Electromagnetic Seabed Logging

Stéphane Sainson

Electromagnetic Seabed Logging

A new tool for geoscientists

Translated from French
By
Linda Sainson M.Sc.

Springer

Stéphane Sainson
Orléans, France

Original French edition published by "Lavoisier", Paris, 2012

ISBN 978-3-319-45353-8 ISBN 978-3-319-45355-2 (eBook)
DOI 10.1007/978-3-319-45355-2

Library of Congress Control Number: 2016960702

Cover illustration: The bottom picture of the cover (EM receiver) comes from the film: [Tony Howell, RSK Communication Services and Peter Taylor Managing Director], Identify new oil prospects with 3D electromagnetic methods, by EMGS, published in 2012. Used with permission.

Printed on acid-free paper

This Springer imprint is published by Springer Nature
The registered company is Springer International Publishing AG
The registered company address is: Gewerbestrasse 11, 6330 Cham, Switzerland

In tribute to Professor Jean Mosnier*, who gave me a taste for research.*
To my mother.

We do not fully understand a science
as long as we do not know its history
Auguste Comte

Preface

Originally, the book published in 2012 under the title *Electrographies de fond de mer: une révolution dans la prospection pétrolière,* edited by Lavoisier, was written for the French public who had no or little knowledge on these new developments of offshore geophysical exploration.

The book was nominated for the French international Roberval's Award in 2013.

It must be said that hydrocarbon direct prospecting has always aroused in France some very unfavorable reactions a priori often relayed by the institutional and economic circles in place. So it was normal despite a few French development attempts in the past that these new techniques would emerge in the Anglo-Saxon and Scandinavian countries in particular.

However, under pressure from my English-speaking colleagues, it seemed appropriate to me to deliver a translation of my work in English with several additions concerning the period 2012 to 2015, making it accessible to all of our small community of geoscientists.

Since the publication of my book, several very good papers, reviews and articles about the history, the theory or the instrumentation synthesis have been published.[1] They can be used as excellent complements about these subjects. Written by the actors and pioneers of this research, these overviews give different points of view, which may epistemologically interest the reader seeking deeper knowledge of this new subject still in development.

I hope this sometimes unskillful translation will meet the expectations of my friends and scientific colleagues, researchers, engineers, technicians and especially

[1]The French edition of my book and the book of Professors Alan D. Chave and Alan G. Jones, *The Magnetotelluric Method*, were simultaneously published in 2012. The latter is only about the magnetotelluric method (MT). A large part is specifically devoted to marine investigations using natural sources (mMT), which are less widely used in oil exploration than controlled source methods (mCSEM). So I recommend to the reader this excellent book for the history, the theoretical developments and the practice of this prospecting method, which was born in France in the 1950s, through the work of Professor Louis Cagniard.

young students preparing for careers in prospecting[2] or more specifically in marine geology. Secondly, I think this book could also interest the community of petroleum geophysicists as a complement to information on marine electromagnetic techniques. This book is also intended for managers and executives of gas and oil service companies whose activities' development is increasingly shifting toward the sea.

This book is rather devoted to the description of *the physical principles* and *the data acquisition* of seabed logging. For this reason, this work is incomplete. Some aspects of theoretical interpretation used as mathematical methods, numerical analysis computations and inversion techniques are reduced to a bare minimum. In the different theoretical parts of my work, I preferred to expose exact formulations using analytical calculations (with the simplest models) but closer to physical realities (directly related to the field measurements). Therefore this may be a good introduction to a future book solely devoted to mathematical modeling and interpretation. On the other hand, I tried to insert qualitative digressions and descriptions rather than equations whenever possible. I think that this approach should probably be more accessible to a larger public readership.

This treatise was therefore designed in a didactic way in order to explain as simply as possible the notions and physical concepts, which in electromagnetism are more complex and certainly less known by professionals in petroleum geoscience than those traditionally used in seismic exploration. Finally, this text also needs to be viewed as an introduction to a new technology and as a tool for reflection, from the historical point of view, on future developments that will undoubtedly enrich this new field of investigation.

Never before, indeed, in the small world of applied geophysics and especially in that of offshore oil exploration has a technology aroused, in so little time, so much interest and enthusiasm. There is not one month, not one week, when an article does not hit the headlines of a specialized professional magazine. It is true that the technique is attractive and the stakes huge. Indeed, if the method reaches a record rate of discovery of nearly 90 %, never equaled so far, it is also environmentally quite remarkable. Directly, unlike the seismic method, it has virtually no impact on marine life, and, indirectly, it severely limits by its excellent results exploration drilling.

In any case we really are today the witnesses of an unprecedented technological breakthrough that could in the future completely change the situation in terms of the energetic, industrial, and commercial aspects as well as the plan of world geopolitics, by extending the territorial marine areas dedicated to mineral exploration, for example.

[2]This is the reason why mathematical developments have been limited, and reduced to the essentials, something quite difficult to do when we know the more and more important part given to interpretation methods especially in electromagnetism. Geophysicists and earth physicists, concerned about further details, will deepen some points with bibliographies located at the end of the chapters and the end of the volume.

Are we attending, for all that, the birth of a new industry? Are traditional upstream companies endangered? Is this finally a revolution in the field of oil prospecting or just a redistribution of the cards in the very closed environment of offshore exploration?

More particularly, technically, this book intends to answer these questions. It aims, after nearly a decade of commercial exploitation, to review this new applied method, for a long time in gestation in university circles and today open to new conquests.

This is the first book devoted to this subject. Despite some parts being treated in a new way, this volume does not claim in any way to be an original work, with much input from specialists around the world, but rather a summary document, which has the merit of offering the public today an exhaustive inventory of the matter. This was relatively difficult to achieve as the number of documents is impressive (see the bibliography at the end of the volume), the authors numerous and the presentations of some of them hardly clear from the point of view of the physics, introducing sometimes real doubts and, even more damaging, controversies. This is one of the reasons why the acquisition data part is more developed than the interpretation part. So I imagine that this work may have many defects yet.

Finally, throughout the text, despite these shortcomings, I hope the reader and particularly geoscience students will find a line of thought and an embryo of an answer to the delicate problem raised by the concept of direct prospecting, which like an old sea serpent has, since the beginning of petroleum geophysics, discreetly haunted the prospector's profession.

Orléans, France Stéphane Sainson

Acknowledgments

I would especially like to thank the companies that have allowed me to report their technologies and reproduce some diagrams and illustrations in this book. Some holders of reserved rights could not be found; may they accept our apologies for these involuntary omissions and be assured that corrections will be made in future editions.

I also want to thank my friend and colleague Christophe Poirmeur, the pioneer of numerical methods of electromagnetic interpretation and general manager of the company Stilog IST, who took time out of his life to reread and correct the typescript.

I owe the colossal work of translation to my wife, Linda. Without her, this version in the English language would not have been able to exist, so neglecting that part of the public that does not read French. In a short time she was able to assimilate this subject that was not familiar to her and overcome the additional work related to her biology–geology teacher's activity, and subsequently to health problems. The translation is not perfect, she says, as English is not her mother tongue, and she is not completely fluent in this language, which she likes very much however. So she apologizes for those mistakes that may survive and hopes to have, nevertheless, respected the original text.

I hope the success of this English version will be worth the effort.

I will be forever grateful to her for that.

I also think of my mother, recently deceased; her most recent drawing appears in the dedication.

Finally, I hope not to have distorted or truncated too much, for the sake of brevity, the thoughts of my colleagues. It is also possible that some illustrations or technical descriptions are not entirely accurate. Similarly inaccuracies, errors, and omissions surely must have slipped into the text. If that is the case, I offer in advance my apologies to the authors. Adjustments will be made in subsequent editions.

Contents

Chapter 1
Introduction

Abstract After a preamble, this introduction first places seabed logging in the global economy. After a succinct reminder of petroleum geology the aspects of its research and discovery are discussed. Indirect (also called structural) prospecting is of course mentioned, thus introducing the benefits of direct prospecting today offered by electromagnetic exploration. The latter uses the electrical properties of rocks and especially the conductivity, which is the best indicator of the facies and lithology and consequently of the presence of hydrocarbons. A history of electromagnetic prospecting techniques is given for information, followed by a study of the seabed logging market and its environmental impact.

Keywords Direct prospecting • Electrical resistivity • Hydrocarbons • Facies • Lithology

1 Prologue

The seas and oceans represent nearly 70 % of the total surface of the earth, about 360 million square kilometers. An important part of the ocean floor is unsuitable for harboring hydrocarbons. However, if we consider the areas likely to contain some, i.e., sedimentary basins of the submerged continental shelf (30 million square kilometers situated less than 500 m deep, the equivalent of the emerged lands) and the associated slopes that border them, farther from the coast but still accessible, we roughly arrive at a number of 100 million square kilometers. This area could contain, according to experts, at least 100 billion tons of oil and gas, the recoverable amount of which is likely to rapidly evolve with advances in exploration techniques. For the future, it is estimated that the oceans might contain 90 % of the hydrocarbon reserves of the world.

Seabed logging, or SBL for short, is one of the very recent and innovative techniques that can meet this expectation. Appearing commercially in the 2000s, it probably is a technological breakthrough regarding the previous prospecting techniques currently used, particularly the marine seismic method whose growth over the past 20 years has been meteoric.

This is largely due to advances in technology (embedded computing for measures, supercomputer use for the interpretation of acquisition data) and to the

S. Sainson, *Electromagnetic Seabed Logging*, DOI 10.1007/978-3-319-45355-2_1

expanding geological and especially petrophysical knowledge of the seabed that the method has provided. However, fairly extensive studies had previously been conducted in several areas of fundamental physical oceanography, marine geophysics in particular, without actually leading to industrial applications. It is only very recently that these principles and methods, extended to geological mapping of the ocean floor, have successfully turned into oil exploration at sea.

2 The Economic Environment

It is expected that overall energy consumption will significantly increase in the coming years[1] to almost double by 2030, with probably a part of growing importance taken by nuclear energy[2] (Bauquis 2001).

In 2014, the annual hydrocarbon consumption already accounted for over 65 % of the total energy demand (14 Gtep, with 4.2 Gtep for oil and 2.9 Gtep for gas), values likely to reverse in the future in favor of the latter. Offshore oil production reaches more than a third of the world production[3] which was around 90 million barrels per day in recent years.[4]

Fifty years ago the energy of one barrel of oil was necessary to extract 100 barrels. Today, with the same amount you can only withdraw 20 barrels in Saudi Arabia and five in the Athabasca oil sands.[5]

A key feature of the production of hydrocarbons is the considerable importance of costs awarded to the research and discovery of new extraction fields whose production is reduced over time (a few decades at most).[6]

With an unprecedented increase in demand for hydrocarbons (with the emergence of countries such as China, India, Brazil, etc.) and the rarefaction of the

[1]According to forecasts by the IEA (International Energy Agency), the global demand for oil will increase by an average rate of 1.2 million barrels per day (mb/d) through 2021 (IEA 2016), with world demand growing from 94.4 mb/d (2015) to 101.6 mb/d (2021).

[2]In the current state of knowledge, the scenario takes into account a growing demand for electricity, an ecological imperative (reducing greenhouse gas emissions) and a significant decline in oil resources. The nuclear electricity production currently accounts for just 5 % of the energy consumed in the world.

[3]Probably more than a half by 2020.

[4]In 2011, the average price of a barrel of crude oil over the year was $US111. Today the production of oils and shale gas (diffuse deposits) has completely changed the game in economic matters in the energy mondial marketplace and in geophysical prospecting investments. After the consumption of oil slowing down and in an overabundance market, oil has recently seen its price drop below US$30 a barrel, well below its cost price. Worldwide, in 2015, the oil industry lost over a third of its workforce. A number of analysts however already agree that this phenomenon is transient and that the price of oil should in late for the future years return to a more realistic level of around US$100 a barrel. Stay tuned… Last minute: the most recent OPEP agreements on the reduction of oil production, mainly supported by Saudi Arabia, are expected to raise oil prices significantly, or at least stabilize prices around sixty dollars for 2017.

[5]An operation that also requires a huge amount of water.

[6]Much less with the current techniques of horizontal drilling.

resources (pick oil),[7] the exploitation of deep oceans (the Gulf of Guinea, Gulf of Mexico, North Atlantic), and Arctic seas[8] (the Barents Sea, Sea of Okhotsk) becomes profitable today with the probable opening of the Northwest Passage[9] (melting ice).

However, the marine seismic prospection,[10] which currently represents about 90 % of the overall budget of the offshore campaigns,[11] has some limitations in these particularly hostile and ecologically fragile areas.

The oil industry spends annually more than $10 billion on geophysical prospecting[12] (Sanière et al. 2010)—more than a half offshore, whose share will probably grow in the coming years.[13] Nevertheless, this development effort will necessarily go hand in hand with a reduction in costs, and among—them, those of the research drillings (named *wildcat* and which might cost $ 25–100 million in deepwater). On average the latter reach two thirds of the exploration expenses and especially deep offshore where they remain the principal expenses since a significant proportion of soundings (about two thirds) remain permanently dry and therefore not redeemable.[14]

To limit the number of failures, which are expensive in terms of resources, the industrials seek above all to minimize the role of hazard in the exploration, trying to

---

[7]We should rather say *oil pick*.

[8]"Arctic oil and gas resources represent the next big chapter in offshore development. Yet, the development of these resources remains challenging in terms of engineering, construction and installation, and related logistics" (Kenny 2011). This is also evidenced by the recent exploration agreement (September 6, 2011) between American companies Exxon Mobil and Russian Rosneft. With this contract Russia seems to initiate a serious investment policy. The USGS estimated in 2008 that undiscovered conventional hydrocarbon reserves exceeded 90 billion barrels of oil, 1669 trillion cubic feet of natural gas, and 44 billion barrels of natural liquid gas (Bird et al. 2008).

[9]Shortages in the relatively short term of our energy resources today push nations that have the means to discover new El Dorado mining (oil and gas). This is particularly the case in countries bordering the Arctic Sea (the USA, Canada, Denmark, Norway and Russia), which recently engaged in a "war" of ownership of the seabed with a very clear challenge to the territorial limits of the international maritime domain (a 2001 request by Moscow to the United Nations to claim the arctic seabed followed in 2007 by an exploratory mission with laying of flags on the seabed). This first geographical stage with the opening of the Northwest Passage actually precedes a no less important exploration phase itself, which will intensely begin in this decade. The Arctic Ocean would contain, according to specialists, a quarter of the global oil reserves (on the Russian side, the arctic subsoil would conceal hundred billion tons of oil equivalent).

[10]Marine seismic acquisition represents a cost five times less than that of land acquisition, which requires more important logistical and human resources. However, the boats' immobilization is more expensive especially in times of economic inactivity (standby).

[11]A magnetic survey (related to volcanic activity) and/or gravimetric survey (linked to salt-bearing diapiric activity) can complete the seismic campaign.

[12]Exploration with well/drilling activities and pipeline transport are the upstream sectors or exploration/production sectors of oil activity (upstream or E & P petroleum sector).

[13]On September 25 and 28, 1945, the US President Harry Truman launched the real policy of development of offshore deposits (Diolé 1951). A few years later (1949), in Azerbaijan, the Soviet government under the leadership of Stalin started oil production in the Caspian Sea (Oil Rocks).

[14]For 5 years, more than 3000 wells per year on average have been drilled offshore, corresponding to a turnover of over $40 billion.

locate boreholes with the maximum chance[15] and accuracy. Modern geophysics must therefore meet these requirements and the constantly improved technics should strive toward this goal.

To date, seabed logging has appeared to meet this dual challenge of reducing costs and decreasing risks as well as the one, no less important, of preservation of the environment, which, for deep areas where it is difficult to intervene,[16] or for sensitive areas such as those found in polar seas, will present more and more hazards in the future.

3 Elements of Geology and Oil Prospecting

Geological and geophysical studies are the fundamental elements of a campaign of prospection, more commonly known as exploration. This precedes the evaluation phase of the field itself, which is technically well testing and reservoir engineering.

3.1 Exploration: Main Phases

Historically, in the beginning, the exploration of oil areas (Lees 1940; Robert 1959) was based on collecting surface clues (shows) and on samples (cuttings) and carrots (drilling cores) collected in the exploration wells (wildcats), without providing a real discovery strategy.

Today, the search for a hydrocarbon reservoir is more complex and essentially comprises three major phases of decision, which are chronologically:

- The choice of the region to explore
- The choice of the location of the exploration wells
- Wider monitoring of the research drilling

These stages of the exploration use specific, appropriate technologies, varied and more or less interconnected with each other, which we usually roughly classify into geological techniques (first phase), geophysical techniques (second phase) and well-logging techniques (third phase), petrophysics then linking these different disciplines.

[15]This concept was first formalized in the 1940s (Gabriel 1945).

[16]Such as the accident at the Deepwater Horizon platform in 2010 in the Gulf of Mexico.

3.2 Marine Geophysics: Preference Domain of the Seismics

Offshore the geological technics at the beginning of all studies[17] are necessarily indirect since the geologist does not have access to rocks neither near (hammer prospecting) nor in the distance (photographs) especially as the omnipresence of unconsolidated marine sediments prevents visual inspection.

In these extremely hostile and unfavorable conditions for observation, the engineer will have to rely solely on information taken from a distance using particularly the indirect methods of applied geophysics. These have as a main purpose the knowledge of invariants of the subsoil by the study at the surface of the effects caused by deep geological structures.

Today, the geophysics now practiced are almost exclusively seismic and primarily focus on methods of seismic reflection, which are quick to implement because they are virtually automated and efficient in determining the shape and structure of the underlying grounds.

But despite steady progress, the seismics that specifically inform on the relative position of the plans of the strata may not provide sufficient information on the horizontal evolution of the properties of the various geological strata, including the changes in facies, characteristics equally as important for the discovery of hydrocarbon deposits as the structural features where this technique excels. For this reason, among others, the wildcat and especially the well-logging investigations that follow are then indispensable (stratigraphic correlations).

Apart from its "pre-history," which saw a proliferation of direct prospecting methods that were still unsuccessful because they were generally based on false concepts (physical or even metaphysical in some cases), geophysical prospecting, since its birth in the 1920s, has been fundamentally based on the geological sciences and especially on those that describe the structure of the subsoil, from which comes its original name of *structural geophysics*, in intimate relation with the tectonics of the sedimentary basins giving birth to the oil accumulation.

3.3 Acquisition Strategy for Geological Data

In this very particular context of applied research, technically it is then the geologist who initiates the first geological studies and who alone should take, on a practical level, the decisions:

- To initiate geophysical surveys
- To interpret the results with the geophysicist
- To execute and control the exploration drillings
- To establish and interpret the logs

[17]Geological studies also follow all stages of the exploration process.

These are the reservoir and oil field engineers who will later have to estimate the reserves and establish an economically viable and profitable extraction and exploitation plan.[18]

Applied geophysics is then a powerful tool available to geologists and petroleum engineers, which today represents one of the largest budget items behind drilling and downhole logging.[19] This is particularly true for investigations at sea where the water depth is still currently one of the major obstacles to the acquisition, sensu stricto, of data and geological information.[20]

If in fact, in its principles, marine geophysics is not fundamentally different from terrestrial geophysics, seismic prospecting in ocean deeps however requires, because of the thick blanket of water, the use of a large amount of energy[21] not always compatible with environmental issues and especially with marine life (see Sect. 9.2).

Moreover, in the Arctic, for example, onshore and offshore seismic techniques, both together, do not generally provide good quality operations, due in the first instance to the presence of ice, which forms a barrier to the penetration of the acoustic waves in the water, and due in the second instance to the streamers' tearing risks on drifting ice. Having the means for exploration completely or partly "at the bottom" therefore seems a good alternative in these very inhospitable environments.

3.4 Oil Fields: Brief Recollections

This section is not intended to replace a course or even less a treatise on oil. There are many pertinent works, and readers will find some excellent ones to satisfy their curiosity (Perrodon 1980; Chapman 1983; Abrikossov and Goutman 1986; Laudon 1996; Selley 1998; Bjorlykke 2010; Vining and Pickering 2010).[22]

---

[18]This is called the evaluation phase, which allows through wells seismic and log data among others to define the deposit in terms of its shape and its size. This is known as delineation of the deposit.

[19]The downhole logs include instant logs (*logging while drilling*, or LWD) executed during drilling with instrumentation located behind the drill bit, and well logs (*delayed well logging*), performed after drilling with instrumented probes connected to the surface by an electric cable. These measures correspond to microvolumetric investigations into the immediate surroundings of the borehole.

[20]Sediment sounders, whether mechanical or acoustic, can also provide information within certain limits.

[21]A seismic source cannot be deeply submerged. Unless it is otherwise (with an implosive source), it has to fight against the increasing hydrostatic pressure with the water level.

[22]Old treatises on petroleum geology available in university libraries (e.g., Levorsen 1954) are interesting in their naturalistic and historical approach but instead should be read with great caution regarding the interpretations of the geological phenomena including the geodynamics presented there.

Everyone knows that[23]:

- The formation of oil results from a thermodynamic process transforming the organic matter trapped in sediments, ranging in time, on more or less large surfaces.
- Its trapping then arises from regional geodynamic processes, inserted into contexts of global tectonics (at the globe scale).

All these phenomena then produce oil on a geological time scale whose unit is a million years.[24]

3.4.1 Oil Formation

Oil is formed after a long maturation that needs special burial conditions (high temperatures and pressures). They cause the original organic material, composed of carbon (including plankton), concentrated in marine sediments located in shallow basins (e.g., delta area), after subsidence, to turn very slowly into oil.[25]

The history of these organic sediments is then the history of the sedimentary rocks.[26] Today we know most of the laws by which the geological facies and their characteristic tectonic forms determine the location of the deposits and the presence of hydrocarbons.

3.4.2 Oil Traps

Essentially the oil that is exploited through wells looks like an essentially mobile mineral fluid. The current position of its accumulation (reservoir rock) is not that of its origin (source rock) and even less that of its formation (diagenesis).[27] This very special place, or trap, which can also contain no hydrocarbons, mainly depends on:

- The structural characteristics of the reservoir rocks (sand, calcareous sandstone or cracked dolomite, etc.)
- Their porosity and permeability resulting from the facies
- The shape, the height and the surface of the closure,[28] thus creating the trap

[23]See the main treatises on petroleum geology mentioned above.

[24]It may be recalled that the earth is about 4.5 billion years old and that the oil deposits discovered to date cover the period from –500 My to –4000 y.

[25]This is the term used in practice by the oil industry instead of the word "petroleum."

[26]The reader will find in the book *La recherche pétrolière en France* an excellent introduction to the origins of oil and a history of the evolution of ideas on the subject (Pelet 1994).

[27]All these physicochemical and biochemical processes lead marine sediments to turn into sedimentary rocks. These transformations (compaction, déshydration, dissolution, cementation, etc.) occur at shallow depths.

[28]Level surface through the vanishing point (the top point of the oil and gas filling).

- The waterproof quality of the cover rocks[29] (clays, marls, evaporites, anhydrites, gypsum, carbonates, etc.) forming the high closure
- The water pressure pushing up the oil, forming the low closure

Oil, a mainly migratory mineral, comes to accumulate at the high points of the reservoir rocks and more precisely in the updip of the porous store layers.

We therefore understand the importance for the *petroleum geologist* of the tectonics and structural knowledge of the basins, and more specifically the places, tiny fractions of the subsoil (fault, anticline, salt dome, etc.) able to stop the migration then to accumulate oil and keep it.

For the *geophysicist* the classification of the deposits thus schematically reduces to a classification of the *forms of the closures* (high and low) and of the *nature of the reservoir rock* (petrophysics).

Finally, given the existence of a geological probability for a particular trap or reservoir to be present in the surveyed area, it is the geologist who objectively has to increase or decrease the value of the conclusions of the geophysicist.

3.4.3 Deposit Topology

Hydrocarbon deposits, depending on their formation and some subsequent tectonic movements, can take on different morphological aspects.

The most representative are structurally concomitant to the presence of an obstacle or a geological barrier, which has local aspects of faults, pinches, salt domes, discordances, end of lens, etc., and more specifically anticlinal folds for deposits of a greater magnitude.[30] We will thus speak of *structural traps* (80 % of all the potential encountered traps) preferably localizable in depth.

Others less important are due to stratigraphic accidents located in geological formations of large extent. This is the case, for example, of the presence of sand channels, lenticular formations, and ancient coral reefs where the oil as it migrates laterally finds itself blocked because of lithological discontinuities. We will

[29]Geological layer of impermeable rock (marl for example) or less permeable rock (dolomite) that the rock store, located at its apex and whose shape prevents any upward movement of hydrocarbons (migration stop). The main qualities ensuring a good seal are the lithological composition, the degree of homogeneity of the rock, the thickness and the formation distribution mode. The greatest power of containment is provided in general by saliferous layers. Depending on their size, we can distinguish the regional coverings governing petroliferous provinces, the zonal coverings filling up several fields and local coverings in one field.

[30]Briefly, these traps usually come from ground movements related to the tectonics of the sedimentary basins, which are themselves derived from geological phenomena on a larger scale as well as orogeneses, which then cause at more or less large distances foldings, breaks, thrustings, etc. A review in detail would require much discussion. For this reason we invite the reader to consult some books about geodynamics.

specifically speak of reef reservoirs or more generally *stratigraphic traps*,[31] mostly laterally localizable.

These traps can take different forms, which are relatively well specified and described with seismic techniques. However they will be much less explicit with electromagnetic methods used alone.

The convergence of all these natural phenomena (biogenic, thermodynamic and finally geological):

$$\text{presence of organic matter} \rightarrow \text{sedimentation} \rightarrow \text{diagenesis}$$
$$\text{catagenesis} \rightarrow \text{geodynamic sequence}$$
$$\text{migration} \rightarrow \text{trapping}$$

is rare and unique. It is often only an epiphenomenon in the local geological history. So many potential traps, when they exist, do not contain hydrocarbons.

For more details on this particular aspect of oil formation and accumulation, the reader is invited to read or reread specific works on petroleum geology (Perrodon 1966, 1980; Bjorlykke 2010) (Figs. 1.1a, 1.1b).

3.4.4 Trap Detection: The Seismics

If the topology of the deposit (form of the geological horizons) is relatively easy to identify and structurally describe by seismic reflection (with surface technics in force), we can obtain additional information under certain favorable conditions of acquisition, thanks to more modern methods of high resolution seismics, and of seismostratigraphy.[32] This information concerns in particular:

- The nature of the contact surfaces between layers such as those that concern the bank and erosion surfaces
- The areas of passages and of stratigraphic transitions
- The limits and lithological discontinuities between benches
- The nature of the cover rocks

These various types of geophysical information can therefore specify, among other things, according to their degree of precision, the geological conditions of the formation of deposits (including the nature of the deposits) and therefore can be a valuable support for decisions in completing the historical and tectonic data.

[31] We can distinguish the primary stratigraphic traps contemporary to the sedimentation, from the secondary stratigraphic traps posterior to it.

[32] The contribution of gravity and magnetic studies is also used in some favorable situations when the traps are associated with volcanism (lava), for example.

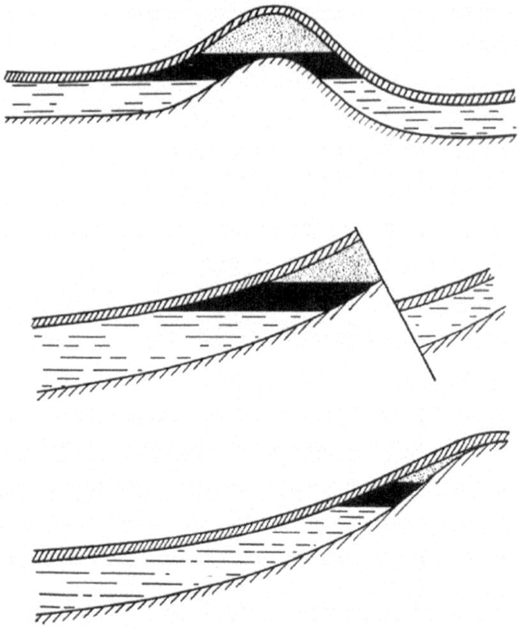

Fig. 1.1a Very simplified diagrams of petroleum traps formed (top to bottom) by an anticline, a fault (structural trap) and a pinch (stratigraphic trap). The oil, in black, is generally "stuck" between the gas (lighter) above and the deposit water (heavier) below. These fluids are immiscible; the phases differ in the reservoir itself. These types of traps are covered with sedimentary layers (not shown), often called "mort terrain" : "or overburden", which often limit investigations by geologists. In areas with the most complex tectonics, some other traps are added due to the presence of salt domes (the Gulf Coast), diapirs or even those due to the presence of volcanism

Fig. 1.1b Photograph of an onshore anticline. In this case, the axis of the fold is substantially vertical

3.5 Physical Properties of Rocks and Hydrocarbons

Most physical properties of the oil are not clearly distinguishable from those of water: density of the same order of magnitude, the same magnetic permeability, almost the same elastic and radioactive properties, etc. Only the electrical conductivity, i.e., the ability to conduct an electrical current, frankly differentiates oil from water. Very simply, one may say that oil is resistant while the deposit water is conductive[33] through the greater or lesser presence of dissolved salts.

But oil rocks do not contain just oil; they often contain water in significant proportion so that it is difficult at first in a normal geological environment (onshore) to distinguish them from aquifer rocks themselves.

If, onshore, this element does not allow us to properly approach the detecting problem, offshore, the significant infiltration of seawater in depth, driven by the hydrostatic pressure to the level of or beyond the deposits, allows us to approach this point with this time much more hope. Indeed, the presence of this seawater, highly conductive, immiscible to oil, then significantly increases the resistivity contrast between the reservoir rock and the surrounding sedimentary series, therefore increasing the sensitivity to facies variations, detail which, as we have seen, is important in oil exploration (cf. Fig. 1.2).

3.6 Indirect Prospecting: A Success in Oil Prospecting

As the geological information is most of the time inaccessible, there were therefore early proposals for the use of devious means and especially physical means. This led in the 1920s to the birth of applied geophysics or exploration geophysics,[34] as the principles of this commercial activity were already known. From this date there were two more or less conflicting concepts: that of direct prospecting, which is older and was abandoned very quickly, and that of indirect prospecting, which has been used until today (see details below in the note history).

Indeed, except in special cases, the results of direct surveys carried out until the 1930s remained well below the expectations, due to the weakness of the phenomena

[33]Freshwater in return is resistant.

[34]Applied geophysics is a relatively young science. If we disregard the divining rod (hazel stick) used by dowsers (Agricola 1556), we can trace the first concrete application of physics to the magnetic detection of minerals in the early seventeenth century (Gilbert 1628). But it was probably with the torsion balance (gravity) of the Hungarian scholar Eötvös (1898) and with the electrical methods of the Electrical Ore Finding Company Ltd (1902) and then of the Frenchman Conrad Schlumberger (1912) that geophysics entered the industrial era. In 1933, offshore exploration began in the Gulf of Mexico (USA) some time after the investigations of Lake Maracaibo (Venezuela) and of the Caspian Sea off Baku (Azerbaijan). At that time it was done to recognize the extension of onshore fields. It was in 1947 that the first subsea well was actually put into production (see note 13). A brief history can be reconstructed by consulting Appendix A1.1 and specifically the following books: Allaud and Martin (1976), Anonyme (1983), Castel D du et al. (1995), Collectif (1966), Iakanov (1957), Kertz and Glassmeier (1999), Sweet (1969), Virchow et al. (1999), Walter et al. (1985) and Ward (1952).

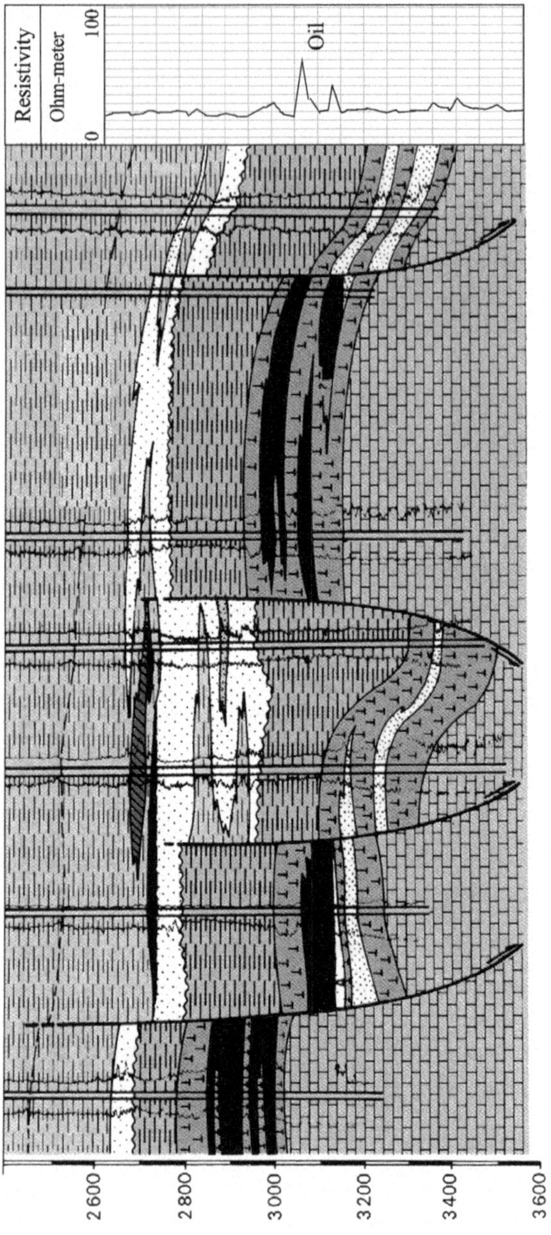

Fig. 1.2 Geological schematic section of a structural unit representing an oil field. Section taken from seismic data, supplemented by downhole information (mud logging, cuttings, corings, etc.) and by logging records: sonic logs for setting the depth of seismic sections, and resistivity logs (*at the top*) to determine the presence of hydrocarbons and correlations between wells. Logging or electric log giving the electrical resistivity of the geological horizons penetrated in the borehole. Up to 2800 meter depth this diagraphic method was the only way to deliver this crucial information for the detection of hydrocarbons and localization of oil-producing layers

to be measured and the difficulty of distinguishing their useful part from all the disturbances caused by irregularities in the constitution and the topology of the subsoil.

However, hydrocarbon deposits are controlled by geological laws that explain their accumulation by their immediate or more distant topological location. Under those laws of the probability of presence that define the existence of the deposits by accurate knowledge of the elements that surround them, the indirect geophysics is then attached not to detection of hydrocarbons as such but the traps likely to contain them; this approach is described as structural or stratigraphic geophysics today.[35] In a such study of the geological environment, where the conditions of deposits are used as the main source of information, only seismic reflection has so far achieved these objectives but with a success rate of only 25 %.

I would like to remark that the relative regularity of sedimentary series particularly along the basin gives a good chance of success to the indirect approach.

In the history, we can say that two main schools of thought have animated the philosophy of geophysical prospecting.

The first, or *direct prospecting,* which comes from magnetic surveys of the nineteenth century, was undoubtedly that which took into account the action of the mineral accumulation in the distance when it was strong enough to affect any detection apparatus.

Due to the smallness of the phenomena to be observed (the target) and the importance of disturbances related to irregularities in the constitution of the subsoil (the geological environment), even in the topography, applied geophysics would have vegetated if it had continued to move in this way even more because oil does not have specific obvious characteristics.

Direct prospecting would have gradually been discredited if a second idea had not come and revived it. Quite naturally, approaching the geology related to deposits where significant progress had just been made (in the early twentieth century) it was possible very early to redirect geophysics this time toward indirect research.

Concentrations of minerals and hydrocarbons are governed by well established geological laws that explain them by their environment and define them precisely in their reception characteristics. It is no longer the product that is researched but favorable conditions for its existence.

It was probably the Frenchman Conrad Schlumberger who was the inventor of *indirect research* by running electrical prospecting in 1912 (already!) in the Iron Basin of the Calvados (in France). This allowed at that time not only determination of the direction of the layers' stratification but also location on the ground, covered with alluvium, of the passage of geological contact between the Calymene schists and the Armorican sandstone (Schlumberger 1920a, b).

For oil exploration, this approach was confirmed for the first time by the Société de prospection électrique teams (Schlumberger methods), by the discovery in 1923 of a salt dome in the Aricesti area, west of the Ploesti field (in Romania).

From that date, other investigative methods were successfully used, such as the discovery in 1924 of several shallow diapirs in the Gulf Coast (USA) by seismic

[35]We then go back to the concept of the geologist who, from his observations on the ground (surface), imagines, by his knowledge of the genesis of the deposits, the arrangement of the different grounds of the subsoil.

refraction (by the German company Seismos) and the Nash Salt Dome in Texas by the Rycade Oil Company using the Eotvos balance, then in 1926 the discovery of deep domes in Oklahoma, this time by seismic reflection (Karcher 1974). Maurice Ewing in 1937 was the first person to apply the seismic refraction method at sea, under the impulse of Richard Field and William Bowie, in a work to study the structure of the Atlantic coastal plain (Ewing et al. 1940) (See also Appendix A1.1).

All these discoveries attached to different physical detection principles then put into evidence the importance of "the architecture of the subsoil" directly resulting from structural geology and therefore from tectonics,[36] where the accumulation of raw material is related among other things to the anomalies among the monotonous layers of the sedimentary basins. But these features and geological accidents (anticline, fault, salt dome, etc.), more commonly known as traps, do not necessarily contain these precious materials (ores, oil). This explains why, despite favorable geological structures, the rate of discovery has always remained very low.

Until now, only drilling has allowed us to confirm the existence of oil and/or gas and downhole well logging to estimate the economic potential. A number of oil companies are then prepared to limit these research campaigns, of a very high cost, at sea particularly, especially when the indirect survey provides only a low probability of discovery.

We can remark that only radioactive minerals (with emission of radiation) and some magnetic rocks (iron ore) are directly accessible to measurement and those with strong contrasts in physical properties with the enclosing environment (electrical conductivity for sulfide masses, density for massive minerals), or even with special effects with the immediate geological environment (spontaneous polarization). Apart from these exceptional cases, all other metal ores, copper, lead, etc., or even oil, are intimately dispersed in the rock and so their intrinsic properties are difficult to detect at a distance.

3.7 *Direct Prospecting: A Marginal Technique Until Now*

Minerals including oil are abnormal by the fact that they are strongly differentiated from most of the surrounding rocks by one or more of their physical properties. The latter by their action at a distance can be utilized for detection using specific instrumentation. This is the case, for example for magnetite (magnetic susceptibility and permeability), sulfur masses (electric conductivity), graphite (density), etc., which can be detected by singular fields of force or by their induced effect when interacting with their immediate environment (polarization). In the metal mining domain, in metallogenic and gitological known contexts, which are very often complex (metamorphism), this direct prospecting, often the only method possible, can be used successfully.

[36]A greater ambition of indirect exploration, but more difficult to satisfy, is to date the marker horizons allowing establishment of a certain chronology.

The idea of using the electrical characteristics of the hydrocarbon reservoir and more particularly its electrical resistivity is not new. US researchers in the 1930s had already realized oil prospecting campaigns by electrical methods in Kentucky, in the Legrand Basin. At that date, the authors had reached or invested a depth of 224 m, but given the sharpness of the results they could have been expected to reach a priori very much greater depths (Lee and Schwarz 1932).

A few years later, other engineers specified that the electrical method for the detection of oil sands was all the more effective when the width of the deposit exceeded several times the burial depth (Cercher and Dermont 1935).

Finally, to strengthen and quantify these results in promising fields, the calculation of charts of interpretation and experiments on scale models were proposed and carried out mainly in the Soviet Union. At this time, we have already pointed out the need to intervene in the interpretative model of external data (Gorbenko 1936) (cf. Fig. 1.3):

> To find the cause of the rise in resistivity, we must know a number of circumstances and in particular the geological structure and the nature of the rocks of the studied district; this information is important especially here...[37]

Shortly afterward, in the years 1940–1950, and more specifically across the Atlantic, electromagnetic prospecting techniques were invented,[38] using high frequencies (radio waves). More or less scientifically supported (a significant *skin effect*), these methods had mitigated success (Fig. 1.3).

But it is probably in the decade that followed that the most founded and most promising hopes for the direct detection of hydrocarbons were noticeable with the shy but certain breakthrough of the magnetotelluric method (MT), a supplement of the seismic method. These expectations were based on the idea that on (already) recognized traps, the substitution of the brine by oil causes a significant variation in the conductance of the reservoir, thus introducing an even greater anomaly when the relative saturation in hydrocarbons, the porosity and the reservoir closing are great.

---

[37]Translated from Russian...

[38]This is for example the case of the Barret Company Inc., which suggested in the 1950s a radioelectrical method (the Radoil method), working in the band of 100 kHz to 500 kHz (Barret 1947, 1949). This controlled source terrestrial technique enabled, by measuring the intensity of the reflected waves in the subsoil, detection of the presence or absence of anomalies of conductivity in depth and thus marked the difference between the salt water, the hydrocarbons and the rock. Use of radio waves more directive indeed than those related to low frequency (and continuous) currents could then make apparent some accurate phenomena of reflection/refraction as observed in seismic prospection (Melton 1933). But it did not take into account the low penetration of energy related to the high frequency used, and the theoretical difficulty, if any (a very important skin effect), which was bypassed straightforwardly by the developers of the method by announcing the existence of selective absorption zones in certain frequency bands (unproven) then favouring penetration at great depth. Furthermore, to overcome significant dissipative effects due to the ground surface, this company devised a transmitter able to generate a surface wave, which, while flowing over it, could chip, thereby forming a number of secondary sources up to the receivers causing a cumulative effect on receiving. Another interesting case was that of the Sorge Company, whose approach at low and very low frequencies was somewhat different and more plausible in its conceptual approach (Calvaresi, 1954, 1957).

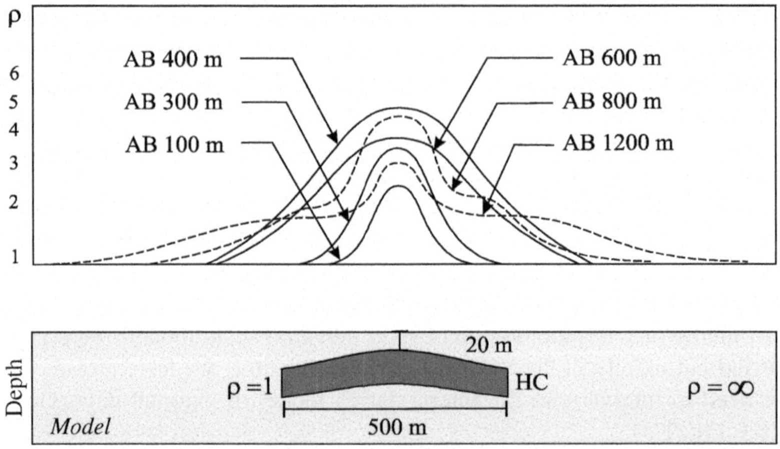

Fig. 1.3 Series of profiles of resistivity ρ over an oil layer (HC) for different lengths of line (AB) varying from 100 to 1200 m, defining the depth of investigation. (According to Gorbenko (1936))

At that time, Professor Louis Cagniard pointed out in his famous article on the magnetotelluric survey (Cagniard 1953a, b):

> If it is a preference domain where electrical survey should have done the most important services, it is certainly in the study of large sedimentary basins, i.e., in research of oil…

Ahead of the very rapid advances in seismic reflection, with its easy use and its meaningful results, accompanied (which is not an insignificant fact) by fierce protectionism in the USA, where most American exploration companies were operating, the MT method remained at an almost experimental stage.

3.8 Geology and Geophysics: Two Indissociable Sciences at Sea

Except where oil crops out,[39] oil exploration has always relied on solid geological knowledge (genesis, history and structure of deposits, etc.), allowing, through

[39]The surface signs were at the base of the first oil onshore discoveries in the mid-nineteenth century (Titusville, Bakou, Grozny, Mossoul, Pechelbron). Today, oil companies try to develop observation submarine devices (ROVs, AUVs equipped with thermal cameras) for physical signs (gas bubbles or droplets of oil escaping from the sediments). What did we know in 1900 about the nature of underwater rocks? Virtually nothing. The engineers had a good idea about the variations in the oceanic relief thanks to the works of laying trans-Atlantic telegraph cables (in the nineteenth century), but had no more information because of the absence of effective means of investigation of the soil and even less sea subsoil exploration at great depth (Delesse 1872; Murray and Renard 1891; Renaud 1902; Termier 1951; Whittar et Bradshaw 1965).

natural laws of occurrence and in particular tectonic and structural laws (Billings 1959), establishment of probabilities then predictions on the presence or absence of hydrocarbons (Suter 1948).

Since the early twentieth century, geological studies have been systematically completed by geophysical studies using the laws of physics, considerably increasing the certainty of the degree of discovery.

Drilling is then the means of confirming (or not) the presence of oil; instantaneous and/or delayed drilling logs with well testing evaluate the nature and economic potential of the discovered deposit.

The offshore exploration like that practiced on land is still largely dominated by seismic methods (over 90 % of total exploration activity) where they can be used on larger areas, then compensating for the lack of geological information caused by difficult underwater access for the field geologist.

Furthermore, geology and petroleum geophysics made a big step forward a few years ago, when it was discovered that:

- Deposits with high potential could exist in the deep zones of the continental slopes under over 1000 m of water (in the Gulfs of Mexico and Guinea).
- Seawater generally could largely penetrate very permeable marine sediments several hundred meters deep, making them widely conductors of electricity.

This has the immediate consequence that the locally trapped oil can be distinguished by its resistivity, which is higher than that of the surrounding marine sediments gorged with conductive seawater (Fig. 1.4).[40]

I would like to remark that unlike methods of natural field measuring (magnetic and gravimetric fields) depending on a single variable (depth z), the electrical methods on two horizontal variables (x, z) and especially seismic and electromagnetic methods have the ability to measure the lateral dimensions (x, y, z). This particularity is partly due to the dynamic nature of the auscultation where the distribution of the artificial fields through the different grounds is after all the result of measurements made at the relative positions of the transmitters and receivers.

4 Electrical Resistivity and Geology

Outside the logs domain (measurements during or after drilling), the electrical resistivity has been a very little used physical parameter in the world of surface oil exploration. Yet this property of the rocks is one of their most interesting and representative features in many aspects.

[40]The resistivity contrast between the reservoir and the marine sediments is more important than the contrast of acoustic impedances encountered in seismic exploration.

Fig. 1.4 Seismic reflection 2D section (time/depth) after treatment, highlighting the geological folded structure of the subsoil (anticline). Seismic acquisition alone does not allow us to exactly stall the model in depth. It is necessary to know the propagation velocities in the different sedimentary strata to make a sonic log, which can only take place in a well. The seismic profile can only detect traps without providing evidence of the presence of hydrocarbons (?)

4.1 Lithology Index and Facies Index: Old and New Reality

Of all the physical rock properties (density, elasticity, etc.) the electrical resistivity is the one that has the greatest variability (Telford et al. 1978), but until now it has only been used in boreholes (in situ measurement) under vertical investigation only. The SBL therefore offers today by its three-dimensional character a new dimension, that of lateral extension.

The electrical resistivity, as has been demonstrated by the extensive use of its measurement (log)[41] during the last half century, is one of the fundamental and essential characteristics[42] of petroleum geology delivering vital information on:

[41]Very briefly, the electric logs allow understanding of a very precise diagnosis of permeable layers. In particular, we can derive information on:

- The value of the porosity
- The water saturation
- The oil saturation
- The amount of oil displaced by water pressure (water drive)
- The precise thicknesses of the producing layers
- The contact of water/oil and often gas/oil

[42]"Resistivity is important because oil is insulating and water is conductive, so there is something natural, deep: it is truly the measure which is needed to identify the oil." Jean-Pierre Causse (Dorozynski and Oristaglio 2007).

- The *lithology*, i.e., the nature of the rocks
- The *facies*, i.e., their appearance and disposition

In addition, the porous and permeable formations that form the sedimentary sequences that may contain hydrocarbons may have resistivities that vary over a large domain according to their content in more or less mineralized fluid. For example, a rock as sand moistened with saline water has low resistivity, while limestone impregnated with oil or gas or even freshwater will increase its resistivity in a very important way.

4.2 Index of Hydrocarbon Presence: A New Concept

Electrical resistivity was so far the only reliable petrophysical characteristic indicative of the presence of hydrocarbons in the subsoil. This unique and sufficiently sensitive property was demonstrated by drilling in the 1930s (Schlumberger 1927) and its measure according to the well depth is today one of the main activities of the world para-oil industry.

It is then not absurd to think at first that an "equivalent" but this time noninvasive device (without the need to drill), located at the surface or more precisely on the bottom of the sea, delivering such a parameter, may advantageously be used as an index of the presence of hydrocarbons.

5 Seabed Logging: Its Place in Modern Oil Exploration

In general we consider a prospecting campaign under two distinct aspects. The first one consists of recognizing the shapes of the grounds; it is the structuralist aspect of the problem solved in particular by the seismic methods. The second one is the determination of the identified structure at each point, of the properties of the sedimentary strata (*porosity* and *permeability* in this case) defined by the facies and determined by downhole logs. EM seabed logging can therefore fit naturally between these two fundamental aspects of exploration and change the approach of it.

In summary, offshore oil exploration can therefore be defined in several major phases: the strategic, tactical and evaluation phases.

5.1 Strategic Phase: The Seismics (Structure)

Seismic reflection can be part of the strategic phase of exploration.[43] It aims to identify favorable structures, selected in a probable geological context (e.g., ancient deltas), i.e., those corresponding to potential traps at a *large scale*. It thus succeeds the preliminary geological studies that have defined the prospect zonation (Sheriff and Geldart 1983).

5.2 Tactical Phase: Seabed Logging (the Presence Index)

Seabed logging however takes place rather in the tactical phase, i.e., the one that at a *smaller scale* detects the reservoir, or more specifically the one that specifies the presence or absence of hydrocarbons in the trap. This approach, which follows the strategic phase (trap limitation), increases significantly the rate of discovery, and of course has as an economic corollary reduction of the risk of dry drilling, then reducing costs on drilling recognition programs and well measures.

5.3 Evaluation Phase: Well Logging (Facies)

Indications of seismics, the option of SBL, depth geology (mud logging, cuttings, corings, measurements while drilling, etc.), whose activity takes place during the drilling, then provide the information necessary for the first petrographic, stratigraphic and geological interpretations.

In the case of positive results in these investigations, the well is then logged (delayed logs). The results of these measurements (log)[44] specify among other things the details of the petrophysical characteristics of the rocks, namely—most importantly—their porosity, permeability and saturation in hydrocarbons (Serra 2000).

When several wildcats have been drilled in the same geographical area (a basin in this case), correlations between wells are established. They are then used to locate more or less precisely the discordances (facies) and geological discontinuities (structures) that are very useful for identifying potentially oil-rich areas.

Apart from some exceptions, SBL cannot, in the current state of knowledge, substitute for seismic and well log campaigns, especially as the electromagnetic interpretation methods, especially using inversion methods inherent to this new technique of investigation, require the input of seismic data (Fig. 1.5).

---

[43]Magnetic studies (in the volcanic context) and especially gravimetric studies (in the context of saliferous tectonics) may also be in keeping with this phase.

[44]Recording of the measurements of physical properties of rocks in relation to depth.

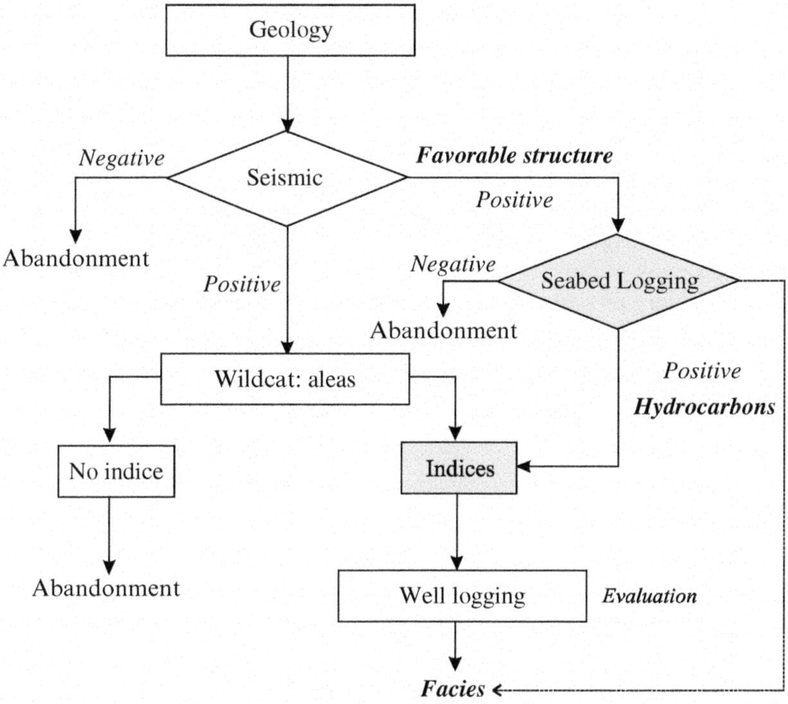

Fig. 1.5 Simplified decisional synoptic for an offshore geophysical exploration campaign with and without an SBL program. The SBL application allows the economy of a drilling program—by far the most expensive and risky component of the exploration

6 Historical Landmarks in EM Seabed Logging

Among all the applications of electricity, one of the most curious is probably that relating to the exploration of the subsoil. This is, strictly speaking, the first "artificial field" technique that was developed for the detection of mineral wealth and especially ore bodies.[45]

Aside from their great extension in the well-logging domain,[46] electrical techniques, more generally the electromagnetic one, remained, despite a promising start

[45]We owe this original technique to the Frenchman Conrad Schlumberger, who applied it as far back as 1912 (Schlumberger 1920a, b). Magnetic prospection is the oldest of all the geophysical methods applied with discernment. However it only affects magnetic and ferromagnetic materials. In 1640, the Swedes tried to find iron mines with compasses. But it was only after 1870 (Brocke in 1873 in the USA and Thalen in 1876 in Sweden) that we clearly realized the disruptive effect of magnetic masses in the earth's magnetic field.

[46]In 1927, Conrad Schlumberger expanded his surface concept to the well by the electrical logging technique known today more commonly as logs (Schlumberger 1927).

in the 1920s, relatively marginal in oil exploration and mostly confined to structural geophysics until the recent advent of EM seabed logging.

Historically, the SBL takes its distant origins in the early electrical submarine and lacustrine soundings conducted in 1930 by Schlumberger school teams (SPE and CGG).[47]

It was much later in the 1970s in real investigations at sea, for the purpose of deep oceanic crust exploration under the major programs of earth physics, that efficient equipment (electrometers and vectorial magnetometers) were developed and validated.

It was only very recently in the 2000s that the technique was used for commercial purposes of oil research. As in many complex techniques, we can say that it is the convergence of different subjects coupled with complementary technologies that have enabled the success of the method[48] (Ridyard 2006).

6.1 Terminology

Seabed logging includes geophysical techniques of representation and imagery of the seabed subsoil obtained by electromagnetic methods (EM) in the broad sense (see Fig. 1.6).

So we can include here the sounding techniques of:

- Continuous or pulsed current, the first to be used and still in effect: marine direct current sounding (or mDC)
- Source controlled AC: marine controlled source electromagnetic sounding (or mCSEM)
- Natural, telluric and/or magnetic source: marine magnetotelluric sounding (or mMT)

To these may be added more marginal techniques such as the ones concerning spontaneous potential measurements: marine spontaneous polarization (mSP) or marine differential magnetic sounding (mDM).

Today, these new EM methods are reported in the technical press as revolutionary because they allow, according to their developers, direct detection of oil or at least give a relatively reliable index of its presence.

---

[47]Société de Prospection Electrique, Procédés Schlumberger and Compagnie Générale de Géophysique.

[48]One could compare the beginnings of seabed logging to the birth of wireless telegraphy where Marconi had the intuition to bring together into a single concept Maxwell's (theory) and Hertz's discoveries (experience), and Popov's (the antenna), Ruhmkorff's (power transmitter) and Branly's inventions (detector: coheror), to make a remarkable communication tool with the commercial success that everyone knows.

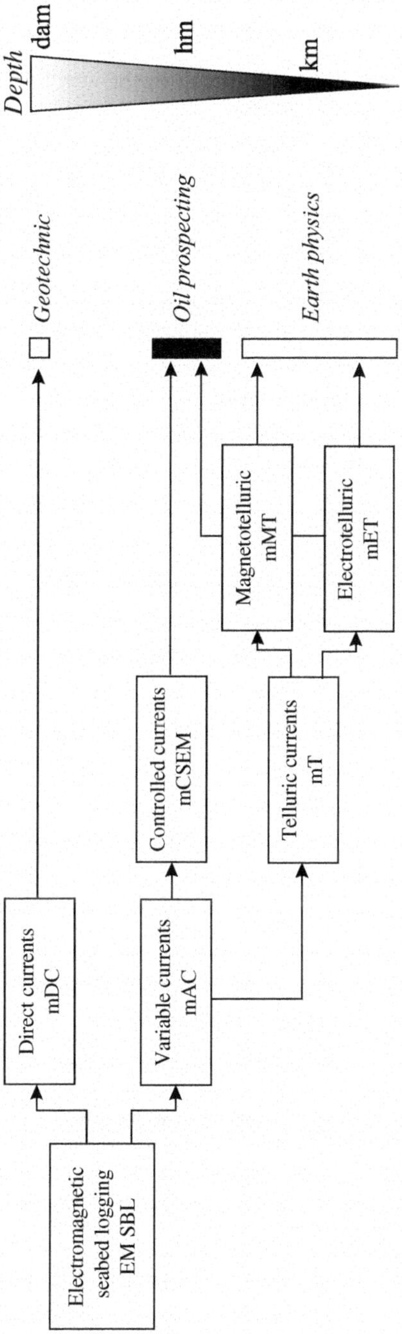

Fig. 1.6 Possible applications of different EM seabed logging techniques, given depending on their respective depth investigation (dam → km)

6.2 *The Precursors: Electrical Prospecting by the French School*

With its experience since the 1920s, it seems that it was a French team from the Société de Prospection Electrique (SPE) who in October 1926 performed the first electric marine sounding on Prien Lake in Louisiana (Baron and Breusse 1926). This first lacustrine experiment, derived from terrestrial processes[49] (pulsed DC electrical profiling) and whose device was placed at the water surface (floating electrodes), driven by a small boat, allowed establishment of the practical feasibility of the method (Fig. 1.7).[50]

These encouraging results brought, a few years later, in Gabon, as the river system is there more convenient to browse than the rainforest, those same engineers to repeat the experience between Pointe Noire (Congo) and Fernand Vaz (Gabon), this time for hydrocarbon exploration (Fig. 1.8), then a few months later, in the Caspian Sea, to study the extension of the Bibi Eibat[51] (Azerbaijan) deposit for the location of a resistant limestone bench.

But it was really with the missions in the harbor of Algiers, at first in the summer of 1932, conducted by the CGG engineers Jabiol and Donner, and the year after by the engineers Bazerque and Leleu, whose aim was to establish the thickness of marine sediments at a variable depth, that the method was improved, consolidated and finally validated with a device located at the bottom of the harbor (Renaud 1933) (Fig. 1.9). With such an arrangement, more stable and reliable, it was then possible to avoid the inductive effects caused by the deformation of the measuring

[49]The idea of using the electrical properties of (nonmagnetic) metallic minerals dates from the nineteenth century (Fox 1830). From that time, researchers have suggested studying the distribution of the electromagnetic fields using high frequency devices (Leimbach and Löwy 1910), low frequency processes (Daft and William 1902) or measurements from DC injection with electrodes planted in the ground (Brown 1903; Mc Clatchey 1900). The first serious investigations began before the First World War (Schlumberger 1912; Wenner 1914) to take, from the 1920s, an economic boom with the creation of societies exclusively devoted to this particular technique of direct prospection (Hedstrom 1930). Only the use of direct current (or pulses) was recommended and able to quickly force itself until today. On land, it has the main advantage of virtually controlling the depth of investigation and theoretically increasing it to infinity (never achieved in practice), but has the main drawback of integrating an excessive volume of land, which does not allow any lateral resolution. This therefore confines the DC method only to an investigation in depth of horizontal geological strata in an often limited number (three or four maximum), and applicable only in some domains of subsurface applied geophysics. "It's certainly by electrical methods that have been obtained, 40 years ago, the first indisputable successes of geophysical prospecting" (Cagniard 1953a, b).

[50]Interesting results had already been obtained in several surveys in marshy ground and in the bayous of Louisiana, where the electrodes were then planted in the ground.

[51]Also written as "Bibi Heybat"—the first oil field reclaimed from the sea south of the Bay of Baku whose remains today show disrepair where pipes and wells are leaking from everywhere.

Fig. 1.7 Photograph of the mission (measurement by potentiometer) and execution scheme designed at the time (1926) by the engineers Baron and Breusse from the SPE for the execution of the first marine electrical marine (lacustrine) sounding on Prien Lake, Louisiana (According to Baron and Breusse (1926))

system (due to movement of water) in the earth's magnetic field (Schlumberger and Leonardon 1934). These investigations remained there until the end of World War II, laying then the theoretical and practical bases without giving rise to real enthusiasm (Fig. 1.9).

During the years 1950–1960, with the appearance of the terrestrial surface methods (electrotelluric and magnetotelluric methods mainly developed in France and the Soviet Union), some researchers took under consideration the possibility of conducting electromagnetic soundings in deep water (Cagniard 1953a, b) with also the possibility to use artificial sources (Cagniard and Morat 1966).

At this time we can note some patents related to electromagnetic methods of offshore exploration (Thomson 1950; De White 1958; Postma 1962) and already some experiments conducted in the Caspian Sea (Sarkisov and Andreyev 1962).

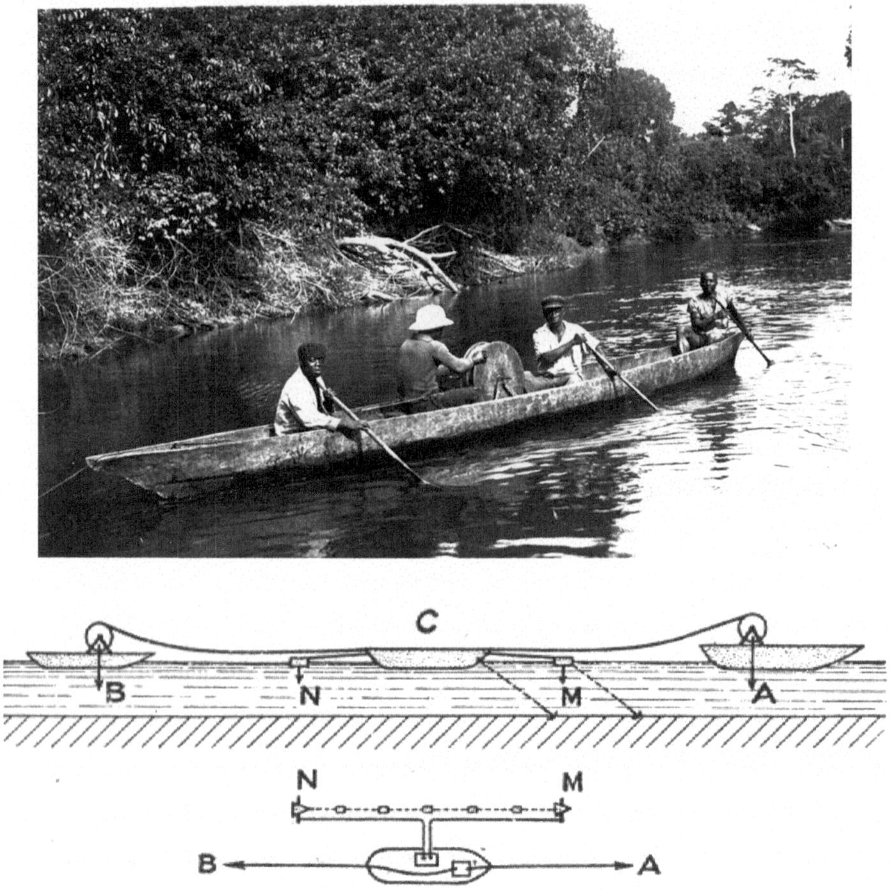

Fig. 1.8 Photograph and execution scheme of the mission in Gabon in 1933 (According to Schlumberger and Leonardon (1934))

6.3 *The Continuers: Researchers and Earth Physics*

The first experimental marine investigations date from 1950 to 1960[52] and were mostly concentrated in the USSR. They were intended, for example, to:

– Determine the deep layers of the globe where the difficulty at the time, for the Soviet scientists, was to send high intensities of direct current in the ground

[52]At that time there was also initiation of the first experiments of underwater detection, of study of the seaboard effect, etc., where specific and confidential materials (magnetic variometers) were developed (Mosnier 1986).

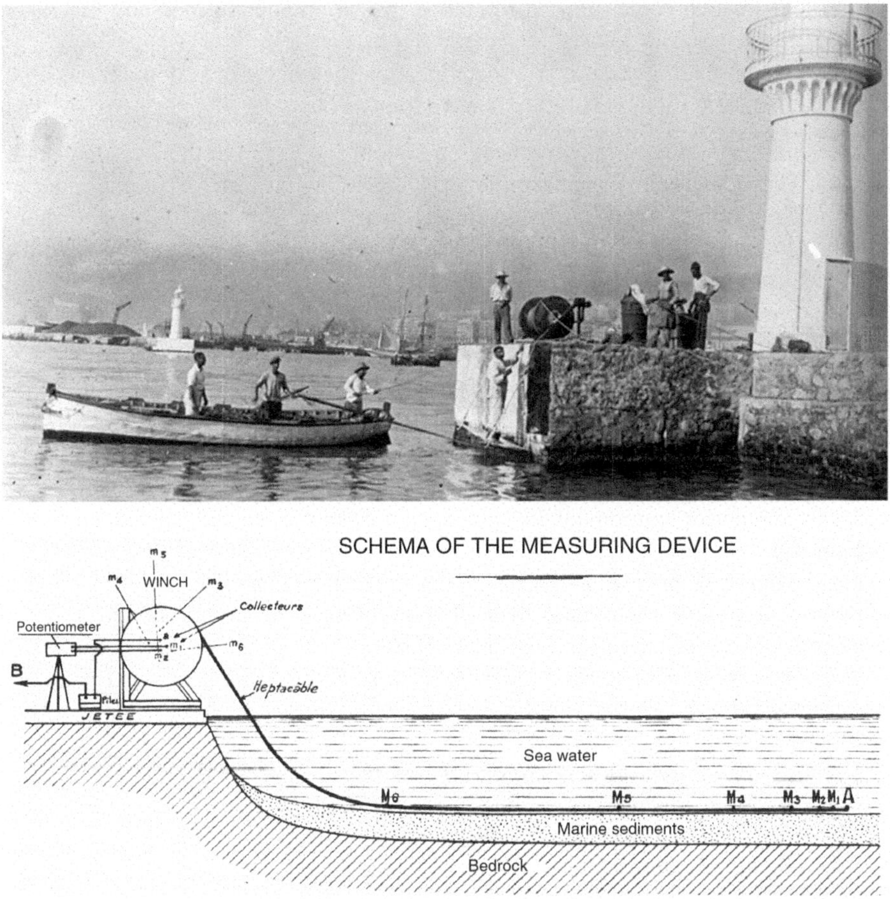

Fig. 1.9 Photograph and execution scheme of the evaluation device used for the second Algiers mission led by the CGG (1933). The measures were effected for the first time in depth with a device towed and placed on the bottom, made of a flute with several electrodes, thereby establishing the first commercial underwater sounding. To avoid the effects of electrode polarization due to salt water (electrolytes), these different investigations were already using pulsed currents requiring a suitable induction corrector. The measurement was done with a potentiometer equipped with a double inverter (measurement/injection) or with a telephone. The interpretation was conducted using abacuses, specially calculated taking into account the system architecture and its shallow submersion (According to Renaud (1933))

(on land), which forced them to practice that in an aquatic medium[53] (Krajev et al. 1948)
- Extend the land variational methods to the maritime domain as the magnetotelluric method (Cagniard and Morat 1966)

Measurements of telluric currents at sea are old and began in the late nineteenth century. They had as aims, for example, the study of the movements of major ocean water masses (tides, ocean currents, etc.). It was only from the 1930s and more specifically in the years 1950–1960 that some investigations were conducted with a geological purpose:

- 1936: the first recordings of telluric currents at sea (Nitronov 1936)
- 1952: the study of sea electromagnetism (Le Grand et al. 1952.)
- 1957: measurement of electrical currents in Lake Baikal (Iakanov 1957; Vinogradov 1957) and examination of the nature of telluric currents on the ocean floor (Ryshkov 1957)
- 1959: large scale mapping of oceanic current circulation (Parkinson 1959)
- 1960: submarine equipment for the study of telluric currents (Bogdanov and Ivanov 1960)
- 1961: studies of the variations of the telluric fields associated with the oscillations of the earth's magnetic field (Fonarev 1961) and effects of the induced currents on sea water in Hel (Kolovski 1961)
- 1962: comparison of telluric currents at sea and on land (Hessler 1962) and quantitative study of the edge of sea effect in the Sea of Azov (Popov 1962; Mosnier 1967)
- 1963: study of the vertical component of the electric field for Lake Baikal (Foranev 1963), calculations of the induced electric currents in the sea (Ivanov and Kostomarov 1963) and study of telluric currents in the Arctic Ocean (Novysh and Fonarev 1963)
- 1964: distribution of electromagnetic variations depending on the depth (Fonarev 1964), examination of the consistent fluctuations between telluric and magnetic fields at sea (Cox et al. 1964) and telluric experiments in the Puerto Rico Trench using the bathyscaphe Archimède (Launay et al. 1964)
- 1965: telluric experiments in the Hellenic Trench with the search for Schumann resonance (Launay et al. 1965).

In the years 1970–1980, following the discovery of plate tectonics (Hess 1962; Wine and Matthews 1963; Le Pichon 1968; Hallam 1976), the international scientific community pretty quickly took an interest in the nature of the oceanic

[53]The system used then was relatively sophisticated: current sendings every 15, 30 and 60 s by a fixed dipole and acquisition when the steady state was established on a mobile dipole. The measures of the current at emission and the voltage at receiving were connected by a radioelectrical telemetry device taking account of the speed of relative movement of the two dipoles and of their spacing. We can say that these early experiences prefigured the current methods (reverse SBL).

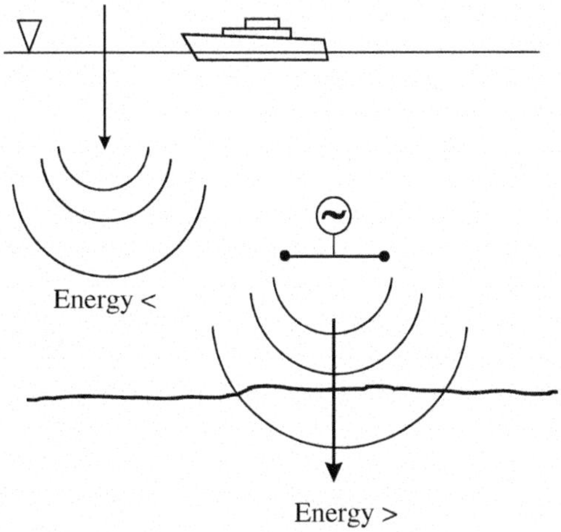

Fig. 1.10 By the 1960s, it was proposed to have artificial sources at the bottom of the sea. Energy, this time controllable, more powerful than telluric sources, no longer has to completely cross the water column. The use of higher frequencies is then possible (According to Jones (1959))

lithosphere in order to conceptualize dynamic models of the globe (Jones 1999; Juteau et al. 2008). It was at this time that the first magnetic soundings (Filloux et al. 1987) and then electromagnetics with a natural source (ELF)[54] (marine magnetotelluric or mMT) or an artificial source (VLF)[55] were experienced for the purposes of scientific research (Mott and Biggs 1963; Coggon and Morrison 1970; Cox et al. 1964, 1973; Vacquier 1972; Filloux 1973; Larsen 1975; Vanyan et al. 1978; Parkinson 1983; Geyer et al. 1983; Ferguson 1988).[56]

The electric conductivity determined by all these authors varied from 10^{-3} to 1 $mho.m^{-1}$, much lower than that found onshore (Palshin 1996; Virchow et al. 1999).

To supplement the lack of power of the natural sources (cf. Fig. 1.10), and to increase the accuracy, the use of artificial or controlled sources was then indispensable (Lawrence 1967).

It was probably during the RISE project (a campaign in the eastern Pacific in 1979) that the mCSEM method was born (Spiess et al. 1980). We then had a real system consisting of an 800 m length horizontal towed antenna (HED)[57] injecting a current at frequencies from 0.25 Hz to 2.25 Hz and with three field sensors. The results (cf. Fig. 1.11), supplementing the mMT previous works (cf. Fig. 1.12), show conductivities of the crust of about 0.004 $S.m^{-1}$ (1 Hz).

From that time, various techniques were proposed and systematically integrated into international research programs (Chave et al. 1981; Chave and Cox 1982; Edwards et al. 1981, 1985; Nobes et al. 1986), with the MOSES method (an acronym for magnetometric offshore electrical sounding) for investigations at

[54]Extremely low frequency (see Chap. 4 Sect. 3).

[55]Very low frequency. Frequency band used in the submarine telecommunications.

[56]The reader will find a comprehensive history of modern development of the methods in the article by Professors Constable and Snrka on mCSEM (Constable and Snrka 2007).

[57]A horizontal electromagnetic dipole.

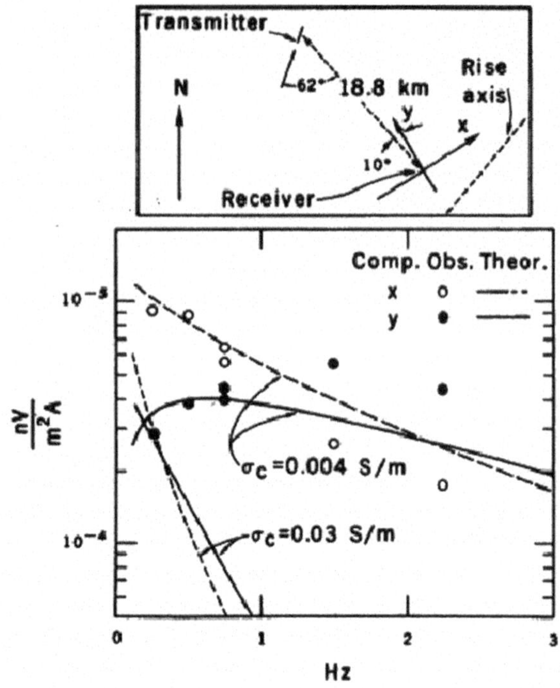

Fig. 1.11 Result of the first survey with a controlled source (HED) performed on a ridge in the eastern Pacific (According to Spiess et al. (1980))

lower water depths (cf. Fig. 1.13). These operations were carried out at low frequency (0.125 Hz) with a vertical transmission or VED (a galvanic source with little or no inductive effect) thus allowing use, to calculate the apparent resistivity, of propagation models with continuous current (a quasistatic approximation where the skin effect is neglected).

It was then in the years 1980–1990 that more convenient and efficient mEM devices were developed for the study of the midocean ridges (Cox et al. 1986; Filloux et al. 1989; Chave et al. 1991; Evans et al. 1994; Yungul 1996; MacGregor et al. 1998, 2001). This specific material (equipment) was developed in England, during this time, at Cambridge University at the instigation of Professor Martin Sinha, forming, so to speak, the first full mCSEM device. Then was joined to detection equipment based on the system of the Scripps Institute of Oceanography, a mobile and submerged transmitter antenna that can be towed (Sinha et al. 1990).

Alongside these studies some researchers proposed advanced ideas on new methods and magnetic exploration with the marine magnetometric resistivity method (mMR) for knowledge of the conductivity of the seafloor or the location of large terrestrial faults, such as Professor Nigel Edwards's group from Toronto University (cf. Fig. 1.14) and Professor Jean Mosnier's team from the Ecole Normale Supérieure de Paris.

To simplify the chronological understanding, a more comprehensive recent history is associated with each chapter. For a complete synthesis on the progress of the various stages of the implementation of the concepts, principles, techniques and instruments, the reader can refer to excellent articles by Professors Alan Chave,

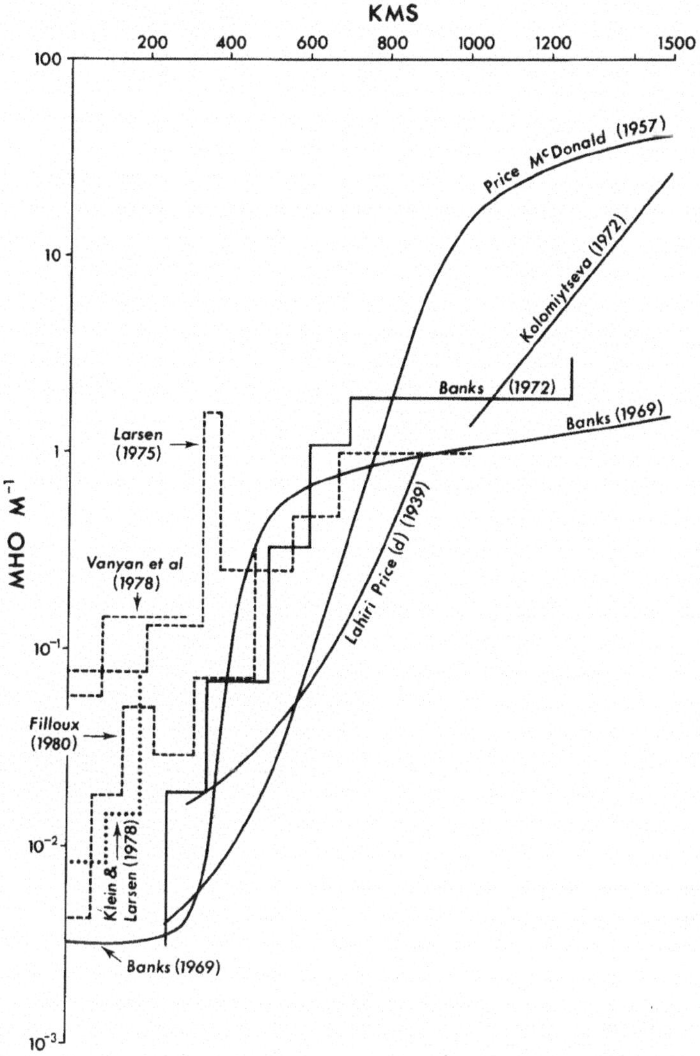

Fig. 1.12 Comparison of electrical conductivity measurements of the earth's and oceanic lithosphere (MT and mMT methods), depending on the depth, performed by different authors: on land (*solid curves*) and offshore (dotted curves) (According to Parkinson (1983))

Steven Constable, Leonard J. Srnka and Nigel Edwards (Chave et al. 1987; Edwards 2005; Constable and Srnka 2007; Chave 2009; Constable 2010, 2013; Key 2011) and the book by Professors Alan Chave and Alan Jones specifically devoted to the mMT method (Chave and Jones 2012).

I would like to remark that in the ocean, measurements of the electrical conductivity of the mantle were conducted mainly in the 1960s by researchers such as Price, Ashour, Zhigalov, Mason, Schmuker, Parkinson and Rikitake (Gaskell et al.

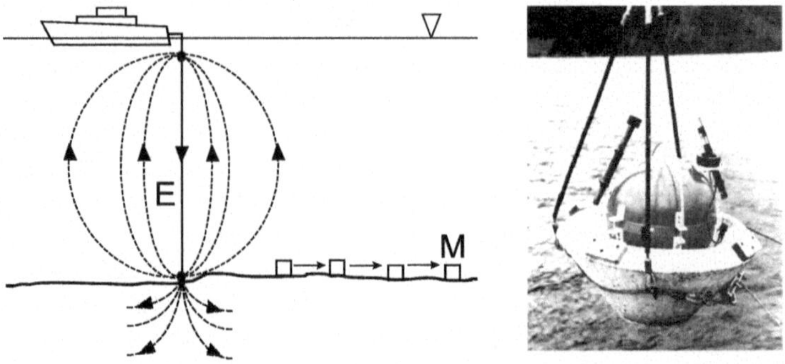

Fig. 1.13 Diagram illustrating the principle of the MOSES method developed in the 1980s for the study of the oceanic crust. The current injection line, symbolized by the electrodes E_1 and E_2, is vertical (VED), and the receivers at right placed on the *bottom* are magnetometers (M) allowing the measurement of the two horizontal components of the magnetic field (According to Chave et al. (1991))

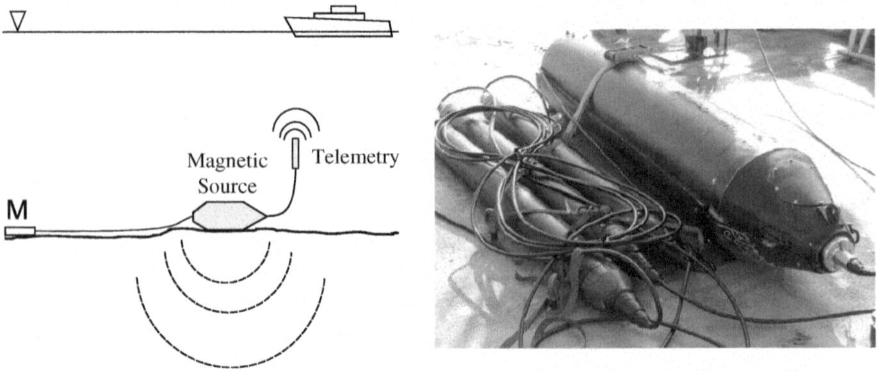

Fig. 1.14 Block diagram of magnetic surveys with an artificial magnetic source (dipole) and a vectorial magnetometer with two horizontal flux gates. The apparent resistivity is calculated from measurements of the transverse components of the magnetic field (According to Cheesmann et al. (1987, 1990, 1991). A rigth a 3 receptors with transmitter submersible magnetic sounding device. According to Edwards (2005))

1967). Due to the very low penetration of the electromagnetic waves (the *skin effect*), the studies to date were done with the only available data, i.e., those of the transient magnetic field variations, associated with the movement of telluric currents in depth. From the record of several years of fluctuations of this field, the discrimination of the parts with an internal and external origin, the analysis of the relations between them and the comparison with more or less sophisticated mathematical models (Coulomb et al. 1976), it was then possible to establish some values of the mantle conductivity ranging from 0.1 to 1 $S.m^{-1}$. This high conductivity value, mainly located in the upper part of the asthenosphere, would be explained by the significant presence of liquid (molten) carbonates directly related to volcanic activity (Gaillard et al. 2008).

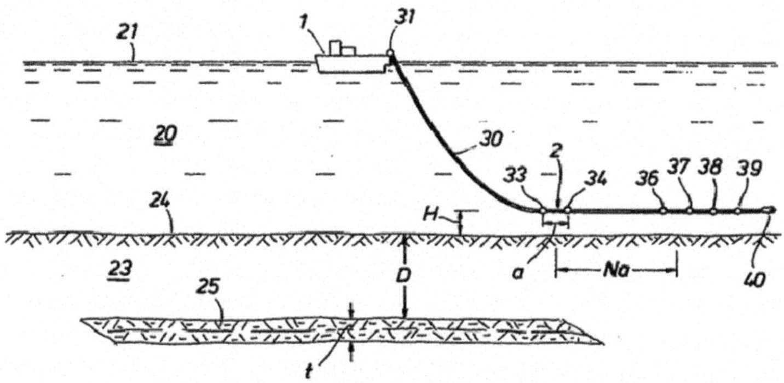

Fig. 1.15 Detail of the patent application filed by Exxon for the detection of hydrocarbons at sea (According to Srnka (1983))

6.4 The Innovators: Offshore Oil Prospecting

From this basic research, it was only a step for the oil industry to expeditiously implement practical applications. Before the innovators, since the 1950s, and more specifically in the mid-1980s when the industry began to get interested in mCSEM, a number of applied geophysicists and technologists (inventors) proposed a lot of solutions to directly map the oil contained in the reservoir structures. One can count more than 200 patents for seabed logging techniques at large. Not all are scientifically substantiated. Thus began a patent battle, as this book partly shows in this brief history.

It is probably to Professors Peter Bannister from the US Navy (Bannister 1968) and Charles Cox with Steven Constable from the Scripps Institution of Oceanography[58] in San Diego that we owe the first conclusive tests (Cox 1981; Fischer 2005; Constable 2010).[59] Some researchers (Hoehn and Warner 1983; Bahr 1988) and oil companies like Exxon/Mobil had previously made a few attempts without reaching commercially successful results (Srnka 1983, 1986) (Fig. 1.15).

On the industrial side, it is ultimately the Norwegian company Statoil that initiated in 1997 in Trondheim the first targeted studies conducted by Doctors Svein Ellingsrud and Terje Eidesmo, considered today as the pioneers of this industry (Eidesmo et al. 2002). For this major step forward these two researchers received in 2007 the very prestigious Virgil Kauffman Gold Medal delivered by the SEG. These studies have been completed by some laboratory experiments in Rotvoll (Norway) and a first demonstration of real size in 2000 in the Gulf of Guinea off the coast of Angola (West Africa), on the Girassol offshore field (1500 m depth) operated by the company Total/Fina/Elf (Total today).

---

[58]The world's largest Marine Research Institute involving several university laboratories, private and military (US Navy).

[59]The reader will find in this last article a comprehensive summary of this epic and a very concise summary of the SBL technique, especially applying to the uninitiated.

Meanwhile other teams, coming from the academic environment, such as the one of Professor Lucy MacGregor, University of Southampton,[60] were already engaged in more basic research, and more recently Professor Anton Ziolkowski's team, from the University of Edinburgh, positioned themselves very quickly in this innovation by offering a service adapted to oil and gas offshore.

Today, as in all sciences requiring large financial resources and heavy investments, these researchers, as shown by the many collective publications, join in common operations of cofinanced demonstrations.

6.5 The Inventors: Seabed Electrometry

Without dating back indefinitely, with among others Faraday's experiments on the river Thames,[61] the seabed electrometric measures, in the broad sense, began during the first half of the twentieth century: first from 1920 (Young et al. 1920) in the context of a dynamic study of the oceans (marine currents, tides, etc.) and then more partially and confidentially within detection activities and underwater discretion with the worldwide magnification of nuclear submarine fleets in particular (Cold War).

The first experiments, as well as those relating to prospecting itself (see Sect. 6.2), did not require elaborate electronic equipment (a potentiometric method). However, the investigations that were then conducted to deepen knowledge of the nature of the deep layers of the oceanic crust (research in global geodynamics) thus necessitated much more efficient equipment.

It was therefore in the early 1970s, with the advent of electronics (transistors and then integrated amplifiers) that such investigations could be carried out (Filloux 1977; Cox 1981). The first electrometers were developed by the Scripps Institution of Oceanography (Cox et al. 1971, 1973, 1978; Filloux 1973). However, these devices, to deal with the electronic noise inherent to the technology of the moment and with instrumental drift, needed to be supplemented to counter these negative effects by more or less effective electromechanical systems (cf. Fig. 1.16a).

It was only in the years 1980–1990[62] that it became possible with the advent of low noise electronic components, and more specifically thanks to instrumentation amplifiers, to measure below the background noise. It was therefore only at this time that the technology became truly effective and ultimately essentially practiced in earth physics (Edwards et al. 1985).

[60]The authorship of these innovations is currently hotly debated. The industrial stakes are high, and several trials are ongoing (see Chap. 2, Footnote 47).

[61]Experience at Waterloo Bridge (Faraday 1832). After laying two copper plates connected to a sensitive galvanometer in the Thames, a short distance from the banks of the river, the English scholar noted in the circuit the forming of a small electric current. The explanation for this phenomenon of induction (a moving conductor in a magnetic field) would be given a few decades later by Lorentz (see Chap. 3 Sect. 6.9.2).

[62]Previously, after the introduction of total field magnetometers in the 1960s (proton or optical pumping magnetometers), sensitive magnetic variometers appeared in the 1970s. These ones (declinometer, H meter and Z meter) offered the possibility of simultaneously measuring the two components of the field (Mosnier 1977).

Fig. 1.16a One of the very first seabed electrometers, built in the 1970s and launched in 1976 (see details in Chap. 4) (According to Chave et al. (1991))

Fig. 1.16b Proposal (French patent) for a method for measuring the field with the aim of detecting the resistivity contrast of the underlying grounds (According to Duroux (1974))

Meanwhile other technologies have been proposed. This is the case, for example, for induced fields measures (Duroux 1974), a method specifically adapted to offshore exploration (cf. Fig. 1.16b),[63] or even for current density measurements, for which the apparatus was developed by the team of Professor Jean Mosnier from the Ecole Normale Supérieure of Paris in the 1980s, for a variety of applications ranging from underwater detection to mining exploration, and is now being introduced as part of SBL (see Sect. 5.3 Chap. 4).

The Duroux acquisition device includes:

- A low frequency towed transmitter (1–100 Hz) constituting a transmitter (loop shown in the figure) or electric transmitter (not shown)

[63]This comes from the Melos process, a magneto-electric method using a surface wave, developed in particular by the Bureau de Recherches Géologiques et Minières (Duroux 1967).

- A receiver on the vessel (electrically isolated) allowing by suitable processes identification of anomalies by measurements of either the dip in the magnetic field (amplitude H_x/H_z ratio or their phase difference) or the impedance (E_y/H_x or their phase difference), more generally by a measure of one of the transverse components of the electric and/or magnetic fields.

6.6 The Industrial Rise: A Very Recent Activity

We can say that the first attempts at prospecting, at an industrial scale, are a credit to the Norwegian oil company Statoil, which then gave to its researchers the responsibility for the development of the service through the company EMGS (for Electromagnetic Geoservices), specifically created for this task. This had as a result the discovery in 2000 of an oil reservoir off Angola (Ellinsgrud et al. 2002), which became a reality 2 years later by the first commercial job (the Ormen Lange field).

Today this activity is all the less marginal compared with the overall activity of offshore prospecting and particularly of seismics. Yet few companies, mostly startups, have entered this market gap, whose turnover increases very quickly (cf. Sect. 8) (Fig. 1.17).

7 Philosophy and Interest of SBL Methods

The term SBL may broadly contain all marine and underwater electromagnetic methods and techniques using towed or untowed equipment, submerged in open water or resting on the bottom of the sea. The philosophy consists of studying the variation of amplitudes and phases in the received fields as a function of the source–receiver offset, with mathematical modeling using geological hypothesis and data.

Roughly the following can be distinguished:

- The methods with controlled sources which are opposed to telluric sources
- The DC and AC principles
- The techniques that apply in the frequency and time domains
- The electrokinetic (electric potentials) and electromagnetic measurement (electric and/or magnetic fields) technologies

We can say that these methods, techniques and technologies can be combined in many possible ways and that those proposed today will certainly be complemented in the future by other probably more efficient processes (Fig. 1.18).

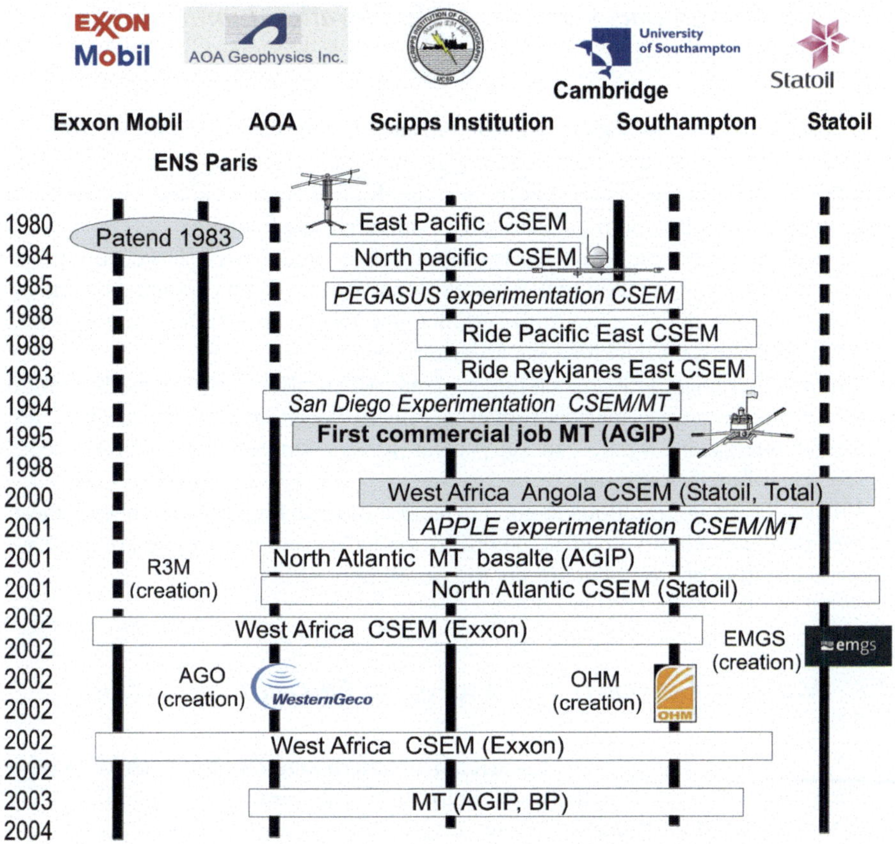

Fig. 1.17 Chronological summary of seabed electrography investigations (mCSEM) since the first measurements in the earth's physics to its economic development with the creation of companies in the 2000s. This diagram does not include more or less confidential experiments conducted in the underwater detection domain and those in studies of the seaboard effect (1972–1989) (According to Constable (2010))

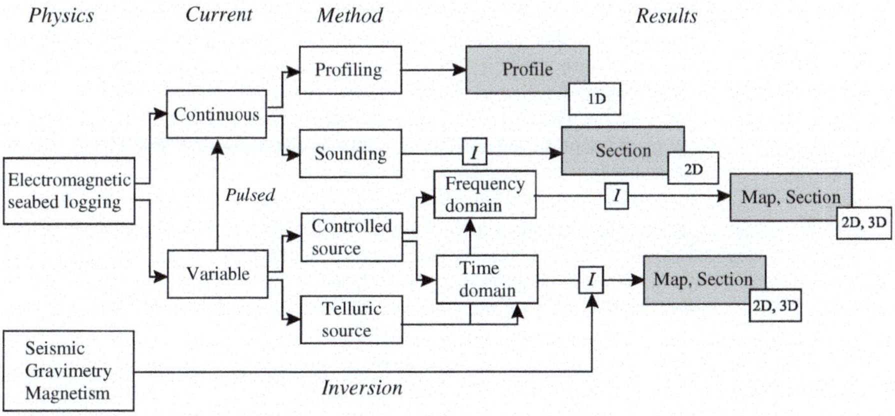

Fig. 1.18 Different EM SBL principles, methods and techniques placed in an overall scheme. Today, the results also tend to be proposed as a mathematically reconstituted image (imagery)

7.1 *The Continuous Current: An Exploration Limited to the Vertical*

The continuous current widely used in subsurface exploration (civil engineering, hydrogeology, metal mining, archeology, materials research, etc.) has serious drawbacks since we are interested in large geological structures and especially in its employment in the marine electrically conductive medium.

Indeed, investigation with DC, by its principle, integrates a large volume of ground confining the method to vertical exploration only, if this volume is made up of horizontal (or under a low dip) sedimentary layers, in a limited number and laterally of a constant thickness (as an assumption).

On the other hand, unlike sounding on land, which offers a theoretically infinite depth of investigation, marine sounding, due to very rapid dissipation of the electrical energy in water, can only offer a very limited investigation depth (a few tens of meters at most).

Finally, DC is unpropitious to good detection of insulating bodies. Indeed, these bodies, unlike conductive bodies, do not let the DC current go through, but instead allow alternative or variable currents to spread more easily.

7.2 *Variable Currents: The Accessible Lateral Exploration*

In oil exploration, the possibilities offered by alternating currents, thanks to their guiding properties, allow exhaustive exploration, both vertical and lateral.[64] They therefore attracted early interest from researchers and were tested and used with more or less success (Horton 1946).[65]

In fact it was with the emergence of the telluric methods, electrotelluric[66] first (Migaux 1948; Porstendorfer 1960; Berdichevsky 1965) then magnetotelluric[67]

---

[64]This is also the case in mineral exploration for the detection of conductive masses.

[65]For example, the low frequency inductive methods (the methods of the spiral or hoop developed by the SPE in the 1930s) for measuring the dips of deep geological structures (an indirect method), or the high frequency methods (radio waves) using the phenomena of reflection and refraction (or absorption) of the waves on the roof of hydrocarbon reservoirs (a direct method). With a lack of conclusive results, these technics proposed by virtually all companies at that time (1930–1950) were quickly abandoned in favor of seismics (an indirect method).

[66]The electrotelluric method developed in the 1930s by the SPE and implemented in France in 1940 by CGG had great success after the war in Europe and North Africa with the discoveries, among others, of the deposits of Saint Marcet (France) and Hassi Messaoud (Algeria). Lack of a sufficient market (especially absent in the USA) and with more restrictive conditions than seismic reflection, which asserted itself more and more, the electrotelluric method was gradually abandoned despite a revival and development in the Soviet Union.

[67]This method, where we simultaneously measure the variations of the electric and magnetic fields, should rather be called the electromagnetotelluric method.

(Cagniard 1953a, b, 1956) that it was a success. This time, use of variable currents at very low frequencies, naturally propagating in nappes in the subsoil, then allowed us to reach significant depths of investigation and acceptable resolutions in relation to the considered geological objects.[68]

Therefore the researchers naturally turned toward this type of current, whether telluric or artificial, to approach the problem of the detection of hydrocarbons at sea.

Covering the range of ultra and very low frequencies (0.01 Hz to 10 Hz), the propagation in a conductor (seawater) environment as well as the detection of insulating objects (hydrocarbon deposits) are thereby made much easier[69] than with DC. In this case, the depth of investigation related to the frequency and conductivity of the traversed media corresponds to a few hundred meters for the controlled source methods and can reach several thousand meters for the natural source methods.

8 Industry of EM Seabed Logging

The global market for geophysics, all techniques together, including sales of equipment and services (acquisition/processing), represents an annual turnover with strong growth in recent years.[70]

In addition to the SBL, promising technologies today are wide and multiple-azimuth seismics, which improve the resolution and particularly the imagery of the subsoil and 4D seismics with permanent seabed systems.

8.1 The Market: A Highly Growing Activity

The SBL industry turnover increased in 5 years from $30 million to $700 million annually,[71] which remains a relatively low percentage compared with marine seismics (92 % of the market) and even more regarding the four major oil companies' income estimated at $800 billion.[72] But, on the other hand, this technology has had exponential growth, which, let us hope, if the results are confirmed, will achieve a much more important turnover in the coming years, comparable to the one due to the emergence and progression of 3D seismics in the 1990s (Fischer 2005; Ellingsrud et al. 2008).

---

[68]Meanwhile geomagnetic soundings were completed (Barsczus 1970).

[69]For a sufficient resistivity contrast.

[70]Rather variable turnover, very sensitive to economic conditions.

[71]With a wide disparity between the different companies ($140 million for EMGS in 2007) and a significant decline in recent years (2009–2010).

[72]Exxon/Mobil, Shell, BP and Total.

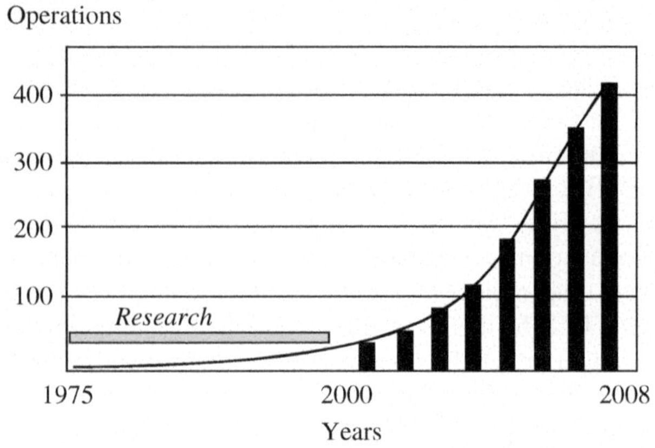

Fig. 1.19 Cumulative evolution of the EM operation numbers at sea since the introduction of the method to the market in 2000

With a similar survey, the price of a seabed logging campaign is slightly higher than that of a 3D seismic campaign. However, these costs are to be compared with the rates of discovery, which are of course at the moment in favor of SBL, but as the latter are integrated into the overall processes of investigation (combined seismic prospecting + EM SBL), the analysis may seem a priori more difficult to conduct. On the other hand, SBL cannot be applied everywhere and the success also depends on the available additional data that are necessary for the interpretation of the data, especially in the inversion processes.

In any case, the economy will be seen in drillings and downhole logs, or, if the discoveries are significant (large deposits), from the start of operations. For example, we can estimate the cost of an operation at $2 million (20 times less than deep drilling), and it can go for a full 3D survey in the Gulf of Mexico to more than $10 million (*Offshore* magazine, 15 July 2010, source EMGS). By comparison, in 40 years, the cost of seismics was multiplied by five and the cost of drilling by ten.

Today this activity is still marginal to say the least. Yet few companies have engaged in this niche. At the end of 2005, nearly 15,000 km of EM profiles had been made. Today about 700 commercial operations have been conducted around the world by a few companies on the international market. Since 2008, the activity has suffered from two crises: one in 2007 because of growing too fast for an immature technology with aggressive marketing, and in 2015 because of the collapse of oil barrel prices with the increase of shale oil production in the USA (Fig. 1.19).

8.2 The Active Companies in this Market

The main service companies in the past and present markets have been very limited. Thus, in alphabetical order, we have:

- EMGS—Electromagnetic Geoservices (Norway)
- Exxon Mobil (USA)

- OHM—Offshore Hydrocarbon Mapping (England)
- MTEM—Multi-Transient Electro-Magnetics (Scotland)
- Schlumberger, with its subsidiary company WesternGeco (USA)
- PGS—Petroleum Geo-Services (Norway)
- Petromarker (Norway)
- Seabed Geosolutions (the Netherlands)

These companies established in the early 2000s (cf. Fig. 1.15) were developed from three different strategies:

- EMGS[73] was founded in 2002 by defectors (Drs. T. Eidesmo, S. Ellingsrud and Ståle Johansen) from the Norwegian oil company Statoil. Today the leader company to exploit the process.
- Exxon Mobil at the origin of the method (US patent Srnka) internally developed for its own service the R3M for reservoir resistivity mapping.
- OHM was created by academics (Drs. Martin Sinha and Lucy MacGregor) from the University of Southampton. This company has now ceased its activity.
- MTEM was created in 2004 by academics (Dr. David Wright and Pr. Anton Ziolkowski) from the University of Edinburgh. PGS acquired this company in 2007.
- WesternGeco (a US subsidiary company of Schlumberger) began its activity from external acquisitions (AGO Geomarine: the marine division of the company AOA Geophysics, which uses patents and technology developed by Dr. Steven Constable of the Scripps Institute of Oceanography, University of California, San Diego).
- PGS, a services company, today bases its strategy on simultaneous acquisition of seismic and electromagnetic data thanks to a new type of streamer developed in collaboration with the University of Edinburgh.[74]
- Petromarker, formed in 2004 by ORG Holding AS, is a private Norwegian services company using transient EM vertical stimulation and measurement.[75]
- Seabed Geosolutions, the youngest, is a joint venture between Fugro (60 %) and CGG (40 %).

All firms together, the business today employs nearly 700 people (full time) worldwide, mostly engineers, technicians and researchers. This very capitalizing domain also includes a fleet of ten high sea specialized ships (heavy duty) and several data centers exclusively dedicated to the tasks of EM data interpretation.

In 2014 the total income of this business was about $450 million, with US $198 million for EMGS, the principal services company (300 employees). With the oil crisis we are experiencing now, certain service companies were forced to temporarily suspend their mEM activity or disappeared completely.

[73]Currently in a joint venture with the operator Fugro.

[74]Probably the most advanced company in this sector.

[75]First job in 2006.

Besides these companies, a small number of university or research laboratories build equipment and perform missions and experiments with an industrial character from scientific works.

9 Environmental Impact

Today, it goes without saying that when a new technology appears on the market, particularly in the field of energy, its direct and indirect impacts on the environment are systematically analyzed and carefully evaluated.

9.1 Effect on the Number of Exploration Drillings

With an oil shortage that is arriving with no surprise in the coming decades, EM SBL appears to be one of the answers to limit the drilling risks in deep water by simply reducing them. Indeed, the large number of dry holes resulting from only seismic exploration speaks for systematic use of SBL campaigns (Fig. 1.20).

9.2 Effect on Submarine Fauna

If marine seismics have some impact on aquatic fauna (BF sound sources such as air guns go up to 250 dB re 1 μPa at 1 m propagating energy hundreds of miles away), it is clear that seabed logging has in return a very small impact.

Specialized scientific studies (Girondot 2007) show that despite the high power electrical sources (especially in amperage: 1500 A) used in SBL, the electric fields possibly able to confuse some species (causing strandings, for example) would virtually have no impact on the behavior of these animals.[76]

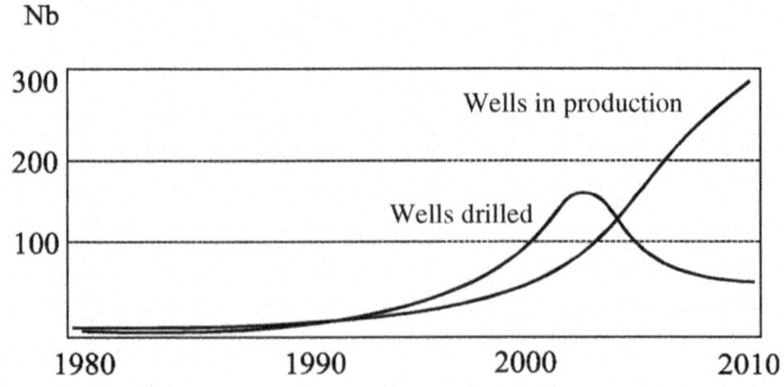

Fig. 1.20 Number per year (Nb) of drillings and producing wells over 1000 m deep in the Gulf of Mexico between 1980 and 2010

[76]See detailed calculations in the Appendix A3.2.

However, intensities of 10 nT (corresponding to the values of a magnetic field 10 km away from the source) could be felt by turtles, cetaceans[77] and, to a lesser extent, sharks more sensitive to the electric field than other animals without a priori a significant change in their behavior.

10 General Presentation of the Book

This book is primarily a teaching and educational work in which the effort is focused on the description of the principles and basic concepts. Repetitions are numerous to avoid references and immediately allow, by bringing together all the elements relating to a matter, better understanding of the studied phenomena.

An important part is devoted to metrology and its corollary: instrumentation, capital science and technology when operational and instrumental limits are reached.

This book only briefly discusses the equally important interpretation because, as pointed out by Jean Goguel in the preface of Professor Muraour's book (Muraour et al. 1970):

> Before thinking about interpretation, we must have the desire to obtain the best possible measures, and treat them so as to highlight the part of useful information they contain.

The remainder of this volume is divided into five parts, chronologically chaptered, and covers all of the main methods used at this time.

Chapter 2 reviews the different detection principles that support the main EM seabed logging methods. Based on the concepts and laws of electromagnetism (Ampère, Faraday, Ohm and Maxwell), this chapter provides an overview of the physical phenomena contributing to the various configurations of the exploration and widely recalls in its appendices the basic relations, theorems and theoretical formulations used in the demonstrations.

Chapter 3 is devoted to metrology. It approaches the question of the propagation and diffusion of electromagnetic energy in a particular environment consisting of seawater and marine sediments, and that of the detection of hydrocarbon deposits, with their constraints and limitations. It also specifies the objectives and orders of magnitude of interesting signals.

Chapter 4 describes the instrumentation, different apparatus, sources and detectors that are now used or are going to be used, the associated treatment techniques, especially those concerning analog electronics, which assumes here an absolutely essential interest (measure of very weak fields). This chapter also briefly discusses

[77]Note that these animals for their orientation are a priori sensitive to changes in continuous magnetic fields (the earth field) or over very long periods (telluric currents). At the present day, the currents used in SBL vary in a frequency band from 0.25 Hz to 10 Hz, which logically should not affect these animals.

the devices, specific methods and technical means of implementing equipment for surveys in deep waters.

Chapter 5, the longest with its appendices, but maybe less comprehensive, gives an overview of the methods of interpretation by mathematical modeling, with a deliberate choice of exposing in a more fully diversified manner the resolution of the forward problem by analytical methods. More intelligible, because they are very close to the physics, these latter will, hopefully, make easier the description and understanding of the physical phenomena, emphasizing the complexity of the problem as the data interpretation a posteriori. Numerical and inversion methods, which are briefly mentioned but remain the foundation of interpretation in different dimensions, may be supplemented, if necessary, by reading articles and monographs (specialized papers) quoted in the text and referenced in the bibliography. Finally, a few words are devoted to experimental analog models (rheostatic tanks) still in force, and to migration and imagery methods, well known by seismicians.

Chapter 6 presents and summarizes some historical examples of exploration (case histories), which the technical press has talked about. The latter discussion, dealing with some different cases, can serve as a reference and interpretative basis for geological applications. The reader is encouraged to continue and complete this work especially with reading of periodicals (generalist newspapers), a list of which is given at the end of the volume.

On a formal level, some chapters are enriched by one or more appendices, which clarify certain concepts briefly discussed in the text, and by a list of references, which allow readers and especially geophysicists to deepen their knowledge on accurate points. The latter can also be complemented by a list of items not referenced in the text at the end of the volume.

Finally, for a broader understanding of the processes, principles, methods, techniques, technologies, etc., small historical notes, improving the one presented in the introduction, have also been added to several developments (chapters). They will, I hope, offer young people a less dogmatic and more vivid presentation of this new discipline by giving an overview of the evolution of ideas and thought.

References

Abrikossov I, Goutman I (1986) Géologie du pétrole. Généralités, prospection, exploitation. Ed. Mir. 301 p

Agricola G (1556) De re metallica. English translation de H C and LH Hoover. Ed. Dover

Allaud L, Martin M (1976) Schlumberger: histoire d'une technique. Ed. Berger-Levrault

Anonyme (1983) A short history of electrical technique for petroleum exploration

Bahr K (1988) On the feasibility of seafloor magneto-telluric soundings in the North Sea. Ocean Dyn Vol. 41, no. 3–6. Ed. Springer. pp 277–281

Bannister PR (1968) Determination of electrical conductivity of the sea bed in shallow water. Geophysics 33:995–1003

Bannister PR (1968) Electromagnetic fields within stratified earth produced by a long horizontal line source. Radio Sci 3:387–390

Baron, Breusse (1926) Essais de résistivités sur le Prien Lake Charles, 29 October 1926, note dactilographiée. 4 p
Barret WM (1947) Method for electromagnetic wave investigations of earth formations. US patent no. 2426,918. Sept. 2
Barret WM (1949) The Radoil method. A geophysical technique for the direct location of oil. 33 p
Barsczus HG (1970) Sondages géomagnétiques: bibliographie. Ed. ORSTOM de M'Bour. 85 p
Bauquis PR (2001) A reappraisal of energy supply and demand in 2050. Oil & Gas Sci Technol Rev. IFP Vol. 56, no. 4, Ed. Technip. pp 389–402
Berdichevsky MN (1965) Electrical prospecting with telluric current method. Traduit du russe par GV Keller. Ed. Quarterly of the Colorado school of mines. Vol. 60
Billings MP (1959) Structural geology. Ed. Prentice-Hall. 514 p
Bird et al. (2008) Circum artic resource appraisal: estimates of undiscovered oil and gas north of the Arctic Circle. USGS fact sheet 3049
Bjorlykke K (2010) Petroleum geoscience: from sedimentary environments to rock physics. Ed. Springer. 340 p
Brown FH (1903) Apparatus for locating metals mineral, buried treasures etc. without digging. US patent no 727,077
Bogdanov, Ivanov (1960) Les courants électriques dans la mer et leur mesure par l'appareil EMIT. Trudy Inst. Okeanol. SSSR, 39. pp 80–84 (in Russian)
Cagniard L (1953a) Procédé de prospection géophysique. Brevet français no. 1025683
Cagniard L (1953b) Principe de la méthode magnéto-tellurique, nouvelle méthode de prospection géophysique. Annales de géophysique, Tome 9, fascicule 2. pp 95–125
Cagniard L (1956) Electricité tellurique. Handbuch der Physik. Geophysik 1. Ed. Springer. pp 406–469
Cagniard L, Morat P (1966) Extension de la prospection magnéto-tellurique à l'exploration offshore. 3 p
Calvaresi MC (1954) SORGE S.p.A. Società organizzazione ricerche geofisiche. Brochure publicitaire en italien. 47 p
Calvaresi MC (1957) Procédé de repérage direct des gisements géologiques et de minerais et appareil pour l'exécution de ce procédé. Brevet français no. 1.180.109. 25 July
Castel D du et al. (1995) Les aventuriers de la terre. CGG : 1931–1990. Ed. La Sirène. 95 p
Cercher, Dermont (1935) La prospection électrique profonde. Bulletin of the American association of petroleum Geologist no. 1
Chapman RE (1983) Petroleum geology. Developments in petroleum science. Vol.16. Ed. Elsevier. 415 p
Chave AD (2009) On the electromagnetic fields produced by marine frequency domain controlled sources. Geophys J Int 29 p. doi:10.1111/j.1365-246X.2009.04367.x
Chave AD, Cox CS (1982) Controlled electromagnetic sources for measuring electrical conductivity beneath the oceans, 1, forward problem and model study. J Geophys Res 87:5327–5338
Chave AD, Jones AG (2012) The magnetotelluric method: theory and practice. Ed. Cambridge University Press, Cambridge, pp 23–24
Chave AD, Von Herzen RP, Poehls KA, Cox CS (1981) Electromagnetic induction fields in the deep ocean northeast of Hawaii: implications for mantle conductivity and source fields. Geophys J R Astron Soc 66:379–406
Chave AD, Constable SC, Edwards RN (1987) Electrical exploration methods for the seafloor. Electromagnetic methods in applied geophysics Vol.2 Applications. Ed. by MN Nabighiam, SEG investigation in geophysics no. 3. pp 931–966
Chave AD et al. (1991) Electrical exploration methods for the sea floor. Electromagnetic methods in applied geophysics. Vol. 2. Chap. 12. Ed. Society of exploration geophysicists. pp 931–966
Cheesman SJ, Edwards RN, Chave AD (1987) On the theory of seafloor conductivity mapping using transient electromagnetic systems. Geophysics 52:204–217
Cheesman SJ, Edwards RN, Law LK (1990) A test of a short baseline seafloor transient electromagnetic system. Geophys J Int 103:431–437
Cheesman SJ, Edwards RN, Law LK (1991) Porosity determination of sediments in Knight Inlet using a transient electromagnetic system. Geo Mar Lett 11:84–89

Mc Clatchey AF (1900) Electrical prospeting apparatus US Patent 681,654
Coggon JH, Morrison HF (1970) Electromagnetic investigation of the seafloor. Geophysics 35 (3):476–489
Collectif (1966) Techniques marines pour la recherche et l'exploitation du pétrole, IFP
Constable S (2010) Ten years of marine CSEM for hydrocarbon exploration. Geophysics 75 (5):75A67–75A81
Constable S (2013) Instrumentation for marine magnetotelluric and controlled source electromagnetic sounding. Geophys Prospect 61(Suppl. 1):505–532
Constable S, Srnka LJ (2007) An introduction to marine controlled source electromagnetic methods for hydrocarbon exploration. Geophysics vol 72, no. 2. Ed. Society of exploration geophysicists
Coulomb et al. (1976) Traité de géophysique interne. Tome 2. Magnétisme et géodynamique. Ed. Masson. pp 138–154
Cox CS (1981) On the electrical conductivity of the oceanic lithosphere. Phys Earth Planet Inter 25:196–201
Cox CS et al. (1964) On coherent electric and magnetic fluctuations in the sea: studies on oceanography. Tokyo University press, Tokyo, p 449–457
Cox CS et al. (1971) Electromagnetic studies of ocean currents and the electrical conductivity below the sea floor. The sea. Ed. Willey & Sons
Cox CS et al. (1973) Plate tectonics and geomagnetic reversals. Ed. Freeman. 702 p
Cox CS et al (1978) Electromagnetic fluctuations induced by wind waves on the deep-sea floor. J Geophys Res 83:431–442
Cox CS et al. (1986) Controlled source electromagnetic sounding of the oceanic lithosphere. Nature 320(6057):52–54
Daft L and William A (1902) Electrical apparatus for detecting, localizing mineral deposits. US patent no 817,736
De White L (1958) Apparatus for fogging the sea floor. US patent no. 2839721. June 17
Delesse AEOJ (1872) Lithologie des mers de France et des mers principales du globe. 2 vol. Ed. Lacroix
Diolé P (1951) L'aventure sous-marine. Ed. Albin Michel. pp 249–252
Dorozynski A, Oristaglio M (2007) Le sens du courant, la vie d'Henri Georges Doll, inventeur. Ed. Le cherche midi. 230 p
Duroux J (1967) Caractères de l'onde électromagnétique de surface engendrée par un dipole magnétique. Application à l'investigation en profondeur de la résistivité électrique du sous-sol. Geophys Prospect 15(4):564–583
Duroux J (1974) Procédé et appareil de prospection en mer par mesure de champ électromagnétiques. Brevet français no. 74 26465
Edwards N (2005) Marine controlled source electromagnetics: principles, methodologies, future commercial applications. Survey in geophysics. Ed Springer. pp 675–700
Edwards RN, Law LK, De Laurier JM (1981) On measuring the electrical conductivity of the oceanic crust by a modified magnetometric resistivity method. J Geophys Res 86(B12):11 609–11 615
Edwards N et al (1985) First results of the MOSES experiment: sea sediment conductivity and thickness determination, Bute Inlet, British Columbia, by magnetometric offshore electrical sounding. Geophysics 50:153–160
Eidesmo T et al. (2002) Sea bed logging (SBL), a new method for remote and direct identification of hydrocarbon filled layers in deepwater areas. First break. Vol. 20.3 March
Ellingsrud et al. (2002) Remote sensing of hydrocarbon layers by seabed logging: results from a cruise offshore Angola. Lead Edge 21(10):972–982
Ellingsrud S et al. (2008) CSEM: a fast growing technology. SEG Las Vegas 2008 annual meeting. Recent advances and road ahead
Evans et al. (1994) On the electrical nature of the axial melt zone at 13°N on the East Pacific Rise. J Geophys Res 99:577–588

Ewing M, Woollard GP, Wine AC (1940) Geophysical investigations in emerged and submerged Atlantic coastal plain, pt IV. Bull Geol Am 51:1821–1840

Faraday M (1832) Bakerian Lectures. Experimental research in electricity. Phil Trans Roy Soc London, part 1, p 176

Ferguson IJ (1988) The Tasman project of seafloor magneto-telluric exploration. Sripps Institution of Oceanography. 436 p

Filloux JH (1973) Techniques and instrumentation for study of electromagnetic induction at sea. Phys Earth Planet Inter 7(3):323–338

Filloux JH (1977) Ocean floor magneto-telluric sounding over North Central Pacific. Nature 269:297–301

Filloux et al. (1987) Geomagnetism. 3 volumes. Ed. Academic Press, London

Filloux JH, Law LK, Yukutake T, Segawa J, Hamano Y, Utada H, White A, Chave A, Tarits P, Green AW (1989) Offshore EMSLAB: objectives, experimental phase, and early results. Phys E Pl Int 53:422–431

Fischer PA (2005) New EM technology offerings are growing quickly. World Oil 226(6):1–9

Fonarev GA (1961) On variations of telluric currents in the sea. Geomagn Aeron 1:417 (in Russian)

Fonarev GA (1964) Distribution of electromagnetic variation with depth in the sea. Geomagn Aeron SSSR 4(6):881–882 (in Russian)

Foranev GA (1963) Sur les courants telluriques verticaux dans la mer. Geomagn Aeron SSSR 3(4):784–785 (in Russian)

Fox RW (1830) On the electromagetic properties of metalliiferous veins in the mines of Cornwall. Philos Trans 120(pt 2):399

Gabriel VG (1945) Mathematical chance of finding oil. Oil weekly. 21 May 1945

Gaillard F et al. (2008) Carbonatite melts and electrical conductivity of the asthenosphere. Science magazine, Nov

Gaskell TF et al. (1967) The earth mantle. Ed. Academic Press, London/New York, p 163–165

Geyer RA et al. (1983) Handbook of geophysical exploration at sea. Ed. CRC Press, Boca Raton, 397–416

Gilbert W (1628) De magnete. Traduction anglaise de 1958. Ed. Basic Books

Girondot M (2007) Impact de champ électromagnétiques sur la faune protégée au large des côtes de Guyane françaises. Ed. Université Paris Sud

Gorbenko A (1936) Les recherches des gisements de pétrole par les méthodes de l'exploitation électrique de surface. Traduit du russe. 15 p

Hallam A (1976) Une révolution dans les sciences de la terre: de la dérive des continents à la tectonique des plaques. Ed. Du seuils Points. 186 p

Hedstrom H (1930) Geo-electrical exploration methods. The Oil Weekly, vol. 58, July 25, pp 34–36

Hess HH (1962) History of ocean basins: Petrologic studies: a volume in honor of A. F. Buddington. Geological Society of America, Boulder, p 599–620

Hessler VP (1962) Characteristics of telluric currents at land and sea based stations. J Phys Soc Japan. 17 supp. Al., part 1. pp 32–34

Hoehn GL, Warner BN (1983) Magneto-telluric measurements in the Gulf of Mexico at 20 m ocean depths. Handbook of geophysical exploration at sea. Ed. CRC Press, p 397–416

Horton CW (1946) On the use of electromagnetic waves in geophysical prospecting. Geophysics 9:505–518

Iakanov D' BP (1957) Nature des courants électriques naturels et leur étude sur le fond des océans. IZV. AN SSSR. Ser. Geofiz, no. 6. pp 800–802

IEA (2016) Medium-term oil market report 2016. 22 February, 12 p

Ivanov VI, Kostomarov DP (1963) Computation of electric currents induced in the sea by Sq Variations. Geomagn Aeron SSSR, 3(6):868–875 (in Russian)

Jones SB (1959). Ocean bottom stratigraphy. US patent no. 2872638. Feb. 3

Jones EJW (1999) Marine geophysics. Ed. John Willey. 474 pp

Juteau T, Maury R, Vewer P de (2008) La croûte océanique: pétrologie et dynamique endogénes. Ed. Vuibert
Karcher JC (1974) The reflection seismograph, its invention and use. The discovery of oil and gas fields. Ed. American Institute of Physics
Kenny S (2011) Webcast: meeting the challenges of arctic development. Offshore. West Africa report. Feb. 8
Kertz R, Glassmeier KH (1999) Geschichte der Geophysik. Ed. Georg Olms
Key K (2011) Marine electromagnetic studies of seafloor resources and tectonics. Springer. Surv Geophys. 33 p
Kolovski M (1961) Short period variations of the earth's magnetic field in Poland. Acta Geophys Pol 2(3):205–226
Krajev AP et al. (1948) Premier essais de sondage électrique très profond de l'écorce terrestre. Traduit du Russe, Université Vestnik de Leningrad URSS
Larsen JC (1975) Low frequency electromagnetic study of deep mantle electrical conductivity beneath the Hawaiian Islands. Geophys J R Astron Soc 43:17–46
Laudon RC (1996) Principles of petroleum geology. Ed. Prentice Hall, Upper Saddle River, 267 p
Launay L, Lichtman SW, Seltzer E (1964) Deep sea magnetic and telluric registrations made in the Puerto Rico Trench with the bathyscaphe Archimède. N.B.S. Special report. Boulder Symposium
Launay L, Lichtman SW, Seltzer E (1965) U.L.F. environment of the ocean floor. N.B.S. Special report. Washington
Lawrence LG (1967) Electronics in oceanography. Ed. Howards W Sams, Indiana Polis, p 52
Le Grand J et al. (1952) Introduction à l'électromagnétisme des mers. Ann Inst Océanogr XXVII:235–329
Le Pichon X (1968) Sea floor spreading and continental drift. J Geophys Res 73:3661–3697
Lee L, Schwarz A (1932). Recherche de pétrole par les méthodes de la résistivité. Technical paper no. 521.
Lees GM (1940). The search for oil. The geographical journal. Vol. XCV no. 1. 18 p
Leimbach G, Löwy H (1910) Verfahren zur Systematischen Erforschung des Erdinnern Gröβerer Gebiete Mittels ElektrischerWellen. Patent DE237944. Filed on 15 June 1910
Levorsen AI (1954) Geology of petroleum. W H Freeman and Company, San Francisco
MacGregor L et al (1998) The Ramesses experiment III controlled source electromagnetic sounding of the Rekjames Ridge at 57°45'N. Geophys J Int 135:772–789
MacGregor L et al (2001) Electrical resistivity structure of the Valu Fa Ridge, Lau Basin, from marine controlled source electromagnetic sounding. Geophys J Int 146:217–236
Melton BS (1933) Electromagnetic prospecting method. US patent no. 2077707 April 20
Migaux L (1948) Une méthode nouvelle de géophysique appliquée: la prospection par courants telluriques. Compagnie Générale de Géophysique
Mosnier J (1967) Application des techniques de pompage optique à l'étude des gradients géomagnétiques. Cas particulier de l'effet « bord de mer ». Doctoral thesis. Laboratoire de géophysique de l'Ecole normale Supérieure de Paris
Mosnier J (1977) Magnétisme et électromagnétisme en géophysique interne. in De la thermodynamique à la géophysique, hommage au professeur Yves Rocard. P. Aigrain and P. Nozières. Ed du CNRS. p 251
Mosnier J (1986) La détection magnétique des navires. Aspect géophysique. Revue : L'armement. NS no. 4 September. pp140–149
Mott H, Biggs AW (1963) Very low frequency propagation below the bottom of the sea. IEEE Trans Antennas Propag AP-11:323−329
Muraour P et al. (1970) Eléments de géophysique marine. Géologie des aires océanique. Ed. Mason. 196 p
Murray and Renard, (1891). Deep sea deposits. Published by order of Her Majesty's Government, 1110 pp
Nitronov AT (1936) Zhurn. Geofiz. J Geophys 6(5):5

Nobes DC, Law LK, Edwards RN (1986) The determination of resistivity and porosity of the sediment and fractured basalt layers near the Juan de Fuca Ridge. Geophys J R Astron Soc 86:289–317

Novysh VV, Fonarev GA (1963) Telluric currents in the Arctic Ocean. Geomagn Aeron SSSR 3 (6):919–921 (in Russian)

Palshin NA (1996) Oceanic electromagnetic studies: a review. Surveys in geophysics 17, pp 455–491. Ed. Kluwer Academic Publishers

Parkinson WD (1959) Direction of rapid geomagnetic fluctuation. Geophys JR Astrom Soc 2:1–14

Parkinson WD (1983) Introduction to geomagnetism. Ed. Scottish Academic Press Ltd. p 327

Pelet R (1994) Les origines du pétrole : évolution des idées de 1850 à 2000 dans La recherche pétrolière en France. Ed. du CTHS. p 403–443

Perrodon A (1966) Géologie du pétrole. Bibliothèque de l'ingénieur géologue. Ed. Presses universitaire de France. 440 p

Perrodon A (1980) Géodynamique pétrolière. Genèse et répartition des gisements d'hydrocarbures. Ed. Masson. 381 p

Popov YN (1962) Erfahrungen bei der Registrierung der mittelperiodischen Variationen des Feldes der tellurischen Ströme auf dem asowschen Meer. Razv. I Promysl. Geofiz. 43. p 80–87 (in Russian)

Porstendorfer G (1960) Tellurik. Grundlagen, Messtechnik and neue Einsatzmoglichkeiten. Ed. Akademie Verlag

Postma GW (1962) Method for marine electrical prospecting. US patent no. 3052836. Dec. 24

Renaud MJ (1902) Note au sujet de la recherche des roches sous l'eau. Annales hydrographiques. Ed. Imprimerie Nationale. 11 p

Renaud (1933) Etude géophysique sous-marine exécutée dans le port d'Alger. Annales des Ponts et Chaussées. Ed. A. Dumas. 15 p

Ridyard D (2006) Seabed logging. It's all about finding hydrocarbons. Onpoin, June. p 1–44

Robert M (1959) Géologie des pétroles. Principes et applications. Ed. Gauthier-Villars. p 131–139

Ryshkov YG (1957) Dan. USSR 113, no. 4. p 787

Sanière A et al. (2010) Les investissements en exploration-production et en raffinage. IFP Energie nouvelle, p 14

Sarkisov GA, Andreiev LI (1962) Results and prospects of marine electrical surveying in the Caspian Sea. Sovetskaya geologiia 2, no. 12, pp. 100–114 (in Russian)

Schlumberger C (1912) Process for determining the nature of the subsoil by aid of electricity. US Patent 1,163,468

Schlumberger C (1920a) Essais de prospection électrique du sous sol. Compte rendu de l'Académie des Sciences. Ed. Gauthier-Villard, March. p 519–521

Schlumberger C (1920b) Etude sur la prospection électrique du sous-sol. Ed. Gauthier-Villars, Paris, p 48–56

Schlumberger C (1927) Communication sur le carottage électrique. II congre's international de 1231 forage, Paris, 16–25 September 1929

Schlumberger CM, Leonardon EG (1934) Electrical exploration of water-covered areas. American Institute of Mining and Metallurgical Engineers. March. Technical publication, contribution no. 71. p 122–134

Selley RC (1998) Elements of petroleum geology. Ed. Academic Press, San Diego, 470 p

Serra O (2000) Diagraphies: acquisition et applications. Ed. Serralog. 586 p.

Sheriff RE, Geldart LP (1983) Traité de prospection sismique. 2 tomes. Ed. Erg. 304 and 373 p

Sinha MC, Patel PD, Unsworth MJ, Owen TRE, Mac- Cormack MRJ (1990) An active source EM sounding system for marine use. Mar Geophys Res 12:59–68

Spiess FN, Macdonald Ken C, Atwater T, Bal R, Carranza A, Cordoba D, Cox C, Diaz V M, Francheteau GJ, Guerrero Hawkins J, Hayi R, Hessler R, Juteau T, Kastner M, Larson R, Luyen B, Macdougall JD, Miller S, Normark W, Orcutt J, Rangin C (1980) East Pacific Rise: hot springs and geophysical experiments. Science 207(4438):14–21 and 14–33

Srnka LJ (1983) Offshore electromagnetic sounding. US patent no. 06 554 032

Srnka LJ (1986) Method and apparatus for offshore electromagnetic sounding utilizing wavelength effects to determine optimum source and detector positions. US patent no. 4,617,518
Suter HH (1948) Relative role of some geological tools in oil exploration. AAPG Bull 33 (11):2127–2139
Sweet GE (1969) The history of geophysical prospecting. Ed. EL De Golyer Memorial, Science Press, Los Angeles
Telford et al. (1978) Applied geophysics. Ed. Cambridge University. 860 p
Termier HG (1951) Les Connaissances actuelles en géologie sous-marine. Ed. Revue scientifique. p 186–192
Thomson RR (1950) Electrical prospecting method. US patent no. 2531088. Nov. 21
Vacquier V (1972) Geomagnetism in marine geology. Elsevier oceanography series, 6. Ed. Elsevier. 185 p
Vanyan LL et al. (1978) On the upper mantle electrical conductivity near the Bermudas. Phys Earth Planet Inter 16:7–9
Vining BA, Pickering SC (2010) Petroleum geology: from mature basins to new frontiers. Ed. Geological Society of London, London, 715 p
Vinogradov P (1957) On the record of electrotelluric field potential gradient at various depths in Lake Baikal. Akad Nauk SSSR Doklady 113(6):1255
Virchow et al (1999) Deep electromagnetic exploration. Ed. Springer, Berlin/New York
Walter E et al. (1985) Through the earth electromagnetic trapped miner location systems: a review. Tuscaloosa Research Center, Tuscaloosa, Alabama. Open filed report 127–85. United States Department of the Interior. Bureau of Mines. 57 p
Ward SH (1952) A theoretical and experimental study of the electromagnetic method of geophysical prospecting. PhD thesis. University of Toronto
Wenner F (1914) A method of measuring earth resistivity. US Department of Commerce, Bureau. of Standards, Washington, DC. Sc. Paper No 258
Whittar WF, Bradshaw R (1965) Submarine geology and geophysics. Ed. Butterworths, London
Wine FJ, Mattews DH (1963) Magnetic anomalies over oceanic ridges. Nature 199:947–949 (07 September 1963); doi:10.1038/199947a0
Young FB et al. (1920) On electric disturbances due to tides and waves. Phil Mag série 6, 40:149–159
Yungul SH (1996) Electrical methods in geophysical exploration of deep sedimentary basins. Ed. Chapman et Hall, London, 208 p

Chapter 2
Principles and Methods

Abstract This chapter first recalls the various laws and equations that govern the propagation of electromagnetic energy in the more or less electrically conductive media that are seawater environments, marine sediments and hydrocarbon deposits. From these theoretical elements, principles and methods of detection of oil and gas can be proposed. They make apparent that the seabed recording of lateral and in-depth variations of the electromagnetic fields (electric and/or magnetic fields) induced by a natural or artificial source, locally modified by the distribution of the electrical conductivity of the subsoils, allows highlighting of the presence of hydrocarbons more resistive in more conductive sediments (because they are saturated with seawater). It is also apparent that the variable current and dipole–dipole type controlled source methods in different configurations (in line and broadside) are then more favorable for the detection of hydrocarbons, which is, if certain conditions are met such as a resistivity contrast, able to be measured with a good signal-to-noise ratio.

Keywords Galvanic • Induction • Conductivity • Electromagnetic wave • Eddy current • Skin depth • Maxwell equations • Diffusion • Propagation

Preamble

I would like to remark about the notation: in the following text, the expression of equations, equally written according to the needs in the time and frequency domains, required the use of different notations. The fields are shown in lower-case letters in the *time domain*, and in upper-case letters in the *frequency domain* (Fourier domain and stationary domain : steady state). The constitutive relations and the passage of a notation to another are recalled and explained in the supplement located at the end of the chapter (Appendix A2.1).

1 Introduction

We may think that the use of electrical energy in a conductive medium such as seawater is a challenge. This is what we imagine at the first glance, considering that this energy dissipates too quickly in the open sea to reach deep geological

S. Sainson, *Electromagnetic Seabed Logging*, DOI 10.1007/978-3-319-45355-2_2

layers.[1] This is partly true because marine sediments, which represent a large part of the ocean floor, are themselves conductors through their pores saturated with seawater.[2] In this very particular context, we can assume, without much difficulty, that a penetration of electric power, even small, must allow extraction with adapted means—that is to say sufficiently sensitive—of enough information on the electromagnetic characteristics of the subsoil to deduce its geological structure or at least its composition.

Until the 2000s, marine geophysics, concerning electromagnetic methods, was developed around two distinct concepts: one using the galvanic effect of the *direct current* (in the field of maritime and coastal engineering in particular) and the other one rather using the *inductive effect* of variable currents whether they were natural or artificial (in a field rather reserved to earth physics). These radically different approaches are due to the nature of the investigated rocks, especially their electromagnetic properties, the means that are used (artificial or natural sources), and also the depth of investigation, concerning the subsoil layers, and the depth of the water, concerning the diffusion of the energy.

Due to the electrical characteristics specific to oil rocks (imperfect dielectrics) and more generally the topological characteristics of oilfields (complex geological structures, limited dimensions of the reservoirs and average depths of traps), the galvanic methods in a strict sense could not be truly effective for the detection of

---

[1]We have already known since the nineteenth century—thanks to hydrotelegraphy experiments (wireless telegraphy) by the American researchers Morse (1842), Graham (1882) and Edison (1885); the English researchers Wilkins (1849), Highton (1852), Stevenson and Preece (1892), Whitehead (1897) and Gavey (1900); the Scottish researcher Lindsay (1854); the German researchers Rathenau and Strecker (1896) and Zenneck and Braun (1901); and the French researchers Bourbouze (1870) and Ducretet (1902)—that it is possible to transmit electrical energy through water over large distances (Fahie 1899). An example is the transmission of signals via the Seine River between the cities of Rouen and Paris during the Commune civil war in 1871 (Fournier 1910; Ducretet 1903; Berthier 1908; Meyer 1972). Since the First World War, we have also known about the use of electrical devices for magnetic mine dredging (Rocard 1956).

[2]The seabed and its relief and lithology have been relatively well known since the nineteenth century through works including the installation of transoceanic telegraph lines: the Challenger (1872–1876), Meteor (1925–1927), Carnegie (1909–1929) campaigns, etc. (Harland 1932; Correnz 1937). However, the physical properties of marine sediments were much less well known (Delesse 1871; Murray and Renard 1891; Thoulet 1907). It was only in the 1930s with the first indirect marine seismic investigations (1935) that the structure of the surface sediment layers was suggested (Jacobs et al. 1959). It was then with the oceanographic campaigns of direct exploration (sampling) of the oceanic crust (the Mohole project) in 1950–1960 (Bascom 1961) and those of the Glomar Challenger (1968–1983), where soundings, corings and finally drilling logs were made, that we possessed more complete information this time about the conductivity of the rocks of the oceanic crust and little indeed about the sediments (Hsu 1992). We have only known since very recently, thanks to the resistivity logs achieved by oil companies, that on the continental shelf seawater penetrates deeply into the underlying sedimentary layers, thus making the latter also highly conductive (Cox et al. 1971; Chave and Cox 1982). Programs of deep-sea drilling are currently continuing through the boat driller JOIDES Resolution (Joint Oceanographic Institutions for Deep Earth Sampling).

hydrocarbons. Consequently, this has led in recent years to the development of new techniques, allowing in this singular geological context more accurate highlighting, and in a differentiated way, the oilfields considered as more resistant in the geological surroundings to be considered as rather conductive.

For a better understanding of the involved physical phenomena, this chapter shows the various principles used in the methods of offshore electromagnetic prospecting. Not all have direct oil applications in the strict sense (commercial or finalized ones). However, they are briefly recalled to express the detection problems and to explain, if necessary, why some of these techniques do not a priori allow us to solve the problem of detection/localization of hydrocarbons as such, and why new principles have been finally proposed.

2 Laws and Physical Principles

We recall here the general and basic laws[3] of the electrical and electromagnetic survey in the broader sense, governing the different physical principles used among others in marine exploration.

2.1 General Laws Applying to Electrical Prospecting

If one focuses on the movement and distribution of electric currents[4] in the subsoil, we can simply say that:

- Electric current can travel by different ways, galvanic and/or induced modes,[5] whether it is a continuous or varying current.[6]
- The continuous current only flows in electricity conductors such as sulfide deposits (electronic conduction), or in sedimentary rocks containing in their porous matrix more or less formation of water, itself conductive (a conductive electrolyte), or more generally in fractured rocks with this type of conduction

[3]We advise readers little experienced in vector analysis to see, for example, the work of Professors Chisholm and Morris (1965).

[4]We call "electric current" every general movement of charged particles or charges moving in a referential.

[5]These concepts are explained later in the presentation.

[6]By this we mean alternating currents (fixed frequencies), periodic currents (modulated frequencies) and variational currents (frequencies or periods varying in time and space). Within the SBL, the term "alternative" will be used for controlled source devices (mCSEM) and the term "variational" for telluric measurements (mMT).

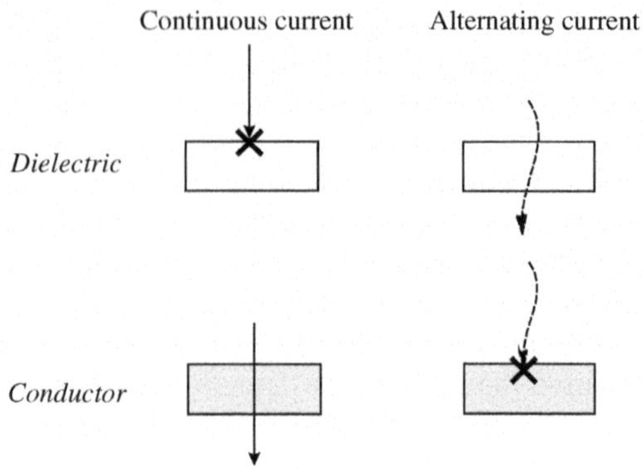

Fig. 2.1 Different behaviors of continuous and alternating currents toward dielectrics and conductors of electricity

- Alternating and variable currents penetrate dielectrics such as rock hydrocarbons, but are more or less stopped by conductors all the more strongly as the frequency and electrical conductivity of the soils[7] are high.[8]
- The movement and distribution of the alternating currents, even at very low frequencies, cannot be completely equivalent to those of the continuous currents unless the latter are variable currents with very long periods, such as telluric currents (Fig. 2.1).

Except for the investigations carried out in earth physics for the recognition of very deep layers of the lithosphere, most of the electromagnetic methods used in conventional geophysical exploration using artificial sources have been studied to highlight:

- Resistivity of horizontal contrasts (superposition of geological horizons) or vertical contrasts (faults, breaking down, etc.)[9] with the use of continuous or low frequency currents (→ galvanic effect; see Sect. 2.4.2)
- Local conductive anomalies[10] (such as mineral ore bodies), buried in more resistive geological environments, using then medium frequency alternating currents (→ inductive effect; see Sect. 2.5.2)

and more recently:

- Resistive anomalies (such as oil reservoirs) in conductive environments (environments invaded by seawater) using in low frequencies (LF) the electrokinetic and electromagnetic joint and combined effects (galvanic + inductive effect; see Sect. 2.5.3)

---

[7]It is to the Italian physicist Carlo Matteucci (1811–1868) from Thouvenel's experiments (Thouvenel, 1792) that we owe the first works (using an electric telegraph) on soil electrical conductivity (Blavier 1857).

[8]This is also true of the radio transmission modes (hertzian and radar).

[9]The "*à coup de prise*" phenomenon, well known by electrical prospectors.

[10]See definition in Appendix A2.5.

2.2 *Electromagnetism and Marine Geophysical Prospecting: Recollections*

Marine electromagnetic surveying (mEM for short) brings together a multiplicity of methods and technics[11] based on somewhat different physical principles; however, it is possible to identify briefly their general outlines. Roughly, these methods can be classified into three categories, depending or not on the time:

- Those that use an artificial continuous current (marine direct current methods or mDC methods)
- Those that are practiced with artificial alternating currents applied as controlled, permanent or transitory currents (marine alternative current methods or mAC methods, generally called marine controlled source electromagnetic methods or mCSEM)
- Those based on natural variable currents, called telluric currents (marine magnetotelluric current methods or mMT methods)

These various investigations, both in their conceptual approach (magnetic, electric or electromagnetic) and in their technological specifications (very different metrological arrangements) are governed by separate laws of current distribution.

These methods are divided roughly into:

- Vertical sounding (vertical or depth investigation)
- Mapping (lateral investigation at a given depth)
- Imaging (lateral investigation at different depths)

2.3 *General Principles of the Electromagnetic Exploration*

These well-known principles in the profession are explained by classical deterministic physics[12] based on the laws of electromagnetism.[13] The interested reader will find more complete developments in the books and monographs dedicated to surface exploration, whose application areas are essentially those of

[11]The first experiments in electromagnetic prospecting took place shortly after the appearance of wireless telegraphy. During the First World War, French troops used a device for telegraphy through the soil. Significant variations in transmission were found on this occasion, attributed to the conductivity changes that affected the different grounds. We owe to the Romanian physicist Sabba Stefanescu, a Conrad Schlumberger collaborator, the theory of the dipole TTS (telegraphy through the soil), the basis of the quantitative interpretation of the first electrical soundings (Stefanescu 1936, 1945). For a complete history on electromagnetic methods, the reader may refer to the recent article by Professor Zhdanov (Zhdanov 2010).

[12]The notions of general relativity or the general theory of the fields are not involved here.

[13]To avoid overloading the following discussion, these laws are contained in the Appendices A2. The reader will therefore be able to refer to it through referrals located in the text.

hydrogeophysics (Patra and Nath 1999; Makoto et al. 2003; Rubin and Hubbard 2006), mining geophysics (Keller and Frischknecht 1966; Telford et al. 1978; Patra and Mallick 1980; Nabighian 1987) and geothermics (Adam 1976). An exhaustive booklist is given in the appendix A1.1

2.3.1 Continuous Current Regime

At steady state, only conduction currents are present in the conductive media. They correspond to the general movement of the electrical charges present in the medium (electrons for the metal conductors, ions for rocks saturated with water, etc.). The electric field $\vec{E}$ is then, in the same direction, directly attached to the current density $\vec{J}$ by the only electrical characteristic representative of the medium, i.e., its conductivity σ (or the inverse: its resistivity).

In this case it is verified that:

- The average charge is zero, and the spatial distribution of $\vec{J}$ does not depend on time and satisfies the equation:

$$\vec{\nabla}.\vec{J} = 0$$

- The field $\vec{E}$ in the conductor is constant too, and satisfies the equation:

$$\vec{\nabla} \wedge \vec{E} = 0$$

In the subsoil, these local phenomena, which are concomitant with the simultaneous movement of the moving charges, i.e., here the ions present in the water in the rock are ruled as homogeneous and isotropic conductors[14] by the law of linear proportionality between $\vec{E}$ and $\vec{J}$, formalized by Ohm's law (cf. Sect. 2.4.4) (Fig. 2.2).

Theoretically, the distribution of the potential fields is comparable to that of electrostatic fields; it is then called static approximation (cf. Sect. A2.10). Except for the limit conditions, it is conveniently given in the absence of a current source by solving the Laplace equation (see Eq. 2.6) and in the presence of a source by the Poisson equation (cf. Eq. 2.8).

The means of bringing it into operation are especially achieved through simple metal electrodes[15] (such as copper or steel) in contact with the ground where the current is injected (+/−) and directly taken from the ground through a quadrupole type device. This technique is finally an ohmic transfer,[16] which places in direct

[14]See definitions in Appendix A2.5.

[15]In resistant grounds a group of electrodes (four or more) Joule is necessary.

[16]Much of the electrokinetic energy is dissipated by a Joule effect (ohmic drop) at the injection electrodes. It is therefore important that the earth or sea plate have as little resistance as possible.

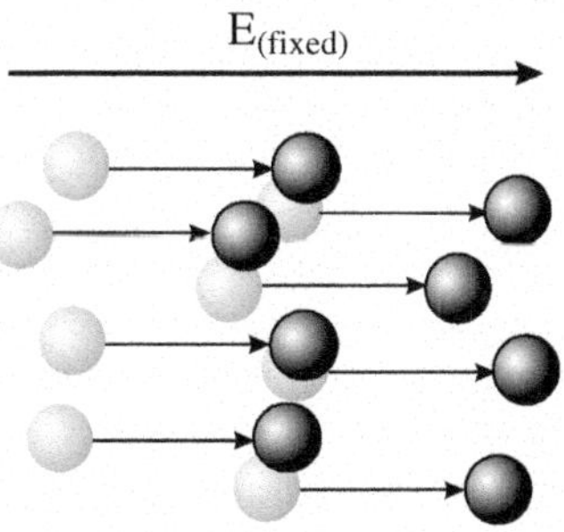

Fig. 2.2 Conduction current concomitant to the movement of all the ions O in the field E (fixed) present in the medium. Electricity does not move through the medium but with the charges of the medium (Lodge 1892)

contact the energy exchanges between the power source and its more or less immediate environment.

The main disadvantage of the continuous current methods, also called conduction current methods, lies in the facts that:

- The continuous current dissipates very quickly in conducting media such as seawater (short circuit).
- The layers of high resistivity, even if they are thin, form a block to any electrical energy penetration, preventing access by the way to any information on these layers and moreover to those more deeply buried in the underlying strata.

These are the reasons why we prefer to use, in these very singular detection conditions of "resistant" layers in a conductive marine environment, variable currents, which are more suitable to propagate, to penetrate these layers, and they can present in addition some directionality according to specific measurement dispositions.

2.3.2 Alternative and Variable Current Regimes

In an alternative or variable regime (a function of the time t), the current distribution depends not only on the electrical conductivity, but also on the dielectric constant and to a lesser extent on the magnetic permeability of the medium of propagation.

Electric and magnetic fields are indissociable and are more or less directly attached through the intervention of cause and effect actions governed by the classical laws of electromagnetism. Five dynamic parameters then define the electromagnetic state of the environment and are represented by the five vectors:

$$\vec{d},\ \vec{h},\ \vec{e},\ \vec{b} \text{ and } \vec{j}$$

These are connected together by an equation couple formed by Faraday's law and by Ampere's theorem.

In any medium, a variable magnetic field of induction $\vec{b}$, generates an induced electric field $\vec{e}$; this is the effect known as induction or an inductive effect, described in its local form by *Faraday's law.*[17]

$$\vec{\nabla} \wedge \vec{e} = -\frac{\partial \vec{b}}{\partial t} \tag{2.1}$$

In an electrically conductive medium, an electric field $\vec{e}$ creates a current of density $\vec{j}$, which in its turn generates a magnetic field $\vec{h}$; this is *Ampere's theorem.*

$$\vec{\nabla} \wedge \vec{h} = \vec{j} + \frac{\partial \vec{d}}{\partial t} \tag{2.2}$$

This second equation then admits two kinds of current:

- one represented by $\vec{j}$ corresponding to a transport of the charges by the current, and called *galvanic current,*
- the other one, a function of the time, equal to $\partial \vec{d}/\partial t$ corresponding then to a displacement of the field $\vec{d}$ and named *displacement current.*

Nearly all electromagnetic prospecting methods use consequences of simultaneous combination of these two interactions. The electromagnetic wave propagation and the distribution of the associated fields therefore depend more particularly on these two fundamental equations (*Faraday's law*[18] and *Ampere's theorem*[19]). Ordinarily, they are incorporated into the system of equations of *Maxwell* (four in total) respectively designated as the first and second equation (Maxwell 1861, 1862), which are supplemented by the constitutive equations. They allow, among other things, with the expression of a priori given restrictive conditions, understanding of most of the problems of distribution of electricity in material media if they are, however, sufficiently basic (cf. Appendix A2.1).

In practical terms, only the electric field $\vec{e}$ (or $\vec{E}$) and the magnetic induction $\vec{b}$ (or $\vec{B}$) are directly accessible to measurement (e, b) (Fig. 2.3).

[17]Expressed by Faraday (1831) and formulated a few years later by Lenz (1834), the law expresses the electromotive force induced in a circuit crossed by a variable magnetic flux. In the nineteenth century, Faraday already sensed the existence of electromagnetic fields in the sea, induced by the dynamic movements of conductive waters in the Earth's magnetic field (Faraday 1832). Local and integral forms are equivalent.

[18]Statement of *Faraday's law*: the time derivative of the magnetic induction flux, sign changed $(-\partial B/(-\partial t)$, through a surface element of a dielectric material, represents the induced electromotive force created along any limit embracing the surface.

[19]Statement of *Ampere's theorem*: the unit of magnetic mass moving along a closed contour enveloping an area crossed by a displacement current flux $(J + \partial D/-\partial t)$ performs work equal to the product by 4π of the displacement current flux, regardless of the medium in which the magnetic mass moves.

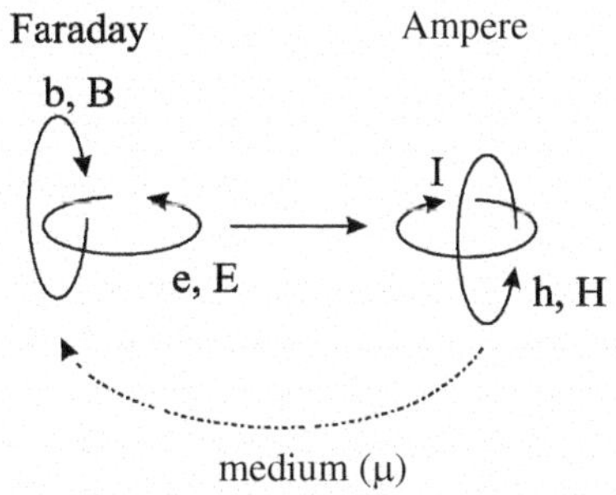

Fig. 2.3 Interactions of the electric and magnetic fields with the currents I, after all formalizing the propagation of the electromagnetic energy through a material medium, derived from Faraday's law and Ampere's theorem. Field $\vec{h}$, $\vec{H}$ and magnetic induction $\vec{b}$, $\vec{B}$ are linked by the intrinsic characteristic of the medium (invariant), namely its magnetic permeability μ (cf. Eq. 2.13). It should be noted here that for Maxwell there is in fact no open circuit but closed current loops; if a circuit composed of conductive bodies is interrupted, the dielectric medium itself at the ends of these bodies closes the circuit

Moreover, these relationships between the fields also present two regimes:

- A regime of *diffusion*, due to the conduction phenomena (introduced by the so-called conduction currents)
- A system of *propagation*, due to the polarization phenomena (introduced by the so-called displacement current: induced effect)

These two phenomena are either simultaneously or separately established. They depend on one hand on the electrical characteristics of the current source (the excitation frequency especially), and on the other hand on the electromagnetic properties of the crossed media (electrical conductivity and permittivity more particularly in the case of sedimentary rocks).[20]

On the theoretical level, the difficulty lies then in the fact that some materials forming the subsoil do not quite behave as conductors of electricity and not as dielectrics.

2.3.3 Fundamental Difference Between Continuous and Variable Regimes

The study of the electric field in a continuous regime is only based on the study of the thin streams of surface currents (ground surface) whose spatial distribution mainly depends on the deep subsurface heterogeneities (→ *vertical exploration*).

In an alternative regime, the electromagnetic field is, conversely, the result of the elementary fields due to all the thin streams of surface currents (including those from the source) and deep currents, and so the study is then one of a total field.[21] It

[20]In the sedimentary (nonmagnetic) layers the magnetic permeability plays a minor role and cannot be used as a deciding factor in detection.

[21]Or, in other words, with addition of the primary field, secondary fields due to induction currents (mutual induction of the current lines between them), which rise from the depths.

therefore has no fixed direction (cf. Fig. 2.13). It has been shown (see below) that, as a result of the difference in the phase presented by its different components at the same point, this global field is largely influenced by the heterogeneities of the subsoil, and so the end of the vector representing the field describes then a polarization ellipse. As a corollary, using separate and oriented measuring devices, it is then possible to obtain directional systems (→ *lateral exploration*).

If in continuous current methods it is sufficient to measure the amplitudes of the electric fields, it is necessary in the variable regime methods to record either their variations or their phases, or most preferably both of them.

Furthermore, another major difference is the fact that the penetration depth of the electric lines of force remains limited for alternating currents (a *skin effect* due to the frequency and the conductivity of the rock formations → formation of *eddy currents*) and substantially infinite for continuous currents—that is, if we can have for them sufficient electrical power and adequate means of measurement (a microvoltmeter with low drift).

Finally, in theory, the formulas that are given for the calculation of the *apparent resistivity* are different—relatively simple for continuous currents (with a direct relationship with Ohm's law) and much more complex for variable currents (resolution of Maxwell's equations system in the frequency and/or time domain).

2.3.4 Geoelectric and Geological Sections: Two Different Realities

Geoelectric sections obtained by any process whatsoever are not necessarily representative of geological sections. This is the case for instance for:

- Successive layers of different ages or facies with the same resistivity
- Impregnation fluid (salt water, oil, etc.) present in two superimposed layers with similar hydraulic characteristics (porosity)
- Low thickness benches interposed between two layers of high power and different resistivity

In these particular situations, the lower and upper limits of the geological layers (wall and roof) do not coincide with those of the measured resistivity, which can lead to an incorrect estimation of the thickness of the different layers or simply an indetermination (a plurivocal problem).

To remedy this disadvantage we can then perform a quantitative calibration when we have stratigraphic sections accessible by drilling for the deep horizons (carrots, samples, cuttings, mud logs, etc.) or by acoustic survey for the superficial layers. A parametric calibration is otherwise performed by applying the principles of equivalence (Bhattacharya and Patra 1968) and suppression (Chapelier 2001).

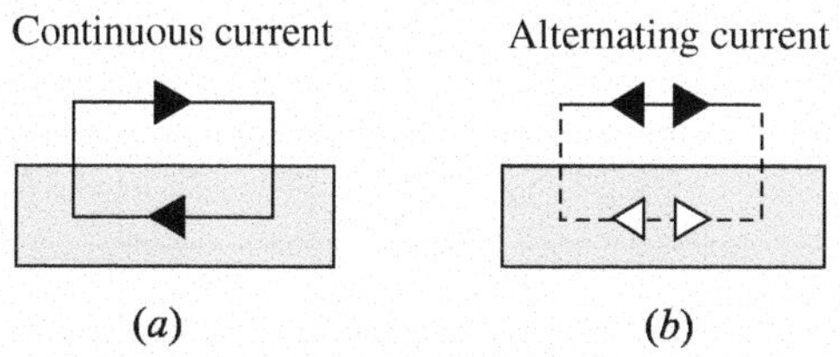

Fig. 2.4 Direct current (DC) only flowing in a closed circuit (*a*) and variable current (AC) able in addition to propagate in an open circuit (free at the ends) coupling to the ground (*shaded* in the figure) by induction by a link that can be immaterial (*b*)

Operating Techniques

There are a wide variety of techniques of electromagnetic exploration processes.[22] They go from injection current methods (permanent or impulsive, artificial or natural, galvanic or inductive) to reception methods (potentials, potential differences, electric and/or magnetic vector fields, total field, etc.).

Whereas the direct current cannot flow in a physically closed circuit, the alternative current can also flow in an open circuit; it is called an antenna (cf. Fig. 2.4).

In continuous current prospecting, the circuit usually closes due to the invested medium (cf. Fig. 2.4a). In the context of variable current prospecting, it can also close over the medium by conduction as for a continuous current, but also spread in an induced manner (cf. Fig. 2.4b).

Controlled Current Injection

Operating techniques for the application of artificial methods or controlled current source methods physically use either:

- Inductive transmitters or inductors (coil,[23] loop,[24] hoop,[25] frames,[26] etc.)[27] isolated from the environment where a periodic current flows
- Transmitters with direct plugs planted in or in physical contact with the environment (land, seawater), where the current is injected at one point and exits through another

---

[22]Compared to other methods, the electromagnetic methods are probably those with the greatest diversity of techniques.

[23]A single turn of wire placed on the ground.

[24]Several turns of wire placed on the ground.

[25]A spire inclinable in all directions placed on a tripod.

[26]A frame whose orientable coils are placed horizontally and vertically (perpendicular to each other).

[27]These devices are also found in reception (receptors).

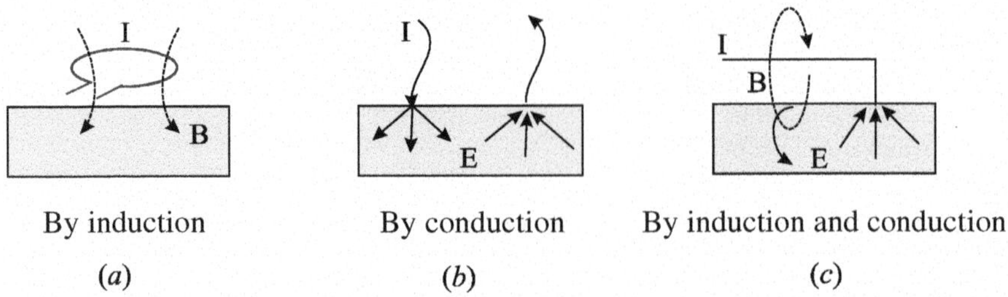

Fig. 2.5 Various technical means for artificially transmitting electromagnetic energy in the subsoil: induction methods (*a*), conduction (*b*) and mixed (*c*) methods. The inductive method is used for investigations in rather resistant ground where the ohmic drop may be important (dry sand for example). However, the technique by conduction, mainly used with direct current, requires a resistance of the electrodes as small as possible, to minimize the ohmic drop at the electrodes. In SBL, the hybrid method has been little used so far. (In the 1920s, surface prospecting technologies using these two effects were tried without success a priori. This was for example the case with Professor Ambronn's old system where, briefly, a small diameter coil (a few feet) was connected to two earth electrodes (Ambronn 1928))

- Electromagnetic loops consisting of a straight horizontal or vertical electric cable and an earth (or sea) connection forming with the ground (reverse current) a type antenna device[28] (Fig. 2.5).

Moreover, some techniques developed for scientific research also use magnetic sources (see Chap. 1, Fig. 1.14), but these techniques, in the context of commercial prospecting, remain relatively marginal today and only are proposed now for deep sea mining exploration (Swidinsky et al. 2012).

Telluric Sources

Natural or telluric current sources have a more complex mode of action to highlight. The origin of these currents, their provenance, their nature, their distribution, their circulation, and their variability in time and space, as well as their concentration and their period, associated with changing scaling factors, do not allow us to establish clear and still less general rules about their specific electromagnetic behavior in the subsoil. However, we admit that, roughly speaking, it can be assimilated into the behavior of the variational and transient currents whose period varies over time and space.

Reception

For direct current, the reception only uses unpolarizable or not potential sockets, and more rarely relative or absolute measure magnetometers.

[28]Interconnected by one or more buried or submerged potential electrodes.

For alternative and variable currents, according to the employed techniques, the sensors are:

- Magnetic (magnetometers, declinometers, variometers, etc.)
- Electric (electrometers)
- Both magnetic and electric (electromagnetometers in development)

In the case of SBL we nearly only adopt vector sensors, i.e., field (electric and/or magnetic) measurers able to register one or several components, which provide for the measure a certain directivity. The so-called scalar or total field sensors are in return reserved for permanent current methods (magnetic or electric methods; cf. Sect. 2.3.3).

2.3.5 Apparent Resistivity Notion

The *apparent resistivity* is in some way an average value, physically and globally integrating the *specific resistivities* (see Chap. 3, Sect. 2.1.3) of the different rock formations crossed by the electric current, if these are considered as a stack of more or less horizontal and homogeneous geological layers.[29] In the case of localized heterogeneities, the apparent resistivity then depends on the *resistivity contrast* (anomaly/surrounding), the size and burial depth of those layers or more simply their relation (Habberjam 1979).

The apparent resistivity also depends on the evaluation method that has been adopted, on the geometry of the acquisition device (the size and layout of the electrodes in relation to the dimensions of the anomalies), on the characteristics of the current and finally on the conditions of its use (hardware and environment).

The apparent resistivity is calculated from the measurements of V, I, B and E combined together and mathematically linked by arrangement parameters, which define among other things the investigation depth.

2.4 Physical Principle of Stationary Current Marine EM Seabed Logging

The principle of stationary current electromagnetic SBL is old (see Chap. 1. Sect. 6.2) and relatively well known by prospectors. It is governed by the laws of electrokinetics principally derived from the laws of electrostatics for the study, for example, of the charge distributions (cf. Sects. 2.4.4 and 2.4.5).

These methods are, with minor exceptions, marine applications of techniques traditionally used on land, such as profile draggings (constant investigation depth),

---

[29]With some corrections or conceptual artifices, we can here tolerate a low dip.

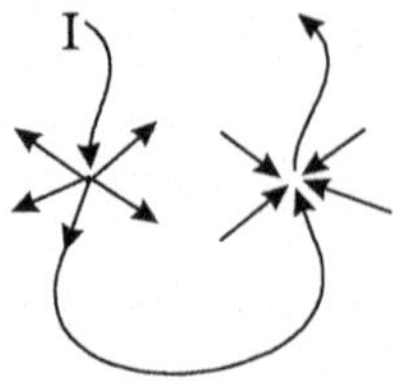

Fig. 2.6 Power line from the entry point of the current I to the exit point (by the electrodes). The electric current flowing in the medium (land or seawater) follows the path of least resistivity (law of least effort), which is not necessarily the shortest path

vertical soundings (variable investigation depth) or even electrical panels, universally used in:

- Hydrogeological prospecting (searches of aquifers, of freshwater resurgence or of infiltration, or for identification of pollution, etc.)
- Structural prospecting (locations of geological lineaments as fractures or faults, etc.)
- Coastal auscultation and harbor geotechnical inspection (evaluation of the sanding or silting thickness, of the stone bedding depth, etc.)
- Building material research (sand, gravel, etc.)
- and to a lesser extent mining prospecting (detection of sulfides masses, veins, stockwerks, etc.).

In DC electrical exploration, except for some special methods, we mainly use the so-called resistivities method that allows through difference of potential measures establishment of a section in depth of the distribution of resistivities (an integrative method). In this case, the energy transfer is always done by conduction; the electrical current then flows from a material point to another (injection/return) by the path of least resistance (Fig. 2.6).

In these processes the dual coupling transmitter/subsoil/receiver remains substantially invariable at each measuring station, i.e., in other words the transmitters and receivers have finite, fixed dimensions, and move together at a constant elevation of the seabed (cf. Fig. 2.20).

To avoid the effects of electric polarization of the electrodes due to electrochemical phenomena of contact (metal immersed in an electrolyte), we very often use variable polarization currents as pulsed, chopped, or inverted currents (regular alternance of positive and negative current) depending on whether one seeks to differentiate a particular type of terrain or geological structure.

2.4.1 Apparent Resistivity and Geoelectrical Section

Generally, at sea, the investigation is carried out using a series of quadrupoles (current injection and capture potential dipoles mutually aligned) fixed on a flute (on the seabed) or arranged in a streamer (in full or surface waters, more rarely), whose different geometries (spacing type Wenner, Schlumberger, Lee partition, double dipole, etc.) allow you to explore the floor at varying depths.[30]

Most investigations at sea, in coastal areas, or in harbor areas correspond to the development and construction of vertical pseudosections—that is to say electrical resistivity section—representative of the various strata or tabular geological structures forming the sedimentary tranche.[31]

To build these sections, the prospector, from a series of recorded measurements, establishes a curve or a pseudosection of apparent resistivity (see Chap. 5, Sect. 3.3.1), then interpretable in terms of the geoelectric model.

2.4.2 Galvanic Effect: Conduction, Detection and Anomaly Location

In the geological electricity conductor media, the flow of direct or low frequency alternating electric currents (where the galvanic effect is also important) can be affected by electrical discontinuities. The latter are due either to horizontal (or vertical) interfaces of different natures (geological layers) or to discrete local heterogeneities whose conductivity represents a contrast with the conductivity of the environment. In these cases, the field lines are then canalized in their spatial distribution (in the case of sedimentary strata) or present a geometric distortion (in the case of conductive or resistive heterogeneities), which then characterize the galvanic effect (static shift).

This singular effect is due to the opposite accumulation of positive and negative electric charges at the boundaries of the different geological media (layers, heterogeneities, etc.) such that:

- Within the anomaly, this accumulation then creates a current generating a secondary field equivalent to the one produced by an electrostatic dipole, in the opposite direction (to the primary field[32] that generates it) for a conductive anomaly, or in the same direction for a resistive anomaly.
- Outside the anomaly (useful information for prospecting), this depolarization current is added for a conductive heterogeneity or subtracted for a resistive heterogeneity.
- Outside the anomaly, we then observe on either side of it a distortion of the force lines, which has the effect of promoting the detection (especially deflection of the electric field) and above all of allowing the location (the intended purpose).

In this case, the total field is treated as an electric dipole. Theoretically, in these anomalous circumstances, the secondary field $\vec{E}_S$, at a distance r from the emission source, is provided by *Coulomb's law* (cf. Appendix A2.10) such that:

---

[30]We suppose that for a given geometry, the depth of investigation is constant. This is partly true because in fact the conductivity of the grounds alters that depth to a lesser extent.

[31]We usually seek to assess the sediment or vase thickness, or to recognize the depth of the bedrock where the imperatives of directivity and great water depth are not required.

[32]This case is not of interest in exploration, except in the technique *mise à la masse*, wherein the current injection electrode is implanted into the accessible ore deposit.

$$\vec{E}_S = \frac{q}{4\pi\varepsilon}\int \frac{\vec{r}}{|r|^3}dv \tag{2.3}$$

where q is the volumetric charge density, which is, depending on the primary field $\vec{E}_p$, on the electrical conductivity σ and on the dielectric permittivity ε of the medium, equal to $q = -\varepsilon E_P . \nabla\sigma/\sigma$ (Fig. 2.7).

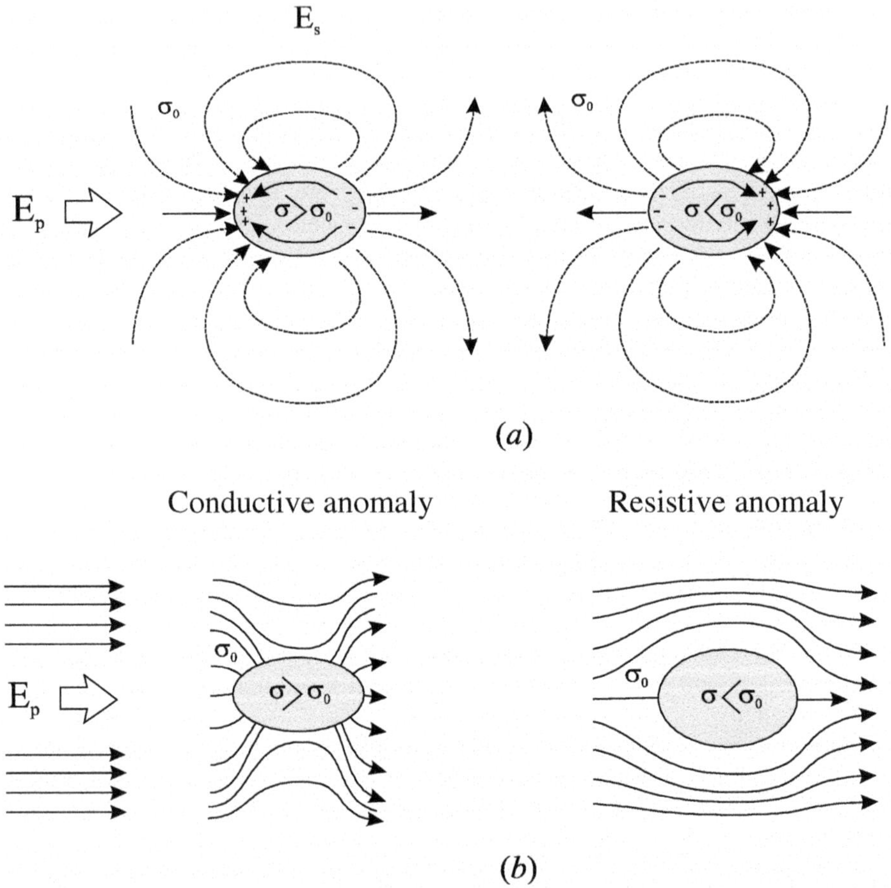

Fig. 2.7 Examples of galvanic effects due to abnormalities of conductivity of different types. (*a*) Polarization mechanisms (accumulation of charges of opposite signs at the ends of the target → creation of an electric current), source of the galvanic effect, for a conductive anomaly and a resistive anomaly. (*b*) In the presence of a conductive (*left*) or resistive (*right*) anomaly, the current lines can bend even more sharply than the contrast of conductivity (or resistivity) is large. The conductive anomalies tend to attract at greater or lesser distance the lines of force while the resistive anomalies tend to repel them. A priori, this local effect is more pronounced for conductive anomalies than for resistive anomalies. The potential lines (not shown in the figures) are perpendicular to these isoanomalous lines and to the limits with the anomaly

Macroscopically, the distortion of the lines of force, of the fields, etc., can be mathematically modeled by either analytical methods for detection itself (Sainson 1984) or by numerical methods to realize corrections (Chave et al. 2004). In all cases, it is necessary to have maximum information on the form and the effective resistivity contrast between the anomaly and the surrounding rock.

2.4.3 Remark About Current Sources

In DC electrical exploration and at low frequency the dipole terminology is suitable for two different physical realities, depending on whether we are interested in the measuring device itself and more particularly the transmitter, or in the anomalous field when it exists in the presence of an heterogeneity [sic] (Fig. 2.8).

In the case of the primary sources, i.e., the transmitters, the dipole corresponds to the injection of an electrical current by two points or poles connected by a cable where the generator is generally inserted, where the current then escapes from one of the poles (the first electrode) to close by the ground on the other pole (the second electrode).

In the case of an anomalous field, caused by a conductivity anomaly, this one then intrinsically plays the role of a secondary power source. This electromotive force is opposed to the primary field in the case of an insulator or is added in the case of a conductor.

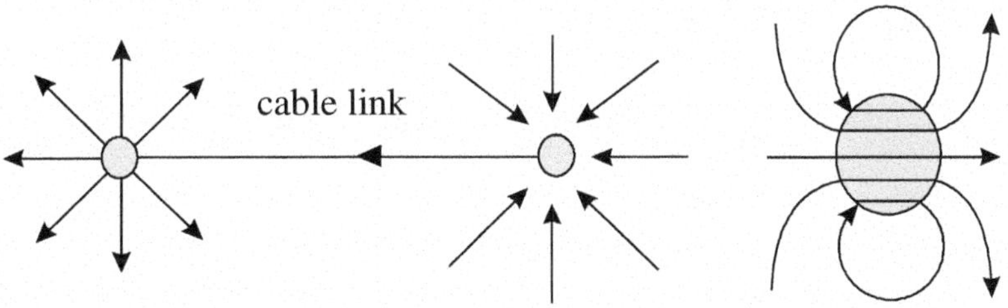

(a) *(b)*

Fig. 2.8 Dipolar sources of different origin and associated current lines: (*a*) corresponding to the primary field of a transmitter with two poles separated but connected by a cable; (*b*) corresponding to the secondary field associated with a conductivity anomaly; the target becomes then a source of electric current

2.4.4 Ohm's Law: Recollections

Techniques using continuous or similar electrical currents[33] are theoretically based on Ohm's law.

According to it, in any point of the system formed by the measurement device and the investigated ground, the current density $\vec{J}$, which corresponds to the current flow per area unit (cf. Appendix A2.2), is proportionally related to the local electric field $\vec{E}$, by the conductivity σ or the resistivity ρ (where $\rho = 1/\sigma$), an intrinsic property of the propagation medium at rest, such that:

$$\vec{J} = \sigma\vec{E} \quad \text{or} \quad \vec{E} = \rho\vec{J} \tag{2.4a}$$

This constitutive law linked to the propagation medium, connecting the currents to the field intensity, adds itself to Maxwell's equations (cf. Appendix A2.3). The charge displacement may also result from an additional external cause, independent of the local electromagnetic field, such as:

- The presence of a power source (transmitter) nearby
- The possible presence of a chemical potential gradient (electrolyte)
- The movement of the conductor (seawater) or the sensor in a present magnetic field (the earth's field for example)

In these cases, to the current density is then added an additional term, such as:

$$\vec{J} = \sigma\vec{E} + \vec{J}^{s} \tag{2.4b}$$

If we consider, at any time, the omnipresence of telluric currents in the soil, this expression seems to be therefore the most appropriate.

2.4.5 Stationary Problem Solution: Laplace and Poisson Equations

Under local conditions, the electric field $\vec{E}$ becomes the gradient of a scalar potential V such as:

$$\vec{E} = -\vec{\nabla} V \tag{2.5}$$

[33]To avoid the polarization of the electrodes, pulsed currents are also used. For example, in transmission, the current is chopped at a given frequency and reconstituted at reception to the same frequency. In the heroic times of electrical prospecting, we used for that common devices such as the pulser/inverter. Today these systems are electronic (IGBT double switcher).

As the direct current is conservative by definition ($\vec{\nabla}.\vec{J} = 0$), then the potential V obeys the Laplace equation (cf. Appendix A2.5)

$$\nabla^2 V = 0 \tag{2.6}$$

or, in Cartesian coordinates,[34] the partial differential equation:

$$\frac{\partial^2 V}{\partial x^2} + \frac{\partial^2 V}{\partial y^2} + \frac{\partial^2 V}{\partial z^2} = 0 \tag{2.7}$$

Crossing a surface separating two media of different conductivities, the voltage V varies continuously (it is the same for the normal component). The solutions of the Laplace equation (cf. Eq. 2.7), giving the distribution of the electrokinetic potentials in space (x, y, z), are mathematically identical, except for the limit conditions, to those giving electrostatic potentials. This approach corresponds to the *static approximation* (cf. Sect. A2.10).

In the presence of an electrical current source of intensity I discharging into the medium, the Laplace equation, which corresponds to an equation without a source, is then endowed with a second member such that:

$$\nabla^2 V = -I\boldsymbol{\delta}\left(\vec{r} - \vec{r}_S\right) \tag{2.8}$$

where $\boldsymbol{\delta}$ is the Dirac function,[35] and where $\vec{r}$ and $\vec{r}_S$ are respectively the directional vectors going from any observation point M to the potential V point and the place where the source S is (cf. Fig. 2.9).

This equation corresponds, in these particular conditions, to the Poisson equation (cf. Sect. A2.6).

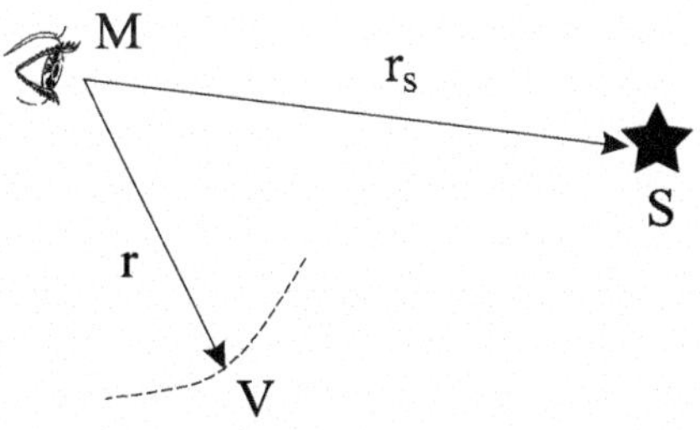

Fig. 2.9 Arrangement of directing vectors ($\vec{r}$, $\vec{r}_S$) relative to an observation point M in the context of solving the Poisson equation

[34] In oil exploration, cylindrical coordinates are also used for anticline modeling, for example.

[35] Also called a pulse function, well known by electricians since Heaviside and reintroduced by P.A.M. Dirac in quantum mechanics (*The Principle of Quantum Mechanics*, 1947, p.58). The response to an excitation of this type generally leads to relatively simple expressions.

2.5 Physical Principle of Alternative Current Marine EM SBL

The principle of alternative current SBL is recent (Eidesmo et al. 2002), distinct from the older DC methods and partly comes from variable current methods previously used in oceanographic research particularly led in the USA (Spiess et al. 1980). Its physics are still apprehended with some questions about its precise mode of action (see Sect. 2.5.8).

The physical principle of SBL detection is based on the study of the specific behavior of electric currents at low frequencies, which is different depending on the environment in which they propagate. Depending on the conductivity of these media, we may be dealing either with a relative predominance of the conduction phenomena, or a preponderance of the propagation phenomena.

2.5.1 Conduction Currents and Displacement Currents: Semantic Recollections

From the macroscopic point of view, we can roughly distinguish two types of movement of electrical currents. They take their difference from their origin, some depending on their mode of production (the nature of the electromagnetic source), the others following from the environment in which they occur.

When the energy propagates because of the displacements of the charged particles under the action of an electric field $\vec{E}$, then we are dealing with conduction currents. Each particle of charge q is then subjected to a electric force $\vec{F}$ such that:

$$\vec{F} = q\vec{E} \tag{2.9}$$

These currents are mainly established in electricity conductor media, solid and liquid, where free charges are absolutely present (electrons, ions, etc.). They both concern continuous and variable fields.

When energy this time propagates because of the motion of electric and magnetic fields, in this case, we are dealing with displacement currents themselves, according to the Maxwell terminology (Webster 1897; Ferraro 1956; Rocard 1956). These currents usually occur in rather resistant environments.

Inside the medium, the current density vector is equal to:

$$\vec{j} = \varepsilon \frac{d\vec{e}}{dt} \tag{2.10}$$

setting (Maxwell):

$$\vec{j} = \frac{d\vec{d}}{dt} \tag{2.11}$$

we obtain:

$$\vec{d} = \varepsilon\,\vec{e} \tag{2.12}$$

which represents the electric displacement vector.

In this state, the displacement currents then ensure the electromagnetic energy transfer and mainly concern the variable fields and dielectric materials.

Schematically, and in absolute,[36] we can say in summary that:

- Only the conduction currents remain in the perfect conductors, as the displacement currents are zero.
- Only the displacement currents[37] remain in perfect dielectrics, as the conduction currents are lacking.

However, the concerned dielectrics (oil reservoirs) are obviously not perfect. In them exist molecules and free charges, which are polarized and move slowly under the influence of electric and magnetic fields. They can therefore counteract the movement of the energy all the more strongly when frequencies are low.

2.5.2 Inductive Effect and Differentiated Propagation

We know (see Sect. 2.3.2) that a variable (magnetic or electric) field involves in the same time the creation of a secondary field, also variable, which in turn creates a new field, complementary to the secondary field, and so on (cf. Fig. 2.10). This phenomenon of "serial cause and effect", so to say of field displacement step by step, whose inseparable intermediaries are displacement currents and electromagnetic induction, is the origin of the phenomenon of electromagnetic propagation itself. It then establishes, during the time, a transfer of energy in the space between the media and especially in the horizontal planes of the geological layers, as discussed later (see Sect. 2.7.4).

In electricity-conducting media, the inductive effects are limited by instantaneous secondary currents precisely called in this case eddy currents. Indeed, the magnetic fields that enter the conductor also induce alternative electric fields. These fields are spontaneously opposed all the more strongly to the primary currents which generated them that the medium is conductive and that the magnitude of the field frequency is high.[38] The "propagation" in these circumstances is limited or even stopped, as in the case of metallic materials, which have very high conductivities (Sainson 2010).

[36]Which is never the case.

[37]On the theory of dielectrics, interested readers may consult, for example, Professor Fröhlich's book (Fröhlich 1958).

[38]The reader will find in my book on corrosion logs a chapter devoted to eddy currents in metals (Sainson 2010).

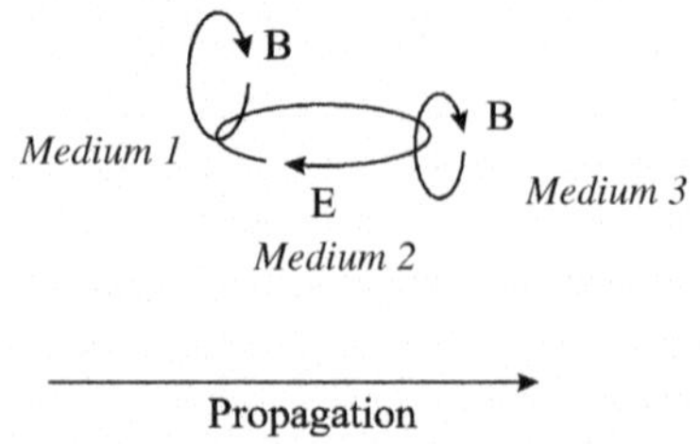

Fig. 2.10 Inductive effect: a variable magnetic field B, which appears in medium 1 creates around it an electric field E in medium 2, which in turn creates a magnetic field B in medium 3, and so on. Magnetic field B has then moved from medium 1 to medium 3 → propagation

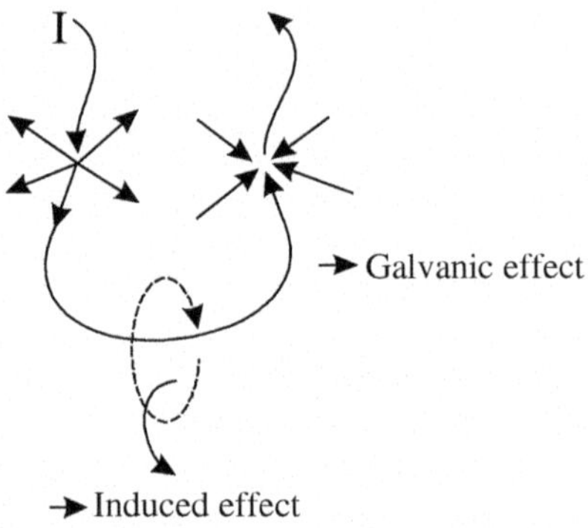

Fig. 2.11 Combined galvanic effect and induced effect due to the variable current circulation through any medium. These effects will be independently more or less pronounced depending on whether the medium is a conductor or resistant. The identification and indirect measurement of such effects (fields) are the basis of the detection system of SBL

However, in dielectric media, inductive currents are then more important and stronger (little or no eddy currents are present). They allow the propagation to be easier.

We can briefly say that at low frequencies, in conductive media, the energy transfer will be essentially due, as will be seen further, to the galvanic effects (→ electrical conduction phenomenon), whereas in dielectric media, proportionately the propagation phenomenon will be predominant (→ electromagnetic induction phenomenon).

This fundamental difference in the behavior of the lines of force and the fields toward different natural environments (the conductive marine sediments and the resistive oil reservoir especially) is the basis of the low frequency alternating current detection of SBL, since we are able to differentiate and quantify both effects separately—that is, in fact, to measure them individually at the same time and with good accuracy (Fig. 2.11).

2.5.3 Galvanic/Inductive Effect: Key of Seabed Logging Detection

As has been indicated above, the presence of alternative or variable currents more or less creates, in any geological environment, secondary currents produced by the variations of the primary magnetic flux.

These vortical derivative currents (*eddy currents*) are even more intense than the medium is conductive, and are consequently opposed to the penetration of the main current, defining then what is conventionally called the *skin effect*, or historically the *Kelvin effect* (see Chap. 3, Sect. 3.3.1).

In contrast, in a dielectric medium, they are weak and just slightly weaken the main current, allowing it to propagate more easily. In this type of resistant environment, the dominant responses are rather the galvanic type (open current). In this case precisely, the vertical component of the electric field changes directions when the current passes from a conductive medium to a dielectric medium and vice versa. This phenomenon is, among others, the basis of the SBL detection devices (measures of the field vector components) (Figs. 2.12 and 2.13).

The innovation of the currently accepted SBL method is therefore based on the fact that the eddy currents, widely present in marine sediments (a conductive medium), are significantly decreased in the presence of a hydrocarbon reservoir (a resistive medium), leading to a specific local behavior and therefore abnormal regarding ambient electromagnetic fields (excluding oil deposits).

More specifically, at the reservoir level, we can say that the resulting field then locally corresponds to the juxtaposition of two concomitant electrical effects:

- One galvanic (*static shift*), superficial, sensitive to the resistive layer, only depending on the depth
- The other one inductive or a vortex (*inductive shift*), insensitive to the resistive layer, depending on both the depth and frequency, as the eddy currents can only be formed moderately

Specifically, in the absence of a reservoir of oil or gas, the electromagnetic currents will thus tend to diminish very quickly (because of the skin effect in marine sediments), whereas in the presence of a reservoir, these currents will then have the ability to penetrate deeper (eddy currents are then much less important) (Fig. 2.14).

At an absolute level, it is relatively difficult to quantitatively assess the contribution of each of these effects. Nevertheless, it is possible to model them separately, as can be done, for example, for the galvanic effect (cf. Appendix A5.4) However, some authors have proposed models to qualitatively estimate the relative share attributable to either of these effects, in the form of a ratio (galvanic/inductive effect), dependent itself on the ratio of electric and magnetic field values (E/H), and for various geoelectric environments (Walker and West 1992).

Accordingly, and practically, if it is possible to simultaneously generate (cf. Fig. 2.15) horizontal currents (→) and vertical currents (↓), and to accurately measure them straight up the resistive anomaly (HC), if it exists, it is then possible

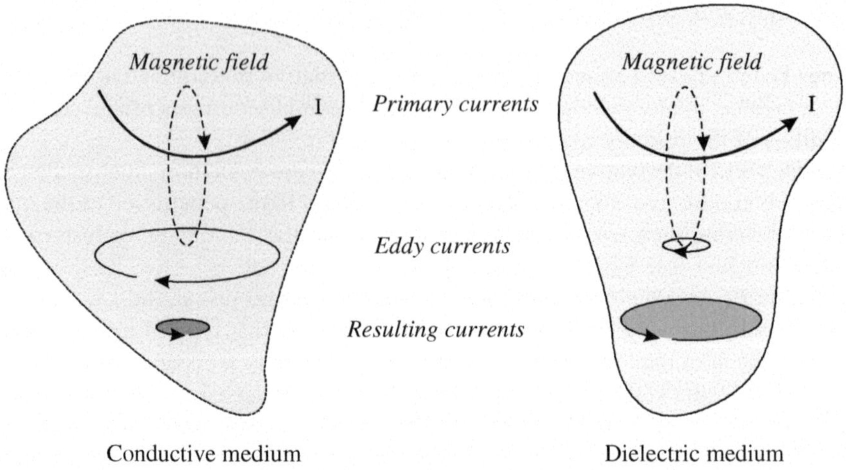

Fig. 2.12 Creation of eddy currents, important in the conductors, low in the dielectrics. The creation of these is effective for both artificial and telluric currents and changes the resulting electromotive force to be lower in the conductors than in the dielectrics

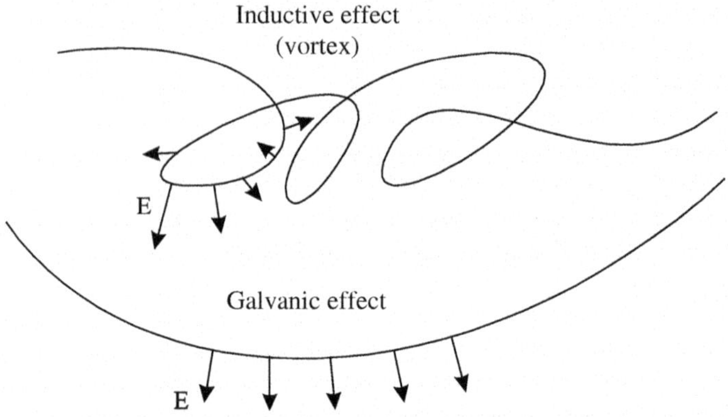

Fig. 2.13 Schematic: galvanic effect (*bottom*) and induction or vortex (*top*). The circulation of electric fields E, along the lines of force (potential), and their topological shape (more or less importantly winding), allow us to differentiate the evolution of these fields in the space. In the presence of a resistivity contrast, the vertical component of the field E changes directions → detection. In geological media, more or less conductive, the electromagnetic classical theory symbolized by the laws of Ohm and Joule (*open current*) coexists with the theory of Maxwell (*closed current*)

to identify these two distinct effects, all the more easily when the conductivity of the reservoir is low and the contrast with the sediment rocks is high.

In these singular detection conditions, in the presence of a hydrocarbon reservoir with a resistivity ρ_{HC} and a thickness t_h, the galvanic effect (cf. Fig. 2.15) comparable to the one encountered in DC will be preponderant in an all the more important manner when the transverse resistance of the reservoir ($R_{TR} = t_h\ \rho_{HC}$) is strong (Constable 2010).

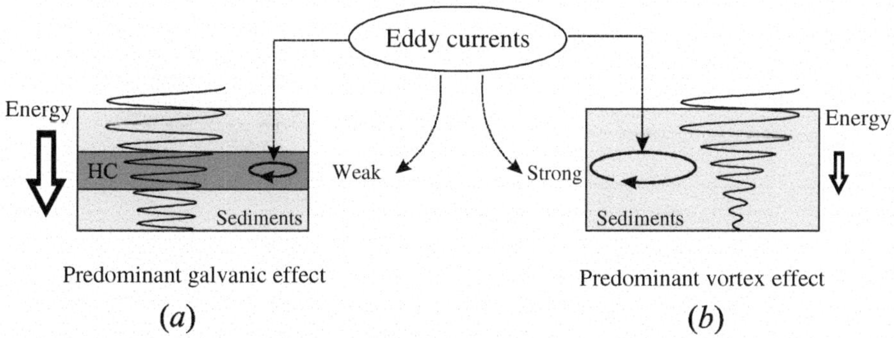

Fig. 2.14 At a similar frequency (LF), the penetration of the force lines ↓ is greater in the presence of an HC reservoir (*a*) than in its absence (*b*)

In other words, the presence of the resistive layers under the EM source combines the increase of the lateral propagation (associated with a larger skin depth) and the galvanic interruption of vertical currents (cf. Fig. 2.15).

2.5.4 Choice of the Method Principle: Emission/Reception

The imperatives of detection described in the preceding paragraph require a specific technology with regard to both the quality of the emission and the quality of the electromagnetic field reception, and their respective coupling with the subsoil.

About the emission, we know (cf. Appendix A3.2) that:

- Galvanic type responses are preferentially excited by galvanic sources too, of an open wire type, or a free antenna type (without an electrical counterweight), operating in the medium by conduction (cf. Fig. 2.5b), rather than by magnetic or inductive sources of a closed wire type or winding and coil (cf. Fig. 2.5a) essentially generating some magnetic field.[39]
- Moreover, a galvanic source supplied more energy to the medium than an inductive source.
- The use of a horizontal galvanic source, parallel in direction to the seabed (cf. Fig. 2.15), can generate both horizontal and vertical currents[40] (Walker and West 1992).

About the reception, we admit that to clearly differentiate the two effects, sensitive to the presence of the hydrocarbon deposit, it will be necessary to:

- Have field vector sensors measuring separately the different components of the electric field and not simply assess potential differences as is done with direct current.

[39]To distinguish and quantify these two effects, we can simultaneously use a galvanic source and an inductive source (see Chap. 4, Sect. 7).

[40]A vertical source can only provide a galvanic response (see the MOSES program).

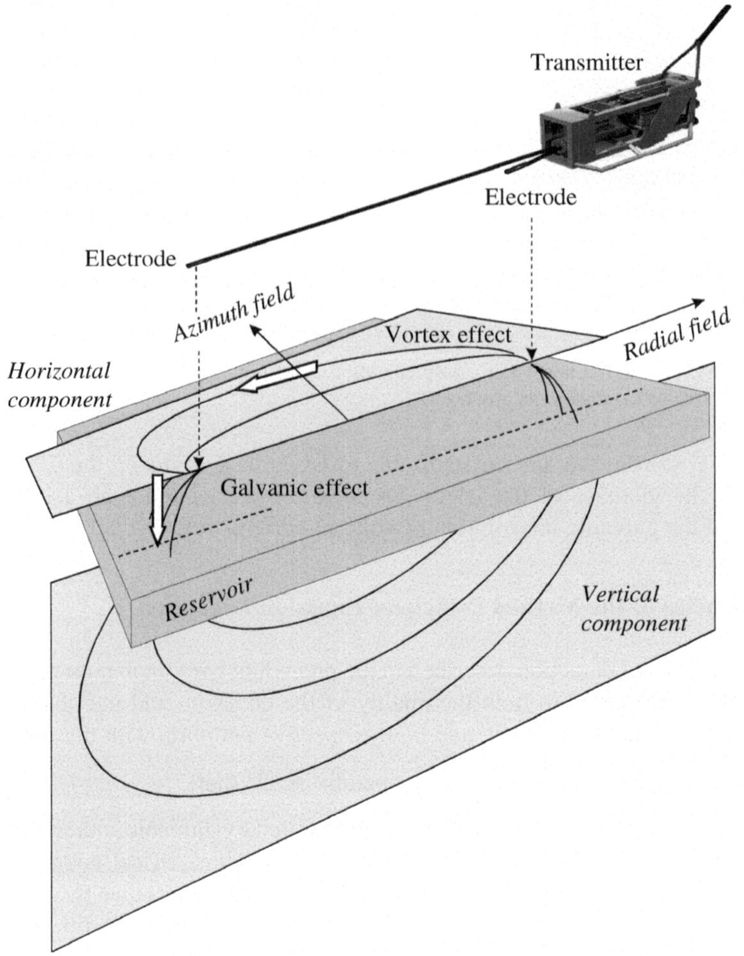

Fig. 2.15 Schematic principle diagram for detecting a horizontal electrical stimulation, which can simultaneously generate responses (fields E) due to galvanic effects (↓) and a vortex or induction (→) in the presence of a resistive layer (oil field) inserted between two electrically conductive media (sea water and marine sediments)

- Measure the components of the electric field on two mutually perpendicular horizontal lines, when the transmitter and receiver are aligned, *in line*, and then when the transmitter and receiver are parallel, *broadside* (cf. Fig. 2.16).

  About the survey, we know that:

- The galvanic effect is even more important when the transmitter dipole is located at the top of the reservoir.

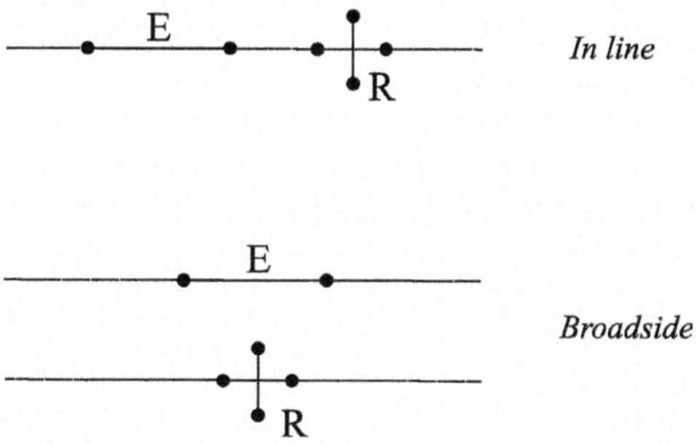

Fig. 2.16 Acquisition geometry *in-line* and *broadside* arrays (seen from above). In the *in-line* configuration the transmitter E and the receiver R (dipoles) are aligned. In the *broadside* configuration, E (HED) and R (horizontal sensors) are in parallel directions

- The inductive effect (attenuation and phase shift) occurs when the skin depths are comparable to the distance over which the electromagnetic energy has traveled.

On the transmitter/subsoil and subsoil/receiver couplings, if we admit a source of transmission common to the receivers (an operational necessity), moving at a constant depth over a fixed set of sensors posed on the seabed, these sensors will then vary at each station (cf. Fig. 2.31).

2.5.5 Acquisition Device Geometry: A Crucial Element of Detection

The data consist of EM amplitude and phase measurements as a function of the source/receiver offset and position. For horizontal electrical excitation and specifically in this case, the response to an anomalous resistivity of the subsoil depends on the magnitude of both effects. Those work, so to speak, in opposition. They preferentially differ in certain geometric configurations (cf. Fig. 2.16).

The *in-line* and *broadside* geometries give different information (Yu and Edwards 1991; Yu et al. 1997).

For example, for a deposit of hydrocarbons with a thin and resistant horizontal geological horizon (MacGregor and Sinha 2000),[41] we know that:

- In the configuration of an *in-line* array (azimuth 0°, in the direction of the source), the galvanic effect prevails and the electric field lines are purely radial and plunge vertically into the subsoil (cf. Fig. 2.15).
- In the configuration of a *broadside* array (azimuth 90° to the direction of the source), the inductive effect is predominant, and the electric field lines are purely azimuthal or horizontal, so they do not produce a galvanic response and the attenuation is primarily due to the skin effect (cf. Fig. 2.15).

[41]This fundamental element had already been stressed by Professor Cox in 1984 in a report from the Scripps Institution (Constable 2010).

The *in-line* configuration (electric dipole–dipole) is sensitive to the vertical resistivity, while the *broadside* configuration is sensitive to the horizontal resistivity.

These important results, corroborated by analog models (rheostatic tank), automatically (analytical and numerical),[42] and by numerous exploration results (field data) are not surprising if we consider that the hydrocarbon reservoir also acts as a guide wave (see Chap. 3, Sect. 3.10) facing a longitudinal excitation (Ellingrud et al. 2002).

2.5.6 Maxwell's Equations: Conditions of the Media and the Emission Source

EM seabed logging physics, considered at low frequencies, are theoretically derived from the Maxwell equations (see details in the Appendix A2.1). The resolution of these equations allows access to different mathematical, analytical or numerical models, then to simulations which will be practiced either:

- A priori in the definition of the acquisition methods through the design of measuring instruments (definitions of the detection limits, systems architecture, sensitivity, resolutions, etc.)
- A posteriori in the formulation of algorithms used in the interpretation of the data of the exploration campaigns.

Moreover, the inherent nature of the propagation media and the topology of the investigative devices will however impose special determining conditions on resolutions.

Specific Conditions Concerning the Propagation Media

Due to the virtually nonmagnetic character of the crossed materials (at least in hydrocarbon exploration),[43] the magnetic permeability μ of the crossed environments is then equivalent to that of the vacuum μ_0 and can thus be considered constant. Under these conditions, with few exceptions, the constitutive relation between the magnetic field and induction vectors (cf. Eq. A2.13) becomes:

$$\vec{B} = \mu \vec{H} = \mu_0 \vec{H} \tag{2.13}$$

[42]Which we shall see examples of in Chap. 5.

[43]Some hydrocarbon deposits may be structurally related to volcanic activity with the presence of magnetic igneous rocks (dykes). Under these conditions the magnetic permeability is different from that of a vacuum.

for all concerned media (seawater, marine sedimentary layers and hydrocarbon reservoirs) involved in the detection process.

Conditions Concerning the Transmitting Source

Because of the proximity of the transmitter and its power, the current density near the receptors acquires an additional term $\vec{j}^s$ related to the source such that we finally have (Ohm's law):

$$\vec{j} = \sigma\vec{e} + \vec{j}^s \tag{2.14}$$

which is to write for the equations[44] considering the constitutive relation (cf. Eq. A2.1.8) that (cf. Eqs. A2.1.5 and A2.1.7):

$$\vec{\nabla} \wedge \vec{h} = \varepsilon\frac{\partial \vec{e}}{\partial t} + \vec{j} \tag{2.15}$$

and

$$\vec{\nabla}.\vec{e} = \frac{q}{\varepsilon} \tag{2.16}$$

In low frequency or quasistatic approximation (cf. Sect. A2.10), as is the case in alternative stimulation SBL technics (cf. Chap. 5, Sect. 3.1.1), we consider that the displacement currents are negligible compared with the conduction currents ($\rightarrow \varepsilon\partial e/\partial t = 0$).

2.5.7 Fundamental Equations of the SBL Detection Principle

Taking $e^{-i\omega t}$ as an expression of the harmonic variation (ω) of the associated electric and magnetic fields, in the unit MKSA or SI system, and replacing the equation (cf. Eq. A2.1.4) by the equality (cf. Eq. 2.13), the above relations including the source term $\vec{J}^s$ then become equivalent in the frequency domain to the following equations:

$$\vec{\nabla} \wedge \vec{E} = -i\omega\mu_0\vec{H} \tag{2.17}$$

and:

[44] In the time domain for the problem of wave propagation to be well established, it is imperative to also establish initial conditions to fix the direction of the time ($\pm\, t$).

$$\vec{\nabla} \wedge \vec{H} - \sigma\vec{E} = \vec{J}^{s} \quad (2.18)$$

These two vector equations are the theoretical basis of the detection principle of the EM SBL method. They govern, under the guise of using well-established restrictive conditions (boundary conditions) and complementary relationships (e.g., Ohm's law), the propagation, the diffusion and the distribution of the electromagnetic fields through various media (seawater, marine sediments and hydrocarbon reservoirs).

2.5.8 Ohm's Law and Maxwell's Equations

Maxwell's equations do not themselves accurately describe the electromagnetic situation in the concerned areas where charge carriers are present, for example. New relationships in relation to the specific nature of these environments must therefore complete these equations.

The detection principle itself, based on a dual effect (inductive/galvanic), imposes, theoretically and at the considered frequencies (LF), the introduction of Ohm's law (cf. Sect. 2.4.4).

Indeed, under certain conditions, Ohm's law can be extended to variable regimes. It completes then the propagation equations as forming one of the constitutive equations of Maxwell's equations (cf. Appendix A2.3).

In the harmonic regime, the electrical conductivity and dielectric permittivity are then considered as complex because they depend on the frequency. In these circumstances, Ohm's law in its complex form becomes:

$$\vec{J} = (\tilde{\sigma} + i\omega\,\tilde{\varepsilon})\,\vec{E} \quad (2.19)$$

where the complex electrical conductivity $\tilde{\sigma}$ and the complex dielectric permittivity $\tilde{\varepsilon}$ are respectively equal to:

$$\tilde{\sigma} = \sigma' + i\sigma'' \qquad \text{and} \qquad \tilde{\varepsilon} = \varepsilon' + i\varepsilon'' \quad (2.20)$$

which is to write (cf. Eq. 2.19) that:

$$\vec{J} = \left(\sigma' - \omega\varepsilon'' + i\omega\varepsilon + i\sigma''\right)\vec{E} \quad (2.21)$$

or that:

$$\vec{J} = (\sigma_e - i\omega\,\varepsilon_e)\,\vec{E} \quad (2.22)$$

with:

$$\sigma_e = \sigma' - \omega\varepsilon'' \qquad \text{and} \qquad \varepsilon_e = \varepsilon' + \frac{\sigma''}{\omega}$$

where σ_e and ε_e are then real numbers, as σ'' and ε'' (depending on the frequency) are negligible in most cases except in clays.

2.5.9 Propagation Physics or Diffusion Physics?

The equation of the electromagnetic waves is the Helmholtz's equation (see Appendix A2.3; cf. Eq. A2.1.27). In writing it in the time domain and involving the propagation velocity c, we obtain:

$$\nabla^2 \vec{e} = \frac{\omega^2}{c^2} \vec{e} + i\omega\mu_0\sigma\vec{e} \tag{2.23}$$

It then shows two terms:

- A *propagation term* (the first member) the so-called Maxwellian term dependent on the speed c explicitly and on the dielectric permittivity ε of the medium implicitly
- A *diffusion term* (the second member) only inherent to the electromagnetic and inseparable properties of the medium (μ_0 and σ), and particularly its electrical conductivity σ

If we now form the ratio of these two terms, replacing c by its value, i.e., $c = (\varepsilon\mu_0)^{-1/2}$, such that we have:

$$\frac{\textit{propagation term}}{\textit{diffusion term}} = \frac{\omega^2}{\omega\mu_0\sigma c^2} = \frac{\omega\varepsilon}{\sigma} \tag{2.24}$$

and considering the magnitude orders of the electromagnetic characteristics of the crossed media, we see that at low frequencies, this ratio is very low and that the diffusion term is then largely preponderant and outweighs the one of propagation.

For example, in the case of SBL, the media have very low dielectric permittivity (see Chap. 3, Table 3.2), slightly higher than that of a vacuum ($\varepsilon_0 = 8.8 \cdot 10^{-12}$ F/m). We obtain in these situations, and more specifically in the sedimentary rocks when σ is 10^9 times greater than ε, for very low and extremely low frequencies (ELF), the values of this ratio, such as (Table 2.1, Fig. 2.17):

$$\text{ELF} \quad \rightarrow \quad 10^{-9} \langle\ \omega\varepsilon/\sigma\ \langle\ 10^{-7} \tag{2.25}$$

This theoretically implies that at low frequencies ($\approx$1 Hz) and very low frequencies (<1 Hz), it has more to do with *diffusion physics* ($\omega\varepsilon/\sigma << 1$) comparable to the transmission of heat (with material support)[45] with *propagation physics* such

[45]The equations are the same as those found by Fourier to represent the propagation of the heat in the media (Fourier 1822).

Table 2.1 Values of the ratio $\omega\varepsilon/\sigma$ in different propagation media, in response to an excitation of frequency 1 Hz

Stimulation at 1 Hz	$\omega\varepsilon/\sigma << 1$
Seawater	$1.5 \cdot 10^{-8}$–$9.5 \cdot 10^{-10}$
Sediments	$3.5 \cdot 10^{-8}$–$4.4 \cdot 10^{-9}$
Reservoir	$1.1 \cdot 10^{-7}$–$1.1 \cdot 10^{-8}$

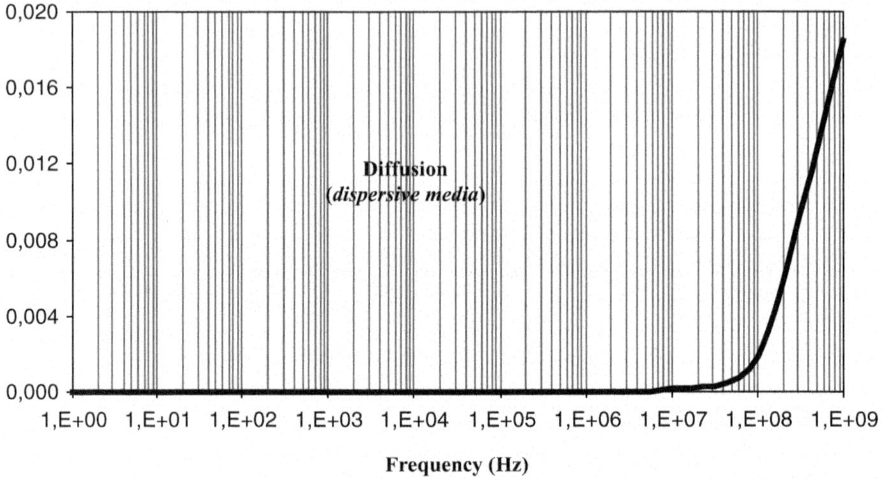

Fig. 2.17 Changes in the ratio $\omega\varepsilon/\sigma$ depending on the frequency, thus theoretically defining broadcast and propagation domains showing that below the GHz, well over the working frequencies ($\approx$1Hz), the diffusion phenomenon is predominant

as those involving the movement of radioelectric waves (without hardware support, such as in a vacuum, for example). In this case, the propagation factor is rather of the form $k = (-i\omega\mu_0\sigma)^{1/2}$ (cf. Eq. A2.1.32).

Nevertheless, we can prove (Loseth et al. 2006) that the electromagnetic energy, despite the low frequencies that are used, can also behave like waves in the limit of the skin depth, then follow the laws of geometric optics,[46] browse preferred paths and also be decomposed in different modes (cf. Sect. 2.7.4.3 and Sect. 3.7 Chap. 3).

It goes without saying that the propagation media should be as homogeneous as possible, so that this theory is valid, which is rarely the case in a reservoir made of different fluids (water, oil, gas) and solid elements (rocks) with among others the presence of fractures, joints, facies transition, discordance or any other remarkable geological discontinuities.

If the theoretical aspect is then important, this does not prevent controversies existing. This is the case between several schools of thought, which are opposed at the present time and which suggest that additional phenomena of propagation of the

[46]Phenomena observed for the first time by the physicists J. C. Bose, A. Righi and H. Hertz.

electromagnetic waves[47] (relaxation, polarization, or others) due to microscopic qualities of geological environments really coexist (Ellis and Singer 2007).

This would particularly be the case at the oil reservoir level, even at its periphery, with, for example, the existence of specific interfaces of more or less pronounced phenomena (Bonner et al. 1996 Loseth et al. 2006)[48] (Fig. 2.18).

In these special conditions, the general wave equation (diffusion/propagation):

$$\nabla^2\vec{e} = \mu_0\sigma\frac{\partial\vec{e}}{\partial t} + \mu_0\varepsilon\frac{\partial^2\vec{e}}{\partial t} \tag{2.26}$$

of which the second order term is negligible (ε very small) therefore reduces to the diffusion equation:

$$\nabla^2\vec{e} = \mu_0\sigma\frac{\partial\vec{e}}{\partial t} \tag{2.27}$$

a "reduced" fundamental equation, which in some way governs the detection principle and especially the distribution of electric fields in the context of the

---

[47]For Dr. Chave, from Woods Hole Institute of Massachusetts, scientific adviser to Schlumberger (WesternGeco), low frequency physics, within SBL, are physics of diffusion (a fact determined a priori). However, for the experts of the Norwegian company EMGS, as well as for Professor Shultz of the University of Oregon, these physics would come under the juxtaposition of the two phenomena at least with regard to the reservoir. For now, the debate is not settled and still remains controversial as evidenced by the discussion that took place at the trial in opposition (EMGS patents), which confronted the two companies in front of the High Courts of Justice in England (England and Wales High Courts, 2009) and Norway.

Remark: The physics of waves generally reveals in the propagation some temporal terms that vary with space (cf. the wave equation). This is for example the case of the signal phase, which may be ahead or behind the original current (natural or artificial).

Phase shift due to the induced effects: The physics of the propagation of currents in conductors and "perfect" insulators is relatively intelligible. In good conductors, the currents will have all the more difficulty in passing if the frequency is high. For good insulators, the opposite happens. In other words, a DC or a VLF current can neither enter nor cross through an insulator while a high frequency current can. Similarly the latter cannot penetrate a very good conductor, as its density is then directly affected by a pellicular effect concentrating the field to the periphery (a *skin effect*). For materials with intermediate conductivity and especially with a heterostructure, such as the reservoir rock, the problem is more complex to understand and the physics are then less clear to establish except in the precise case of DC. For the latter, the propagation is only affected by an energy dissipation essentially due to the *Joule effect* (heating at the current injection), the conductivity of the rock the presence of the ions (chemical energy transfer) and the distance (a geometric or *spherical divergence* phenomenon, which affects the current density: $\partial I/\partial S$). For alternating currents, in addition to the characteristics of the rock, we have to take into account the frequency and its attendant effects on the electrical resistance, capacity and self-induction of the crossed fields. For VLF currents, we can assume that the resistance is very much the same as for DC. However, we can say that the other electrical features then play a significant role in the propagation giving the phase of the current intensity different values at each point of the field.

[48]The reader may also follow a more complete demonstration in the frequency domain for a stratified medium.

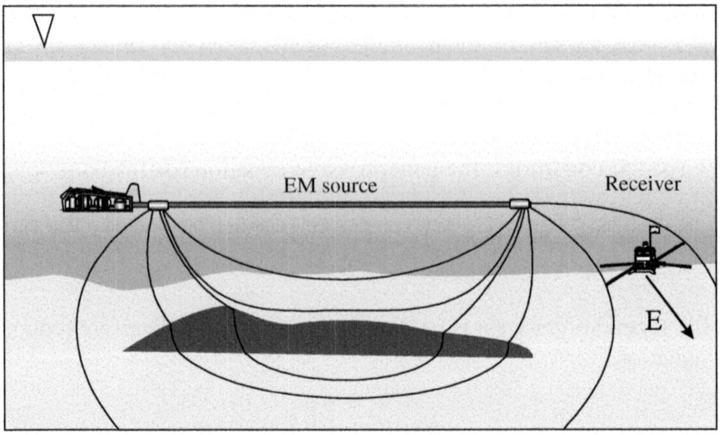

Fig. 2.18 Principle of measuring electric fields (E): details of the current lines near and passing a HC reservoir where inductive and galvanic effects simultaneously combine. These are more marked laterally and differently at the apex and at the periphery of the oilfield

exploration by SBL,[49] taking then into account the electrical conductivity of the rocks (with $\mu_0 = \text{constant}$).

2.5.10 Simple Solutions of the Diffusion Equation

In a homogeneous and isotropic medium, the solutions of the diffusion equation (see Eq. 2.27) for a monochromatic plane wave of angular frequency ω (with a frequency of 0.1–10 Hz or lower for the mMT method) affecting the amplitudes of the electric and magnetic fields in the direction of z are of the form:

$$\begin{aligned} \mathrm{e}(z) &= \mathrm{e}_0 \mathrm{e}^{-\beta z} \mathrm{e}^{-\mathrm{i}\alpha z} \mathrm{e}^{\mathrm{i}\omega t} \\ \mathrm{h}(z) &= \mathrm{h}_0 \mathrm{e}^{-\beta z} \mathrm{e}^{-\mathrm{i}\alpha z} \mathrm{e}^{\mathrm{i}\omega t} \end{aligned} \tag{2.28}$$

where e_0 and h_0 are the values of fields at the distance $z = 0$, where β and α are the respective terms of attenuation and the electromagnetic characteristics of the propagation media.

When the conduction currents are dominant (a conductive medium), the coefficients α and β are then dependent on the electrical conductivity σ and the magnetic permeability μ_0 such as:

[49] In prospecting using geological radar where frequencies are important (≈100 MHz), the second term (propagation) can no longer be neglected (cf. Chap. 5, Eq. 5.9).

$$\alpha = \beta = \sqrt{\frac{\omega\sigma\mu_0}{2}} \tag{2.29}$$

The components of the e_x and e_y fields in the horizontal plane (perpendicular to z) then have constant amplitudes (a plane wave).

The above equations can also be expressed as discussed later (see Chap. 3, Sect. 3.1.1) in term of the skin depth δ. We have then $\beta = \alpha = 1/\delta$.

2.5.11 A Signal-to-Noise Ratio Very Favorable to Detection

The measurement of the very low amplitudes of the electric fields induced in the water, and their variations (in the order of a few $nV/m/\sqrt{Hz}$) would not have been seriously considered if the ambient electromagnetic noise present at the bottom of the sea had been important. But this one, in exploration areas[50] sufficiently distant from the coasts, remains remarkably low (of the order of a few tens to a few $pV/m/\sqrt{Hz}$) or below in some areas (Cox et al. 1978).

This consequently leads to a very acceptable, even comfortable, signal-to-noise ratio (a factor ranging from 100 to 1000), if you have a measuring device that is sufficiently sensitive and with little noise so that measures can be performed with high precision in such a hostile environment.

2.6 Physical Principles of Variable Current Electrography

We say here variable currents for telluric currents, i.e., the natural electrical currents continuously flowing through the oceans and the lithosphere.[51] These currents with long periods may be used as an electrical and/or magnetic source. Their spatial and temporal (frequential) variability advantageously allows investigation of the subsoil at different depths (a variable skin effect). The very low frequencies that characterize them therefore allow us to obtain great depths of investigation, which make these techniques more suitable for the study of earth physics (crustal and mantle structures).

The physical principle is similar to the one governing alternative or periodical currents. In contrast, the transitory and variable (unpredictable) characteristics of the telluric source then require proper treatment of the information, which is more complex to achieve, notably using correlation or comparison of the fields (magnetic

[50]It is quite different in estuaries, harbor facilities and generally coastal areas, where anthropogenic sources nearby and the shallow depth of water concentrate the electromagnetic noise.

[51]Vagabond currents from industrial sources are also variable. With few exceptions these have been used for prospecting or for earth physics, especially employing electrified lines for the railway (Meunier 1975).

and/or electrical); the techniques with methods that use this principle give more relative results than those obtained by techniques using artificial currents (predictable).

2.7 *Acquisition, Disposition and Measurement Principles*

Commercially, DC methods were previously used only in shallow water and are very easy to implement. Currently, they increasingly have to compete with more and more powerful sediment echo sounders and high resolution seismics (Trabant 1984).

Alternating current techniques (controlled source methods) and variable current techniques (telluric source methods) intended for investigation of deeper grounds under deep water require, however, much more important means comparable to marine seismics.

2.7.1 Direct Current: Resistivity Electrical Methods

In the DC regime, the most widely used method is the resistivity method, i.e., one that allows us to obtain, from current and potential measurements, after calculations, the resistivity values of the different soil layers forming the seabed. This technique is not without resemblance to the surface methods from which it is directly derived (see the history in the Introduction). Nevertheless, if its efficiency is real in fresh water (rivers and lakes), it must be theoretically considered differently due to the immersion of the acquisition device in seawater, which is conductive.

Measuring Device Topology

For the convenience of the operations, current injection and potential measurement electrodes are mounted on the same material support constituting a unique line of electrodes (flute) allowing the entire acquisition system to be simultaneously pulled by a boat.

Since the 1930s, when the initial investigations were conducted (cf. Sect. 6, Chap. 1), several geometries have been tested with either:

- The injection points at the end and the measurement points at the center (quadrupole type Wenner or Schlumberger arrays)
- Dipole–dipole type layouts, or
- Focused devices logging on the bottom (Witte 1958; Jones 1959; Postma 1962) or floating with guard electrodes (Mayes 1965)

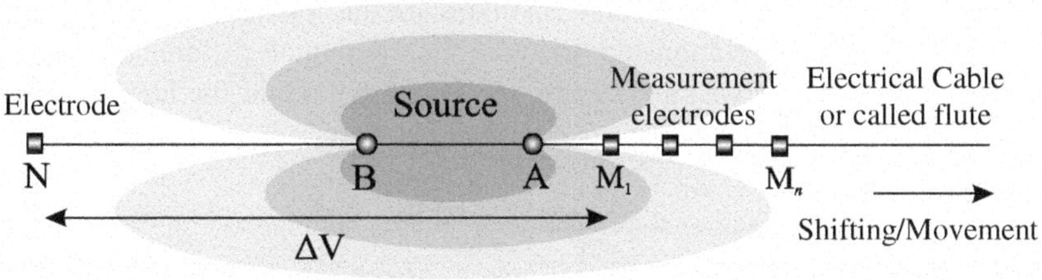

Fig. 2.19 Apparatus for measuring the resistivity (immersed flute) developed from a group of electrodes (two for current injection A, B and several for potential sampling N, M_n) aligned and mounted on a cable (electrically isolated) towed by a boat

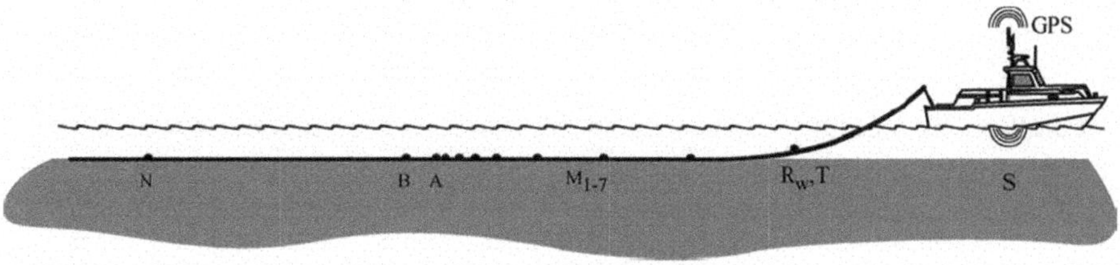

Fig. 2.20 DC device using a cable submerged in shallow water, composed of one or more quadrupoles (interdependent transmitters and receivers) dragging on the seabed (electrical profiling or sounding). The method has no lateral directivity and only provides information according to the vertical at a constant depth (AB fixed) or variable depth (different AB). This technique is widely used in lake exploration (According to Baron and Chapelier 2003)

However, it seems established that some spreads are more suitable than others, as is the case, for example (cf. Fig. 2.19), when the injection dipole (A, B) is at the center of the receiving device (N, M_1,. . .M_n) (Fig. 2.20).

The resistivity method is not sensitive to three-dimensional anomalies as well as to those with limited dimensions (Telford et al. 1983).

However, it is possible to highlight heterogeneities of small dimensions when juxtaposing several profiles side by side. We can also laterally make the position of the measurement dipole vary, using a fixed injection dipole (measuring in a rectangle) and thus obtain a field profile.

Resistivity Evaluation (Continuous Current)

In a marine environment, even if the technique is old, electrical prospecting by DC has only been studied in recent decades by a very small number of engineers, led in France, for example, by Professor Lagabrielle (Lagabrielle 1983, 1992; Lagabrielle et al. 2001), in the USA, Canada, Switzerland (Baumgartner 1996; Baron and Chapelier 2003) and also in Russia (Vishnyakov et al. 1992).

The resistivities of the rock formations and their thicknesses are obtained simply:

– By measurement (in the water) and calculation of the apparent resistivity[52] given by the well-known relationship between the potential difference measured between two electrodes M and N (the receiver dipole) and the injected current I (at point A) such that:

$$\rho_a = K_n \frac{\Delta V_{MN}}{I} = K_n \frac{V_M - V_N}{I} \tag{2.30}$$

where K_n is a geometric factor dependent on the n electrode layout (cf. Fig. 2.19) and (when the electrodes are aligned) equal to:

$$K_n = \left(\frac{1}{AN} + \frac{1}{AM_n} \right) / \left(\frac{1}{AN^2} + \frac{1}{AM_n^2} \right) \tag{2.31}$$

– Parallelly by measurement in situ (still in water) and calculation of the electrical conductance of the seawater such that:

$$C = -\frac{h}{\rho_w} + \frac{K_n}{AM_n} \tag{2.32}$$

where h and ρ_w are respectively the depth and the resistivity of the sea water.[53]

Then from these apparent resistivity measurements, either we compare (a forward problem) or calculate (an inverse problem) the effective resistivity values of the different geological layers forming the subsoil (cf. Chap. 6) (Fig. 2.21).

2.7.2 DC Investigation Limitations

From the apparent resistivity measures,[54] obtained thanks to the transmitted currents and the voltages collected for different geometric spreads (spacing variations between the electrodes in particular → variable depth investigation), we obtain specific resistivity sections depending on the depth. Under these conditions of vertical scanning, investigation by DC offers no horizontal directivity and therefore no lateral resolution,[55] and so the fluctuations of the measured apparent resistivity only depend on the longitudinal conductance[56] of the piling of the different

[52]Not to be confused with the specific resistivities (see Chap. 3, Sect. 2.1).

[53]For this calculation we also measure the water temperature.

[54]True or specific resistivity is a physical quantity that is difficult to understand. For the same material, it may be different and vary depending on the environment in which the measurement is made (laboratory, well logging, surface prospecting, etc.).

[55]In some very rare cases (e.g., a lateral mismatch), the intersection of profiles (perpendicular) between them provides a lateral extension to a certain limit (low resolution).

[56]The longitudinal conductance or the conductance of a layer of thickness h is equal for a portion of a field of a unitary section to h/ρ (the Dar Zarrouk parameter).

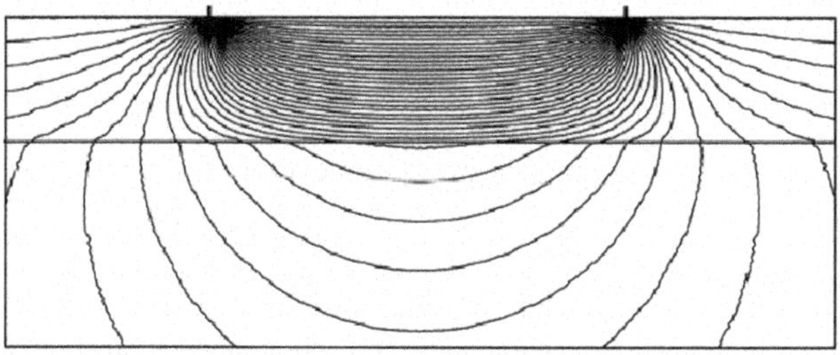

Fig. 2.21 Topology (in section in the vertical plane) of the current lines from the two injection points at the surface of the seawater, which then propagate in less conductive soil (sediment)

conductors (an assumption). Moreover, even if the technique is performed on the seabed, it seems to be limited by the thickness of the water layer (about 10 m) so much more strongly that the ratio of the conductivity of marine sediments to the conductivity of the seawater is low (Lagabrielle 1983, 1992).

For example, assuming in the best case a ratio of 0.03/0.3 or 0.1, it is then necessary for the detection of conductivity variations greater than 10 % to perform measurements with a precision of at least 0.3 %.

Moreover, the technique is limited by the relief of the seabed, or more exactly by changes in the topography whose size must exceed in distance the length of the injection line. Finally, the attenuation will be much stronger as the instrumented cable will be away from the bottom (the largest energy dissipation in water) then limiting surface, subsurface or open water (streamer) techniques and thus making the logging technique more difficult to implement.

Almost all the commercial surveys carried out by these methods in aquatic surroundings aim to obtain information about:

- The location of the roof of geologically resistant formations in the electrical and mechanical sense of the term (consolidated ground, non-cracked, hard rock called bedrock), lying generally below a light and more conductive sedimentary substratum
- The evaluation of the thickness of the first layers of marine sediments (vases, mud, silt or more or less consolidated sediment layers)
- The nature of the detrital rocks forming the bottom of the sea
- The formation, evolution and movement of sand bars (silt)
- The location and measurement of the thickness of submarine permafrosts
- Research into building materials (gravel)
- The identification of layers or horizon marks (present for example in the rift valleys)
- The existence or absence of hidden stones
- The marine resurgence of subsurface freshwater sources
- The hydrogeological evaluation of terrestrial aquifers on the seaside (extension, drainage, overdeepening, etc.)
- The development of accurate maps of the estuary (geographical, hydrological and bathymetric, etc.)

- The mapping of marine or lacustrine archaeological sites
- Leak detection on large structures (dams, dikes, lifting, etc.)
- Identification of polluted areas

In the future, offshore metal deposit prospecting (deep sea mining) could also be created through this relatively simple-to-implement technique.

For oil exploration, in addition to the disadvantage of using very high emission powers (much higher than those of variable currents), the relatively limited nature of the oil concentration, a consequence of which is that they are badly fixed during the measurements, is also a very unfavorable condition for their detection by a continuous current. Even in the case where the oilfield shows high resistivity regarding its immediate geological environment, it is virtually impossible to determine whether the high apparent detected resistivity is due to the presence of hydrocarbons or to another cause (adverse reliefs, heterogeneous rocks, facies variations, irregular thickness, etc.).

2.7.3 Alternative Currents: Resistivity Magnetic Methods

This was one of the first marine investigations to be proposed and conducted with magnetic observation submarine equipment existing at the time (1970) as the OBM (for Ocean Bottom Magnetometer).

Measuring System Topology

The mMR methods (for marine magnetometric resistivity) were primarily used with vertical injection devices, VED (see Fig. 1.13). In these situations, as the electric field is vertical, the magnetic measurements were performed on horizontal or azimuthal B_ϕ components of the associated magnetic field (Chave et al. 1991; Edwards et al. 1981, 1984) such that we can write:

$$B_\phi = \frac{\mu_0 I \sigma_1 h}{4\pi\sigma_w r^2}\left(1 - i\omega\ \mu_0\sigma_1 \frac{r^2}{2}\right) \tag{2.33}$$

where I is the intensity of the injection current, h the height of water, r is the distance from the source to the receiver (offset), and σ_w, σ_1 are respectively the electrical conductivity of the seawater and the subsoil, and μ_0 is the magnetic permeability of the vacuum equivalent for all horizons.

Apparent Resistivity Evaluation

In quasistatic approximation (cf. Eq. 2.33) the resistivity curves are identical to those of a Schlumberger survey (Chave et al. 1991). The apparent resistivity ρ_a is then given according to B_ϕ by the relation:

$$\rho_a = \frac{\mu_0 I h}{4\pi\sigma_w r^2 B_\phi} \tag{2.34}$$

The vertical current injection in this precise case can only highlight the galvanic effect, a phenomenon that fits well with the conductive media investigation but is still insufficient to detect resistivity contrasts between marine sediments and the oil reservoir (conductive/resistant contact). Nevertheless, this specific conformation may limit the surface waves (cf. Chap. 3, Sect. 3.13), which can be significant for instance for surveys realized in shallow waters.

2.7.4 Alternative Currents: Electromagnetic Field Methods

After the detailed establishment of the method principle the question of course is the one that concerns hydrocarbon detection itself and more particularly the identification of the double galvanic/vortex effect (cf. Sect. 2.5.3).

Briefly, to implement this method, the current technology is based on:

- Measurement of the amplitudes of the components (mainly horizontal) of the electromagnetic field (electric especially or magnetic in some cases), in different spatial configurations (in line and broadside)
- If it's possible, identification of the difference in the phase between the received fields in relation to the excitation signal (electromagnetic source)
- Combination and comparison of measurements between them at different points in the horizontal plane (seabed), organizing what we call the prospect

all evaluations being completed in situ and a posteriori by acquisition data correction systems.

Importance of Measure Topology

The device comprises a mobile EM transmitter (source), which moves over a set of vector receptors placed on the bottom of the sea. In such an arrangement, the horizontal electromagnetic excitation (electric dipole) creates in the invested space both horizontal and vertical electric fields, which can be measured according to different geometric configurations:

- *In line*, i.e., when the transmitter and the receivers are aligned (cf. Fig. 2.16). Then both galvanic and inductive effects are present. This configuration is rather sensitive to the *vertical resistivity*, where the difference of conductivities is then important.
- *Broadside*, i.e., when the transmitter and the receivers are in parallel directions (cf. Fig. 2.16). Thus only the inductive effect is operative (horizontal currents). This configuration is more sensitive to a *horizontal resistivity*, which is then

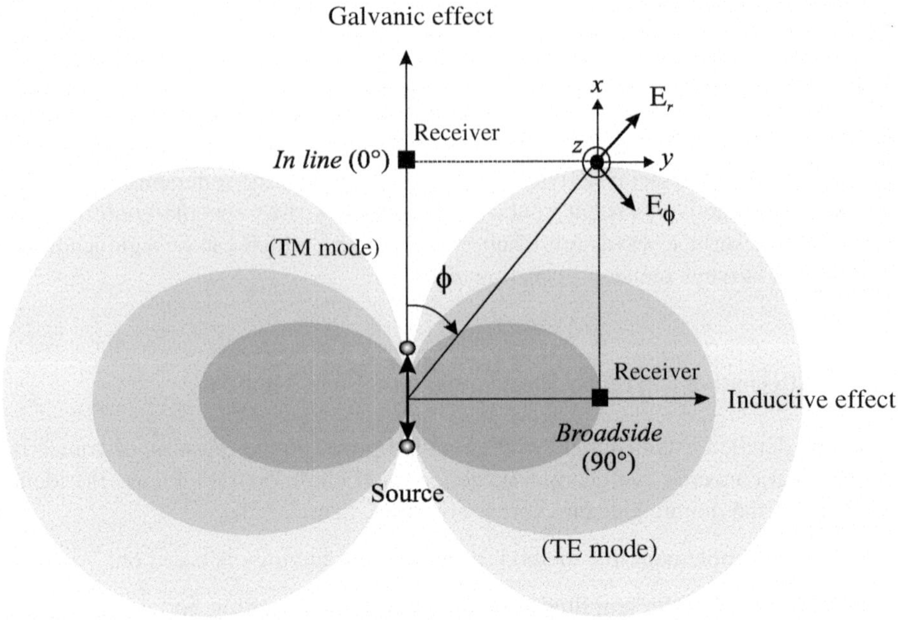

Fig. 2.22 Decomposition in a meridian plane passing through the axis of the dipole of the electric field in the horizontal plane (x, y) by *in-line/broadside* field measures and polar representation (E_r: radial component and $E\phi$: azimuthal component) at a point of the acquisition device (electrometer) allowing us to identify significant effects (galvanic) or not (inductive) on the resistive layer (oil reservoir) (According to MacGregor and Sinha 2000)

interesting, and even determining to detect laterally more or less defined structures.

Electric Field Evaluation

Among the six electromagnetic vectorial components measurable on the sea floor, some are more interesting than others to record in terms of amplitude and phase.[57]

Let us consider a measurement performed in the background, on homogeneous sediment slightly more resistive than seawater, thus separating the propagation medium into two half spaces (Chave et al. 1991; Constable and Srnka 2007).

At a distance r from the EM power source, the radial E_r and azimuthal E_ϕ components of the electrical field (cf. Fig. 2.22), solutions of the propagation equation, are under these conditions equal to:

[57] Statement based on the present state of knowledge.

$$\mathrm{E}_r = \frac{P\rho_w}{2\pi r^3}\cos\phi\left[(k_w r + 1)\,\mathrm{e}^{-k_w r} + (k_r^2 r^2 + k_r r + 1)\,\mathrm{e}^{-k_r r}\right] \quad (2.35)$$

$$\mathrm{E}_\phi = \frac{P\rho_w}{2\pi r^3}\sin\phi\left[(k_w r + 1)\,\mathrm{e}^{-k_w r} + 2(k_r r + 1)\,\mathrm{e}^{-k_r r}\right] \quad (2.36)$$

with a complex wave number for each medium (seawater k_w and sediments k_r):

$$k_w = \sqrt{i\omega\mu_0/\rho_w} \quad \text{and} \quad k_r = \sqrt{i\omega\mu_0/\rho_r} \quad (2.37)$$

where μ_0 is the magnetic permeability of a vacuum, ρ_w and ρ_r are respectively the resistivity of the seawater and the marine sediments, and finally P the dipole moment is equal to the product of the current by the length of the antenna: $\mathrm{I} \times L$ (see Chap. 4, Sect. 4.4.4).

By simultaneously performing measurements of electric and/or magnetic fields in two horizontal directions, perpendicular to each other, we can then distinguish the two pure effects, one galvanic, the other one inductive.

Under these favorable conditions, we thereby obtain the polarization of the average field at the measuring point, which then directly depends on the degree of lateral anisotropy (horizontal plane) of the apparent resistivity of the underlying rock formations. We can therefore, by an appropriate spatial disposition of field sensors with transverse components, laid on the seabed, detect more or less precisely the presence or absence of a resistive anomaly.

By combining the different *in-line* and *broadside* measurements, and considering if necessary some modes (cf. Sect. 2.7.4.3), we now obtain, as discussed later in some modeling, a multidimensional picture of the spatial distribution of the conductivity of the subsoil.

Field Transversal Components: TE and TM Modes

In an homogeneous, isotropic and infinitely extended space, either slightly conductive[58] or dielectric, the electromagnetic wave, whatever its orientation, is in its modal form of the electric and magnetic toroidal type, abbreviated as TEM. This means in other words that the electric field and magnetic field vectors, orthogonal between them, move together perpendicularly to the propagation direction, as if these waves would propagate in the vacuum (see Fig. A2.1). The TEM mode is therefore both a transverse electric mode and a transverse magnetic mode.

In a configuration of a wave guide or more generally in a homogeneous stratified medium, limited then by two distinct horizontal interfaces, we theoretically show (Morse and Feshbach 1953; Collin 1960) that the electromagnetic field propagating

[58]It is not the same in conductors with high conductivity (metals). The electric field and magnetic field vectors are not perpendicular then and also have a phase difference.

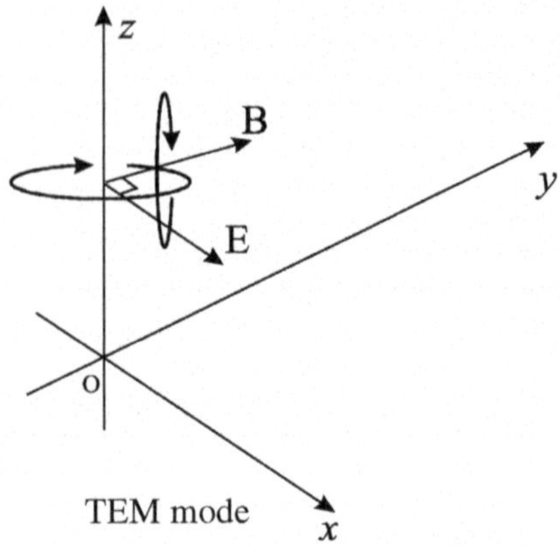

Fig. 2.23 In the TEM configuration, the electric field E and magnetic field B, orthogonal to each other, move simultaneously in the direction of propagation (z) of the electromagnetic wave. In highly diffusive media (e.g., metals), the field vectors B and E are not perpendicular anymore

inside this medium may be decomposed into two different transverse modes, one electric (TE) and the other magnetic (TM).

In practice, we can also show, which is an important point, that the sensitivity to these two modes may depend (not always) on the presence or absence of hydrocarbons in the trap. In these conditions:

- The transverse electric field (TE mode) is originally characterized by a current loop, which flows around the z axis (horizontal plane), which coupling in the propagation medium is achieved by induction (cf. Fig. 2.23).
- The transverse magnetic field (TM mode) is originally characterized by a current loop inscribed in a z-direction plane (vertical plane) and perpendicularly intersecting the separation plane with a medium of different electromagnetic properties (cf. Fig. 2.24).

We can note that the TM and TE modes respectively produce no electric and magnetic fields in the propagation direction[59] (O z here), which means in the case of the figure (cf. Fig. 2.23) that the vertical components $\vec{E}_z$ and $\vec{B}_z$ of the electric and magnetic fields are zero in the vertical direction z.

In theory, based on Maxwell's equations in the frequency domain, where the displacement currents are neglected:

$$\vec{\nabla} . \vec{B} = 0 \tag{2.38}$$

$$\vec{\nabla} \wedge \vec{E} + i\omega\vec{B} = 0 \tag{2.39}$$

[59]It is demonstrated using the Hertz potential, which verifies the wave equation in a rectangular coordinate system, that the transverse components of the electric and magnetic fields (TE and TM modes) are zero in the direction of propagation (Galejs 1969).

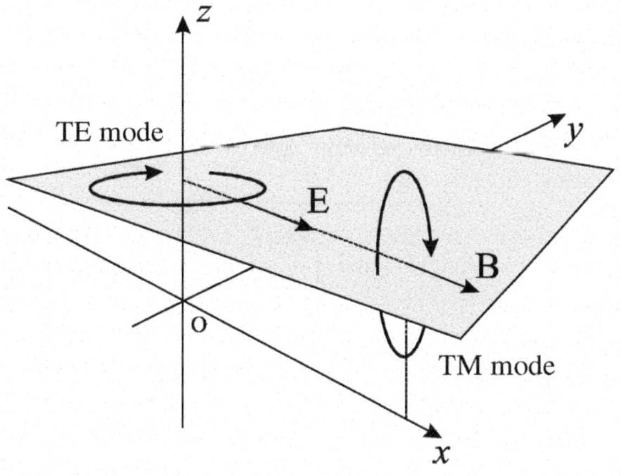

Fig. 2.24 Topology of the magnetic and electric transverse modes TE and TM with regard to the plane of separation between two media with different electromagnetic properties. The electric field E and magnetic field B are included in the plan of separation of two media of different conductivity and propagate in a plane perpendicular to the *z* axis

$$\vec{\nabla} \wedge \vec{\mathrm{B}} - \mu_0 \sigma \vec{\mathrm{E}} = \mu_0 \, \vec{\mathrm{J}} \tag{2.40}$$

considering a Cartesian coordinate system, and also based on the vectorial decomposition[60] theorem, we form the equation (Chave and Cox 1982):

$$\vec{\mathrm{T}} = \vec{\nabla}\phi + \vec{\nabla} \wedge (\psi \, \hat{z}) + \vec{\nabla} \wedge \vec{\nabla} \wedge (\chi \, \hat{z}) \tag{2.41}$$

This allows us to write, using the Eq. (2.38) and the above equation, that the magnetic induction vector is equal to the Eq. (2.41):

$$\vec{\mathrm{B}} = \vec{\nabla} \wedge (\Pi \, \hat{z}) + \vec{\nabla} \wedge \vec{\nabla} \wedge (\psi \, \hat{z}) \tag{2.42}$$

where Π (the first term) and ψ (the second term) then respectively represent the TM and TE modes.

By applying the above equation (cf. Eq. 2.41) to the current source $\vec{\mathrm{J}}^{S}$ and separating the vertical component of the field such that:

$$\vec{\mathrm{J}}^{S} = \mathrm{J}_z^{S} \hat{z} + \vec{\nabla}_h T + \vec{\nabla} \wedge (Y \, \hat{z}) \tag{2.43}$$

where *T* and *Y* are functions that satisfy the Poisson equation (see Eq. A2.1.44), we obtain:

$$\nabla_h^2 T = \vec{\nabla}_h . \, \vec{\mathrm{J}}_{\mathrm{h}}^{S} \tag{2.44}$$

and

[60]Theorem for decomposing a vector field in a combination of three scalar fields.

$$\nabla_h^2 Y = -\left(\vec{\nabla}_h \wedge \vec{J}_h^S\right) . \hat{z} \tag{2.45}$$

The differential equations for Π and ψ are obtained by replacing equations (cf. Eq. 2.42) and (cf. Eq. 2.43) in the equations (cf. Eq. 2.39) and (cf. Eq. 2.40), leading directly to:

$$\nabla_h^2 \Pi + \sigma \partial_z \left(\frac{\partial_z \Pi}{\sigma}\right) - i\omega\, \mu_0 \sigma \Pi = -\, \mu_0\, J_z^S + \mu_0 \sigma \partial_z \left(\frac{T}{\sigma}\right) \tag{2.46}$$

and

$$\nabla^2 \psi - i\omega\, \mu_0 \sigma \psi = -\mu_0 Y \tag{2.47}$$

where the electric field is then equal to:

$$\vec{E} = \frac{1}{\mu_0 \sigma} \vec{\nabla} \wedge \vec{\nabla} \wedge (\Pi\, \hat{z}) - \frac{1}{\sigma}(J_z^S + \nabla_h T) - i\omega \vec{\nabla} \wedge (\psi\, \hat{z}) \tag{2.48}$$

The current source is thus decomposed into exactly two parts: the TM mode with J_z^S et $\nabla_h T$ and the TE mode with $\vec{\nabla} \wedge (\psi\, \hat{z})$.

At each horizontal border (the roofs and walls of the geological formations), the limit conditions of the tangential components of the electric and magnetic fields and those of the normal component of the currents must be checked (Fig. 2.25).

The reader will find in the rest of the description, or in other authors' books (Loeset and Amundsen 2007), the field values separated in the TE and TM modes, this time for different configurations of grounds (oil prospects), calculated from the above expressions (see Chap. 3, Sect. 5.2).

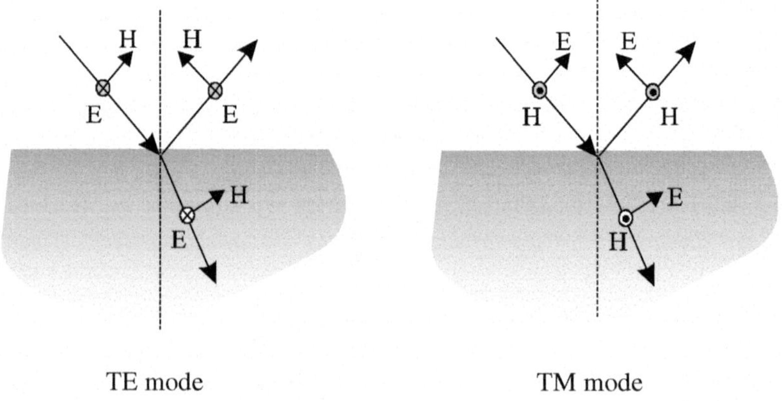

Fig. 2.25 Transverse electric (TE) mode and transverse magnetic (TM) mode at the horizontal interface between two media with different electromagnetic properties for the reflected and transmitted waves (see Chap. 3, Sect. 3.7)

Modal Considerations: In-Line and Broadside Array Answer Optimization

In the case where the galvanic effect is dominant, the magnetic field, compared to the hydrocarbon reservoir, must be transversely polarized (TM). In the TM mode, the electric field then penetrates into the reservoir under a critical angle, which allows the electromagnetic wave to propagate all along the reservoir (limit refraction). Thus, in this way, the reservoir somewhere acts as a kind of a wave guide (cf. Chap. 3, Sect. 3.10).

→ the TM mode then presents the best answer in the *in-line* configuration.

In the case where the inductive effect is predominant, the electric field, regarding the hydrocarbon reservoir, must also be transversely polarized (TE). In the TE mode, the energy is reflected, and is dependent on the source offset. At a certain offset, the electric field coming directly from the transmitter then predominates over the one coming from the underlying grounds. This distance then determines the investigation limit of the method (cf. Chap. 3, Sect. 3.3.2).

→ The TE mode then presents the best answer in the *broadside* configuration.

Geometric Considerations: In-Line and Broadside Configurations

What has been mentioned above and will be developed further in the following chapters demonstrates the extreme need to realize an effective survey (a collection of a maximum of useful and reliable information), to place the measuring instruments and the mobile source in a particular manner following specific geometry.

A priori, so far, one of the most effective ways to get the best results is to:

- Have sensors capable of measuring the values of the horizontal transverse components (T_x, T_y), i.e., perpendicular between them and placed in the horizontal plane.
- If possible, realize combined measures (electric and magnetic) on these transverse components.
- Place these sensors over a wide operation area.
- Simultaneously realize in-line and broadside measurements, i.e., to move the source in different spatial configurations allowing highlighting of the different (galvanic/vortex) effects and modes (TM/TE).

Reciprocity Theorem Consequence

Under the reciprocity theorem (cf. Sect. A2.2), the vertical electric field produced by a horizontal dipole (HED) is equivalent to the azimuthal electric field generated by a vertical dipole (VED), and so the HED always generates greater fields for a given frequency and dipole moment (cf. Chap. 4, Sect. 4.4.4).

3 Different Methods and Associated Devices

EM seabed logging itself gathers under this generic term a set of somewhat different methods, techniques and technologies. Using the laws and principles outlined above, we can roughly distinguish active methods, called controlled source methods, and passive or telluric natural source methods.

3.1 The Controlled Source Methods

Commonly called mCSEM, these methods use artificial sources of energy, whose characteristics of power emission and also the frequency can be remotely controlled. They offer geophysical investigation effective discriminatory detecting means in the sense that the signals are determined in advance both downstream and upstream from the acquisition devices, stable over time and therefore clearly identifiable and recognizable at any time during the job.

Apart from the measurements in the continuous domain, the research performed with the alternating currents may operate either in the frequency domain (variable current but continuous during time) and in the time domain (transient variable current), or simultaneously in the two domains.

3.1.1 Measurements in the Continuous Domain

Measures in the continuous domain were the first techniques to be implemented from 1930 onwards, first in fresh water. They are still in force for works in maritime and coastal engineering in shallow seawater. The injected signal is continuous or better pulsed to reduce the polarization effects on the measuring electrodes when they are immersed in an electrolyte (seawater).[61]

This investigation technique, which can be performed at constant depth (profiling) or variable depth (sounding) is designed to assess the thickness of the conductive layers (marine sediments, vases, etc.) overcoming a resistive horizon (or *bedrock*) of which the lower limit is often considered infinite. This device is particularly sensitive to vertical resistivity contrasts. The common equipment consists of a flute (a multistrand cable dragged along the seabed), more rarely with a streamer (floating cable) supporting the injection current and potential receiving electrodes.

The data interpretation is then performed:

[61] Electrochemical reactions that create potentials occur at the electrodes. For a pair of electrodes, the potential differences may exceed the wanted values.

- Either by solving the forward problem using very simple methods such as the one with electrical images (still in force despite its seniority) or by solving analytically the Laplace equation, or more recently by using numerical methods for geological structures that are a little more complicated geometrically.
- Or, if we have additional qualitative and even better quantitative information, by solving the inverse problem (data inversion processing) for more complex environments.

The results are then presented in the form of profiles (dragging) or resistivity sections (sounding with expanded spreads) or both of them (electrical panels). As noted previously, the use of DC in an aquatic medium[62] is limited in lateral resolution (Telford et al. 1983), investigation depth and also water depth. For these reasons it is preferred to employ the variable current methods usable in the frequency and time domains.

3.1.2 Measures in the Frequency and Time Domains

The frequency and time domains are equivalent. In these domains the waveforms correspond to a permanent periodic signal, sinusoidal or nonsinusoidal (square, rectangular, triangular, etc.). Whatever, the signals are generally described by their amplitude and phase or by a specific wave signature. One of their particularities is that the delays in the time domain are related to phase changes in the frequency domain.

The work frequencies, modulated or not, can be on a broad spectrum, limited in one part by the conductivity of the crossed media (a skin effect) and secondly by the desired spatial resolution. The interpretation of the field data (forward problem) is performed by solving the diffusion equation in the frequency domain (ω):

$$\nabla^2 \vec{E} - i\omega\mu_0\sigma \vec{E} = 0 \tag{2.49}$$

or if necessary in the time domain for complementary studies of the phase, the passage from one to another theoretical expression then being performed by the Fourier transform. The results from these models are then in the form of 2D models or 3D maps.

3.1.3 Measures in the Transient Domain

In the strictly temporal or transient domain, abbreviated as tmCSEM, the principle consists of continuously transmitting to the ground, via the transmitter, a large energetic sequence limited in time (a pulse or sweep)[63] perfectly identified (Ursin

[62]This is not the case in surface prospecting (in an air medium).

[63]The pulses and sweeps are among others characterized by a rate or a recurrence frequency (time between the emission of two consecutive pulses).

1983)[64] and then receiving the signal backscattered on judiciously placed receptors.[65] In this case, there is no primary field; the received signal is measured (with continuous pulse sampling) after the emission. An another advantage is that a lot of signals can be stacked (i.e., a summation of the responses during a determined sequence) to reduce the ambient noise. Anyway, it is more difficult to design circuitry for transient signals than for sinusoidal waveforms including those of the frequency bandwidth (a wide band pass) and the variable field adapted recurrence frequencies.

One other technique (Thomsen et al. 2007), also widely used in seismic reflection, is particularly suitable for the detection of lateral resistivity contrasts and can be conducted by means of digital signal processing very efficiently, and now tested by a correlation technique. The transit time, i.e., the time it takes for the signal to pass through the various environments, directly depends (a linear function) on the nature of these surroundings. The higher the resistivity, the shorter will be the time of flight (ToF) of the reflected or refracted wave.

In a homogeneous and isotropic medium, for a simple reflection, the theoretical path of the wave by applying the Pythagoras theorem is:

$$R = \sqrt{t^2 c^2 + L^2} \tag{2.50a}$$

where t is the arrival time of the event, c is the propagation speed[66], i.e., the signal displacement speed, also called the group speed (see Chap. 3, Sect. 4.4.1), and L is the offset distance (S/R) (see Fig. 2.26b).

Due to the dispersive nature of the phenomenon, it is necessary to add to this path a value R_0. This one is defined from a value of speed c_0 (see Chap. 3, Eq. 3.60), calculated from a reference frequency (ω_0) centered in the reception band, such as:

---

[64]The study of pulse propagation, even brief, shows that it is not inconsistent with the formulation of the skin depth (see Chap. 3, Eq. 3.21). This is explained by the fact that the very high attenuation is related to a large dispersion rather than to energy absorption (see Sect. 2.5.9).

[65]This is probably the first physically interesting concept described in electrical prospecting, in 1907 (Lowy and Leimbach 1911; Lowy 1912), which used radio wave reflection on conductors of electricity (minerals, groundwater). At the time, the authors of the patents could not claim that they precisely measured the transit time (because of an unreliable time base: quartz had not been discovered yet). They proposed then to precisely evaluate the depth of the mirror from the evaluation of the interferences caused by the juxtaposition of the waves transmitted directly from the transmitter and the reflected waves from the reflecting surface (Lowy and Leimbach 1912). It was with the arrival of geological radar (100 MHz to 2000 GHz) in the 1980s that the technique became effective, despite some interesting proposals in the 1940s (Horton 1946).

[66]That is, the displacement speed of the signal (speed group). The first experiments on the determination of the velocity of electricity propagation in water were due to the physicist M. Wheatstone (Blavier 1857).

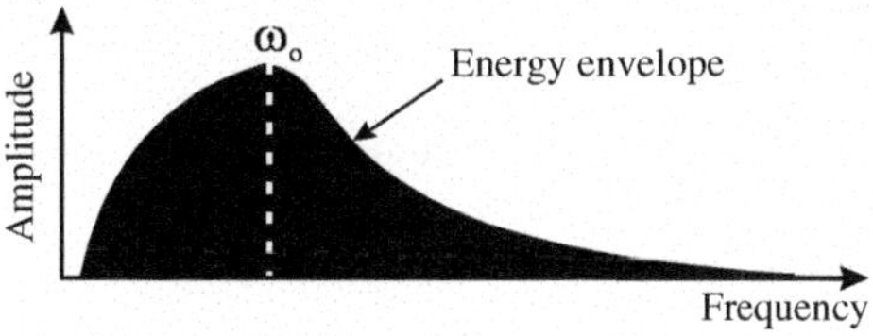

Fig. 2.26a Centering in reception (amplitude wave) of the frequency ω_0 in the spectral band for a receiver (R) (According to Thomsen et al. 2007)

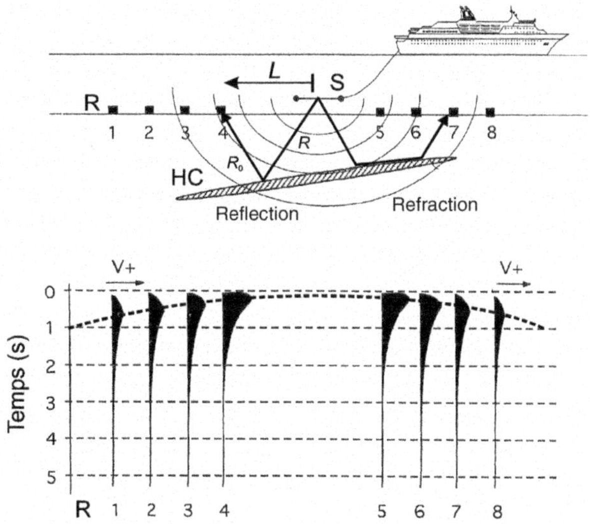

Fig. 2.26b Details of the BP patent showing the acquisition principle in the time domain (t-mCSEM). Technically comparable to the seismic reflection where the receptors R are located on the bottom of the sea and the source S is mobile (According to Thomsen et al. 2007)

$$R_0 = \sqrt{t^2 c_0^2 + L^2} \quad \text{with} \quad c_0 = \sqrt{\frac{2\omega_0}{\mu_0 \sigma}} \tag{2.50b}$$

Knowing *t, L* (measures) for each receiver and *c* (cf. Chap. 3, Eq. 3.61), and the energetic envelope of the signal (cf. Fig. 2.26a), i.e., ω_0, ($\rightarrow c_0$), then we can know the depth of the hydrocarbon reservoir (cf. Fig. 2.26b).

The treatment can be realized as in seismic reflection (Mari et al 1997) especially in using offset compensators[67] (reflected waves), in pointing the first arrivals (time), in stacking[68] (amplitudes) and convoluting the waves[69] (1, 2, 3, etc.) with the

[67] AVO type (an abbreviation for amplitude variation with offset). This changes, in reception, the amplitude as a function of the distance from the EM source; this is different from VGA (variable gain amplifier), which does not take into account the offset source.

[68] Signal summation operation on the same acquisition channel to enhance the amplitude.

[69] Most measuring instruments, due to their metrological imperfections (presence of a convolutional noise in the sensor) often provide blurry, distorted or degraded images of the studied phenomenon. The convolution operator, which can be compared to a mixer function, is therefore intended to restore and reconstruct by the calculation a sharper image of the target.

source signal (frequency) then, if appropriate, in migrating (phases) these answers (see Chap. 5, Sect. 5.5.1).

In heterogeneous and anisotropic media, the relations are then more complex and can be determined by modeling. This[70] and the interpretation of the data that follow can be realized either:

- By numerical methods in calculating the solutions of the diffusion equation in the time domain
- By inversion data methods, combining the direct digital models (previously mentioned) with deterministic and/or stochastic optimization processes

The results of these models are also in the form of 2D or 3D maps for complicated cases.

The waves in marine sediments and hydrocarbon reservoirs "move" at different speeds. Provided we obtain a critical angle, the velocity field can then be studied in refraction (cf. Fig. 2.26b).

3.2 Controlled Source Techniques Under Development

We are currently studying different instrumental topologies allowing either improvement of the performances of the method, especially the investigation depth, the accuracy and the resolutions (Børven and Flekkøy 2009), or increases in the efficiency (Werthmüller et al. 2014) and the quality of the survey (curtailment of the duration, diminution of the operational resources, etc.) especially in shallow water (Ziolkowski and Wright 2008). This concerns, for example, the cases of:

- The power emission by vertical electric dipoles (VED)
- The reception of the vertical component of the field
- The simultaneous displacement of the power source and the field receivers (in a streamer)
- The use of transient currents (t-mEM or mTEM)[71] and multitransient currents (mMTEM)
- The use in addition of downhole sensors or logging tools
- Or the use of combined transient/frequential techniques (Strack 2009)

According to the authors of these studies (Edwards 2005; Scholl and Edwards 2007; Stalheim and Olsen 2008; Holten et al. 2009; Fan et al. 2012), these devices then would present increased sensitivity compared with the actual techniques but would however require for the last one (a streamer) significant miniaturization of the field instrument. This one is only possible with a new measurement technology

[70]The most rigorous approach to study the wave propagation (without the source), and more particularly the reflection phenomena, remains solving the wave equation.

[71]For the *marine transient electromagnetic method.*

and especially the invention of new electrometers with a very small dimension and high sensitivity. Assuming that the difficulties with the quite complex correction systems are overcome, these technologies could in the near future replace those proposed currently (cf. Chap. 4, Sect. 5.4).

3.3 *Telluric Source Methods*

Very often, to compensate for the lack of power of the artificial or controlled source methods, and to increase the investigation depth, we substitute for those methods the telluric source methods, which are lighter to implement because they do not use a source to tow. These techniques, used in early academic research and sometimes in applied geophysics (geothermics), have been receiving renewed interest in recent years and are currently under development in the oil industry.

The telluric (electrotelluric and magnetotelluric) methods use natural exogeospherical and endogeospherical excitation sources, inducing in the ground wandering[72] currents, which flow through the earth lithosphere[73] over large areas. These currents, of very low frequencies, which continuously move across the globe, are partly due to solar activity (magnetic storms associated with solar flares) and to some causes internal to the planet and still poorly identified (Kaufmann and Keller, 1981).

The great period of these currents allow us to explore important depths of ground (a reduced skin effect), which predestined these techniques early in their use to the study both onshore and offshore of the deep composition of the earth's crust. These methods employ identical sensors but with more complex electronics taking into account the spectral diversity of the measurements (large frequency bandwidths). Compared to the control source methods, the telluric methods (especially mMT) have not the resolution of the mCSEM and cannot be used to directly detect hydrocarbons but can help to delineate geological trap structures, particularly below salt domes, and to survey specific reconnaissance.

3.3.1 Electrotelluric Method

The telluric current content in a frequency allows us to investigate the geological layers of variable depths like artificial source electric soundings, for which current

[72]Not be confused with industrial and entropic type currents (60 Hz), which also circulate in the subsoil.

[73]The existence and movement of telluric currents were first suggested by Sir Humphrey Davy in 1821, reinforced 10 years later by the studies of Faraday, evidenced in English telegraph lines by Barlow in 1847 (Rooney 1939; Rougerie 1942; Cagniard 1956) and confirmed a few years later by Matteucci (Matteucci 1862) and several years later calculated by Lamb (1883). The first systematic observations date from the late nineteenth century at differents observatories (Monaco, Berlin, Greenwich, Parc de St Maur, Ebro, Pavlovsk, etc.).

injection devices can be expanded up to a certain limit (Cagniard 1956; Berdichevsky 1965; Keller et Fischknecht 1966). However, with this passive method, we do not reach the vertical staggering of the rock formations, but the horizontal variations due to the structures.

The electrotelluric method has not a priori been used as such at sea for the purpose of prospecting or even observation in earth physics.[74] However, some experimental studies have demonstrated the effectiveness of this method, which has not yet been used in prospecting activity itself, supplanted at this time by the magnetotelluric methods. Indeed, these last methods allow more refined quantitative analysis (theory) using simultaneously and jointly electric and magnetic data, and also accept anthropic currents.

3.3.2 Magnetotelluric Method (Strict Sense)

Very little work has been carried out on the opportunity to explore using a magnetic method in the strict sense because the magnetic fields associated with telluric currents are very weak (cf. Appendix A2.3). However, some tracks have been explored. This is the case, for example, of the measurement of the total (cf. Fig. 2.27) and transient magnetic field gradient,[75] developed in the 1970s by Russian (Trofimov 1975) and French teams (Mosnier 1970; Durand and Mosnier 1977) from facts observed in the 1950s in Germany (Meyer 1951; Bartels 1954; Kertz 1954; Wiese 1954, 1962, 1965; Siebert 1958) and Japan (Rikitake 1950; Rikitake et al. 1952).

In an observation point P, the varying magnetic $\vec{F}$ field at time *t* can be decomposed into two separate fields, one normal $\vec{F}_n$, corresponding to an isotropic ground or to a distribution of layered resistivity, the other one $\vec{F}_a$ abnormal, created by lateral inhomogeneities (Schumcker 1970). Then we have:

---

[74]Method and techniques invented in the late 1930s by the Schlumberger brothers (Conrad and Marcel) and their collaborator Georges Henri Doll and developed particularly in France, North Africa and Italy in the years 1940–1950 by the Compagnie Générale de Géophysique (Migaux 1948). The method was to simultaneously measure the temporal variations of the values of the two horizontal components of the electric field, between a fixed base (E*x*, E*y*) and a mobile station (E'*x*, E'*y*). The comparison of these different records then provided information on the current circulation between the base and the different positions of mobile stations. This could then be related more or less directly to the underlying geological structure (between the base and the stations). This technique was unable to be further developed in the USA, the largest market for oil at this time, and dwindled despite some success (discovery of the field of Saint Marcet in France and of Hassi Messaoud in Algeria). Furthermore, it is surprising that there are only two books in foreign languages, devoted to these techniques (Porstendorfer 1961; Berdichevsky 1965). For more details (history) see Appendix A2.4.

[75]More commonly called *deep geomagnetic survey* (Wait 1982; Berdichevsky and Zhdanov 1984), using long period magnetovariational fields (*geomagnetic deep sounding* or GDS) according to the discovery of geomagnetism induction in the nineteenth century (Schuster and Lamb 1889) and the *Gauss separation* mathematical technique (Gauss 1839).

Fig. 2.27 First submarine optical pumping magnetometer for total magnetude measurement of magnetic field (Ecole Normale Supérieure of Paris, Mosnier 1970)

$$\vec{F}(P,t) = \vec{F}_n(P,t) + \vec{F}_a(P,t) \tag{2.51}$$

Theoretically, we cannot therefore distinguish at one point the anomalous field.

However, if a differential measurement is performed between two points that are geographically invariant, it is then possible to isolate this anomalous field. For example, studying the polarization of the horizontal field components in perpendicular directions, then it is possible to separate the time and space variables. This then allows the detection of lateral conductivity anomalies. In this case the scalar magnetometer cannot be used.

In practical terms, this method uses moving magnet variometers able to move around a vertical axis (α rotation) to measure horizontal components: D-meter and H-meter (cf. Chap. 4, Sect. 5.7.3). It is then easy to establish the relations that link the variations of the transient fields X_T and Y_T to the north–south H_T and east–west D_T variations, such as:

$$\begin{aligned} H_T &= Y_T \cos\alpha - X_T \sin\alpha \\ D_T &= X_T \cos\alpha + Y_T \sin\alpha \end{aligned} \tag{2.52}$$

Today, to know the deep layers of the crust and for shorter period currents we use triaxial tilt-compensated fluxgate magnetometers (Berdichevsky et al. 1984; Tarits 1986).

We would like to remark that in a large, highly original theoretical study in the 1950s the use of displacement of magnetic storms on a large scale was proposed as a source of underground investigation. The author (Matschinski 1952, 1955), by considering a theoretical tri-sheet-like model (ionosphere, atmosphere, land) with use of Maxwell's equations, proposed the possibility to practically use (in prospecting) the propagation velocities of magnetic waves at different frequencies, expressed in the form of tensors to precisely represent the geological underlying structure.

3.3.3 Electromagnetotelluric or Magnetotelluric Method (Broad Sense)

It was the German geophysicist Johann von Lamont who in November 1859 realized the first electromagnetotelluric terrestrial survey from telluric and magnetic recordings made simultaneously at the Astronomical Observatory in Munich (Lamont 1862). At the time, his calculations of the thickness of the ground were only based on a galvanic approach (DC interpretation analytical model), but the author, for lack of a better theory, already evoked an inductive response, as Maxwell's equations had only been published in 1865.[76] It was then the physicist J. Terada who, by measurements of electrical and magnetic fields, linked these variations with the conductivity of the subsoil (Terada 1917).

Most electromagnetotelluric methods, more commonly known as magnetotelluric methods,[77] (MT for short) are based on the study of the relations between the tangential components of the horizontal electric field $\vec{E}_x$ (the TE mode) and the horizontal magnetic field $\vec{H}_y$ (the TM mode) on the soil surface (Simpson and Bahr 2005). The ratio of their energy, which largely depends on the electrical structure of the subsoil in the different spectral bands, is determined by the distribution of the resistivities as a function of depth.

This method was proposed in the 1950s by Professor Tikhonov (from the USSR) who showed that at low frequencies, the amplitude of the derivative of the magnetic field component H_y is then proportional to the orthogonal component of the electric field E_x (Tikhonov 1950). In the same period[78] in Japan, the physicists Kato, Yocoto and Kikuchi associated the period of the electromagnetic phenomenon with the conductibility of two successive layers by introducing phase measures (Kato and Kikuchi 1950). In France, Professor Louis Cagniard, assuming a plane wave telluric field (the concept of the remote source), developed a mathematical formulation between the E_x and H_y fields, thus introducing the MT methodology to the rank of a prospecting technique on its own (Cagniard 1953a, b, c).[79]

[76]After this date, the English physicist George Biddell Airy, one of the fathers of the theory of isostasy, from recordings made at the Astronomy Observatory in Greenwich, clearly showed the relation between telluric currents and changes in magnetic fields (Airy 1868).

[77]The electromagnetotelluric name, to our knowledge, has hardly ever been used: we prefer the contraction "magnetotelluric"—an expression created by L. Cagniard in 1950, which may lead to confusion with the purely magnetic methods.

[78]It was at this time that the first studies of the deep layers of the Earth were conducted (Rikitake 1950).

[79]Thereafter, we owe to various foreign scientists the unification of the theories of interpretation (stratified media, heterogeneity), by a single theory of the telluric and magnetic associated waves (Sejman 1958; Berdichevsky 1961). Today, the theory developed by Tikhonov and Cagniard considering a plane wave source is partly questioned for earth physics (Chave and Jones 2012).

The MT method is based on the simultaneous measurement of micropulses[80] (less than 1 Hz) of the terrestrial electric and magnetic fields in phases and in different bands or frequency spectra, the lowest of which correspond to the deepest investigations (a low skin effect).

MT is probably the most elegant geophysical technique and the most elaborate that has been developed until today. Unlike electrotelluric methods,[81] it does not need a base station and thus records in a continuous regime. Convenient to implement and light (no source), it allows us to explore both subsurface areas (MT audio) and deeper areas of the lithosphere (ULF-MT).

At the theoretical level, from the Maxwell equations, concerning the intensities E and H of the electric and magnetic fields on their respective transverse components x for E and y for H, and also knowing the permeability of the media μ (usually equal to μ_0), we extract, according to the different selected frequencies (ω), the variations of the apparent electrical conductivity σ_a of the subsea medium, such that:

$$\sigma_a(\omega) = \omega\mu_0 \left|\frac{H_y}{E_x}\right|^2 \tag{2.53}$$

where the ratio H_y/E_x represents the admittance for a tabular structure (the inverse then corresponding to the impedance).

Thereafter, Professor Cagniard and his colleagues gave this method several controlled source variants intended for marine exploration (Cagniard 1953a, b, c; Cagniard and Morat 1966; Morat 1970).

Quite rapid development by the USSR took place in the Caspian Sea (Sochelnikov 1977). In the early 1970s (Hermance 1969) and 1980s, the method was further developed by Professors Jean Filloux[82] (1980, 1982, 1983) and Alan Chave (Chave et al. 1991) in the USA, and a few years later by Professor Tarits in France (Tarits 1986).

General Theory of the mMT Method

This very general theory is to obtain from the electric and magnetic field amplitude measurements, the apparent resistivity of a homogeneous and isotropic medium (Patra and Mallick 1980).

The calculation of the apparent resistivity in the marine MT (mMT) context, for a laminate subsoil in a marine medium, is solved in Appendix A5.5 of Chap. 5.

---

[80]These latter include ongoing, regular and sporadic variations. The name "pulse" was given in 1897 by Van Bemmelen to the small rapid variations of the magnetic earth field (Coulomb 1954; Selzer 1972).

[81]See Appendix A2.4.

[82]In the 1960s, MT investigations were conducted on the California coast (Filloux and Cox 1964).

From the Maxwell equations, we can form the wave equation such as:

$$\nabla \vec{e} = \mu\sigma \frac{\partial \vec{e}}{\partial t} + \mu\varepsilon \frac{\partial^2 \vec{e}}{\partial t^2} \tag{2.54}$$

Assuming that the components of the electric field $\vec{e}_y$ and $\vec{e}_z$ are zero as the derivations ($\partial/\partial x = \partial/\partial y = 0$), the above equation with x is then written:

$$\frac{\partial^2 \vec{e}_x}{\partial z^2} = \mu\sigma \frac{\partial \vec{e}_x}{\partial t} + \mu\varepsilon \frac{\partial^2 \vec{e}_x}{\partial t^2} \tag{2.55}$$

Under these conditions, in the frequency domain, the value of the electric field $\vec{E}_x$ varying harmonically satisfies, for z (depth), the equation:

$$\frac{d^2 E_x(z)}{dz^2} = k_n E_x(z) \tag{2.56}$$

where $k_n^2 = i\omega\mu(\sigma + i\omega\varepsilon)$ is the wave number in the considered n medium.

The solution of this equation, for Re $(k) > 0$, is of the general form:

$$E_x(z) = Ae^{kz} + Be^{-kz} \tag{2.57}$$

where A and B are some constants dependent on the rocks electromagnetic properties, evaluated as a function of the limit conditions, in particular those concerning the continuity of the tangential components of the electric and magnetic fields at the interfaces separating the different soil layers.

In a homogeneous medium, the field E_x disappears when z tends to infinity. Under these conditions, the constant A becomes zero, and then only the second term remains, where:

$$E_x(z) = Be^{-kz} \tag{2.58}$$

If we now consider that E_x generates a magnetic harmonic field H_z orthogonal to E_x, we have the relation (Fig. 2.28):

$$\vec{\nabla} \wedge \vec{H}_y = \vec{H}_0 \exp(i\omega t) \tag{2.59}$$

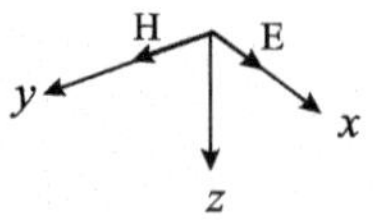

Fig. 2.28 Picture to express the relation between the electric field E and magnetic field H generated in the chosen axis system (x, y, z)

Since the vertical component of the magnetic field H_z does not exist, according to the Maxwell equations, and assuming that $\mu = \mu_0$ (nonmagnetic areas), we thus obtain:

$$\vec{\nabla} \wedge \vec{E} = -i\omega\mu_0 \vec{H} \tag{2.60}$$

which gives:

$$\frac{\partial E_x}{\partial z} = -i\omega\mu_0 H_y \tag{2.61}$$

By differentiating the Eq. (2.58) we have:

$$H_y = \frac{k}{i\omega\mu_0} B e^{-kz} \tag{2.62}$$

Finally, by forming the ratio of the two fields, an impedance known as Cagniard impedance is defined for the n layer such as:

$$Z_{xy} = \frac{E_x}{H_y} = \frac{i\omega\mu_0}{k_n} \tag{2.63}$$

At low and very low frequency, when $k_n^2 = i\omega\mu_0\sigma$, the impedance of Cagniard reduces to:

$$Z = \sqrt{\omega\mu_0\rho_a}\, e^{i\pi/4} \tag{2.64}$$

The apparent resistivity ρ_a and its apparent conductivity σ_a are then derived as:

$$\rho_a(\omega) = \frac{1}{\omega\mu_0}\left|\frac{E_x}{H_y}\right|^2 \quad \text{or} \quad \sigma_a(\omega) = \omega\mu_0\left|\frac{H_y}{E_x}\right|^2 \tag{2.65}$$

In fact, for horizontal layers (where $H_z = 0$), these equations are substituted by the following relation:

$$\rho_a = 0,2T\left|\frac{E_x}{H_y}\right|^2 \tag{2.66}$$

where T is the period of the telluric currents.

For investigations at sea, Professor Cagniard (1953a, b, c) offers, for an electrical device laid at the bottom of the sea, an expression of the apparent resistivity ρ_a and the apparent phase ϕ (the phase shift between E and H) such that at a depth h as a function of δ we have:

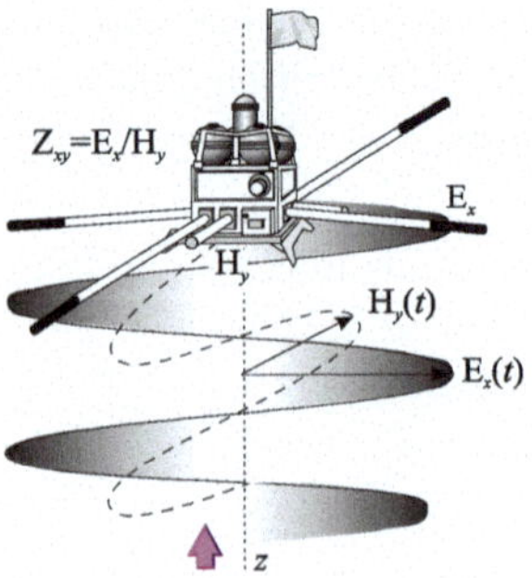

Fig. 2.29 Principle of marine magnetotelluric detection by simultaneous measurements (*t*) in phase and amplitude of a magnetic field H (magnetometer) and an electric field E (electrometer) in perpendicular directions *x* and *y* (horizontal plane)

$$\sqrt{\rho_a}e^{-i\phi} = \frac{2\pi\delta}{\sqrt{T}}e^{-i\pi/4}\frac{\frac{(E_x)_1}{(E_x)_2}\sinh\left(\sqrt{2}e^{-i\pi/4}\frac{h}{\delta}\right)}{1-\frac{(E_x)_1}{(E_x)_2}\cosh\left(\sqrt{2}e^{-i\pi/4}\frac{h}{\delta}\right)} \quad \text{with} \quad \delta = \frac{1}{2\pi}\sqrt{10\rho_a T} \quad (2.67)$$

More interesting information on the theoretical basis for electromagnetic induction are given in the book (Chap. 3) by Professors Alan D. Chave and Alan G. Jones (2012).

mMT Practical Prospecting

Obtaining the resistivities of the different layers, taking into account the water column, can be conceived in different ways. An example of analytical calculation can be found for 1D tabular models in the Appendix A5.5. Other authors, to locate subsurface heterogeneities, presented more elaborate theories with 2D models (Jegen and Edwards 2000).[83]

In practical terms, the survey mMT consists in a very general case of:

- Measuring and recording together the components $E_x(t)$ and $B_y(t)$ or $E_y(t)$ and $B_x(t)$
- Subjecting the records to a spectral analysis to deduce, for the different frequencies (ω), the value of the impedance Z(ω)
- Reversing Z(ω), at different depths *z*, to obtain the layer's resistivity (Fig. 2.29)

The mMT method generally allows us to define with reasonable accuracy the total conductance[84] of the sedimentary rocks. This, when several fields overlap, is equivalent to the sum of a part of the partial conductances of these fields. Therefore it can be assumed that any relatively large variation observed in the conductance

[83] In this article, the authors give a complete theory of the physics of the MT method in a marine environment (mMT).

[84] In mMT we call the total conductance of the grounds the ratio of the overall thickness of the grounds t_h on the apparent resistivity (t_h/ρ).

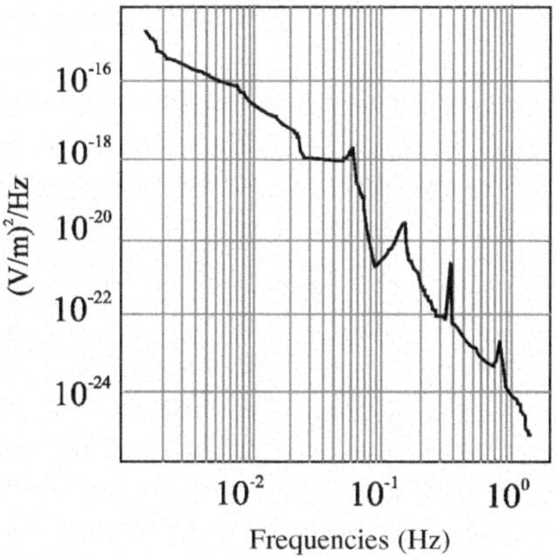

Fig. 2.30 Spectrum of the telluric field (electric) due to the ionospheric and microseismic activities (peaks) in the Pacific Ocean 200 miles off San Diego and in 3700 m water depth (According to Chave et al. 1991)

may have its origin either in strong local thickness variations of the grounds, or in the presence of conductivity anomalies induced by the presence of a hydrocarbon reservoir.

The indeterminacy can rise quickly when we have enough information about the variations of the thickness of the geological layers, i.e., when we have in particular a seismic survey that accurately determines the structural geological environment.

The geoelectric conditions favorable to the detection and to the study of oil reservoirs are defined in relation to the electrical characteristics of the reservoir, i.e., in this case the resistivity (see Chap. 3, Sect. 2.1.3). Above 1 Hz the technique is not applicable beyond 1000 m water depth. It is for this reason, among others, that the mMT method is used at frequencies below 0.01 Hz up to periods T of several minutes.

The mMT survey is particularly suitable for the exploration of deep structures, for areas where there are seismic masks (sub-outcropping limestone slabs, basalt plateaus, salt domes, karsts, anhydrites, evaporites, carbonates, rock with gas, gas hydrate recovery, etc.) and any other configuration with resistant horizons or heterogeneities. The mMT technique is sensitive to lateral facies variations (Fig. 2.30).

At the operational level, the mMT method is easier to implement than other techniques (no artificial source to deploy, less bulky receivers for the expected investigation depths) and can provide more comprehensive results by adapted treatments. This was the case, for example, for the Russian research that developed this method more specifically for offshore use (Berdichevsky et al. 1989) with treatments by inversion (Berdichevsky and Dmitriev 2002, 2008) and for its use in the USA to image salt deposit structures (Hoversten et al. 2000; Key et al. 2006).

A variant of the mMT method was proposed in the 1970s in France (Duroux 1974), which was identical in all aspects to that of L. Cagniard, this time replacing the natural source with an artificial source (a magnetic vertical dipole).

mMT is, for now, still the most interesting method of investigation EM in earth physics and especially in tectonophysics (Ferguson 1988).

3.3.4 Spontaneous Polarization Method

In the industry of downhole logging (well logging), recording the spontaneous polarization (SP) inside an oil well as a function of depth (SP log) played a key role in the oil industry in addition to the log resistivity[85] in the in situ detection of hydrocarbons (Pirson 1981; Desbrandes 1982; Serra 2004). One would think that this property of the geological layers despite a different investigation configuration (missing hole probe, but presence of the fluids migration) would have attracted the attention of researchers, but in fact very few experiments have been conducted in the seabed. We can however note a few studies (Corwin et al. 1970; Corwin 1973, 1975) and those done later by the Scripps Institute (Heinson et al. 1999).

The appearance of the SP is concomitant with several separate phenomena:

- When two media with different electrolytic properties are in contact. This is, for example, the case with seawater and marine sediments whose respective electrical redox potentials vary from +200 to +400 mV and −100 to −200 mV.
- When an electrolyte passes through a porous layer. This is the electrofiltering potential, which results in the creation of the electromotive forces at the interfaces, translatable into a potential difference of up to several tens of millivolts. This effect is particularly marked in the borehole (vertical phenomenon), in the presence of differential pressure, or more generally in the presence of a topographic relief (a horizontal phenomenon present during a surface onshore survey). If this effect can occur laterally, due to the flow of some fluids through faults, for example, it is then possible to imagine seabed devices able to detect hydrocarbons. Yet the question remains.
- When an electrically conductive object passes through geological layers of different electrolytic properties with differential aeration (caused by a hydrostatic level, for example). This results (the self-potential mechanism) in the creation of current *per ascensum* (a galvanic cell) detectable on the surface by potential differences at the apex of the anomaly (Telford et al. 1978).

The detection of the SP is one of the simplest techniques, at least in principle and in practice (Brewitt-Taylor. 1975). It requires virtually no specific and elaborate instrumentation, and can amount to a continuous electronic voltmeter with high input impedance completed sometimes by a rejection filter (60 Hz and harmonics) and a bottom line equipped at these extremities with two unpolarizable electrodes (Ag/AgCl). In contrast, the interpretation is much more difficult to achieve.

The authors of various studies on the subject have highlighted quite a number of ghostly geological anomalies without giving their causes and much less their origins, roughly correlatable with those recorded by a magnetometer immersed

[85]Including the distinction of permeable horizons in impervious areas.

on the surface. Yet this type of investigation has had no industrial application itself.[86]

With a greater chance of success, the method could also be used in metalliferous deposits in underwater exploration (deep-sea mining) or in the detection of submerged and/or partially buried metal bodies, or even of any other polarizable metal structure with the faculty in its immediate environment to establish a localized cell effect. In these cases, the interpretation is then relatively easy.

3.4 Comparison of the Main Methods

DC technics, currently used on the seabed, can only be a priori applied to isotropic and laminated structures (tabular models with horizontal to subhorizontal monotonous layers) and are therefore primarily used in profiling and survey.[87] However, these methods are not suitable for detecting even less locating transverse conductivity contrasts with lateral limits (integrative method → no directivity). Finally, DC methods have the sensitivity of the so-called potential methods (gravity and magnetic technics)—relatively low compared to that of variable current methods (cf. Chap. 5, Eqs. 5.6, 5.7, and 5.8).

To meet these discovery requirements, the measurement of transverse fields of the variable currents seems the most interesting solution to the problem of detection and location of marine subsoil heterogeneities. It provides a certain directivity and moreover may in contrast to DC methods directly evaluate the magnetic field associated with the currents that flow in the subsoil.

However, regarding the quantities of hydrocarbon deposits, with the significant surfaces to explore, field excitation as uniform as possible is then required. It can only be produced by an ad hoc mobile source, sufficiently homogeneously sweeping the whole surface of the prospect, or by a telluric source of regional extent, this one then offering only limited opportunities. Using the displacement of a great width magnetic source such as that advocated by Dr. Matschinski could also be interesting for certain types of prospect (see Sect. 3.3.2) (Fig. 2.31, Table 2.2).

We can note that the greatest depths of investigation are obtained by geomagnetic surveys.

[86]In earth physics, it is planned to use, among others, measures of SP on the midocean ridges (hydrothermal vents) to determine the dynamics of the system (the Elecmomar project).

[87]Practically, the devices "at the bottom" (cables or flutes) or surface (streamer) used in these methods (DC) are limited by the variability of the submarine relief and by the bathymetry. The latter do not allow investigations in a large immersion under a great water depth. It is for this reason that these techniques are mainly used in the field of geotechnical engineering, civil engineering (harbor) and coastal geology.

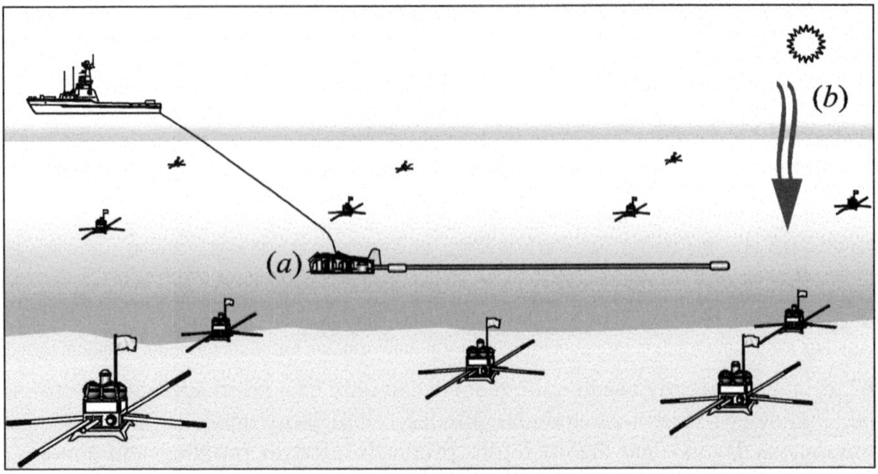

Fig. 2.31 Complete seabed logging survey (mCSEM and mMT) in action. (*a*) Controlled source device (mCSEM). As the EM transmitter goes by, towed by the boat, the electric field transverse components are measured by a plurality of field sensors (electrometers with two or three components) arranged on the bottom of the sea. (*b*) Telluric source device (mMT). The source is natural (magnetic storms, atmospheric electric discharges). The measurement and recording of the components of the electric and magnetic fields are carried out on a frequency spectrum that is more or less wide depending on the desired depth of investigation

Table 2.2 Summary and comparison of different seabed logging methods

SBL methods	mDCS[1]	mCSEM[2]	mMT[3]
Type of investigation	Artificial currents	Artificial currents	Natural currents
Method	Active	Active	Passive
Nature of the current	Continuous (=)	Alternative LF (0.1–10 Hz)	Variable VLF (<0.1 Hz)
Precision	In depth (z)	In depth (z) and laterally (x, y)	In depth (z) and laterally (x, y)
Sensibility (horizon)	Conductor	Dielectric	Conductor
Resolution	Low	High	Middle
Investigation depth*	Low (*decametric*)	Middle (*hectometric*)	High (*kilometric*)
Bathymetric limitation	Little water depth	No limitation**	Little limitation**
Geophysical application (*scale*)	Apparent resistivity (*littoral*)	Resistivity contrast (*reservoir*)	Structure and lithology (*basin*)
Means to implement	Low (*light boat*)	Very important (*offshore boat*)	Important (*offshore boat*)

[1]*Marine direct current sounding*
[2]*Marine controlled source electromagnetic*
[3]*Marine magnetotelluric*
*Actually accessible
**Impractical in little water depth

4 Operational Limitations: Detection Problem

For now, and for the reasons explained above (cf. Sect. 3.4), exploration is carried out by devices with fixed receivers and a mobile transmitter. This is not then without consequences for the shape of the signals picked up by receivers at greater or lesser distances from the emission sources.

Apart from the conventional attenuation phenomena, due to the dispersion of electromagnetic waves and their absorption by the medium, and the presence in receipt of direct waves, the signals will also undergo alterations and distortions essentially due to:

- The more or less stratified nature of the propagation medium inducing reflection and refraction phenomena, or diffusion and diffraction on the different interfaces (air, water, sediments, reservoir)
- The differential removal of the source from the receiving devices, and it will be all the more important that the different measurement channels are large

4.1 *In Terms of Time: The Phenomena of Reflection/ Refraction*

Overall, the information recording is performed in a conductive medium that is not homogeneous. It consists of a succession of strata of very different resistivities where the propagation of electromagnetic waves is carried out accordingly with different velocities (see Chap. 3, Sect. 4.1).

The energy at the level of the source radiates in virtually all directions and will therefore generally be redistributed on different interfaces either by reflection or by refraction. Obviously, during a survey only the energy emerging from the reservoir is to be considered and the energy of the waves reflected and refracted by the discontinuity surface water/air should not be taken into account. This is for example the case in prospecting in a low water depth where this phenomenon is particularly important and where a separation of the signals (up and down) is then necessary.

4.2 *In Terms of Frequency: Limitations Due to the* Offset

If we now focus on the distance between the source and the receptors (offset), when it reaches values of the order of magnitude of the wavelengths in use (1.5 km to 20 km approximately), it causes along axes O*x*, O*y* and O*z* some significant amplitude differences (I_a, I_b, I_c) and phase differences (φ_x, φ_y, φ_z). These gaps are also sensitive to the orientation and dimensions of the vector sensors, as well as to the frequencies (wavelengths).

In this case, for a current of emission form cosωt, of initial intensity I_o, such that:

$$J_0 = I_0 \cos \omega t \tag{2.68}$$

the three components of the current density (it is the same for the field) for a given remote point with coordinates (x, y, z) are equal under these circumstances:

$$\begin{cases} J_x = I_a \cos\ (\omega t - \varphi_x) \\ J_y = I_b \cos\ (\omega t - \varphi_y) \\ J_z = I_c \cos\ (\omega t - \varphi_z) \end{cases} \tag{2.69}$$

values that then introduce biases in the evaluation of the current densities and thus in the evaluation of the electric and magnetic fields (amplitude and phase offsets).

4.2.1 Amplitude Offset: Limit of Detection

Some studies (Eidesmo et al. 2002) showed that the acquisition data (especially the amplitudes of the electric fields), for large offsets (>4 km), were lower than the synthetic answers (1D modeling with HC) and that in these conditions they could possibly be confused with a geological model without HC (cf. Fig. 2.32).

At these "extraordinary" distances, these disadvantages are raised if the transmitter and the receivers move simultaneously, i.e., when they are together. This configuration requires, however, an increase of the intrinsic dimensions of the acquisition devices (a streamer, for example).

4.2.2 Phase Offset: Phantoms?

The phase offset is essentially due to the space orientation of the sensors, which is different, which we do not control during the installation of the instruments and which can be ignored during the acquisition.

During each period the extremity of the vector $\vec{J}$ with rectangular components (J_x, J_y, J_z) no longer describes a straight segment of a sinusoidal motion,[88] but an ellipse then comparable to the phase shifts maybe due to induced effects expressed as a phantom reservoir.

To overcome this major drawback, some authors (Mittet et al. 2007) propose embarked correction algorithms taking into account the instrumental positions, environmental characteristics and operational parameters (see Chap. 4,

[88]It is the same for the electric field components.

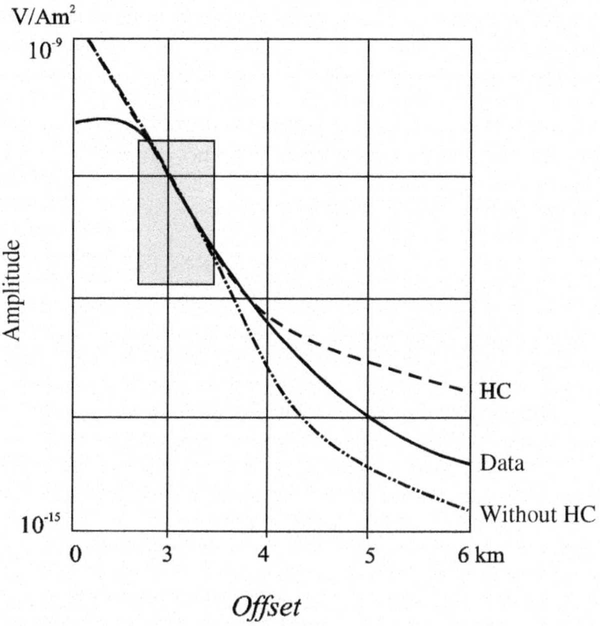

Fig. 2.32 Variations of the amplitude of the electric field according to the offset. From about 4 km, calculations show that the data acquisition in the presence of a reservoir (HC) is lower than the 1D modeling conducted from downhole resistivity values (well log)

Sect. Phase Corrector (Offset and Orientation Effects) in addition to the values of the electric and magnetic fields measured under certain conditions.

4.3 Problem of Anisotropy and its Treatment

It goes without saying that a method sensitive to lateral conductivity will inevitably be sensitive to the transverse anisotropy of the geological rocks and layers. We can then treat the forward problem as a special case of the isotropic problem (Newman et al. 2010)—that is, if:

- we consider that the conductivity can be put in the form of 3 × 3 matrices,
- we can set then some particular limit conditions (Dirichlet conditions for the conductive anomaly).

In this case, we have for the electric field ($\vec{E} = \vec{E}^s + \vec{E}^b$), the equation:

$$\vec{\nabla} \wedge \vec{\nabla} \wedge \vec{E}^s + i\omega\,\mu_0\,\boldsymbol{\sigma}\vec{E}^s = -i\omega\,\mu_0\left(\boldsymbol{\sigma} - \boldsymbol{\sigma}^b\right)\,\vec{E}^b \tag{2.70}$$

with as a conductivity matrix $\boldsymbol{\sigma}$ and anisotropy matrix $\boldsymbol{\sigma}^b$:

$$\boldsymbol{\sigma} = \begin{pmatrix} \sigma_h & 0 & 0 \\ 0 & \sigma_h & 0 \\ 0 & 0 & \sigma_v \end{pmatrix} \quad \text{and} \quad \boldsymbol{\sigma}^b = \begin{pmatrix} \sigma_h^b & 0 & 0 \\ 0 & \sigma_h^b & 0 \\ 0 & 0 & \sigma_v^b \end{pmatrix} \tag{2.71}$$

where σ_h and σ_v are respectively the conductivities in the horizontal and vertical planes.

The inverse problem[89] is then solved by considering the forward problem solutions $\vec{d}^p$ and by comparing these solutions with the data acquisition $\vec{d}^{obs}$ such that we have for the values of the horizontal $\vec{m}_h$ and vertical $\vec{m}_v$ gradients the equality:

$$\vec{\nabla}\varphi_d(\vec{m}_h, \vec{m}_v) = -\mathrm{Re}\left\{ \left(\mathbf{D}\mathbf{J}^T\right) \mathbf{D} \left(\vec{d}^p - \vec{d}^{obs}\right)^* \right\} \tag{2.72}$$

expression which then requires a Jacobian matrix $\mathbf{J}^T$ taking into account the horizontal and vertical conductivities.

The anisotropy effect is obtained:

- Considering the error minimization function Φ, such that:

$$\begin{aligned} \Phi = &\frac{1}{2}\left\{ \mathbf{D}\left(\vec{d}^p - \vec{d}^{obs}\right)^{T*} \right\} \left\{ \mathbf{D}\left(\vec{d}^p - \vec{d}^{obs}\right) \right\} \\ &+ \frac{1}{2}\lambda_h\{\mathbf{W}\vec{m}_h\}^T\{\mathbf{W}\vec{m}_h\} + 1/2\lambda_v\{\mathbf{W}\vec{m}_v\}^T\{\mathbf{W}\vec{m}_v\} \end{aligned} \tag{2.73}$$

where **D**, **W** are respectively the matrix regularization (compensation of noise measures and compensation approximations of the Laplacian operator ∇^2 numerical resolution) and $\lambda_{h,v}$ parameters of horizontal and vertical regularization.

- Then applying the additional stress $\vec{m}_h \geq \vec{m}_v$ or $\vec{m}_h = \alpha\vec{m}_v$ with $0 \leq \alpha \leq 1$

3D imagery allowing us to highlight the anisotropy and isolate it uses the *in-line* and/or *broadside* array data. It can be done using a finite differences code (forward problem) and a reverse approach by least squares or other methods (Morten et al. 2010; Ramananjaona et al. 2011). This numerical investigation requires significant computing means (parallel computation) (Fig. 2.33).

Under certain conditions, other treatments (see Chap. 5) inserted into more sophisticated interpretation models can be advantageously used (Wiik et al. 2010; Ogunbo 2011; MacGregor and Tomlinson 2014). The effects of anisotropy are still poorly understood and still pose problems of interpretation today.

4.4 The Oilfield Topologic Problem

The obtained values for electric fields depend not only on the values of conductivities of the rocks and their contrast but also on the shape, extent and strength

[89]The resolutions of the forward and inverse problems are discussed with more details in Chap. 5: Interpretation.

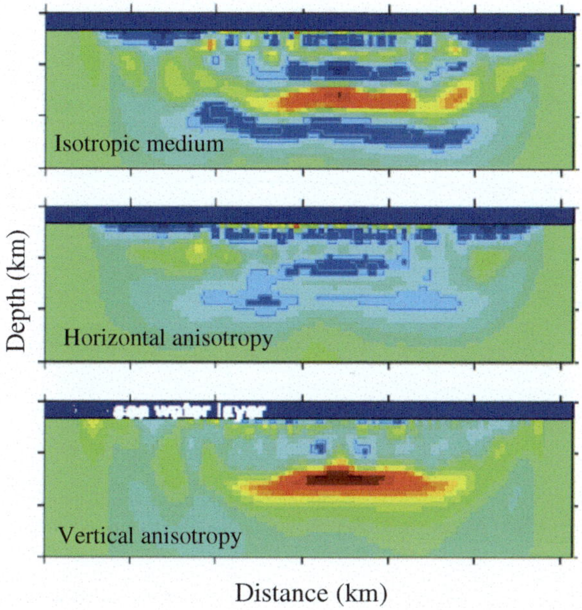

Fig. 2.33 Isotropic and anisotropic effects (vertical and horizontal) brought out on a geoelectric pseudosection (According to Newman et al. 2010)

of the layers formed by these rocks. In some circumstances, these effects may be potentiated with the presence of oil; in others, conversely, they can escape or literally vanish. This is what happens when the trap (anticline, salt dome, etc.) has high resistivity or conversely when it partly contains saline water.

5 Advantages and Disadvantages with Regard to the Seismic Method

If we can do so, with certain reservations (different geological aims), seabed logging methods must be directly compared with seismic reflection and refraction methods, which are currently the only techniques to be used commercially and effectively in the offshore oil exploration. At a geophysical level we could also compare the evaluation of the resistivity (an in situ parameter) with that of the local gravity (gravimetry) in the case, for example, of detection of salt domes, a lower-density geological object.

5.1 In Geophysical Terms: Saturation Indicator

In terms of pure geophysical investigation, seabed logging seems to be now more a means of detecting or locating hydrocarbon targets (oilfields) than a means of structural, lithological or faciological description of geological traps as understood in the context of seismic exploration (Constable 2005). However, SBL appears to

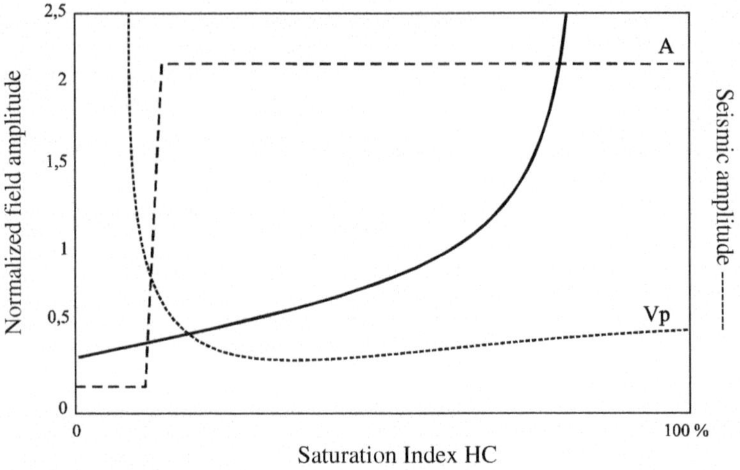

Fig. 2.34 Comparison of the electromagnetic (field amplitude) and seismic (velocity Vp and amplitude A) responses depending on the saturation index HC

be a much better indicator of hydrocarbon saturation than seismics (cf. Fig. 2.34) thanks to the characteristic resistivity values (see Chap. 3, Fig. 3.8).

At the operational level, the present technique may be more difficult to implement than certain existing seismic technologies (immersion and picking up the equipment). However, we can expect in the near future that the use of electromagnetic acquisition streamers, as happens in seismics, will then be more convenient and less expensive.

5.2 *In Environmental Terms*

Electromagnetic investigation broadly appears to be far less aggressive toward marine fauna than seismic methods, whose acoustic sources such as air guns or sparkers can reach powers of more than 210 dB at 1 m, i.e., the equivalent of 250 g of TNT[90] in a 10 m depth, and for which the propagation of the vibrations may extend over hundreds or even thousands of kilometers (in the SOFAR[91] network, for example), thereby necessitating the establishment of HSE standards and controls (Tidy et al. 2011).

[90]These mechanical sources have their energy decreasing very rapidly with the water depth under the influence of hydrostatic pressure. An air gun, for example, no longer delivers 10 kJ at 2000 m depth.

[91]A water layer (SOFAR hydroacoustic channel) located at a depth of 500–1000 m used by the US Navy during the Cold War to detect the presence of submarines. The range of the immersed devices in this channel was several thousand kilometers. The old stations are now used for oceanographic research in sectors as diverse as the study of ocean currents or volcanic eruption forecasting.

In contrast, the high electrical conductivity of seawater causes very rapid attenuation of the electrical emitted power, then advocating this technology. Considering the field values (see Eq. 2.35), neglecting the effects due to the frequency and supposing that cos $\phi = 1$, we obtain the maximum field value (half space: seawater) such as:

$$E_r = \frac{I\rho_w L}{2\pi r^3} \quad (2.74)$$

In these situations, for an injection current of 1500 A (I) and an antenna 200 m long (L), this amounts, at approximately 100 m away from the source (r), to a maximum electric field of 15 mV/m. At a distance of 10 km, the signal is only 15 nV/m (cf. Chap. 4, Sect. 4.1), an intensity comparable to that of telluric currents in a narrower band.

Recent studies conducted by independent marine biology laboratories have a priori shown the harmlessness of this technique to marine fauna (Girondot 2007). Furthermore, if we look this time at telluric methods, where the current is permanently present in the marine environment, they have in fact no human impact on the environment.[92]

Finally, logically, rationalizing resources, SBL must also allow us to significantly limit the number of exploration drillings where the consumption of water and mud is considerable.

6 Combined and Hybrid Methods

Here we touch on the rapid and undeniable success of underwater seabed logging as it stands today, in the sense that this method, in its problematic interpretation, as will be seen later (data inversion), strongly uses multiple, diverse and varied information coming from other acquisition systems and more particularly those acquired yet by seismic means. We can therefore say that the electromagnetic processes of exploration are at the very center of a sophisticated device that requires the use of other methods and that without these methods would probably quickly lose its effectiveness. It is reasonable to assume that in the future other sources of

[92]It may also be recalled that ships produce at a few meters from their hull some electric fields of the order of 1 mV/m and that river fishing societies, as well as associations for the protection of nature, practice electrical fishing for the capture, counting and identification of the fishery resource. This time the value of the fields that put fish in shock is of the order of 1 V/m (sampling rod: 200–500 V, 0.8 A). See also the report of the Association of Geophysical Contractors (EPCM) prepared by LGL Ltd. Environmental Research Associates on the safety of EM sources on marine wildlife.

information will complement the system, such as, when available, data from geology and downhole logs[93] that are now only partially used.

7 Conclusion and Synthesis

That is, if we can summarize the physical principle of this method to a simple causal effect, we can say in brief that:

- The main cause that allows us to detect hydrocarbons buried in the seabed is the existence of a noticeable (detectable) conductivity contrast between marine sediments and the hydrocarbon reservoir eventually present.
- The result of this unique metallogenic and particular geoelectric situation is then the local decrease in the subsoil of the eddy currents induced by the variable electric currents flowing therein, whether they are natural (terrestrial) or artificial (controlled source).

Ultimately, this has as a corollary, at the reservoir, an increase in the values of some diffuse electromagnetic fields, which, in a normal environment (without oil), are generally significantly lower.

Moreover, the principle of the method posed as a basic concept is that we have to jointly use the *galvanic* and *vortex* separate effects in different modes (TE and TM) and under different spatial configurations using *in-line* and *broadside* arrays to then ensure an optimal detection.

This new concept, in metrological terms, accordingly results in a spatial measure of the fields on the two horizontal components at least.

Moreover, it is the result after all:

- Of the discovery of the high conductivity of deep sea sediments infiltrated by seawater, and especially of its lateral variations in the presence of more resistive hydrocarbon deposits
- Of the existence in most cases of a level of very low ambient electromagnetic noise that exists on the ocean floor, of which the amplitude is well below detectable anomalous values
- Of the progress, as it will be seen later, of the interpretation methods (mathematical models) and more particularly of the data inversion, favored by the calculating power of modern super computers

However, the actual measurement techniques of electrical and magnetic fields and, more particularly, their horizontal components are older and have been proposed, for example, and experienced in the past (in the 1980s) in earth physics,

---

[93]The resistivity given by the electric logs does not match that assessed on the seabed. This significant difference can be attributed to the anisotropic nature of the lithology (lateral extension, and wider field of inquiry).

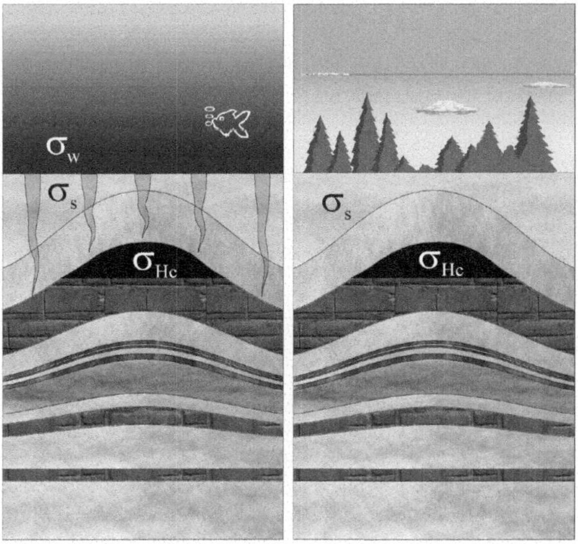

Fig. 2.35 Offshore, the infiltration of sea water (hydrostatic pressure) increases the conductivity of marine sediments and allows oil discovery ($\sigma_S > \sigma_{HC}$). Onshore, it is quite different; the contrast of conductivities between sedimentary layers and the hydrocarbon reservoir is not sufficient to consider any detection ($\sigma_S \approx \sigma_{HC}$)

in submarine detection, and even in mineral exploration, where low noise electronics had already at that time greatly increased (Fig. 2.35).

References

Several references in this following list correspond to appendices A2.1 to A2.5

Adam A (1976) Geoelectric and geothermal studies. Ed. Akademiai Kiado, Budapest, 752 p

Airy GB (1868) Comparison of magnetic disturbances recorded by the self-registering magnetometers at the Royal Observatory Greenwich with magnetic disturbances deduced from the corresponding terrestrial galvanic currents recorded by the self-registering galvanometers of the Royal Observatory. Philos Trans R Soc Lond 158:465–472

Ambronn R (1928) Elements of geophysics as applied to explorations for minerals oil and gas. Ed. McGraw-Hill, New York, p 176–177

Baron L, Chapelier D (2003) Application des méthodes électriques lacustres et terrestres à l'étude d'un aquifère côtier. 4ème colloque GEOFCAN

Bartels J (1954) Erdmagnetische erschliessbare lokale inhomogenitaten der elektrischen Leitfahigkeit in Untergrund. Nachr. Akad. Wiss. Gottingen II Math-Phys K1 2A. pp 95–100

Bascom W (1961) A hole in the bottom of the sea. The story of the Mohole project. Ed. Doubleday, Garden City, pp 154–157

Baumgartner F (1996) A new method for geo-electrical investigations underwater. Geophys Prospect 44:71–98

Berdichevsky MN (1961) Das magneto-tellurische Feld über geneigten Medien. Prikl Geof Vyp 28:70–91

Berdichevsky MN (1965) Electrical prospecting with the telluric current method. Ed. Colorado School of mines, Golden

Berdichevsky MN, Dmitriev VI (2002) Magnetotellurics context of theory of ILL—posed problems, Investigations in Geophysics, 11. Society of Exploration Geophysicists, Tulsa

Berdichevsky MN, Dmitriev VI (2008) Models and methods of magnetotellurics. Ed. Springer, New York, 561 p

Berdichevsky MN, Zadanov MS (1984) Advanced theory of deep geomagnetic sounding. Elsevier, Amsterdam, 408 p

Berdichevsky MN et al (1989) Marine deep geoelectric. Ed. Nauka, Moscow, 80 p

Berthier A (1908) Téléphonie et télégraphie sans fil. Ed. H. Desfores. pp 5–9

Bhattacharya PK, Patra HP (1968) Direct current geoelectric sounding. Method in geochemistry and geophysics. Ed. Elsevier, Amsterdam, pp 61–63

Blavier EE (1857). Cours théorique et pratique de télégraphie électrique. Ed. De Lacroix Comon, pp 413–417 and pp 406–412

Bonner et al (1996) Resistivity while drilling. Images from the string. Oil field review. Document Schlumberger. Ed. Elsevier, Amsterdam, pp 44–54

Børven JM, Flekkøy EG (2009) New EM technology uses vertical rather than horizontal electrical lines. Using a vertical dipole source and vertical receivers should increase horizontal resolution of anomalous resistors in offshore EM exploration. World Oil, September. pp 41–45

Braun F (1901) Drahtlose Telegraphie durch Wasser und Luft. Ed. Leipzig, Verlag von Veit & Comp. 68 p

Brewitt-Taylor CR (1975) Self potential prospecting in the deep oceans. Geology 3:541–543

Burrows ML (1978) ELF communications antennas. Ed. Peter Peregrinus, Stevenage, pp 25–29

Cagniard L (1953a) Basic theory of the magneto telluric-method of geophysical prospecting. Geophysics 18, p 606 and p 633

Cagniard L (1953b) Procédé de prospection géophysique. Brevet français no. 1025683

Cagniard L (1953c) Principe de la méthode magnéto-tellurique, nouvelle méthode de prospection géophysique. Annales de géophysique, Tome 9, fascicule 2, pp 95–125

Cagniard L (1956) Electricité tellurique. Handbuch der Physik. Geophysik 1. Ed. Springer. pp 406–469

Cagniard L, Morat P (1966) Extension de la prospection magnéto-tellurique à l'exploration offshore

Carson JR (1923) Reciprocal theorems in radio communication. Proc Inst Radio Eng 17:952–956

Chapelier D (2001) Cours de géophysique. Prospection électrique de surface. Ed. Université de Lausanne. pp 88–91

Chave AD, Cox CS (1982) Controlled electromagnetic sources for measuring electrical conductivity beneath the oceans. 1. Forward problem and model study. J Geophys Res 87 (B7):5327–5338

Chave AD, Jones AG (2012) The magnetotelluric method. Theory and practice. Ed. Cambridge University Press, New York, pp 9–10, pp 19–49

Chave AD, Constable SC, Edwards RN (1991) Electrical exploration methods for the sea floor. Electromagnetic methods in applied geophysics, vol 2. Chap. 12. Ed. Society of exploration geophysicists, pp 931–966

Chave AD et al (2004) Correction of motional electric field measurements for galvanic distortion. J Atmos Ocean Technol 11:317–330

Chisholm JSR, Morris RM (1965) Mathematical methods in physics. Ed. W B Saunders company, Philadelphia, 719 p

Claerbout JF (1976) Fundamentals of geophysical data processing with applications to petroleum prospecting. Ed. McGraw-Hill, New York, pp 174–177

Collin RE (1960) Field theory of guided waves. Ed. McGraw-Hill, New York, 606 p

Constable S (2005) Hydrocarbon exploration using marine EM techniques. Offshore technology conference, Houston. OTC 17068

Constable S (2010) Ten years of marine CSEM for hydrocarbon exploration. Geophysics 75 (5):75A67–75A81

Constable S, Srnka LJ (2007) An introduction to marine controlled-source electromagnetic methods for hydrocarbon exploration. Geophysics 72(2):WA3–WA12

Correnz CW (1937) Die Sedimente des äquatorialen atlantischen Ozeans. Ed. De Gruyter. 298 p

Corwin RF (1973) Offshore application of self-potential prospecting. Ed. University of Colombia, Berkeley, 606 p

Corwin RF (1975) Offshore use of the self potential method. Geophys Prospect 24:79–90

Corwin RF et al (1970) A self potential detection system for the marine environment. Proceedings of Offshore technology conference Paper OTC 1258
Coulomb J (1954) Les pulsations du champ magnétique terrestre. Annuaire du Bureau des longitudes pour l'an 1954. Ed. Gauthier-Villars. 23 p
Cox CS, Kroll N, Pistek P, Watson K (1978) Electromagnetic fluctuations induced by wind waves on the deep-sea floor. J Geophys Res 83:431–442
De Becherrawy T (2012) Electromagnétisme: équations de Maxwell, propagation et émission. Ed. Hermes/Lavoisier
Delesse A (1871) Lithologie des mers de France et des mers principales du globe. Ed. E Lacroix. 2 vol
Desbrandes R (1982) Diagraphies dans les sondages. Ed. Technip. 575 p
Ducretet P (1903) Traité élémentaire de télégraphie et de téléphonie sans fil. Ed. Librairie militaire R Chapelot. pp 71–76
Durand P, Mosnier J (1977) Magnétomètre sous-marin pour l'étude des composantes horizontales du champ transitoire. Annales de géophysique. Tome 33, fac. 4, pp 519–526
Duroux J (1974) Procédé et appareil de prospection en mer par mesure de champ électromagnétiques. Brevet français no. 74 26465
Edwards N (2005) Marine controlled source electromagnetics: principles, methodologies, future commercial applications. Surv Geophys 26:675–700
Edwards RN et al (1981) On measuring the electrical conductivity of the oceanic crust by a modified magnetometric resistivity method. J Geophys Res 86:11609–11615
Edwards RN et al (1984) Offshore electrical exploration of sedimentary basin: the effects of anisotropy in horizontally isotropic layered media. Geophysics 49:566–576
Eidesmo T et al (2002) Sea bed logging (SBL), a new method for remote and direct identification of hydrocarbon filled layers in deepwater areas. First break. vol 20.3 March
Ellingrud S et al (2002) Method for electric dipole configuration on the seabed. Brevet international PCT no. GB01/03473. 14 August 2000
Ellis DV, Singer JM (2007) Well logging for earth scientists. Ed. Springer, Dordrecht, pp 216–218 and 77–78
Fahie JJ (1899) A history of wireless telegraphy, 1838–1899: including some bare-wire proposals for subaqueous telegraphs. Ed. William Blackwood and Sons, Edinburgh
Fan Y et al (2012) Increasing the sensitivity of controlled-source electromagnetics with synthetic aperture. Geophysics 77(2):135–145
Faraday M (1832) Bakerian lectures. Experimental research in electricity. Phil Trans Roy Soc Lond Part 1. pp 163–177
Ferguson IJ (1988) The Tasman project of seafloor magneto-telluric exploration. Sripps Inst Oceanogr 436 p
Ferraro VCA (1956) Electromagnetic theory. Ed. The Athlone Press, London, p 450
Filloux JH (1980) Magnetotelluric soundings over the Northeast Pacific may reveal spatial dependence of depth and conductance of the asthenosphere. Earth Planet Sci Lett 46:244–252
Filloux JH (1982) Magnetotelluric experiment over the ROSE area. J Geophys Res 87 (B10):8364–8378
Filloux JH (1983) Seafloor magnetotelluric soundings in the Mariane island arc area. Geophys Monogr 27:255–265
Filloux JH, Cox CS (1964) Magneto-telluric sounding under the California coast. Symposium on magnetism of the earth' interior, Pittsburgh, November 16–20, p 74
Fourier J (1822) Théorie analytique de chaleur. Ed. Firmin Didot. 670 p
Fournier L (1910) La télégraphie sans fil. Ed. Librairie Garnier. p 161
Fröhlich H (1958) Theory of dielectrics: dielectric constant and dielectric loss. Ed. Clarendon press, Oxford
Galejs J (1969) Antennas in inhomogeneous media. Ed. Pergamon press, Oxford, pp 6–7
Gardiol F (1976) Electromagnétisme. Traité d'électricité. vol 3. Ed. Georgi
Gauss CF (1839) Allgemeine Theorie des Erdmagnetismus. In: Gauss CF, Weber W (eds) Resultate aus den Beobachtungszahlen des magnetischen Vereins im Jahr 1838. Dieterische Buchhandlung, Gottingen, pp 1–57

Girondot M (2007) Impact de champ électromagnétique sur la faune protégée au large des côtes de Guyane française. Publication du Muséum national d'histoire naturelle de Paris, Laboratoire des reptiles et amphibiens, université Paris sud. 31 p
Habberjam GM (1979) Apparent resistivity observations and the use of square array techniques. Geopublication associates. Ed. Gebruder Borntraeger, Berlin, p 31
Harland PJ (1932) The last cruise of the Carnegie. Ed The Williams and Wilkins Company, Baltimore, 333 p
Heinson G et al (1999) Marine self potential exploration. Explor Geophys 30:1–4
Hermance JF (1969) Resolution of ocean floor magnetotelluric data. Ed. J Geophy Res, vol 74, no 23, pp 5527–5532
Holten TE et al (2009) Vertical source, vertical receiver, electromagnetic technique for offshore hydrocarbon exploration: first break, 27
Horton CW (1946) On the use of electromagnetic waves in geophysical prospecting. Geophysics XI(1–4):505–517
Hoversten MG, Constable SC, Morrison HF (2000) Marine magnetotellurics for base-of-salt mapping: Gulf of Mexico field test at the Gemini structure. Geophysics 65:1476–1488
Hsu KJ (1992) Challenger at sea. A ship that revolutionized earth science. Ed. Princeton University Press, Princeton, 417 p
Hulin N et al (1993) Equations de Maxwell. Ondes électromagnétiques. Ed. Dunod
Jacobs JA et al (1959) Physics and geology. Ed. McGraw-Hill, New York, pp 247–248
Jegen MD, Edwards RN (2000) On the physics of marine magnetotelluric sounding. Geophys Int 142, 35 p
Jones SB (1959) Ocean bottom stratigraphy surveying. US patent no. 2872638. February 3
Kato Y, Kikuchi T (1950) On the phase difference of earth currents induced by the changes of the earth's magnetic field. Part 1. Geophysics 2(1):139–141
Kaufmann AA, Keller GV (1981) The magnetotelluric sounding method. Ed. Elsevier, Amsterdam, 596 p
Keller GV, Frischknecht FC (1966) Electrical methods in geophysical prospecting. Ed. Pergamon Press, Oxford, 517 p
Kertz W (1954) Modelle fur erdmagnetischen induzierte elektrische Strome in Untergrund. Nachr. Akad. Wiss. Gottingen Math. Phys. Klasse, Abt., 11A., pp 101–110
Key KW, Constable SC, Weiss CJ (2006) Mapping 3D salt using 2D marine magnetoluuric method: case study from Gemini Prospect. Gulf of Mexico. Geophysics 71:B17–B27
Kraichman MB (1976) Handbook of electromagnetic propagation in conducting media. Ed. Navy Department Bureau of Ships, Washington, pp 1.7–1.8
Lagabrielle R (1983) The effect of water on direct current resistivity measurement from the sea, river or lack floor. Geoexploration 21:165–170
Lagabrielle R (1992) Génie civil, génie côtier : apport de la géophysique. Reconnaissance géotechnique côtière. Section 2
Lagabrielle R et al (2001) Reconnaissance par prospection électrique du sous-sol marin pour le nouveau port du Havre. 3ème colloque Geofcan, 25–26 September. Orléans. pp 150–153
Lamont J von (1862) Der Erdstrom und der Zusammenhang desselben mit dem Magnetismus der Erde. Leipzig
Landau LD, Lifchitz EM (1969) Electrodynamique des milieux continus. Ed. de Moscou. pp 376–377
Landau LD, Lifchitz EM (1971). Mécanique des fluides. Ed. de Moscou. pp 364–367
Lodge OJ (1892). Modern views of electricity. Ed. Macmillan, New York, p 79
Loeset LO, Amundsen L (2007) On the signal propagation in marine CSEM. EGM international workshop. Innovation in EM, gravity and magnetic methods. A new perspective exploration. Capri Italy, April 15–18
Loseth LO et al (2006) Low frequency electromagnetic fields in applied geophysics: waves or diffusion? Geophysics 71(4):W29–W40
Lowy H (1912) Method of locating subterranean strata. US patent no. 1045575. November 26
Lowy H, Leimbach (1911) A method of indicating the presence and determining the position of veins of metallic ore or subterranean water level. British patent no. 11,737

Lowy H, Leimbach (1912) Interference method for ascertaining the presence of minerals in the earth or the underground level. British patent no. 8716
MacGregor LM, Sinha MC (2000) Use of marine controlled source electromagnetic sounding for sub-basalt exploration. Geophys Prospect 48:1091–1106
MacGregor LM, Tomlinson J (2014) Marine controlled-source electromagnetic methods in the hydrocarbon industry: a tutorial on method and practice. Interpretation 2(3):SH13–SH32
Makoto T et al (2003) Land and marine hydrogeology. Ed. Elsevier, Amsterdam, 199 p
Mari JL, Glangeaud F, Coppens F (1997) Traitement du signal pour géologues et géophysiciens. Ed. Technip. 457 p
Matschinski M (1952) Les équations de Maxwell pour le tri-feuille et leur application à la théorie du déplacement d'un orage magnétique. Ed. Les éditions de la revue scientifique. pp 91–103
Matschinski M (1955) Some methods of interpretation of the magnetic fields. Geof Pura E Appl 30 (1):68–85
Matteucci C (1862) Sur les courants électriques observés dans les fils télégraphiques. CRAS 55:264–266
Maxwell JC (1861, 1862) On Faraday lines force. Part 1: the theory of molecular vortices applied to magnetic phenomena, vol 21, pp 161–175. Part 2: the theory of molecular vortices applied to electric currents, vol 21, pp 281–291. Part 3: the theory of molecular vortices applied to static electricity, vol 23, pp 12–24. Philosophical magazine
Mayes FM (1965) Surface electrical prospecting apparatus utilizing current focusing electrodes means. US patent no. 3182250. May 4
Meunier J (1975) L'apport des sondages a courant continu dans l'étude de la résistivité des couches profondes de la terre. Comparaison avec les résultats magnéto-telluriques. Doctoral thesis, Université Louis Pasteur de Strasbourg. pp 36–42
Meyer O (1951) Uber eine besondere Art von erdmagnetischen Baystorungen. Deutsch. Hydrogr. Z.4, Heft ½, pp 61–65
Meyer HW (1972) A history of electricity and magnetism. Ed. Burndy, Norwalk, p 199
Migaux L (1948) Une méthode nouvelle de géophysique appliquée: la prospection par courants telluriques. Tiré à part de la Compagnie générale de géophysique. 31 p
Mittet R et al (2007) On the orientation and absolute phase of marine CSEM receivers. Geophysics 72(4):145–155
Morat P (1970) Magnéto-tellurique marine. 22 p
Morse PM, Feshbach H (1953) Methods of theoretical physics. Ed. McGraw-Hill, New York. Chap.13. pp 1759–1767
Morten JP et al (2010) Hydrocarbon reservoir thickness resolution in 3 D CSEM anisotropic inversion. SEG technical publication
Mosnier J (1970) Les magnétomètres amèliorent notre connaissance du champ magnétique terrestre. Ed. Gauthier-Villards, 5, pp 75–80
Murray J, Renard AF (1891) Deep sea deposits. Ed. Longmans & Co., Endinburgh, 525 p
Nabighian MN (1987). Electromagnetic methods in applied geophysics. Theory. vol 1. Ed. Society of exploration geophysicists, Tulsa
Newman GA et al (2010) Imaging CSEM data in the presence of electrical anisotropy. Lawrence Berkeley National Laboratory, Berkeley, Ref. LBLN-2967E
Ogunbo JN (2011) CSEM modeling of VTI layered medium with effective anisotropic medium. Université de Delft. Mémoire de Master. 68 p
Patra HP, Mallick K (1980) Geosounding principles, 2. Time-varying geoelectric soundings. Methods in geochemistry and geophysics. Ed. Elsevier, Amsterdam
Patra HP, Nath SK (1999) Schlumberger geoelectric sounding in gound water: principle interpretations and applications. Ed. Balkema, Rotterdam, 153 p
Pirson SJ (1981) Geologic well log analysis. Gulf Publishing Company, Houston
Poincelot P (1961) Le théorème de réciprocité de l'électromagnétisme théorique. Ann Telecommun 16(7–8):154–162
Porstendorfer G (1961) Tellurik. Grundlagen, Messtechnick und neue Einsatzmoglickeiten. Ed. Akademie Verlag

Postma GW (1962) Method for marine electrical prospecting. US patent no. 3052836. September 4
Ramananjaona C, MacGregor L, Andreis D (2011) Inversion of marine electromagnetic data in a uniaxial anisotropic stratified earth. Geophys Prospect 59:341–360
Reitz JR, Milford FJ (1962) Foundations of electromagnetic theory. Ed. Addison-Wesley, Reading, pp 300–304
Rikitake T (1950) Electromagnetic induction within the earth and its relation to the electrical state of the earth's interior. Bulletin of the earthquake research institute. Université de Tokyo, vol 28, pp 268–283
Rikitake T, Yokoyama I, Hishiyama Y (1952) A preliminary study on the anomalous of geomagnetic variations of short period in Japan and its relation to the subterranean structure. Bulletin of the earthquake research institute. Université de Tokyo, vol 30, pp 207–221
Rocard Y (1956) Electricité. Ed. Masson & Cie. pp 64–157
Rooney WJ (1939) Earth currents. Physics of the earth. Tome 8, Terrestrial magnetism and 1938 electricity. Ed. McGraw-Hill, New York, p 270
Rougerie P (1942) Contribution à l'étude des courants télluriques. Annales de l'Institut de Physique du Globe. Tome 20. Ed. PUF. p 61
Rubin Y, Hubbard SS (2006) Hydrogeophysics. Ed. Springer, Dordrecht, 528 p
Sainson S (1984) Etude d'une méthode de détection électrique d'anomalies conductrices voisines d'un forage. Doctoral thesis in geophysics. Université d'Orléans. 177 p
Sainson S (2010) Les diagraphies de corrosion. Acquisition et interprétation des données. Ed. Lavoisier. p 299 and p 324
Scholl C, Edwards RN (2007) Marine down hole to seafloor dipole–dipole electromagnetic methods and the resolution of resistive targets. Geophysics 72(2):WA39–WA49
Schumcker U (1970) Anomalies of geomagnetic variations in the south western United States. Bull Scrip Inst Ocean 13
Schuster A, Lamb H (1889) The diurnal variation of terrestrial magnetism. Philos Trans R Soc Lond A210:467–518
Sejman SM (1958) Über die Möglichkeit der Ausnutzung der Felder der tellurischen Ströme und von Radiostationen für di geologische Kartierung. Tr. Vses. In-ta metodiki I techniki razvedki Gostoptechizdat
Selzer (1972) Variations rapides du champ magnétique terrestre. Handbuch der Physik. Geophysik III Teil IV. Ed. Springer-Verlag. pp 231–394
Serra O (2004) Well logging, vol 1 data acquisitions and applications. Ed. Technip, Méry Corbon. 688 p
Siebert M (1958) Die Zerlegung eines lokalen erdagnetischen Feldes inausserenund inneren Anteil mit Hilfe des Zweidimensionalen Fourier theorem. Ahandlungn der Akad. Wiss. Gottingen, Math. Phys. Kl. Beitr. Int. Geophys. Jahr. 4, pp 33–38
Simpson F, Bahr K (2005) Practical magnetotellurics. Ed. Cambridge University Press, Cambridge
Skilling HH (1942) Fundamentals of electric waves. Ed. Wiley, New York, pp 114–120
Sochelnikov UV (1977) Marine magnetotellurics. Acta Geod Geophys Montan 12(1–3):163–176
Spiess FN et al (1980) East Pacific Rise: hot springs and geophysical experiments. Science 207 (4438):1421–1433
Stalheim SO, Olsen PA (2008). Sea to borehole application of marine controlled source EM method: 3rd Saint Petersburg conference and exhibition, EAGE, Extended Abstracts, C009
Stefanescu S (1936) Sur les fondements théoriques de la prospection électromagnétique par courant alternatif à très basse fréquence. Beitrage zur angewandten Geophysik
Stefanescu S (1945) Le champ électromagnétique de la télégraphie par le sol. Doctoral thesis in science. Université de Bucarest N° d'Ordre 211, br. 11 July 1945. 89 p
Strack KM (2009) Method for conbined transient and frequency domain electromagnetic measurements. US patent no. 7474101 January 6
Stratton JA (1961) Théorie de l'électromagnétisme. Ed. Dunod. 702 p
Swidinsky A, Hölz S, Jegen M (2012) On mapping seafloor mineral deposits with central loop transient electromagnetics. Geophysics 77(3):E171–E184
Tarits P (1986) Conductivity and fluids in the oceanic upper mantle. Physics of the Earth and Planetary Interiors. Ed. Elsevier vol 42, n°4, pp 215–226

Telford et al (1978) Applied geophysics. Ed. Cambridge University, London, 860 p
Telford et al (1983) Géophysique appliquée. tome 3. Ed. Erg. p 62
Terada J (1917) On rapid periodic variations of terrestrial magnetism. J Coll Sci. Imperial University of Tokyo. vol 37, pp 56–84
Thomsen et al (2007) System and method using time distance characteristics in acquisition, processing and imaging of tCSEM data. US patent no. 7,941,273. May 10
Thoulet J (1907) Précis d'analyse des fonds sous-marins actuels et passés. Ed. Librairie militaire Chapelot. 220 p
Thouvenel P (1792) Résumé sur les Expériences d'Electrométrie Souterraine faites en Italie et dans les Alpes depuis 1789 jusqu'en 1792. Pour servir de suite aux Mémoires publiés en 1780 et 1783, sur les rapports qui existent entre les phénomènes du magnétisme, de l'électricité, et de la baguette divinatoire. Ouvrage physique et polémique divisé en deux parties. Ed. Brescia, dalle stampe Bendiscioli. 230 p
Tidy J, Goncalvez L, Pujol E (2011) Envionmentally sound offshore operations. International Oil and Gas engineer. Febuary pp 4–5
Tikhonov AN (1950) Determination of the electrical characteristics of the deep strata in the earth's crust. Dok. Akad. URSS, 72,2. pp 295–297
Trabant PK (1984) Applied high resolution geophysical methods. Offshore geo-engineering hazards. Ed. International human resources development corporation. 265 p
Trofimov I (1975) Experiment of gradient magnetic sounding in Pacific Ocean. Geomagnetizm I Aeronomiya 15(1):181–183
Ursin B (1983) Review of elastic and electromagnetic wave propagation in layered media. Geophysics 48:1063–1081
Van Nostrand RG, Cook KL (1966) Interpretation of resistivity data. Ed. US government printing office, Washington, p 10 and p 38
Vishnyakov AE et al (1992) Detailed mapping of deep sea sediments with deep towed geophysical system. (*in Russian*). Dokladi Akademii Nauk 32:77–80
Wait JR (1982) Geo-electromagnetism. Ed. Academic press, New York, 268 p
Walker PW, West GF (1992) Parametric estimators for current excitation on a thin plate. Geophysics 57:766–773
Webster AG (1897) The theory of electricity and magnetism being lectures on mathematical physics. Ed. Macmillan, London, pp 507–508
Werthmüller D, Ziolkowski A, Wright D (2014) Predicting controlled source electromagnetic responses from seismic velocities. Interpretation vol 2, No. 3 August, pp SH115–SH131
Wiese H (1954) Erdmagnetische Baystorungen und ihr heterogener in Erdinnern induzierter Anteil. Z Meteorol 8:77–79
Wiese H (1962) Geomagnetische Tiefentellurtik. Die Streichrichtung der Untergrundstrukturen des elektrischen Widerstandesn erschlissen aus geomagnetischen Variationen. Geofis Pura Appl, 122:52–83
Wiese H (1965) Geomagnetische Tieffentellurik. Deutsche Akad. Wiss. Berlin. 146 p
Wiik et al (2010) TIV contrast source inversion of nCSEM data. Norvegian University of Science and Technology, Trondheim
Witte L de (1958) Apparatus for logging the ocean floor. US patent no. 2839721. June 17
Yu L, Edwards RN (1991) The detection of lateral anisotropy of the ocean floor by electromagnetic methods. Geophys J Int 108:433–441
Yu L, Evans RL, Edwards RN (1997) Transient electromagnetic responses in seafloor with tri-axial anisotropy. Geophys J Int 129:300–306
Zhdanov MS (2010) Electromagnetic geophysics: notes from the past and the road ahead. Geophysics, vol 75, no. 5. September–November, pp 75A49–75A66
Ziolkowski A, Wright D (2008) Shallow water multi transient EM surveys in the North Sea. Petroleum Geo-Services. 3rd international oil and gas symposium in western Newfoundland marble mountain ski resort. 14 Sept

Chapter 3
Metrology and Environment

Abstract This chapter describes the physical environment where measurements will be made. It then presents the electromagnetic properties of the concerned media (electrical conductivity, magnetic permeability and dielectric permittivity) depending on the characteristics of seawater and subsoil rocks (facies, lithology). It then discusses the frequency and temporal aspects of the detection method depending on the propagation media and the background noise in the deep sea (several $\mathrm{pV/m/\sqrt{Hz}}$). It defines in substance the skin effect, the energy attenuation, the investigation depth, the magnitude of the amplitudes of the fields accessible to measurement (about 1 $\mathrm{\mu V/m/\sqrt{Hz}}$ or, if normalized, about 10^{-12} V/A.m^2), the signal-to-noise ratio and the modes and propagation/diffusion conditions in the presence or absence of oil. Then it proposes data acquisition systems to establish the intrinsic characteristics of the measuring instruments and especially the power of the transmitters and the receptor sensitivity. This chapter ends with a description of optimal conditions of detection and favorable field procedures.

Keywords Electromagnetic properties • Signal to noise ratio • Attenuation • Field amplitude • Depth of investigation

1 Introduction

Once the physical principle is laid down (see Chap. 2), the question is then to establish the optimal conditions for obtaining the information (measurements) needed to retrospectively interpret the acquired raw data.

It is the purpose of metrology which has on one hand the obligation to fix the order of magnitudes of the signals coming from the investigation and on the other hand the obligation, according to them, to identify and define the means of acquisition and especially of the equipment adapted to the environment.

Metrology, if it uses the actual measurement techniques, must take into account for its definition all the aspects of the energy transfer (→) occurring within and between the different concerned sectors, from the emitting source, whether artificial or natural, in the subsoil (?) to the receiving device (cf. Fig. 3.1).

These energy transfers are directly dependent on the electromagnetic properties of the crossed media (water, sediments, rocks, etc.) and are linked by propagation

S. Sainson, *Electromagnetic Seabed Logging*, DOI 10.1007/978-3-319-45355-2_3

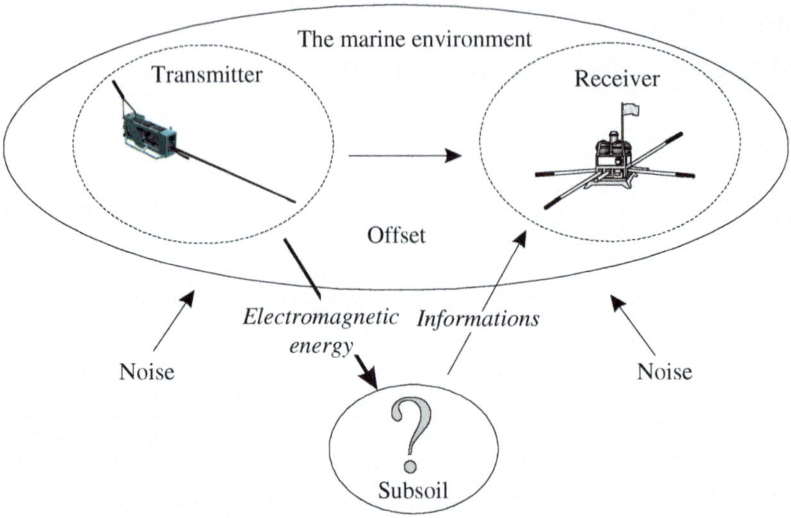

Fig. 3.1 Synoptic of the detection system (transmitter/receiver) in its marine geoelectric environment

laws also dependent on the electrical characteristics of the emission and especially the power and frequency of the source. Metrology therefore requires the study of these energy transports, which result in transfer functions simultaneously affecting grounds (physical properties) and measuring equipment whose intrinsic qualities must be in perfect harmony with the desired characteristics (Fig. 3.2).

Perhaps more than any other method of geophysical investigation, electromagnetic seabed logging acutely sets the double problem of technology and science.

So, technically, the question is to isolate with the greatest possible clarity the desired physical quantity (electric and magnetic fields in particular) representative of the parameters to evaluate (electrical conductivity) and whose amplitude and variations can be very low. The solution to this technological problem then depends on the intrinsic quality of the measuring devices (electrometers), most particularly their sensitivity, accuracy and stability, knowing that they must operate in a most hostile environment (noise).

Besides the problem of the data acquisition itself, the very difficult question of the representativity and interpretation of measurements is then raised. These latter must establish the precise relationship between the various measures corresponding to the observed perturbations, related or not to the inaccessible subsoil heterogeneities (the aim of the exploratory research). These relationships are generally established through very general laws of physics. However, the latter, most often because of their differential formulation, are unfit for the study of integral problems such as those that generally arise in the prospecting context. This is especially true for electromagnetic seabed logging, where, if you specifically wish information on the shape, depth, and nature of the anomalies, the mathematical study remains

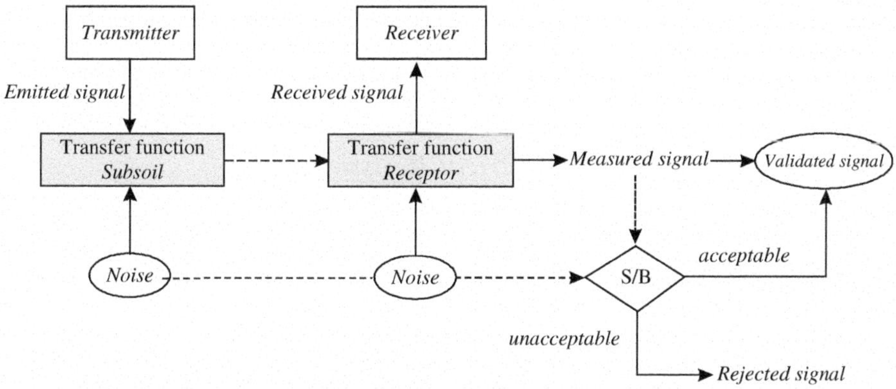

Fig. 3.2 The measure requires at least two transfer functions: the one that is sought, which concerns the subsoil and is based on electromagnetic characteristics of the crossed grounds, and the one that is known, which affects the receptor and must comply with certain intrinsic qualities. The entire acquisition chain must also have an acceptable signal-to-noise ratio (S/B) for the measure to be validated

singularly complex to be implemented in alternative or variational regimes and thus comes under a real theoretical challenge.

If the technical problem (sensor) has so far received satisfactory solutions,[1] the mathematical study of the interpretation still today substantially progresses in offering a more and more extensive and eclectic choice of conceptual approaches and geophysical models (see Chap. 5).

2 Electromagnetic Properties of Propagation Media

In a macroscopic view, we can simply say that the media crossed by electromagnetic energy have relatively homogeneous properties that can differentiate themselves. This is, for example, the case with electrical conductivity, which generally presents a wide contrast between sediments impregnated with seawater and oil-containing rocks, or the case with dielectric permittivity, which more specifically affects insulators such as oil. Nevertheless, the magnetic permeability, a property existing only in ferromagnetic materials, will play a minor role in the differentiation of these media.

These different electromagnetic parameters will consequently affect electric fields in magnitude and in direction on one hand, and the speed of wave propagation in various media on the other hand. The variability of these physical characteristics is, among other things, the basis of the detection systems in EM seabed logging. In

[1] An increase in accuracy (see note below) can only be effective if the quantity to be measured is well defined at each point where the determination is made.

terms of detection itself, we will therefore successively discuss the means allocated to frequency and time domains.

2.1 Conductivities and Electrical Resistivities

Of all the physical properties of the rocks present on the globe (the lithosphere), electrical conductivity offers the widest range of values, approaching a record factor of 10^{20} (Grand and West 1965).

The electrical conductivity of any material is its ability to allow more or less electrical current to cross and it is thanks to this quality that energy transfer can take place. It is to the physicist G. S. Ohm that we owe this discovery (Ohm 1827).

In terrestrial materials, if we only focus on conduction current, this can roughly flow in two modes[2]:

- By the displacement of electrons; conductibility is then known as electronic or metallic
- By the movement of ions; the conductibility is then called ionic or electrolytic

Unlike ores,[3] or more generally metals, the latter of which only have their own electronic conductivity, the electrical conductivity of rock is not an intrinsic property. It is rather of the electrolytic type but *grosso modo* depends on the porosity and its quality (closed or open), on the charge in ions of the fluids contained in its pores and more especially on the constitution of the formation or infiltration of water, or on the interstitial water more or less rich in chlorides and dissociated minerals, located in sedimentary strata[4] (Abrikossov and Goutman 1986).

In these configurations, we then speak of specific resistivity, a value that takes into account all of these parameters, and which already allows us to understand the importance of its estimation in the detection of hydrocarbon deposits which are poorer in water.

In general, the term *resistivity* is dedicated to the rocks considered as rather resistant, such as oil reservoirs for example, and the term *conductivity* to those that are less resistant such as marine sediments that seawater has extensively infiltrated.

---

[2]Minerals such as arsenides and sulphides have the two conductivity types, which do not make them good conductors.

[3]The conductivity of minerals is more complex to understand.

[4]Porous and permeable rocks have resistivities that may very significantly differ depending on the nature of the fluids contained therein. For example, sands containing oil or gas will have a much larger resistivity than those containing salt water. Moreover, if we consider that the resistivity is part of a more macroscopic concept (volume including the notions of stratum, bench, layer, etc.), the resistivity at a given temperature then depends on three basic factors, which are the lithologic character, the amount of present pore water and the mineralization of the latter.

The resistivity ρ is expressed exactly in Ω.m^2/m or more simply in Ω.m (a very largely used unit but which can be confusing). The conductivity σ, which represents the inverse of ρ, is translated into mho.m^{-1} or (ohm.m)$^{-1}$ or S.m^{-1} (International System: IS).

$$\sigma_{[S/m]} = 1/\rho \tag{3.1}$$

In dry air and materials considered as insulating media (pure dielectrics), the electrical conductivity is zero.

The electric resistivity ρ is defined by the relation that links the electric resistance R of a solid as well as a liquid material, to its geometric characteristics. For a rectangular prism of section *A* and length *l*, the resistivity (Ellis and Singer 2007) is then given by the well-known expression:

$$\rho_{[\Omega.m]} = R_{[\Omega]} \times \frac{A_{[m^2]}}{l_{[m]}} \tag{3.2}$$

This definition is of course valid only when *A* and *l* are finite. Furthermore, the Ohm × meter (amount taken as a unit) where 1 Ω.m corresponds in absolute terms to a cube of an homogeneous and isotropic matter with 1 m sides (IS unit), the electrical resistance R taken on two opposite sides *A* (1 m^2), whatever the direction, is equal to one Ohm (1 Ω).

In practice the above formula (3.2) is difficult to apply because the instruments measuring the resistance R (ohm-meter, quotient-meter, Wheatstone bridge, etc.) in general introduce additional resistances and more particularly contact resistances. In a laboratory, its measurement is thus performed more accurately by using a quadrupole of a current/voltage type (see Chap. 4, Fig. 4.39) and its value depending on the geometry of the adopted device defined by the factor *K* is calculated using the formula:

$$\rho_{[\Omega.m]} = K_{[m]} \frac{U_{[V]}}{I_{[A]}} \tag{3.3}$$

In this case, the resistive term (R) is no longer present. This is the same approach that is used in electrical prospecting by direct current (cf. Chap. 2, Sect. 2.7.1.2). In those circumstances where current injection (I) system and voltage measuring circuit (U) are decoupled, it is not correct to say that: U/I = R.

2.1.1 Electric Conductivity of Seawater

Seawater can be considered as an electrically conductive liquid and its conduction mainly comes from its content in dissolved salt, decomposed into ions, especially sodium chloride (Na^+, Cl^-), i.e., in other words, its salinity, which is about 35 g of

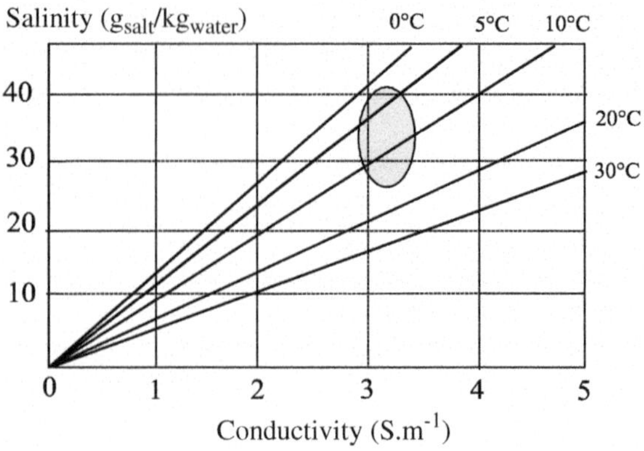

Fig. 3.3a Conductivity/salinity diagram at different temperatures between 0 and 30 °C, and conductivity field in the marine environment ($35g_{NaCl}/kg_{water}$ on average, that is about 0.5 mol.l^{-1} NaCl)

salt per kilogram of water ($35g_{NaCl}/kg_{water}$). The conductivity of seawater is about 4000 times greater than that of fresh water and varies only weakly with the hydrostatic pressure[5] (Horne and Frystinger 1963).

We would like to remark that quantitative chemical analysis does not provide representative results regarding salinity (evaporation of some salts at the end of the chemical metering).[6] However, it is possible to indirectly obtain them (g_{salt}/kg_{water}) by evaluating the salinity s (mg/l) by electrical conductivity measurements such as we have, as a function of experience parameters and especially temperature T:

$$\sigma_{w[S/m]} = 1.6 \times 10^{-4} \times s_{[mg/l]} \times \left(1 + \frac{(T_{[°C]} - 25)}{50}\right) \tag{3.4}$$

A change in salinity of 1 g_{salt}/kg_{water}, for example, gives a variation of the conductivity of 2.5 % under normal conditions of temperature (20 °C) (cf. Fig. 3.3a).

The electrical conductivity of different waters is given as an indication in the form of a chart in this chapter (cf. Appendix A3.1).

More generally, the electric conductivity of seawater varies depending on the temperature (which itself decreases with depth), and decreases with the pressure (Hamon 1958 and Horne and Frystinger 1963) (cf. Fig. 3.3b). A good approximation of the electrical conductivity (Becker 1985), taking into account the average conductivity in the ocean (approximately 3 $S.m^{-1}$), is given by the formula:

$$\sigma_{W[S/m]} = 3 + \left(\frac{T_{[°C]}}{10}\right) \tag{3.5}$$

[5]The hydrostatic pressure is given by the formula: $p = h\rho g$ where h is the height of water, ρ is the volumic mass of water (1028 kg/m^3) and g is the value of the acceleration of gravity (9.81 m/s^2). Depending on the case, it is expressed in bars, Pa or psi. Whenever it sinks 10 m, the pressure increases a bar. At 4000 m deep, the pressure is equal to 40 MPa (5800 psi).

[6]However, it is possible to achieve consistent results through the application of Knudsen's dosage method (Knudsen 1901; Thomsen and Menache 1954).

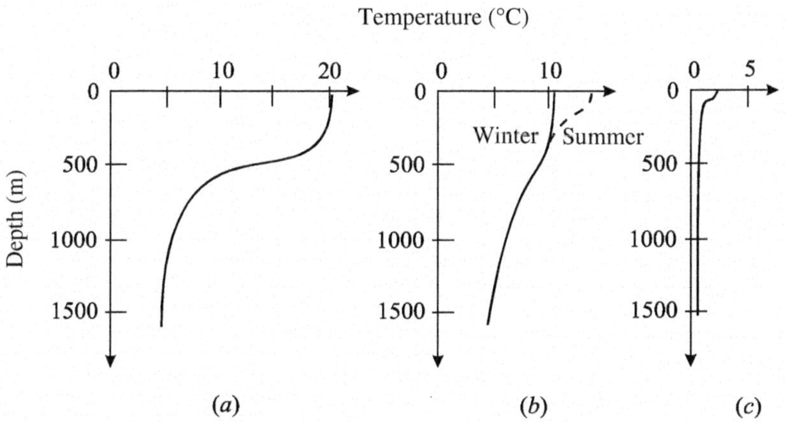

Fig. 3.3b Evolution of the average ocean temperature as a function of depth at low (*a*), medium (*b*) and high latitudes (*c*) (By Ingmanson and Wallace 1989)

Fig. 3.4 Changes in the conductivity of seawater as a function of water depth (By Chave and Luther 1990)

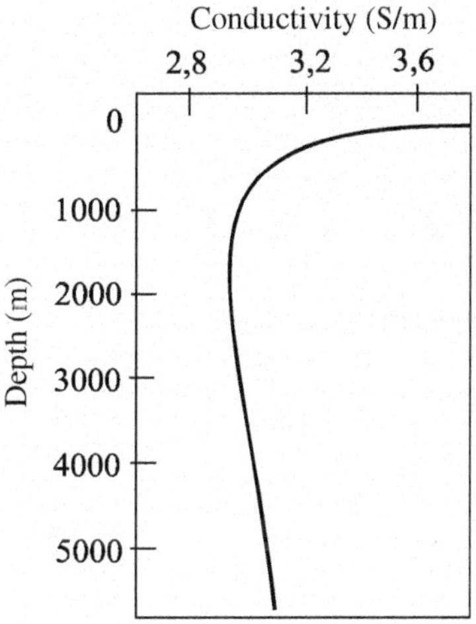

At great depth the temperature is around 3 °C[7] except in arctic regions, where it can reach −2 °C. It is more or less constant over the globe, and leads in this situation to an average conductivity of 3.2 $S.m^{-1}$. Geographically, the latter therefore varies from 2.5 $S.m^{-1}$ at the geographic poles to 5 $S.m^{-1}$ in the tropics.

To take into account at the same time all these changes (pressure, temperature, salinity), some authors (Fofonoff and Millard 1983) have proposed algorithms directly expressing electrical conductivity as a function of water depth (cf. Fig. 3.4).

[7]The first temperature measurements at sea and in deep water date from the end of the eighteenth century (1785–1788 campaigns of the Venus).

2.1.2 Physical Properties of Rocks: Generalities

For geophysicists, sedimentary rocks are heterogeneous composite materials, anisotropic and even multiphasic for those containing hydrocarbons composed of:

- A matrix, solid material of the rock, more or less structured, made of grains and/or cement, generally stable but possibly mobilizable (leaching, etc.)
- Pores, which are globally expressed by the porosity and correspond in volume to the percentage of the present vacuum (no solid matrix)
- Fluids such as water, oil and gas, which can more or less fill the pores

These different parameters are tightly interrelated and may vary in proportion over time.

When porosity is open and pressure differentials exist, the fluid in the rock can then move or flow, providing greater or lesser hydraulic permeability (Guegen 1997).

This is the case, for example, when the hydromechanical conditions are met, from the migration of oil from the source rock to the reservoir rock, or during the extraction of oil by pumping in production wells (a reverse process) (Fig. 3.5).

These various factors will on one hand influence, for the solid parts, the mechanical properties of the rock (density, elastic constants, hydraulic permeability, etc.) and on the other hand, for the fluid elements (oil, water, gas), the electromagnetic properties. Of the latter, it is the electrical conductivity that has the wider variability (Fig. 3.6).

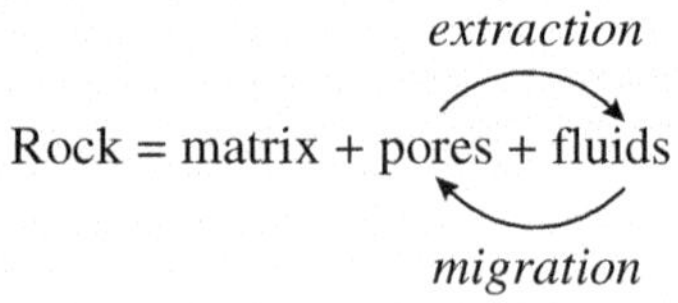

Fig. 3.5 Main components of rock, static (matrix and pores) and mobile (fluids), which will directly influence its electromagnetic properties

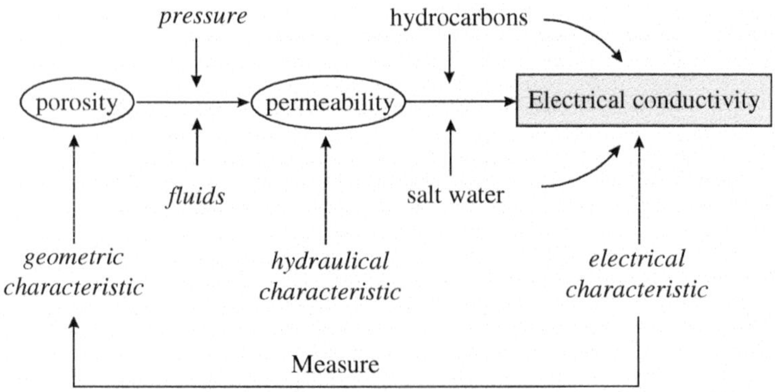

Fig. 3.6 Electrical conductivity, a privileged witness to the porosity and hydraulic permeability of sedimentary rocks and of the presence or absence of hydrocarbons

One of the peculiarities of underwater rocks and more specifically marine sediments is their saturation in salt water. This is due among other things to the combined effect of the hydrostatic pressure under a certain water depth and the large hydraulic permeability, related to the rocks' porosity and its quality, which enable the seawater to largely infiltrate it at great depth or at least in the first few hundred meters of the subsoil. The porosity ranges from 40 to 80 % in the subsurface and gradually decreases with depth.

2.1.3 Electrical Conductivity Specific to Geological Formations: Reminders

The conductivity of geological formations cannot be reduced to a sum or even less to an average of electrical conductivities of the various constituents of the rock. In reality the so-called specific electrical conductivity of a rock depends not only on the solid and liquid elements the rock is made of, but also very strongly on its lithology, structure and more particularly its porosity and its intrinsic quality, which is to say open or closed porosity.

In summary, as will be detailed a little further on, the specific conductivity of a rock depends mainly on:

- The facies
- The lithology
- The interstitial water content and its composition of minerals
- The volume occupied by the communicating pores

A very small variation in one of these four parameters or formation factors can significantly alter the intrinsic resistivity. Furthermore, experience has shown that it can vary even if a direct examination reveals no change in the structure and facies.

Ultimately, we can say that the variation of the electrical conductivity is a very good indicator of the likely lithology and fluid content of the geological formations at large.

2.1.4 Electrical Conductivity of Marine Sediments

Marine sediments have an electrolytic conduction, which was empirically defined by G. E. Archie (Archie 1942)[8] and is mainly applicable to water-saturated rocks.

[8]The physicists Gray and Wheeler in 1720 were the first to measure the electrical conductivity of some rocks (Seguin 1971), followed in the nineteenth century by some others (Becquerel 1834, 1847; Bruck 1841).

$$\sigma_S = \frac{\phi^m}{a} S^n \sigma_w \overset{\text{sea water}}{\rightarrow} \sigma_S = \frac{\phi^m}{a} \sigma_w \tag{3.6}$$

where:

- σ_S is the electrical conductivity of the sediments.
- ϕ^m is the sediment porosity and m is the form or cementation factor ranging from 1.3 to 1.5 for unconsolidated sediments.[9]
- S^n is the fraction of pores infiltrated by water, where S is the saturation (1 for seawater) and n a constant that depends on the fluid[10] ($\approx$2 for water).
- σ_w is the electrical conductivity of seawater (average: 3 $S.m^{-1}$).
- a for the tortuosity factor is a constant for marine sediments between 0.6 and 1.3 (Parkhomenko 1967).

Numerical application: For example, for marine sediments, by strictly applying the Archie formula (cf. Eq. 3.6) and considering the values for the usually encountered unconsolidated sediments (cf. Table 3.1), we precisely obtain an electrical conductivity equal to:

$$\sigma_S = \frac{0.25^{1,37}}{0.88} \times 3 \approx 0.5 \text{ S.m}^{-1}$$

a value greater than the one found in reality.

The problem with this formula is that it uses coefficients that are only accessible by in situ measurements, that is to say well logging (borehole measurements) and often, as it is the case for the value of the porosity, they are not known from the operator before the SBL campaigns. However, some authors (Bahr et al. 2001) have proposed for porosity an approximate formulation according to the depth z such as:

$$\phi^m(z) = 0.05 + 0.6\, e^{-z/1500} \tag{3.7}$$

This formula is more realistic and then usable in various models (Key 2009).

Ordinarily, this electric conductivity, for sediments infiltrated with water, ranges from 0.03 to 0.25 $S.m^{-1}$ (average value: 0.1 $S.m^{-1}$). It is 10 to 100 times lower than that of seawater ($\approx$3 $S.m^{-1}$).

The conductivity of marine sediments also changes, as the one of seawater, with temperature T_S (Constable et al. 2008) such that:

$$\sigma_S = 2.903916 \times \left(1 + 0.0297175\ T_S + 0.0001555\ T_S^2 - 0.00000067\ T_S^3\right) \tag{3.8}$$

[9]This factor is to link the shape of the pores and their degree of connection. For round grains (sand), m is equal to 1.4, while for the tabular grains (clay), the m value increases to 1.9. For fractured rocks, m can rise to 2.5.

[10]For oil and gas, the n values respectively are 2.08 and 1.162.

Table 3.1 Summary table of values *a* and *m* to use in the Archie formula for various sedimentary, clastic and volcanic rocks

Rocks (description)	Porosity (%)	*a*	*m*
Weakly cemented detritus rock	25–45	0.88	1.37
Moderately cemented sedimentary rock	18–35	0.62	1.72
Well-cemented sedimentary rock	5–25	0.62	1.95
Volcanic rock	20–80	3.5	1.44
Rock with very low porosity	<4	1.4	1.58

By Chouteau and Giroux (2006)

2.1.5 Electrical Resistivity of Oil Reservoirs

Hydrocarbons in themselves are considered as fluids, with very high electrical resistance (resistivity of 2.10^{14} Ω.m on average).[11] In geological reservoirs, it is quite different. Oil and gas are part of a complex heterogeneous structure where the different constituents of the oil rock are intimately intertwined with each other and bound by other also varying elements. As for sediments, that is the so-called interstitial water, containing a certain amount of dissolved salts, which is responsible for the electrical conductivity of the reservoir rock. In its upper part, it can take up to 10 % of the pores' volume and in any case rarely less than 5 % of this volume.

So that there is a flow of electric current by electrical conduction (cf. Chap. 2), it is also necessary that the porosity is open, i.e., the pores are in communication with each other, and the fluids are able to move or not. The volume of conductive water is then at the most equal to the effective volume of the pores and is obviously lower when the rock contains oil and gas.

Furthermore, in the fields, the percentage of interstitial water increases toward the sides and the bottom points of the geological structure, until the water completely occupies the volume of the pores in the region of the so-called marginal water. The presence of interstitial water and its particular distribution in rocks are mainly due to the action of the capillary pressure and surface tension between the present fluids and solids.

- In the upper part of the structure (the reservoir itself), this water forms a thin film around the grains and thus completely fills up the gaps of capillary dimensions. The oil is found only in the free space left by the water in the voids of larger dimensions, and this under conditions of local pressure.
- In the lower parts of the structure, conditions are such that there is more oil present. The percentage of water in the aquifer level then depends on certain conditions.[12] In a gas reservoir, this percentage is close to 100 %, while in an oil reservoir, it can be much lower.

[11]They are also used in industry as an electrical insulator (e.g., in transformers).

[12]The deposit water contains salts and ions in solution (Cl^-, Na^+, Ca^{2+}, Mg^{2+}, K^+), dispersed colloids and dissolved gases (N_2, CO_2 and CH_4). The main characteristics of the deposit water come from their primary and secondary origin (Robert 1959).

Finally, the mathematical relation connecting the main constituents and the physical characteristics of the rock, especially the electrical conductivities to the mechanical properties, is given by the formula:

$$\sigma_{HC} = a\sigma_{ef}(1 - S_{HC})^n \phi^m \tag{3.9}$$

where:

- ϕ^m is the porosity of the rock and m is the cementation factor, with a value greater than 1.5 for unconsolidated sediments.[13]
- n is the saturation factor.
- a is the tortuosity factor.
- S_{HC} is the fraction of pores infiltrated by oil.
- σ_{ef} is the electrical conductivity of the pore water.

Under the conditions of the deposit, the formation resistivity ($1/\sigma_{HC}$) then falls to relatively low values, which can vary from 30 to 300 Ω.m (cf. Figs. 3.7 and 3.8).

To study the behavior of electromagnetic energy in geological media, some authors have attempted to translate by analog models the electrical effects of currents on the porous matrix. The various elements are simulated in an analog way by resistors and capacitors inserted in a particular network (Bitterlich and Wöbking 1972).

2.1.6 Resistivity, Facies, Lithology, Structure and Anomaly

Electrical prospecting in a large sense would not have developed if we had not noticed early on that the specific electrical resistivity of the same geological layer does not vary laterally. Indeed, many experiments on the spot have shown that resistivity can be considered as an extremely stable indication of facies (cf. Chap. 1, Sect. 4.1). However, this statement needs to be nuanced for sands or for any porous rocks whose resistivity varies with the interstitial fluid. However, from the practical point of view, oil-bearing rock is actually another rock, discernible from the identical rock this time invaded by both fresh and salted water, and so also in this case, the electrical resistivity then follows the facies. Under these conditions the "electrofacies" may be considered then as one of the most remarkable characteristics of the reservoir.

Whatever the geological circumstances, the specific electrical resistivity is therefore an overly sensitive indicator of facies, often finer than direct, that is to say stratigraphic, examination. We then conceive the enormous advantage that

[13]The Archie and Humble formulas, used for the exploitation of electric logs, admit higher values for exponents m and n (close to 2). This leads us to assign to the reservoirs resistivity values much lower than those observed on electrical logs (Serra 2004). The anisotropy of the geological layers is one of the reasons partly invoked to explain these differences. The factors m and n are described in detail in the technical literature, and more particularly the one concerned with logs.

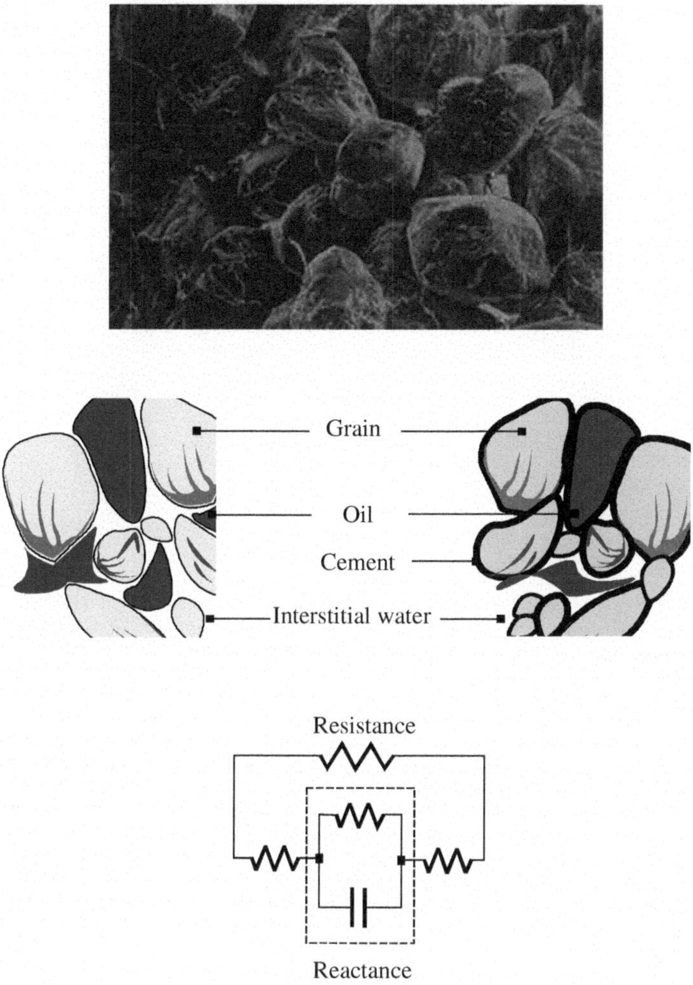

Fig. 3.7 Grain structure of a reservoir rock (microphotography) without and with cement (diagrams), and an electrical equivalent (circuit) taking into account the resistive (resistance) and inductive (with dotted reactance capacity) aspect of the oil in its porous matrix (By Pirson 1950)

electrical prospecting methods or more generally electromagnetic exploration methods have over other methods of geophysical investigation, provided that we can extract by measuring and then by calculating the values of this resistivity. Moreover, we can say that only sedimentary grounds are virtually conductors and that resistivity variations, however small they may be, (where the porosity and the lithology can be regarded as sensibly invariable) then result from the structure of the subsoil only and particularly its layers. Finally, we can also say that the most important changes in resistivity are due in some way, in this specific context, to anomalies characterizing the presence of oil or gas.

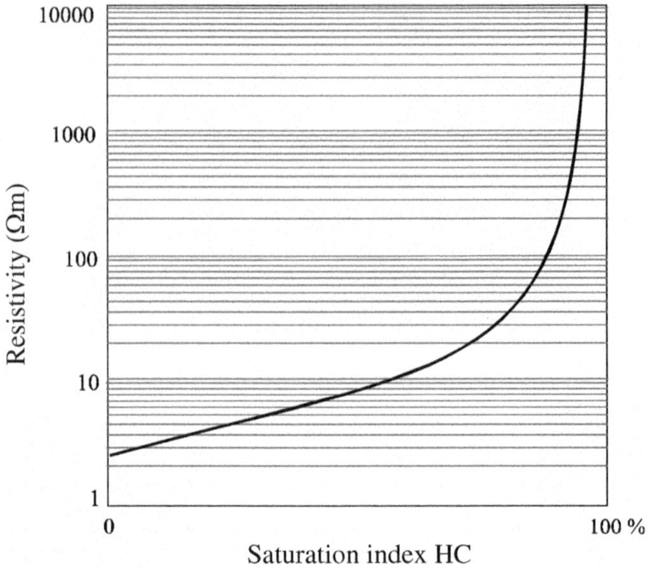

Fig. 3.8 Resistivity as a function of the saturation index in hydrocarbons (HC) for a typical reservoir ($\sigma_{ef} = 0.1$ Ω.m, $a = 0.65$, $m = 1.8$ and $n = 2$). For water saturation of 100 %, the resistivity is 2 Ω.m. For hydrocarbon saturation from 40 to 90 %, the resistivity varies from 30 to 300 Ω.m

As an indicator of facies and lithology, and for sensing favorable structures and revealing defects or heterogeneities as resistive fluids, in these various aspects electrical resistivity is probably the measurable physical property of rocks that delivers the greatest amount of useful geological information for the prospector. This is especially true in comparison with density or elasticity, the main source of petrophysical data currently used in modern oil research (seismic and gravimetry, incidentally).

2.1.7 Reservoir Resistivity and Vortex Effect: Debate?

Measurement and study of the vortex effect is the basis of current detection systems (see Chap. 2) and can be compared with the values of electrical resistivity and more specifically with their lateral variations.

Virtually nonexistent in good conductors of electricity (mainly because of the abundant presence of eddy currents), the vortex effect mainly occurs in the dielectrics whose very important resistivity generally exceeds 10^4 Ω.m.

However, it remains less obvious for the reservoir rocks, which have significantly lower values, in the order of about 300 Ω.m. This suggests at this level that another concomitant action, probably of a different origin, would then add to this effect. This would explain in some way the current controversy that fuels the debate on the physical nature of the phenomenon (see Chap. 2, Sect. 2.5.9).

2.2 Magnetic Permeability

The magnetic permeability μ of a material is expressed by the product of the permeability of the vacuum μ_o (Henry/meter), which is a universal constant whose value is $4\pi \times 10^{-7}$ $H.m^{-1}$ (or 1.257×10^{-6} $H.m^{-1}$) and the intrinsic or relative magnetic permeability μ_r (dimensionless) is:

$$\mu = \mu_0\mu_r = 4\pi \times 10^{-7} \times \mu_r \tag{3.10}$$

With the exception of certain rocks (volcanic rocks and minerals), sedimentary rocks and seawater are nonmagnetic. Their permeabilities with $\mu_r = 1$ are then equivalent to μ_0.

2.3 Dielectric Permittivity

The permittivity is also known as the dielectric constant and physically corresponds to the polarization of the bound charges (electrons, ions, dipolar molecules, space charges, etc.) when the material is traversed by an alternating electric current, or more generally a variational current. It is the product of the dielectric permittivity of vacuum ε_0 (Farad/meter) which is $10^{-9}/36\pi$ or 8.854×10^{-12} $F.m^{-1}$ and the relative permittivity ε_r (dimensionless) is dependent on the environment, such as:

$$\varepsilon = \varepsilon_0\varepsilon_r = \frac{1}{36\pi} \times 10^{-9} \times \varepsilon_r \tag{3.11}$$

The permittivity of seawater is variable and depends on local conditions, i.e., salinity, depth (pressure) and temperature (cf. Table 3.2). That of the rocks is highly dependent on their water content. The relative permittivity of marine sediments more particularly results at a low frequency from the part of the present volume of water β_v (Rubin and Hubbard 2006). It is equal to:

$$\varepsilon_r = 3.03 + 9.3\beta_v + 146\beta_v^2 - 76.6\beta_v^3 \tag{3.12}$$

The dielectric permittivity of the hydrocarbon reservoir is more difficult to estimate. It is among other things dependent on the structure and the various constituents of the rock and more particularly on the oil content and thermodynamic conditions (cf. Table 3.2).

Table 3.2 Summary table of average electromagnetic properties of different media crossed by the electromagnetic waves

Media	Electrical resistivity Ω·m	Electrical conductivity $S{\cdot}m^{-1}$ou $(\Omega{\cdot}m)^{-1}$	Dielectric permittivity × $8.8{\cdot}10^{-12}$ F/m	Magnetic permeability H/m
Dry air	∞	0	1	$4\pi{\cdot}10^{-7}$
Seawater (19 °C)	0.2	5	78–80	"
Seawater (0.5 °C)	0.3	3	"	"
Marine sediments	4–30[a]	0.03–0.25[a]	20	"
Reservoir (oil/gas)	30–300	0.003–0.03	6	"

[a]Values that fluctuate with depth

2.4 *Summary of the EM Properties of Propagation Media*

The characteristics of the electromagnetic properties of the crossed media—namely, air, seawater, marine sediments, and reservoir rocks—are summarized in the Table 3.2. Some of them have significant variabilities (marine sediments, reservoir especially). Others, however, remain relatively constant (air, seawater). We can see that the resistivity of underwater subsurface rocks is much lower than that usually found onshore (cf. Table 3.3). This primordial particularity is the original source of the detection method, which can only be effective at sea.

In addition, comprehensive experimental works have been done on the electromagnetic properties of rocks at depth (Nover 2005), in particular those concerning the terrestrial crust and earth mantle, showing their variability depending on the pressure and temperature (cf. Fig. 3.9).

3 Frequential Aspect of the Method

The use of alternating currents implies an interaction of electric energy with matter (Harrington 1961), which is different depending on whether we are interested in seawater, marine sediments or even in the hydrocarbon reservoir. In seawater, the propagation of electromagnetic waves was particularly studied in the years 1950–1960 by American researchers from the US Navy (Kraichman, Bannister, Hart, Jakson, Walters, Wait, Campbell, etc.),[14] succeeding the early fundamental works in the late nineteenth century (Whitchead 1897; Heaviside 1899).

[14]Research led in the context of underwater detection and published as monographs by the US Navy Underwater Sound Laboratory also relating the EM studies (Bannister 1980).

Table 3.3 Electrical resistivities of some earth materials (rocks and sediments) according to their water content and their mineralization

Rocks and sediments	Electrical resistivity Ω·m
Clay and marl	6–40
Shales	40–250
Chalk	100–300
Limestone	100–5000
Sandstone	500–10,000
Sand	30–10,000
Crystalline rock	>5000

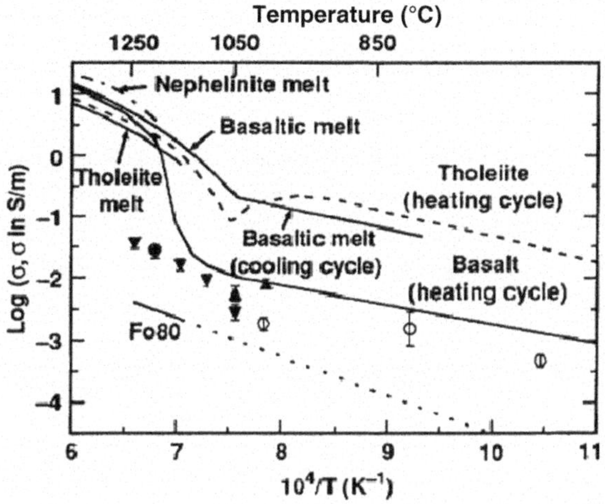

Fig. 3.9 Conductivities (Log) of the deep layers of the crust and mantle (By Nover 2005)

3.1 Conductor Media: Eddy Currents and the Skin Effect

In conductors of electricity like seawater and marine sediments, this interaction results in the medium by creating secondary currents, also called *eddy currents*. They are opposed to the propagation of energy even more intensely when the frequency and the conductivity are high. At this stage the energy dissipated by the *Joule effect* (heat) by the chemical energy transfer (ions) and the propagation is then characterized by a concentration of the field lines on the surface (*skin effect*), thereby limiting the depth of penetration of the strength lines. This effect is the consequence of Faraday's law (second Maxwell equation):

$$\vec{\nabla} \wedge \vec{e} = -\frac{\partial \vec{b}}{\partial t} \tag{3.13}$$

which theoretically shows that variations of the magnetic field induced by the flow of an alternating or variational current in a conductor of electricity in turn create a secondary magnetic field, which is thus opposed to the primary field.

3.2 Dielectric Media: Low Attenuation

In dielectrics or similar materials (air, oil and gas reservoirs), if at high frequencies (≈1 GHz RADAR investigations) additional phenomena of relaxation (the Maxwell–Wagner effect) or of interfacial polarization are present, at low frequencies (Nover 2005; Ellis and Singer 2007), electromagnetic waves only suffer a slight attenuation essentially attributable to the conduction effect of the electrical charges in the interstitial fluid (Razafindratsima et al. 2006) and to the transfer of electrical energy due to displacement currents (see Chap. 2, Sect. 2.5.1).

3.3 Investigation Depth and Penetration Depth

The depth of investigation and the depth of penetration of the electromagnetic waves are concepts that directly arise from the skin effect (Stratton 1961). However, they are probably the most complex concepts to theoretically understand in geophysical prospecting because they depend in practice on factors that fluctuate.

3.3.1 Investigation Depth of the EM Waves in Seawater

Seawater and marine sediments may be regarded as electrically conductive media. The equation of propagation/diffusion, or the Helmholtz equation, is given by the expression:

$$\nabla^2 \vec{\mathrm{E}} + k^2 \vec{\mathrm{E}} = 0 \tag{3.14}$$

and is characterized by the wave number k such that:

$$k^2 = \omega^2 \mu\varepsilon - \mathrm{i}\omega\mu\sigma \tag{3.15}$$

In a dispersive medium, as is the case with seawater and involved rocks (ε is very small for the different media), k is reduced to:

$$k^2 \approx -\mathrm{i}\omega\mu\sigma \tag{3.16}$$

or to:

$$k = \sqrt{-\mathrm{i}}\sqrt{\omega\mu\sigma} \tag{3.17}$$

Using the following relation:

$$\sqrt{-\mathrm{i}} = \frac{1-\mathrm{i}}{\sqrt{2}} \tag{3.18}$$

and also knowing that the transmitted wave is progressive in the positive direction, from which:

$$\mathrm{Re}\{k\} > 0 \tag{3.19}$$

and setting that:

$$k = \frac{1 - \mathrm{i}}{\delta} \tag{3.20}$$

we consequently obtain:

$$\delta_{[\mathrm{m}]} = \sqrt{\frac{2}{\omega\mu\sigma}} \tag{3.21}$$

which represents the *skin depth*, still called the *depth of investigation*. It corresponds to the distance at which the magnitude of the fields is attenuated to $1/e$ or 0.368 (e being the base of natural logarithms equivalent to 2.718).[15]

Seawater and marine sediments are nonmagnetic materials. The magnetic permeability of the latter[16] is therefore equivalent to that of the vacuum, that is $\mu = \mu_0 = 4.\pi.10^{-7}\mathrm{H.m}^{-1}$, from which finally comes a depth of investigation roughly equal to:

$$\delta_{[\mathrm{m}]} = \frac{1}{\sqrt{\pi f \mu_0 \sigma}} = \frac{503.25}{\sqrt{f\sigma}} \approx 500\sqrt{\frac{\rho}{f}} \tag{3.22}$$

This formula shows that the electric current passes along the conductors even more easily at the surface when the frequency f is high, thereby designating a skin effect (*Kelvin* effect).

At this stage (δ), we can schematically admit that there is no more propagation beyond this depth, that the wave is then almost "dead" and that δ is a jointed characteristic of the crossed medium and of the frequency of the propagating wave.

We can equally say that the current penetration is even more important when the product ρT is great. Accordingly, we have on a purely metrological plan the possibility of separating surface anomalies from deeper anomalies by then very simply varying the value of the period T.

---

[15]Distance at which the amplitude of the wave is equal to 36.8 % of the amplitude E_o, ie $E_o = \mathrm{e}^{-\frac{k}{\sqrt{2}}\delta} = E_o/e$ being the base of natural logarithms and corresponding to $\ln(e) =$ or to $\exp(1) = e$. This number may be defined as: $e = \lim_{n\to\infty} e_n = 2,718281....$ with $e_n = 1/0! + 1/1! + 1/2! + ... + 1/n!$

[16]In various nonmagnetic sedimentary grounds, μ_r varies from 1 to 1.00001, and can be then considered as a constant (≈1).

We would like to remark that in the field of earth geophysics, i.e., the one that particularly concerns the great depths (mMT), the prospector in practice prefers to use a literal formula depending on the period T (in seconds) such that we have:

$$\delta'_{[\mathrm{km}]} = \frac{1}{2\pi}\sqrt{10\rho T} \approx \frac{1}{2}\sqrt{\rho T} \tag{3.23}$$

where the investigation depth is then given in kilometers (Kunetz 1966; Benderitter and Dupis 1969).

In the expression above, as in the previous one, the relative magnetic permeability does not appear anymore, because not only does it hardly vary from one field to another, but also it is equal to one unity in most of the cases encountered in oil exploration (very small variations from 1 to 1.00001 in the nonmagnetic sedimentary layers). For example, for a phenomenon during 500 s, the depth is around 5 km.

More generally, we can also set the *skin depth* based on the wavelength λ in the medium such that:

$$\delta''_{\sigma[\mathrm{m}]} = \frac{\lambda_{[\mathrm{m}]}}{2\pi} \tag{3.24}$$

a formula where the electrical conductivity is of course implicit (contained in λ) (Fig. 3.10, Table 3.4).

3.3.2 Penetration Depth of EM Waves in Seawater and Sediments

Generally we theoretically admit that at the considered frequencies (low and very low frequencies), the controlled source (mCSEM) system is effective when the depth of investigation is half less than the depth of penetration.[17] Practically, we accept that for investigations through the entire substrate (marine sediment conductors of electricity), a skin depth at least equal to the effective depth of the latter ($\delta \geq h$) is sufficient to obtain significant results (Fig. 3.11).

In mMT prospecting, where the source of energy is more versatile, it is estimated that the maximum allowable frequency according to the water depth h would in reality correspond to $\delta = 3h$ or at:

[17] Ampere/Maxwell's law implicitly involves a duality between two types of current. At low frequencies, in the conductors, conduction currents are predominant, whereas in the higher frequencies ($\sigma/\omega\varepsilon \ll 1$), i.e., those that are above the light spectra, the movement currents predominate. In DC, investigation depth and penetration depth are then equivalent (no skin effect) and among other things depend on the geometry of the acquisition device (the distance between the electrodes of the injection device particularly). These concepts were defined for the first time in 1938 (Evjen 1938) then supplemented by many authors (Guérin 2007).

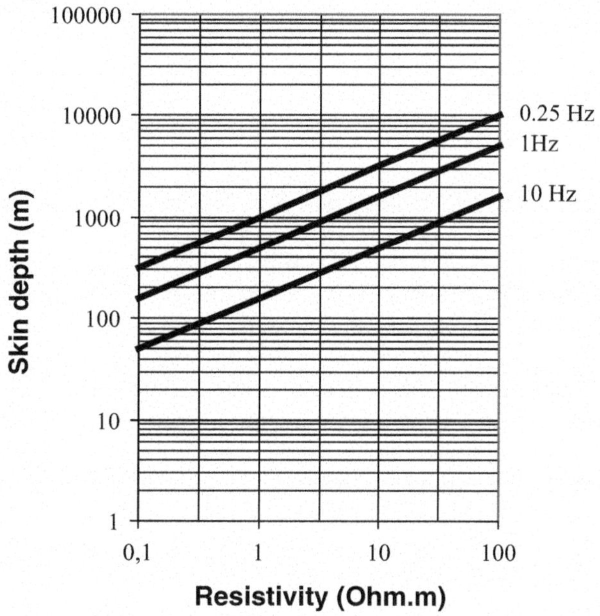

Fig. 3.10 Evolution of the *skin depth* as a function of the excitation frequency and the resistivity of the crossed media

Table 3.4 Table of *skin depths* in seawater and in marine sediments for different frequencies (0.1–15 Hz)

	Frequency			
Media (ρ)	0.1 Hz	1 Hz	5 Hz	15 Hz
Seawater (0.3 Ω·m)	866 m	275 m	123 m	71 m
Marine sediments (1 Ω·m)	1580 m	500 m	224 m	129 m

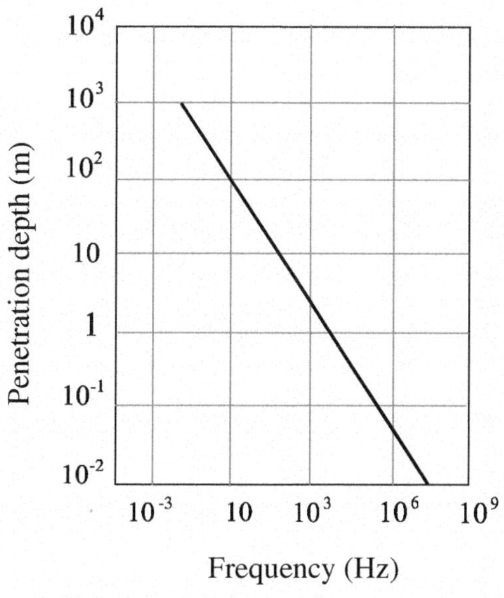

Fig. 3.11 Penetration depth of EM waves in seawater according to the frequency and for an attenuation of 40 dB

$$f_{max} = \frac{1}{\sigma}\left(\frac{500}{3h}\right)^2 \tag{3.25}$$

3.3.3 Investigation Depth and Resolution

In general, the resolving power (ability to separate two very close objects) or the resolution and depth of investigation vary inversely. However, this result needs to be qualified, because the methods of investigation are different (broad frequency spectrums, geometric arrangements more or less directive, etc.) and are also to be compared with the target dimensions, their electrical conductivity and especially the contrast they offer with regard to their immediate geological environment.

For example, survey by DC has in the best conditions (very good resistivity contrast) a vertical resolution of 2–4 % for a very low depth investigation (at sea), while it has no lateral resolution (nondirectional device).

For the alternating current methods with a controlled source (directional device for vector measures), the lateral and vertical resolutions, which strongly depend on the accuracy of the measurement system positioning (cf. Chap. 4, Sect. 6.2), are estimated at 5 % at best.

For telluric methods, with great integrative power, the resolution is poorer and depends on the conditions of acquisition, especially the periods of the natural currents and their variability in time and space. Unlike investigation depths, which can be calculated with more or less accuracy, resolution reasonably can only be estimated (d'Arnaud Gerkens 1989).

We can note that many scientific papers reveal patterns using rays as a convenient way of visualizing the propagation of electromagnetic waves in different media (cf. Figs. 3.20 and 3.21). Although this representation is not completely wrong (we will also use it later in our discussion), it can be confusing when compared, without caution, with that of light rays or even seismic rays, which implicitly bear the resolution of the method. Indeed, if we strictly consider the wavelengths (in relation to geological objects considered as targets), the only electromagnetic technology comparable with seismics (10–200 Hz) is geological radar (*ground-penetrating radar* or GPR), which is used in the range of 100 MHz. Under these conditions, the ray representation is entirely justified and the Snell–Descartes laws of reflection/refraction perfectly valid. These are even more accurate when the second term of the radar equation (cf. Chap. 5, Eq. 5.9) is significant. In the case of SBL, on one hand the propagation term of the above equation is zero, and on the other hand the wavelengths can be much greater than the dimensions of the observed objects (cf. Table 3.5). In this case, it is an extremely important fact that variable current SBL, with all methods combined, cannot of course have the resolution of seismic reflection, but the resolution is higher than that of continuous current methods.

Table 3.5 Summary table of the maximum frequencies f_{max} as a function of the water depth h, in mMT prospecting

h (m)	f_{max} (Hz)
500	0.03
1000	0.007
2000	0.0017
3000	0.00078

3.4 *Phase Difference*

The phase difference measurement, or more precisely the phase delay, is with the measurement of the amplitudes of the fields one of the keys to detection. However, this phase variation may be due, among other things, to several concurrent phenomena such as those established, for example, in conducting media (presence of eddy currents), or even those due to the propagation itself over quite large distances.

3.4.1 Phase Difference: Inductive Effect Due to the Presence of Eddy Currents

In the harmonic regime, in low frequency approximation (far field criterion)[18] in a homogeneous and isotropic conductive medium where the dielectric permittivity is neglected (seawater and marine sediments), the value of the current density J (uniform field) varies depending on the distance and time[19] (cf. Fig. 3.12*a*). If at the surface ($z = 0$), the components of the current density J are given by:

$$J_x = J_0 \cos \omega t \quad \text{and} \quad J_y = J_z = 0 \tag{3.26}$$

At the depth ($z = h$), they are given, taking into account the simplified conditions on limits in the direction of z (depth), by the expressions:

$$\begin{gathered} J_x(h, t) = J_0 e^{-h\sqrt{\pi f \mu \sigma}} \cos\left(\omega t - h\sqrt{\pi f \mu \sigma}\right) \\ J_y = J_z = 0 \end{gathered} \tag{3.27}$$

where:

- J_y and J_z are the values of the current density in the directions y and z.
- J_0 is the value of the original field (source: electric dipole).
- h is the depth.
- ω is the pulsation or angular frequency ($=2\pi f$).
- f is the frequency.
- t is the time.

---

[18]Unique field independent of the distance.

[19]The electric field is distributed according to the same law.

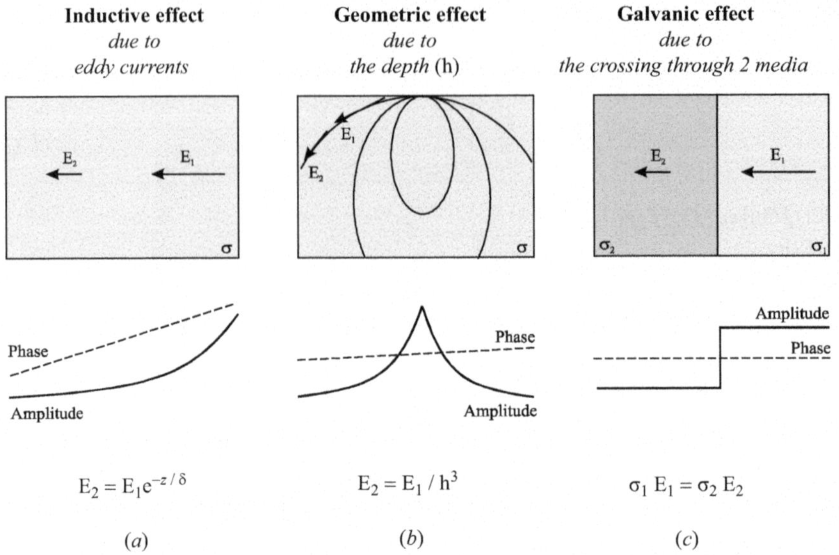

Fig. 3.12 The key to the detection is in the identification of the three effects, which have different behavior with respect to the phase and amplitude E of the electric fields. Only the inductive effect causes a significant increase in the phase (By Constable 2010)

- μ is the magnetic permeability of the medium.
- σ is the electrical conductivity of the medium.

The expression (cf. Eq. 2.27) is a particular solution (real part) of the diffusion equation:

$$J_x(h, t) = J_0 \mathrm{Re}\left[e^{-h\sqrt{\pi f \mu\sigma} + i\left(\omega t - h\sqrt{\pi f \mu\sigma}\right)}\right] \quad (3.28)$$

it is only valid for a plane wave, and shows substantially:

- An attenuation term ($J_0 e^{-h\sqrt{\pi f \mu\sigma}}$), sometimes called damping, caused by a loss of energy by absorption transferred to the matter, which results in chemical energy transfer and in heating (Joule effect) and which then causes an exponential decay of the current density with distance, with an energy flow that weakens gradually
- A phase term ($h\sqrt{\pi f \mu\sigma}$) increasing almost linearly as a function of the energy distribution, which corresponds to the phase difference between the invariable phase of the emission current (ωt) and the variable phase of the resulting induced currents

Theoretically, from the above expressions (cf. Eqs. 3.26 and 3.27), in the conditions of the far field (plane wave and uniform current threads, etc.), we can still deduct after integration of the current density J_x, the values of the horizontal components of the magnetic field H and the phase difference ($\pi/4$) that follows such as:

$$\begin{aligned} H_x &= 0 \\ H_y &= 4\pi \int_0^{+\infty} J_x dz = 2\sqrt{\frac{\pi}{\sigma\omega}} \cos\left(\omega t - \frac{\pi}{4}\right) \end{aligned} \tag{3.29}$$

3.4.2 Offset Phase Shift Due to the Distance from the Source

In a homogeneous and isotropic medium, for a plane wave, where the propagation velocity is then constant, the phase gradually and slightly increases linearly with the distance (cf. Fig. 3.12*b*).

Considering the phase term (cf. Eq. 3.27) for a diffusion of the electromagnetic energy in the seawater for example, then the phase difference varies:

$$h\sqrt{\pi f \mu_0 \sigma} = 10^3 \sqrt{\pi \times 1 \times 4\pi \times 10^{-7} \times 0.3}$$

i.e., about 1 rad/km (or even 57°/km).

3.4.3 Phase Shift: Galvanic Effect

Passing two media of different conductivities σ, if the amplitude of the electric field abruptly changes at the surface of separation (conservation of the charges), the phase, however, remains unchanged (cf. Fig. 3.12*c*). Proportionally, this effect is comparable to that of a direct current.

3.4.4 Conclusion

In short and schematically, we can say that only the *inductive effect* due to the existence of eddy currents (*vortex effect*) has a significant change in phase (cf. Fig. 3.12*a*) detectable on a greater or lesser distance (skin depth).

We would like to remark that if at significant water depths the information about the amplitude of the field is fundamental, it also shown that for shallow investigations, the information about the phase now appears to be more appropriate for detection (Mittet 2008) (Figs. 3.13 and 3.14).

3.5 Effect of the Frequency on Detection

The mathematical models, including 1D, in the frequency domain, allow the evaluation of the detection sensitivity depending on the frequency (Eidesmo et al.

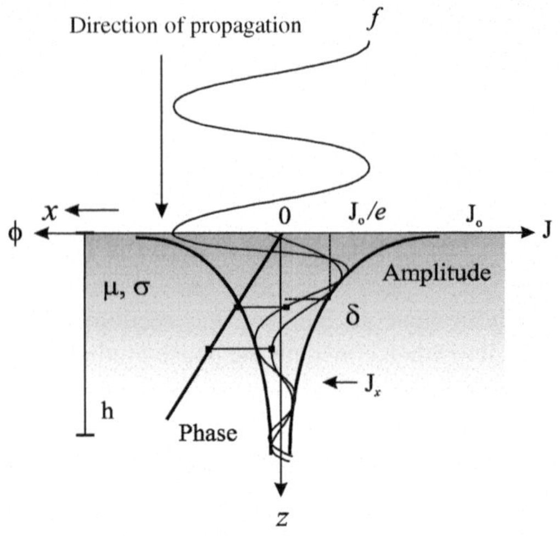

Fig. 3.13 Variation in amplitude and phase (inductive effect) of an electromagnetic wave passing through a more conductive medium (air/seawater, for example). By entering this medium (μ, σ) the wave is attenuated (exponential law) and also undergoes a linear sliding of its phase

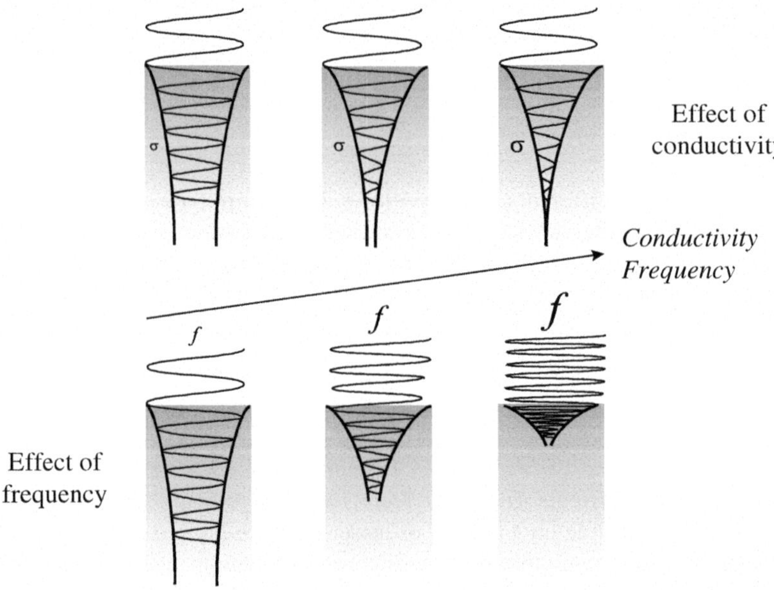

Fig. 3.14 Effect of conductivity (*top*) and frequency (*bottom*) on the penetration of electromagnetic waves in a conductive medium

2002). They show, for example, within a certain distance (small offset), that the amplitude of the fields then increases with the frequency (cf. Fig. 3.15).

The increasing frequency also results in an improved accuracy in detection, shown by the spreading of the curve being less pronounced for the highest frequencies (cf. Fig. 3.15).

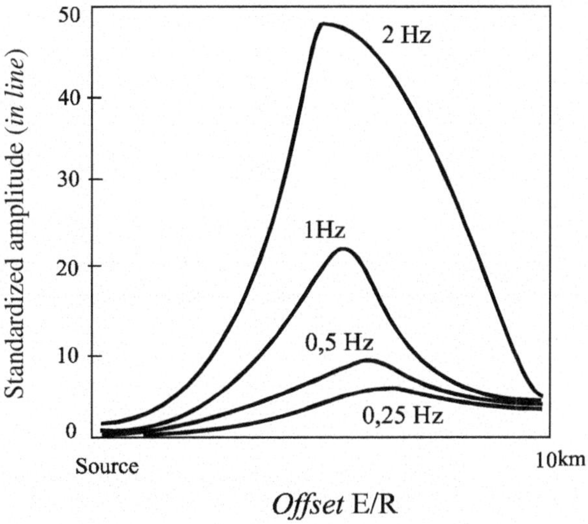

Fig. 3.15 Synthetic, interpretative curves, in response to a resistive reservoir-type anomaly for four different frequencies (0.25–2 Hz) in the *in-line* configuration (By Eidesmo et al. 2002)

3.6 *Sensitivity to Changes in Resistivity, Bathymetric and Topographic Effects: Three Examples of Operational Constraints*

Besides the purely instrumental aspects (used frequencies and power), and the propagation phenomena themselves, other important factors may limit the efficiency of the detection method. This is the case, for example, of the impact of variations in resistivity on the sensitivity, the effects of bathymetry, or even that of the submarine relief.

3.6.1 Sensitivity to Variations in Resistivity: Fréchet Derivative

Sensitivity can be mathematically estimated, for example by Fréchet derivative (MacGillivray and Oldenburg 1990). This allows us to know how a vertical variation of the resistivity $\rho(z)$ of the subsoil will influence the field values at the level λ of the receivers. The higher the value of this function, the more the influence will be considered important. For example for a 1D pattern, the sensitivity function is equal to the following integral:

$$h_k(\lambda, 0) = \int_0^{\infty} \mathbf{G}^*(\lambda, z)\,[\delta(z - z_k) - \delta(z - z_{k\text{-}1})]\frac{\mathrm{d}}{\mathrm{d}z}(\lambda, z)\,\mathrm{d}z \tag{3.30}$$

with as Green functions:

$$\begin{cases} \mathbf{G}^*(\lambda, z)\big|_{z\to\infty} = 0 \\ \mathbf{G}^*(\lambda, z) = -\dfrac{2\pi}{\lambda \mathrm{I}}\,\dfrac{h(\lambda, z)}{\rho(z)} \end{cases} \tag{3.31}$$

With this type of function, the investigation depth with a current source I may be then quantitatively determined with greater or lesser precision.

3.6.2 Bathymetric and Topographic Effects

Shallow water remains today an obstacle to the proper execution of a survey and may in some cases limit the investigations or even make them impossible when the electromagnetic noise is also important (in the case of surveys in coastal areas).

The Bathymetric Effect and Its Consequences

The effect of bathymetry is mainly present in shallow water where the proximity of the surface (air) intervenes in the wave propagation. The latter is somehow in direct competition with the resistant horizon that the deposit represents if it is buried to a depth comparable to the height of the water. At these depths, it is therefore essential to make the acquisition with a minimum offset (cf. Table 3.6) so that the surface waves do not obscure the waves from the hydrocarbon reservoir. The shallow water then makes the method more sensitive to changes in operating parameters, namely:

- The frequency, geometry, offset source and direction of the measurement devices
- The conductivity contrast of the marine soil with the seawater
- More especially the submarine relief (Fig. 3.16)

We will see later in the presentation how to remedy this problem in the context of surveys in shallow water (cf. Sect. 3.13).

Table 3.6 Source/receiver distance (km) depending on the depth and frequency at which the surface waves (*air waves*) are dominant

	Frequencies			
Depth (m)	0.25 Hz	0.5 Hz	1 Hz	2 Hz
500	4.8	4.0	3.4	3.0
600	5.2	4.3	3.9	3.5
700	5.7	4.7	4.3	3.8
800	6.1	5.0	4.6	4.1
900	6.5	5.4	4.9	4.5
1000	6.9	5.8	5.4	4.9
1200	7.6	6.7	6.1	5.6
1400	8.5	7.5	6.8	6.2
1600	9.3	8.3	7.5	7.1
1800	10.1	9.0	8.3	7.8
2000	11.0	9.8	8.9	8.4

According to Eidesmo et al. (2002)

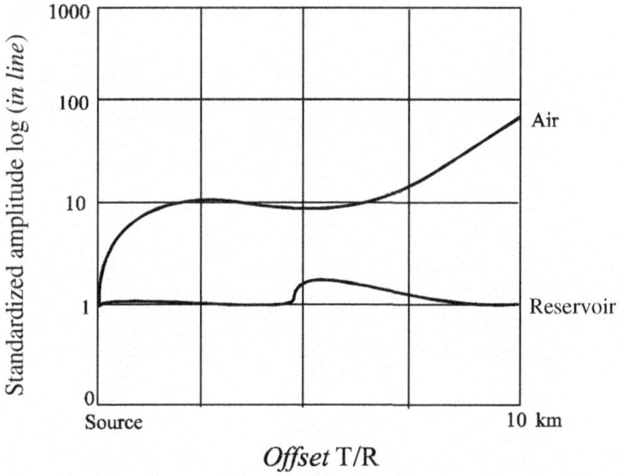

Fig. 3.16 Bathymetric effect in shallow water. The waves from the surface (*air waves*) then prevail

Topographic Effect

We consider here the position of the electromagnetic sensor. If the latter is placed on the seabed at an angle β with the horizontal (*y* axis, for example), the values of the components of the electric field $E_{y,z}$, and magnetic field $H_{y,z}$, in Cartesian coordinates, according to the parallel (//) and perpendicular (⊥) fields to the topographic slope, form the following equation systems:

$$\begin{cases} E_y = E_{//} \cos\beta + E_{\perp} \sin\beta \\ E_z = E_{\perp} \cos\beta + E_{//} \sin\beta \end{cases} \quad (3.32a)$$

and:

$$\begin{cases} H_y = H_{//} \cos\beta + H_{\perp} \sin\beta \\ H_z = H_{\perp} \cos\beta + H_{//} \sin\beta \end{cases} \quad (3.32b)$$

Some authors have determined the effect of this slope on the values of the components $E_{y,z}$ and $H_{y,z}$, in the presence or absence of subsoil anomalies of conductivity, to propose then some corrections.

Without going into too much detail on the 2D modeling (FDM),[20] where the field is divided into primary and secondary fields, calculations have shown that, among other things:

- The distortion of the EM fields generated by variations in seafloor relief also has both galvanic and inductive effects (Jiracek 1990).
- These effects differently affect the electric and magnetic fields in the in-line and broadside configurations, by some 10 % at most, for a slope of ten degrees, and

[20]See Chap. 5.

can be differentiated from an anomalous field of the type hydrocarbon reservoir (Li and Constable 2007).
- The vertical component of the field is less or not at all affected by this topographic phenomenon.

Other authors (Kiyoshi and Nobukazu 2002) have proposed, for example, for the mMT method to include mathematic patches in 3D interpretation models by introducing precise cartographic data.[21]

3.7 *Reflection/Refraction of EM Waves*

As concerning relatively composite, heterogeneous, anisotropic media, the theories based on the laws of reflection/refraction, which occur among other things in geometric optics, may not be accurate. However, they may be applicable if and only if the wavelength λ is lower than the characteristic geometric dimensions l of the posed problem (Landau and Lifchitz 1969).

$$\lambda \ll l \tag{3.33}$$

Related and complex phenomena must be then considered. In particular, it may be interesting to recall some basics, provided that they can be applied in this very special case of propagation. Here, we will give some of them.

If quantitatively the electromagnetic waves suffer all through their journey differentiated attenuations (cf. Sect. 3.11), qualitatively they will also support significant changes.

We prove (Loseth et al. 2006) that, when passing in different media, characterized in particular by their electrical conductivity, electromagnetic waves, considered as plane waves, will suffer, except for deformation, some changes of direction, polarization (rotation) and phase, especially in the presence of a hydrocarbon reservoir, which makes then one of the essential principles of EM detection.

- In the absence of a hydrocarbon reservoir, the propagation will be characterized by reflections and refractions governed by the laws of *Snell–Descartes*, which will be effective only to a certain depth, corresponding at most to the skin depth δ.
- In the presence of a hydrocarbon reservoir, the laws of geometric electromagnetism provide under a certain incidence of the transmitted wave ($\vec{E}_1$, $\vec{H}_1$) a phenomenon of limit refraction driving the energy ($\vec{E}_2$, $\vec{H}_2$) through the deposit, the energy coming out again at its periphery by the opposite phenomenon (resisting medium $\rightarrow$ conductive medium). The fields of the refracted wave then undergo polarizations and well-known changes of phase, which may be detected, captured,

[21]See Chap. 4.

and stored. These are different from those that could be detected without the presence of the reservoir.

We would like to remark that these geometric properties of waves can be particularly utilized in the context of surveys performed in the temporal domain (*transient electromagnetic technics*), where the differentiated propagation times are then used as in seismic prospection (see Chap. 2, Sect. 3.1.3).

At the interface between two homogeneous media with different electromagnetic properties, the tangential components of the electric $\vec{E}_{1,2}$ and magnetic fields $\vec{H}_{1,2}$ check for a plane wave (Stratton 1961):

$$\vec{n} \wedge \left(\vec{E}_2 - \vec{E}_1\right) = 0 \quad \text{and} \quad \vec{n} \wedge \left(\vec{H}_2 - \vec{H}_1\right) = \vec{K} \tag{3.34}$$

where $\vec{K}$ corresponds to the surface current, which is zero when the conductivities are finite, and $\vec{n}$ corresponds to the normal vector to the surface of separation (cf. Fig. 3.17).

When the conductivity of the media is finite, the tangential components are continuous. In the domain defined by the skin depth δ, the rays follow the law of *Snell–Descartes*, such that, as a function of the respective wave numbers $k_{1,2}$, and angles (see Fig. 3.17), we have:

- On reflection:

$$\sin\theta_1 = \sin\theta_1' \tag{3.35a}$$

- In refraction:

$$k_1 \sin\theta_1 = k_2 \sin\theta_2 \tag{3.35b}$$

Considering the incident field vectors, reflected and transmitted, we have:

$$\begin{aligned} \vec{n} \wedge \left(\vec{E}_1 + \vec{E}_1'\right) &= \vec{n} \wedge \vec{E}_2 \\ \vec{n} \wedge \left(\vec{H}_1 + \vec{H}_1'\right) &= \vec{n} \wedge \vec{H}_2 \end{aligned} \tag{3.36}$$

or by using the coefficients of reflection/refraction involving the electromagnetic properties μ and ε of the media (Born and Wolf 1964):

$$\begin{aligned} \vec{E}_1' &= \frac{\mu_2 k_1 \cos\theta_1 - \mu_1 k_2 \cos\theta_2}{\mu_2 k_1 \cos\theta_1 + \mu_1 k_2 \cos\theta_2} \vec{E}_1 \\ \vec{E}_2 &= \frac{2\mu_2 k_1 \cos\theta_1}{\mu_2 k_1 \cos\theta_1 + \mu_1 k_2 \cos\theta_2} \vec{E}_1 \end{aligned} \tag{3.37}$$

and:

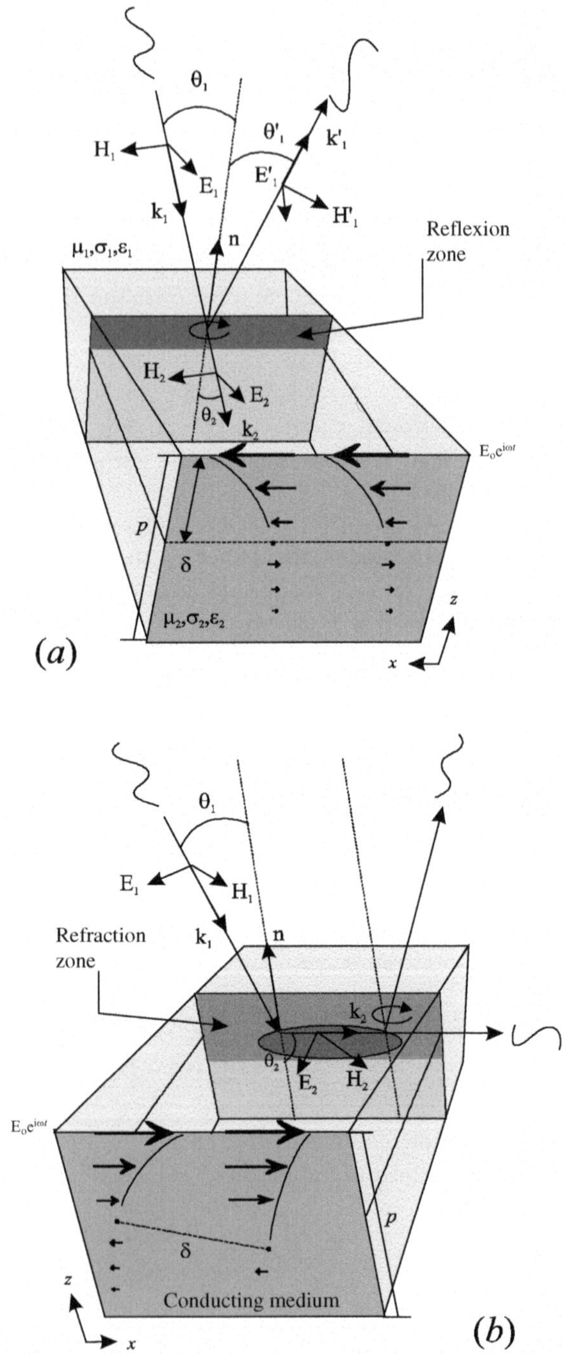

Fig. 3.17 Block diagrams: topological changes in electric $\vec{E}_1$ and magnetic $\vec{H}_1$ field incident vectors in TEM mode (wave vector $\vec{k}_1$). (*a*) In the absence of a deposit, the refracted waves will gradually reduce (significant *skin effect*) and get lost in the subsoil, and the reflected waves will be returned to seawater. (*b*) In the presence of a deposit, the refracted waves will somehow funnel in the latter (*wave guide*) just to come out at its periphery (inverse refraction) and decompose at the interfaces in TE and TM mode

$$\begin{aligned}\vec{H}_1' &= \frac{\tilde{\varepsilon}_2 k_1 \cos\theta_1 - \tilde{\varepsilon}_1 k_2 \cos\theta_2}{\tilde{\varepsilon}_2 k_1 \cos\theta_1 + \tilde{\varepsilon}_1 k_2 \cos\theta_2} \vec{H}_1 \\ \vec{H}_2 &= \frac{2\tilde{\varepsilon}_2 k_1 \cos\theta_1}{\tilde{\varepsilon}_2 k_1 \cos\theta_1 + \tilde{\varepsilon}_1 k_2 \cos\theta_2} \vec{H}_1\end{aligned} \tag{3.38}$$

formulas that can be reduced when the magnetic permeabilities $\mu_{1,2}$ are equivalent ($\approx 4\pi.10^{-7}$) and that give the polarization transverse modes TE and TM of the electric and magnetic fields, allowing us then to define the most suitable methods for the detection (see Chap. 2, Sect. 2.7.4.3).

3.8 Refraction Conditions: Wave Lengths and Frequencies

Some authors (Ellingsrud et al. 2004) have given the limit refraction conditions in the subsoil. These depend on the depth of burial of the deposit p_R and the wavelength λ of the wave propagating in the sedimentary formation as may be defined from k (cf. Eq. 3.16, Table 3.7).

With:

$$\lambda_{[m]} = \frac{2\pi\sqrt{2}}{\sqrt{\omega\mu\sigma}} \tag{3.39}$$

and to ensure proper detection, the investigation process must follow:

$$0.1 p_R \leq \lambda \leq 5 p_R \tag{3.40a}$$

The choice of frequency must also accord limiting factors such as the skin depth and the distance L or offset between the transmitter and receiver, as also checking:

$$0.5\lambda \leq L \leq 10\lambda \tag{3.40b}$$

Table 3.7 Frequencies and wavelengths λ in the sediments (3200 $m{\cdot}s^{-1}$, cf. Sect. 4.1) for different burial depths

Depths (m)	Minimal λ (m)	Maximal λ (m)	Minimal frequency (Hz)	Maximal frequency (Hz)
250	25	1250	0.008	0.4
500	50	2500	0.01	0.8
1000	100	5000	0.03	1.6
2000	200	10,000	0.06	3.1
3000	300	15,000	0.1	4.7

The working frequency will therefore preferably be between 0.1 and 3 Hz. At these frequencies, the depths of investigation are respectively 5000 m and 900 m (cf. Eq. 3.24).

For example, for an excitation at 1 Hz, a commonly used frequency, the offset shall not be less than 1.6 km ($0.5 \times 1 \times 3200$) and theoretically not exceed 32 km ($10 \times 1 \times 3200$), the distance to which attenuation is then very important. Moreover, at these distances, the length of the antenna ($\approx$200 m) can be neglected, and can therefore be assimilated into a simple electromagnetic dipole, then to a point source. To a certain extent this approximation is even more accurate than the considered distance is important.

3.9 Propagation of EM Waves in the Hydrocarbon Reservoir

At low frequencies, the reservoir acts as an imperfect dielectric material ($\sigma \neq 0$ with $\sigma/\omega\varepsilon \ll 1$). The permittivity ε is then complex and depends on the frequency of excitation (ω). The electromagnetic waves penetrate into it, regardless of the frequency used and all the more easily if it is high.

3.10 The Hydrocarbon Reservoir: A Resistive Wave Guide

Under certain conditions (including wave length), the reservoir may behave toward electromagnetic waves as a wave guide allowing the latter to propagate at more or less great distances along these longitudinal boundaries (channeling effect). Under certain limits (diffusive media) mCSEM can be assimiled to refraction seismic technics.

If, for example, we are interested in refracted waves entering the reservoir and agreeing with the theory of guided waves in plates (Kong 1986), which takes into account the dimensions of the wave guide, width (l) and thickness (t_h), the wave number k_z in the z direction, depending on the propagation factor k (cf. Eq. 3.15), is written:

$$k_z = \sqrt{k^2 - \left(\frac{m\,\pi}{l}\right)^2 - \left(\frac{n\,\pi}{t_h}\right)^2} \tag{3.41}$$

where m and n are the integration parameters.

For the same mode (that is to say, for $m = 1$ and $n = 0$) the previous expression is consequently reduced to:

$$k_z = \sqrt{k^2 - \left(\frac{\pi}{l}\right)^2} \tag{3.42}$$

In all cases k_z is smaller than k. When l is largest that the wavelength in the reservoir, we have then: $k_z \approx k$. In these unique conditions, the reservoir can be considered as a high pass filter whose cutoff frequency is then dependent on the dimensional parameters, especially l.

When considering that the dimensions of the reservoir are finite in space (very good resistivity contrasts), it can be assumed that the electromagnetic waves are within a volume that can be proportionally compared to a resonator (Jackson 1998). In this case, the multiple receptions could generate a stationary wave system[22] (resonance frequency), which would somehow have the advantage to amplify the signal or more precisely to minimize energetic losses and thus provide in the same way additional information on the geometric characteristics of the reservoir (in relation to the frequencies) and in particular on its thickness.

3.11 *Attenuation of the Electrical Fields, of the Direct, Reflected and Refracted Waves*

In fact, the skin depth has no real physical significance because the propagation does not stop abruptly. In reality, the magnitude of the field E (it is the same for H) decreases with the penetration of the lines of force such that for a value of the initial field E_0, as a function of the skin depth δ, we have at depth z:

$$E(z) = E_0 \, e^{-\frac{z}{\delta}} \, e^{i\left(\frac{z}{\delta} - \omega t\right)} \tag{3.43}$$

where the real and imaginary parts respectively indicate an attenuation (frequency effect), and an increase of the phase difference (time effect).

At depth z, the attenuation α may be then commonly expressed in decibels per meter (dB/m) such that:

$$\alpha_{[dB/m]} = 20 \log\left(\frac{E}{E_0}\right) = 20 \log e^{-\frac{z}{\delta}} = -\left(20 \frac{z}{\delta}\right) \log e = -8.685 \frac{z}{\delta} \tag{3.44}$$

hence, at depth δ ($z = \delta$), a specific attenuation of –8.685 dB.

For example, in seawater, more precisely, according to its intrinsic properties (σ_w, ε_w and μ_0), the attenuation as a function of the angular frequency ω is given by the formula (Meyer et al. 1969)[23]:

---

[22]Phenomenon not yet exploited. In seismic exploration, in the 1960s, a similar technique was proposed to directly assess the thicknesses of the sedimentary layers.

[23]See also Kraichman (1976).

$$\alpha_{[\mathrm{dB/m}]} = 8.685\omega\sqrt{\frac{\mu_0\varepsilon_w}{2}\left[\sqrt{1+\left(\frac{\sigma_w}{\omega\varepsilon_w}\right)^2}-1\right]} \quad (3.45)$$

In marine sediments the attenuation is two times lower (lower conductivity), and in the reservoir it finally becomes negligible (cf. Fig. 3.18, Table 3.8).

When the source moves near the bottom, and the water depth is great enough, we prove (Bannister 1987)[24] that the electromagnetic field, on the assumption that the wavelength is less than the radial distance r, is divided into three types (cf. Fig. 3.20a):

- The direct waves (corresponding to the real source)
- The waves reflected by the interface (corresponding to the image source)
- Waves refracted along an interface (sea/air or sea/sediments/deposit)

Their respective attenuations follow, depending on the depth of immersion h_s of the source, exponential decreasing laws whose shape in polar coordinates[25] can be summarized as:

- For direct waves (1):[26]

$$e^{-kR_0} \quad \text{with} \quad R_0 = \sqrt{r^2 + (z - h_s)^2} \quad (3.46)$$

- For reflected waves (2):

$$e^{-kR_{-1}} \quad \text{with} \quad R_{-1} = \sqrt{r^2 + (z + h_s)^2} \quad (3.47)$$

- For refracted waves (3, 4, 5):

$$e^{-k(z+h_s)} \quad (3.48)$$

with $k = \sqrt{\omega^2\mu_0\varepsilon - i\omega\mu_0\sigma}$ representing the propagation factor (see Sect. 3.3.1) dependent on the angular frequency ω and on the electromagnetic properties μ_0, ε, σ of the crossed media.

In contrast to direct and reflected waves, which attenuate very quickly (inverse to the squared distance), the attenuation of the refracted wave does not depend on r. It therefore propagates further than the other contributions. This more distant propagation is the base of the EM submarine surveys where (guided) interface waves then decrease less quickly than volume waves (spherical or geometric divergence).

[24]Theory of images modified in a infinite conducting half space.

[25]Polar coordinates (r, θ) reported in Cartesian coordinates.

[26]For dialing (1, 2, 3, 4, 5) see diagram (see Fig. 3.19*a*).

Fig. 3.18 Attenuation (dB/m) of electromagnetic waves in seawater as a function of frequency

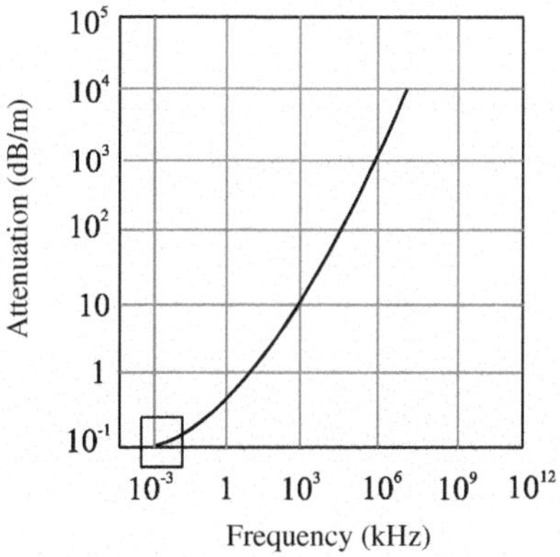

Table 3.8 Summary table of attenuation of EM waves at 1000 m for different involved horizons ($\neq \delta$, cf. Table 3.4), including the reservoir

Media	Attenuations (1 Hz, at 1000 m)
Seawater	−32 dB
Marine sediments	−17 dB
Reservoir	−2 dB

3.12 *Nature of the EM Waves: Upwaves or Downwaves*

At reception, in the presence of a conductivity anomaly (existence of a refraction), the amplitude of the total field E_T received by the receiver, direct waves excepted, is then composed of the amplitudes of two electrical fields (E_{down} and E_{up}) distinctly coming from:

- *Down waves* composed of the direct waves coming from the source (set higher than receptors), and of the refracted and reflected waves coming from the air interface
- *Up waves* composed of reflected and refracted waves coming from the reservoir

such that we finally have:

$$E_T = E_{down} + E_{up} \tag{3.49}$$

In these circumstances, only the fields concerning the up waves E_{up} coming from the reservoir must be taken into consideration. It is therefore necessary to isolate them by some adapted discrimination processes.

Schematically, the differentiation between a trap filled or not by oil is made with the existence or not of *up waves* (Fig. 3.19 and Fig. 3.20*b*).

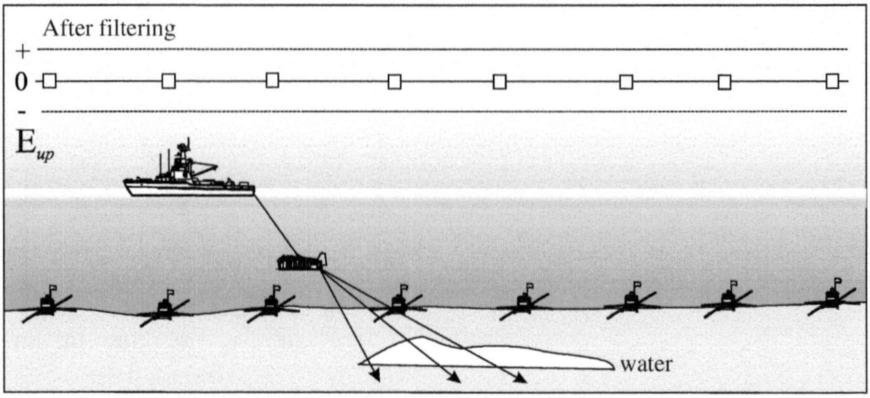

Fig. 3.19 In the presence of a water reservoir, waves and energy are directly transmitted ($\sigma_e \approx \sigma_s$). No wave is refracted and reflected. There is no upgoing wave; the transmission is complete

3.13 Separation of the Up and Down EM Waves

For a given receiver, placed on the bottom of the sea, the waves refracted by the reservoir correspond to *up waves* (*up*), the only interesting thing; *down waves* (*down* or *d*) coming from other places must then be eliminated (cf. Fig. 3.20). It is therefore essential to do a discrimination in order to isolate the up waves. Fortunately, these latter have a certain number of more or less specific identifiable features, distinguishable from those of down waves such as:

- Propagation at different speeds (cf. Sect. 4.1)
- Stronger attenuation for the direct and reflected waves (cf. Sect. 3.11)
- A quasilinear offset phase shift (cf. Fig. 3.13)
- Specific polarization (cf. Fig. 3.17)

Theoretically, these effects can be derived from Maxwell's equations and some authors (Admundsen et al. 2006)[27] offer from these latter some processing algorithms using measurements of the horizontal components of the electric fields (E_1, E_2) and magnetic fields (H_1, H_2). This allows us to obtain in favorable cases the values of the up fields (E^{up}) such that, combining these fields, we have:

$$\begin{cases} E_1^{up} = \dfrac{1}{2}\left(E_1 - \dfrac{1}{\tilde{c}\,\tilde{\varepsilon}} H_2\right) \\ E_2^{up} = \dfrac{1}{2}\left(E_2 - \dfrac{1}{\tilde{c}\,\tilde{\varepsilon}} H_1\right) \end{cases} \tag{3.50}$$

[27] An overview of this process is presented in the thesis of L. Loseth (2007).

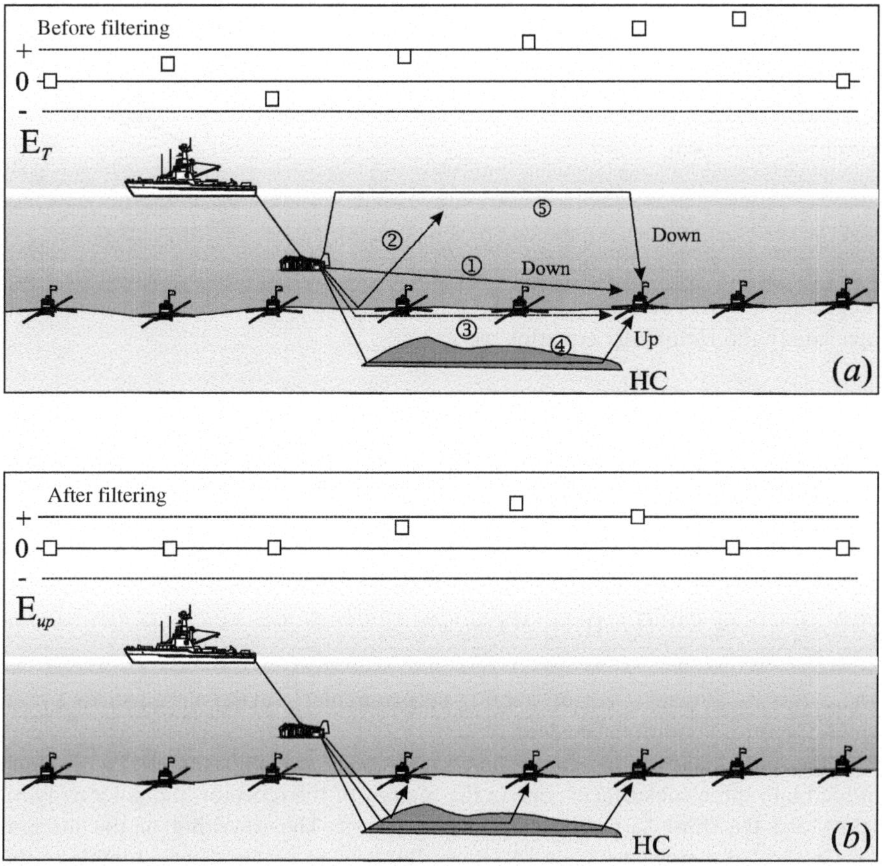

Fig. 3.20 Records of the variations of the electric field before (*a*) and after filtering *up/down* waves (*b*). At an oblique incidence (*critical angle*), the waves transmitted in the reservoir are refracted. The "useful" waves correspond to *upgoing* phenomena whereas the waves from the *downgoing* phenomena are undesirable. The direct and reflected waves are more attenuated than the refracted waves (presence of eddy currents). At low offsets (<2 km), direct waves (*1*) predominate. Beyond this distance (>2 km), the dominant contribution is due to the surface waves (*5*) and to those from the reservoir (*4*). In reality, there are many reflections on the different interfaces (not shown)

where $\tilde{c}$ and $\tilde{\varepsilon}$ are respectively the complex speed ($\tilde{c} = \sqrt{\omega / i\mu_0\sigma}$) and the complex permittivity, which, for a diffusive field ($\omega \ll \sigma/\varepsilon$), is equal to $i\sigma/\varepsilon$

Under these conditions, the dielectric permittivity ε no longer depends on the electric conductivity σ (Fig. 3.21).

Other authors, this time from the three components of the electric and magnetic fields, use convolution filters (Rosten and Admundsen 2008). These are obtained by

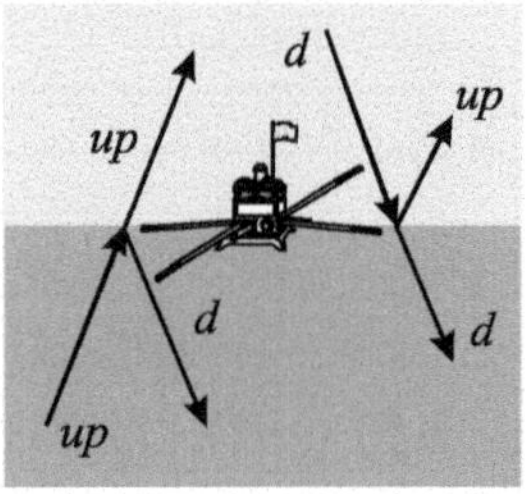

Fig. 3.21 Decomposition of electromagnetic waves in upgoing waves (*up*) and downgoing waves (*d*) at ground level (According to Admundsen et al. 2006)

integrating the Maxwell equations by Green's functions[28] (scalar or tensor), which then satisfy the Helmholtz equation as:

$$\left(\nabla^2 + k^2\right) \mathbf{G}(x, x', \omega) = -\boldsymbol{\delta}\,(x - x') \tag{3.51}$$

For example, at the point with coordinate x', different from x, the electric field is given by the integral of the surface such that:

$$\begin{aligned} \mathrm{E}(x') = \; & \oint_S \mathrm{d}S(x) \left[\left[\vec{\Pi} \wedge \vec{\mathrm{E}}(x)\right] \wedge \vec{\nabla}\mathbf{G}(x, x') + \left[\vec{\Pi}.\vec{\mathrm{E}}(x)\right] \vec{\nabla}\mathbf{G}(x, x') \right. \\ & -\zeta(x) \left[\vec{\Pi} \wedge \vec{\mathrm{H}}(x)\right] \vec{\nabla}\mathbf{G}(x, x')] \end{aligned} \tag{3.52}$$

where $\vec{\Pi}$ is the Poynting vector[29] acting perpendicularly to the closed surface S on which the limit conditions are laid.

In the spatial domain, to obtain a discrimination of the fields, the surface S is then divided into three subsurfaces, one in the plan S_1 of the receptor, the other S_2 lying below and the third S_r normal to the two others. The rewriting of the integral (cf. Eq. 3.52) for surfaces S_1 and S_2 thus allows us to obtain filters suitable to the selection of *down waves* (*d*) for S_1 and *up waves* (*up*) for S_2.

Practically, in x' the field corresponding to *up waves* with regard to the reservoir is obtained by subtracting E(x') as calculated above (cf. Eq. 3.52) from the actually measured field E_T such as:

[28] We call Green's function, denoted **G**, the elementary solution of a linear differential equation or a partial derivative equation with constant coefficients. In electromagnetism, the solution is obtained using a single source (pulse or Dirac delta or $\boldsymbol{\delta}$). The general solution corresponding to the actual source is then equivalent to the superposition of impulse responses, that is to say, corresponding to the Green functions. These functions may take varied forms as, for example, analytic functions when the solution of the homogeneous differential equation is known, or an infinite series of orthogonal functions then satisfying the boundary conditions when the solution of the equation is unknown.

[29] Vector that indicates the direction and the sense of propagation of an electromagnetic wave. The modulus of the Poynting vector ($\mathrm{P} \approx \mathrm{E} \wedge \mathrm{H}$) corresponds to a flux, power per area unit (Skilling 1942).

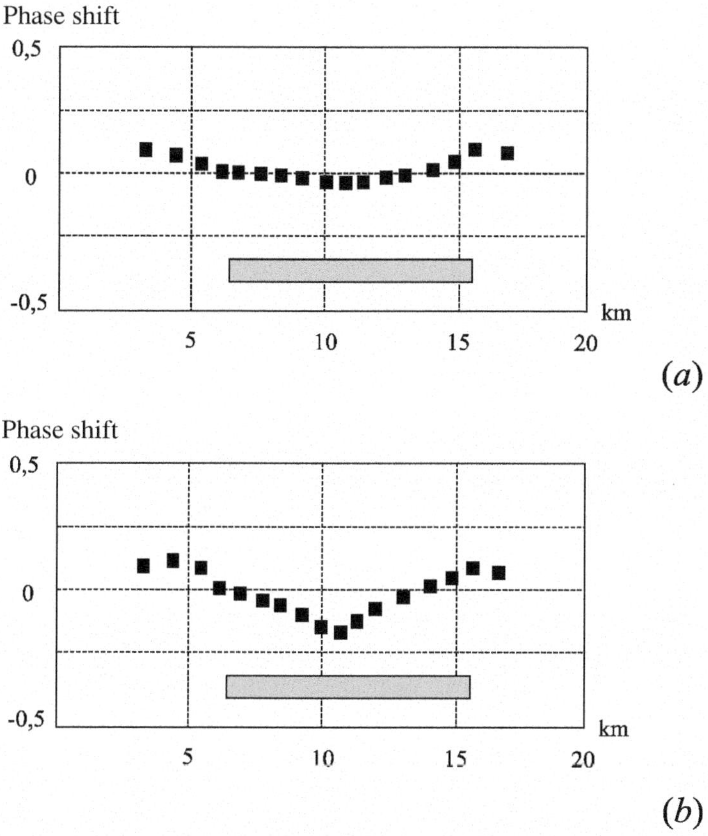

Fig. 3.22 Example of records of the phase shift over a conductivity anomaly, before (*a*) and after filtering (*b*) the *up waves* coming from the reservoir (According to Rosten et al. 2005)

$$E^{up}(x') = E_T(x') - E(x') \tag{3.53}$$

The convolution filters are then assigned to each measuring channel ($\times 2$ or $\times 3$).

The separation of the waves is particularly important in the case of prospecting in shallow water because the waves coming from the interface between the air/sea first arrive on receivers. In this case it is also preferable to work on the phase measurements rather than on those of the amplitudes (Rosten et al. 2005), or to work in the time domain (Weiss 2007; Li and Constable 2010) (Fig. 3.22).

Finally, other researchers (Tompkins et al. 2004), to locate the surface waves (*air waves*), propose to establish the ratio of the wave numbers (see Eqs. 3.15, 3.16, 3.17, 3.18, 3.19, 3.20, and 3.21) in different media such as:

$$\frac{\mathrm{Re}\,\{k_{\mathrm{water}}\}}{\mathrm{Re}\,\{k_{\mathrm{Sediment}}\}} = \frac{\sqrt{\frac{\mu_0 \sigma_w \omega}{2}}}{\sqrt{\frac{\mu_0 \sigma_s \omega}{2}}} = \sqrt{\frac{\sigma_w}{\sigma_s}} \tag{3.54}$$

which gives, for example, for mean values of conductivity a ratio of 1.73 or $\left(\sqrt{3/1}\right)$. In this case, this means that the energy from the surface may be a priori positioned to a depth greater than 1.73 times that from marine sediments. In other words, if we consider a water depth of 1800 m, it sets the surface waves at about 4900 m, i.e., $1800 + (1.73 \times 1800)$, and its first multiples (order 2) at approximately 8000 m, i.e., $1800 + (1.73 \times 1800 \times 2)$. For a value this time of 1.76 (water thickness of 1000 m) the maximum depth of investigation will then be 1760 m.

Many efforts are currently being made to treat and eliminate down waves in the treatment process (Chen and Alembaugh 2009), especially in the case of surveys conducted in shallow water (removal of *air waves*).

3.14 Phase Shift of the EM Waves and Estimation of the Depth of the Reservoir

If the amplitude measurements, taken separately, allow us to assess the lateral position of the conductivity anomaly, they cannot provide in any event sufficiently reliable knowledge about the depth of the deposit. Nevertheless, it is possible in some cases, with additional measurements of signals, in particular those concerning the phases, to obtain this information, which, in the context of the interpretation (solving the inverse problem in particular), will then be very useful.

For example, for a sine wave in $\sin(\omega t)$ (where $\omega = 2\pi/T$ is expressed in rad/s and T in s is the period), the phase difference $\Delta\phi$ (*phase shift*) between the transmitted signal and the received signal for a period of time Δt (*time shift*) is equal to:

$$\Delta\phi_{[\text{rad}]} = 2\pi\frac{\Delta t}{T} \tag{3.55a}$$

Furthermore, the additional field E_S (cf. Fig. 3.23) is affected on its horizontal components (x, y), relative to the signal from the source E_P, by:

- An *offset* phase shift ϕ_0 due to the distance separating the transmitter and receiver
- An anomalous phase shift ϕ_a concomitant to the presence of a resistant horizon (the hydrocarbon target in this case)

such that finally:

$$\Delta\phi = \phi_0 + \phi_a \tag{3.55b}$$

Whatever the position of the measurement $\Delta\phi$, ϕ_a remains constant. This allows us, by the evaluation of the latter, to locate more or less precisely the heterogeneity. The greater ϕ_a is, the more remote the anomaly is from the surface. ϕ_a is then simply calculated by subtracting the *offset* phase shift ϕ_0, which is itself obtained by

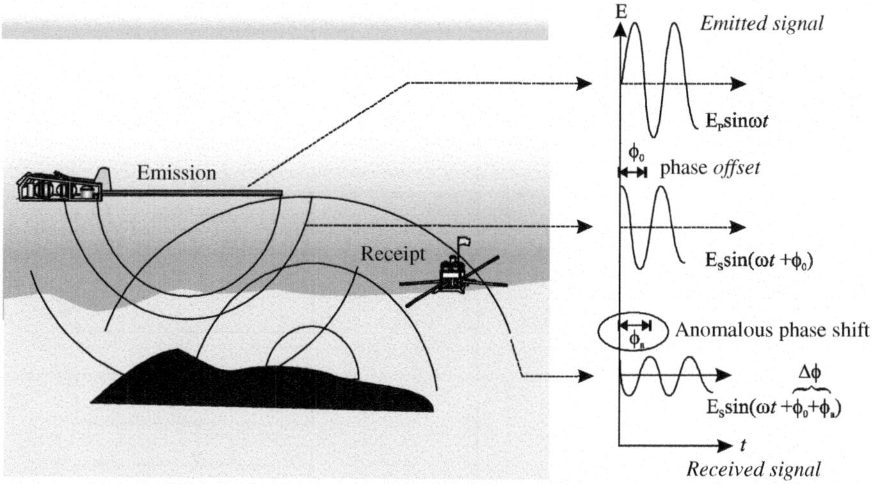

Fig. 3.23 In the presence of a conductivity anomaly, the diffracted field has a phase lag $\Delta\phi$ compared to the emitted field. This anomalous phase shift ϕ_a must also be subtracted from the phase offset ϕ_0 due to the remoteness of the source relative to the receiver. It is the same for the other used waveforms (*rectangular*, *triangular*, etc.)

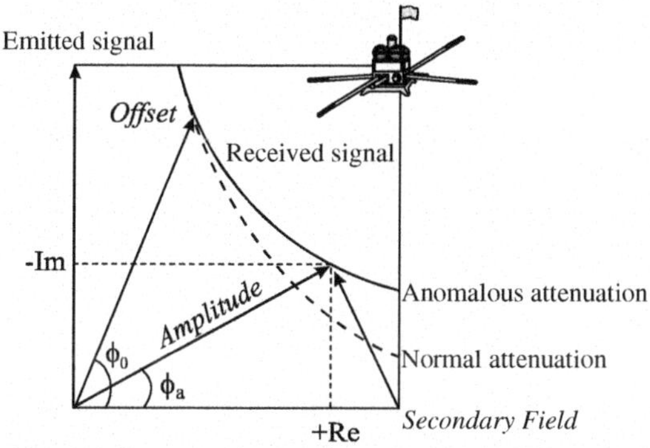

Fig. 3.24 Vector relations between the primary, secondary and resulting fields (the secondary field opposes the primary field) and graphic representations of the amplitude and phase of the reception signals, in polar coordinates (A, $\phi_{0,a}$) and in the complex plane (+Re, −Im). From these two representations, it is then possible to establish the simple trigonometric relations (cf. Sect. 5.4) connecting angles and real part + Re (in phase signal) and imaginary part −Im (in quadrature signal)

calculation (see Sect. 3.4.2) knowing the *offset of source* or by measuring the waves directly coming from the transmitter (Figs. 3.20, 3.23 and 3.24).

To increase the detection accuracy, we can also theoretically use the Helmholtz equation:

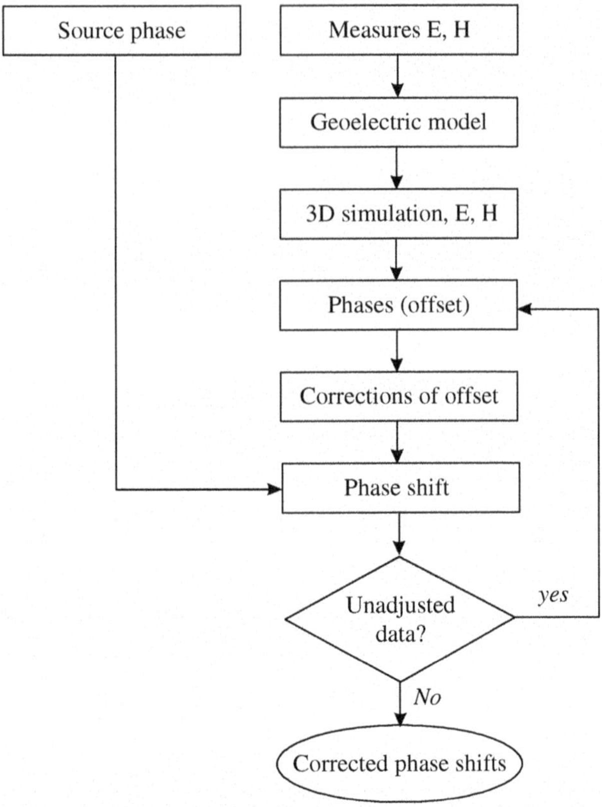

Fig. 3.25 Example of offset correction algorithm for the evaluation and calculation of the phase shift between the transmitter and receivers

$$\nabla^2 \vec{E} = \frac{\omega^2}{c^2} \vec{E} + i\omega\mu\sigma\vec{E} \tag{3.56}$$

It is then possible by solving it (cf. Appendix A2.4) to assess the depth of the deposit. At this level, it is necessary to make precise adjustments to take into account the *offset phase*. Several algorithms fulfilling this function are described in the technical literature. Some of them use methods of errors assessment (Pavlov et al. 2009). On the other hand, to get the (horizontal and vertical) position, as in seismic methods, we also practice migration techniques (Fig. 3.25).

3.15 Amplitude Normalization of the EM Fields

To view more precisely the anomalous field values caused by the presence of the hydrocarbon reservoir, an amplitude normalization is performed for this purpose on a scale of 1 to 4 (convention). This curve (C) is obtained by dividing the field

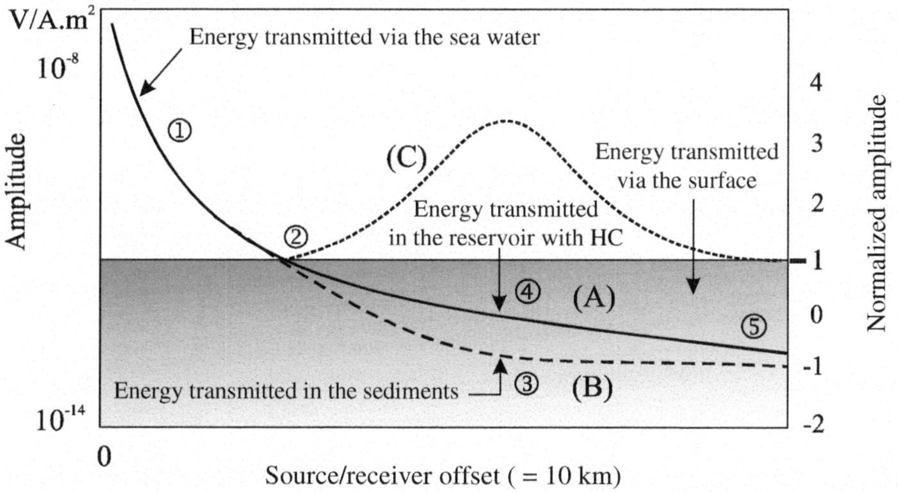

Fig. 3.26 Diagram of the evolution of the electric field (amplitude in V/A.m^2 and unitless normalized amplitude) as a function of the distance T/R (offset) for a given receiver. The long *dashed curve* (*bottom*) corresponds to a response without hydrocarbons (*left scale*). The bell curve with short dots (*top*) corresponds to the normalized response (*right scale*). The numbers ① and following refer to the previous figure (cf. Fig. 3.20)

measurements (curve A) by the values of the so-said "oil free" (OF) fields (curve B) (Fig. 3.26).

$$(A)_{HC} \div (B)_{OF} = (C)_{Nor} \tag{3.57}$$

For investigations in shallow water (40 m or less), this relatively simple technique cannot be effective, because of the faster surface waves that arrive first on the sensors. To overcome this major drawback, a decomposition (Admundsen et al. 2006) followed by a separation of up and down waves are required in advance (Mittet 2008), or work in the time domain.

4 Temporal Aspect of the Method

Another aspect of the principle of the method is that applied this time in the time domain. In itself, this principle is not commonly used at the moment. It accompanies certain acquisition techniques, in particular those concerning operations in shallow water.

In the time domain the analysis of the signals can be enhanced by very powerful digital processing means such as the use of detection windows for example. In a heterogeneous medium of propagation, formed from successive layers (air, seawater, sediment, reservoir) with very different electromagnetic properties, the radiated

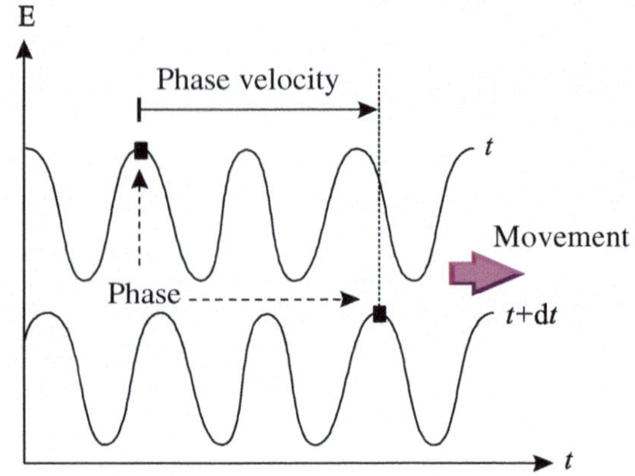

Fig. 3.27 The velocity of propagation of an electromagnetic wave corresponds to the speed of movement of the phase

energy will propagate in space at different speeds. This will have the effect, depending on the route taken by the electromagnetic waves, of a delayed arrival of the different amounts of energy at the receptors. It will be then even faster when the speed of propagation is high and the journey is short.

4.1 *Speed Propagation of the EM Waves in the Different Media*

An electromagnetic field in a harmonic regime can never be completely localized in either space or time. We must therefore admit an arbitrary definition of its speed (Stratton 1961). This is generally defined as the rate at which the wave phase propagates, provided that the field is periodic in space, that it is not deformed and that the wave train has an unlimited duration (Fig. 3.27).

4.1.1 Speed in Conductive Media

For electrically conductive media such as seawater and marine sediments (dispersive media), the speed is affected by the magnetic permeability μ_0 (cf. Sect. 2.2) and especially by the electrical conductivity σ.

If the state of the medium is represented by the function ψ (z, t), the surfaces of the constant phase (plane wave) as a function of the pulsation ω and the wave number k are defined by:

$$kz - \omega t = \text{Cte} \tag{3.58}$$

These surfaces then propagate up to speed:

$$c = \frac{dz}{dt} = \frac{\omega}{k} \tag{3.59}$$

By replacing the value k with $1/\delta$ (cf. Eq. 3.21), we finally obtain the expression of the phase velocity:

$$c = \frac{\omega}{\frac{1}{\sqrt{\frac{2}{\omega\mu_0\sigma}}}} = \sqrt{\frac{2\omega}{\mu_0\sigma}} \tag{3.60}$$

For information, the group speed $d\omega/dk$ distinct from a wave train limited in space is here twice the phase velocity, in this case $2c$. It then represents the speed of transmission of energy through the medium.

4.1.2 Speed in Dielectric Media

For dielectric media such as air and imperfect dielectrics such as rocks containing hydrocarbons, the speed is always affected by the magnetic permeability μ_0, less or not at all by the electrical conductivity. However, roughly, it no longer depends on the frequency, but on the dielectric permittivity ε such that:

$$c' = \frac{1}{\sqrt{\mu_0\varepsilon}} \tag{3.61}$$

4.1.3 In Short

Practically, it is difficult to obtain these speeds precisely, especially in sediments because the variation of the electromagnetic in situ parameters can be very large and dependent on the depth of burial. It may, however, be possible, thanks to the above formulas, to give an estimation (cf. Table 3.9). Theoretical calculations also show that:

- In conducting media (seawater, marine sediments), the speed of propagation of electromagnetic waves is even faster than the electrical resistivity is high ($c_S > c_w$).
- In more or less resistant media (air, oil), the speed of propagation of electromagnetic waves is even faster ($c_{air} > c_{HC}$) than the dielectric permittivity is low.
- Electromagnetic waves propagate with greater velocity in the air and in the reservoir (rather resistant media) than in seawater and sediments (rather conductive media).

Table 3.9 Average speeds of EM wave propagation in SBL (ULF) at the frequencies used and in the different concerned media

Media	Propagation speed ($m \cdot s^{-1}$)
Air[a]	3.10^8
Seawater[b]	1600
Marine sediments	3200
Reservoir[c]	22,000

[a]Speed of light in a vacuum
[b]In seawater and some sediments the propagation velocities of the electromagnetic waves at these frequencies are substantially equal to those of the sound in these components
[c]Speed, which can be very different from one reservoir to another

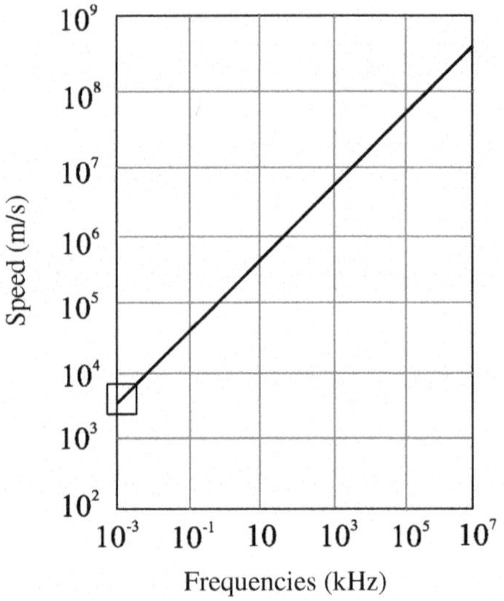

Fig. 3.28 Variation of the propagation speed of EM waves (VHF-ULF) in seawater as a function of frequency (dispersive medium)

– In a wave guide (reservoir) the phase velocity is different from that of a plane wave (infinite medium) (Fig. 3.28).

This has the immediate consequence that the refracted waves in the presence of a hydrocarbon reservoir, on condition that the water depth is sufficient, arrive first at the receivers (cf. Fig. 3.29).

4.2 *Time Courses of EM Waves According to the Distances*

Depending on travel D and propagation speeds c in the media, the different waves will arrive against each other in more or less delayed times.

It is obvious that under these conditions, suitable signal processing in the time domain has an advantage in discriminating the different signals and more

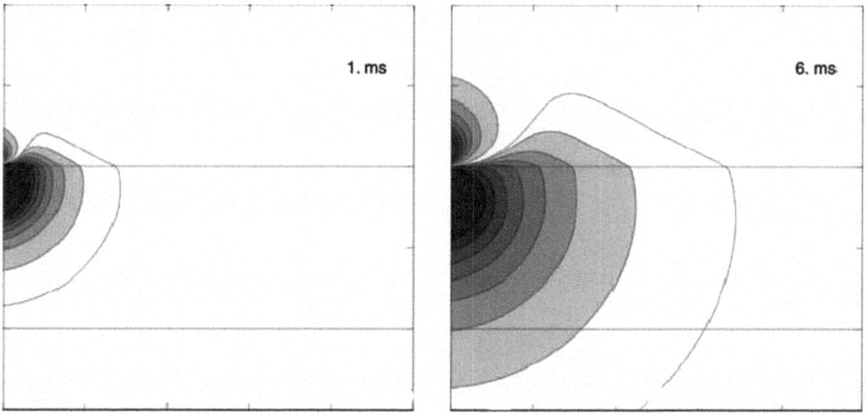

Fig. 3.29 Diffusion model of energy (1–6 ms), which spreads faster in the marine subsoil (0.5 S/m) than in seawater (3 S/m) (According to Edwards 2005)

Table 3.10 Relative magnitude of the travel time of electromagnetic waves in different natural environments according to travel *D*

Media (stimulation at 1 Hz)	Propagation time (*D*/*c*)
Seawater (direct wave)	1 s
Marine sediments (direct wave)	0.5 s
Reservoir (refracted wave)	0.1 s

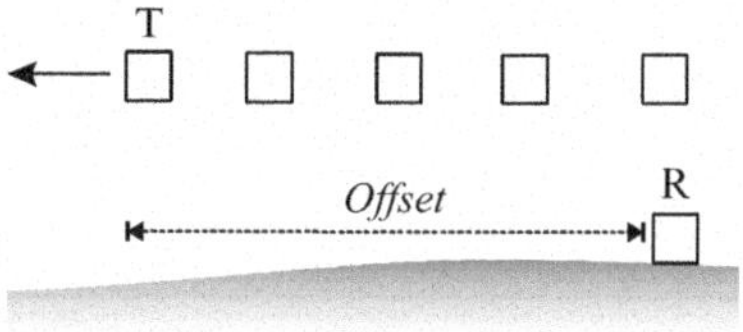

Fig. 3.30 Variation of the propagation time as a function of offset changes due to the movement of the source T (R fixed receiver)

particularly those coming from the reservoir and those arriving from the surface (*air wave*) through the seawater (see Table 3.10).

4.3 Propagation Time Depending on the Offset

The distance between the transmitter and receiver (called the *offset*) varies because during a survey the source is mobile and the sensor is stationary. The relative speed between the two devices is that of the transmitter, i.e., that of the boat towing it at the surface. In a homogeneous medium, the propagation times increase with the *offset*. In a heterogeneous medium, with gradual changes in conductivity even with discontinuities, it is quite different because the waves then propagate at different speeds (Fig. 3.30).

In reception, it is therefore necessary to correct the measurements of the effect of the *offset* to finally keep the useful signals only, i.e., those involving only the propagation.

When the signals can be specifically described in the time domain (*pulse*, *sweep*, for example) then this correction can be easily performed by synchronized clocks present in the instruments (T/R).

5 Theoretical Evaluation and Measurement of EM Fields

The definition of the evaluation devices and their metrological characteristics must fit jointly with the orders of magnitude of the signals that may be collected and the ambient noise (background noise). This determines the power to be injected (transmitter) according to the received powers (receptors). This power ratio then defines the signal-to-noise (S/N) ratio. To determine this ratio, it is desirable to a priori assess it theoretically.

The theoretical evaluation of the electric $\vec{\mathrm{E}}$ and magnetic $\vec{\mathrm{B}}$ fields due to a distant source mathematically corresponds to Maxwell's equations being solved in the marine environment.

5.1 Maxwell's Equations: Positioning Problem

If we consider that, for a harmonic stimulation (source) in $\mathrm{e}^{-\mathrm{i}\omega t}$, the displacement currents are negligible compared to the conduction currents, Maxwell's equations are written:

$$\begin{gathered}\vec{\nabla} \wedge \vec{\mathrm{E}} = -\mathrm{i}\omega \vec{\mathrm{B}} \\ \vec{\nabla} \wedge \vec{\mathrm{B}} = \mu\sigma\vec{\mathrm{E}} + \mu\vec{\mathrm{J}}^{\mathrm{S}}\end{gathered} \tag{3.62}$$

where the current density $\vec{\mathrm{J}}^{\mathrm{S}} = \vec{\mathrm{I}}\boldsymbol{\delta}\,(r - r_0)$ is required by the source distant of r_0.

Using the potential magnetic vector[30] $\vec{\mathrm{A}}$, the magnetic field is at once written:

$$\vec{\mathrm{B}} = \vec{\nabla} \wedge \vec{\mathrm{A}}$$

The electric field, from the preceding equations, becomes equal to:

$$\vec{\mathrm{E}} = \mathrm{i}\omega\vec{\mathrm{A}} + \frac{1}{\mu\sigma}\vec{\nabla}(\vec{\nabla}.\,\vec{\mathrm{A}}) \tag{3.63}$$

[30]The potential vector is a mathematical tool that allows us, by introducing additional functions, to simplify the calculation procedures for the evaluation of magnetic and electric fields. For example, the fields are calculated from the potentials (specified sources), solutions of the Helmholtz equation.

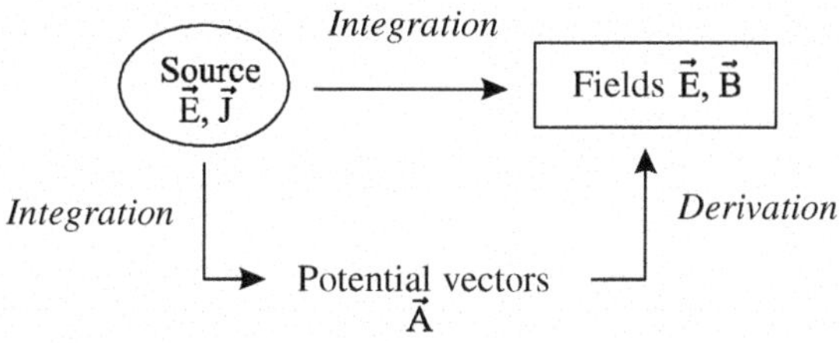

Fig. 3.31 Calculation of the electromagnetic fields $\vec{E}$, $\vec{B}$ created by a remote source can be done either directly by integration or indirectly by derivation using the vector potentials $\vec{A}$

and where the Lorenz gauge[31] has been specified (cf. Fig. 3.31).

5.2 *Calculations of EM Answers in TE and TM Modes*

From the equations developed above (see Chap. 2, Sect. 2.7.4.3), the values of radial electric and magnetic fields (E_r, B_r), azimuthal fields (E_ϕ, B_ϕ) and vertical fields (E_z, B_z) were analytically or more often numerically calculated by some authors (Chave and Cox 1982; Andreis and MacGregor 2008; Chave 2009)[32] depending on their electric transverse components (TE) and magnetic transverse components (TM), in a typical configuration of exploration (a canonic thin HC layer model).

As an indication, in a finite medium of height h (see Chap. 2, Fig. 2.4), of electromagnetic properties ρ_w and μ_0 (seawater), located above a resistant thin structure (HC), the fields corresponding to the model of the figure (cf. Fig. 3.32) are given as a function of the dipole moment P (or $I_a \times L$) by the following relations:

- For the electric field:

$$\begin{aligned} E_r = \frac{P\rho_w}{4\pi}\cos\phi \int_0^\infty \Bigg[& \left(-\frac{\beta_0 k J_0(kr) - J_1(kr)/r}{1 + R_L^{TM} e^{-2\beta_0 h}} + i\omega\mu_0 \frac{J_1(kr)}{r\beta_0\rho_w\left(1 - R_A^{TE} R_L^{TE} e^{-2\beta_0 h}\right)} \right) e^{-\beta_0|z-z'|} \\ & + \left(\frac{\beta_0 k J_0(kr) - J_1(kr)/r}{1 + R_L^{TM} e^{-2\beta_0 h}} R_L^{TM} + i\omega\mu_0 \frac{J_1(kr)}{r\beta_0\rho_w\left(1 - R_A^{TE} R_L^{TE} e^{-2\beta_0 h}\right)} R_L^{TE} \right) e^{-\beta_0|z+z'|} \\ & + \left(-\frac{\beta_0 k J_0(kr) - J_1(kr)/r}{1 + R_L^{TM} e^{-2\beta_0 h}} + i\omega\mu_0 \frac{J_1(kr)}{r\beta_0\rho_w\left(1 - R_A^{TE} R_L^{TE} e^{-2\beta_0 h}\right)} R_A^{TE} \right) e^{\beta_0|z+z'-2h|} \\ & + \left(\frac{\beta_0 k J_0(kr) - J_1(kr)/r}{1 + R_L^{TM} e^{-2\beta_0 h}} R_L^{TM} + i\omega\mu_0 \frac{J_1(kr)}{r\beta_0\rho_w\left(1 - R_A^{TE} R_L^{TE} e^{-2\beta_0 h}\right)} R_A^{TE} R_L^{TE} \right) e^{\beta_0(|z-z'|-2h)} \Bigg] \times dk \end{aligned} \tag{3.64}$$

[31]The Lorenz gauge decouples differential equations on the vector and scalar potential and then gives rise to a general solution using Green's functions.

[32]Other authors have developed solutions in an infinite medium (Chave and Cox 1982).

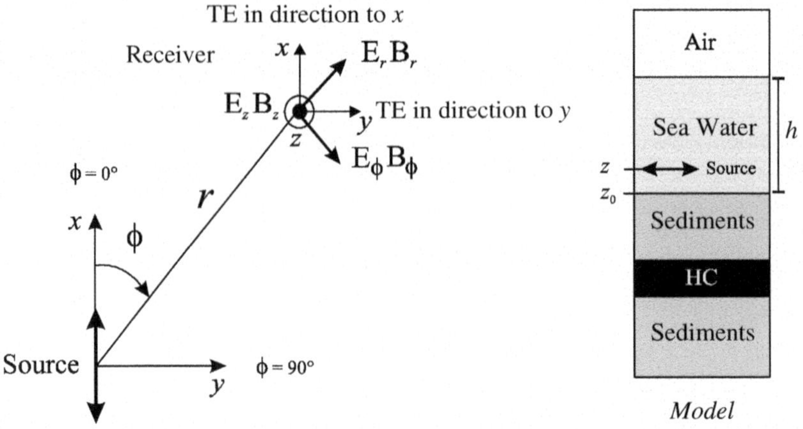

(*a*) (*b*)

Fig. 3.32 Pattern in the horizontal plane (*a*), section in the vertical plane (*b*) (According to Andreis and MacGregor 2008). (*a*) Decomposition of the field into its various components (horizontal and vertical) and transverse modes (TE). (*b*) Geological cross-section or pseudosection currently used in the calculation of synthetic fields, also called a canonic model

$$\begin{aligned}
\mathrm{E}_\phi = {} & \frac{P\rho_w}{4\pi}\sin\phi\int_0^\infty\Bigg[\left(\frac{\beta_0 J_1(kr)}{r\left(1+\mathrm{R}_\mathrm{L}^\mathrm{TM}\mathrm{e}^{-2\beta_0 h}\right)}+\mathrm{i}\omega\mu_0\frac{kJ_0(kr)-J_1(kr)/r}{\beta_0\rho_w\left(1-\mathrm{R}_\mathrm{A}^\mathrm{TE}\mathrm{R}_\mathrm{L}^\mathrm{TE}\mathrm{e}^{-2\beta_0 h}\right)}\right)\mathrm{e}^{-\beta_0|z-z'|} \\
& +\left(-\frac{\beta_0 J_1(kr)}{r\left(1+\mathrm{R}_\mathrm{L}^\mathrm{TM}\mathrm{e}^{-2\beta_0 h}\right)}\mathrm{R}_\mathrm{L}^\mathrm{TM}-\mathrm{i}\omega\mu_0\sigma_w\frac{kJ_0(kr)-J_1(kr)/r}{\beta_0\left(1-\mathrm{R}_\mathrm{A}^\mathrm{TE}\mathrm{R}_\mathrm{L}^\mathrm{TE}\mathrm{e}^{-2\beta_0 h}\right)}\mathrm{R}_\mathrm{L}^\mathrm{TE}\right)\mathrm{e}^{-\beta_0|z+z|} \\
& +\left(\frac{\beta_0 J_1(kr)/\mathrm{r}}{r\left(1+\mathrm{R}_\mathrm{L}^\mathrm{TM}\mathrm{e}^{-2\beta_0 h}\right)}-\mathrm{i}\omega\mu_0\frac{kJ_0(kr)-J_1(kr)/r}{\beta_0\rho_w\left(1-\mathrm{R}_\mathrm{A}^\mathrm{TE}\mathrm{R}_\mathrm{L}^\mathrm{TE}\mathrm{e}^{-2\beta_0 h}\right)}\mathrm{R}_\mathrm{A}^\mathrm{TE}\right)\mathrm{e}^{\beta_0|z+z'-2h|} \\
& +\left(-\frac{\beta_0 J_1(kr)/\mathrm{r}}{r\left(1+\mathrm{R}_\mathrm{L}^\mathrm{TM}\mathrm{e}^{-2\beta_0 h}\right)}\mathrm{R}_\mathrm{L}^\mathrm{TM}+\mathrm{i}\omega\mu_0\frac{kJ_0(kr)-J_1(kr)/r}{r\beta_0\rho_w\left(1-\mathrm{R}_\mathrm{A}^\mathrm{TE}\mathrm{R}_\mathrm{L}^\mathrm{TE}\mathrm{e}^{-2\beta_0 h}\right)}\mathrm{R}_\mathrm{A}^\mathrm{TE}\mathrm{R}_\mathrm{L}^\mathrm{TE}\right)\mathrm{e}^{\beta_0(|z-z'|-2h)}\Bigg]\mathrm{d}k
\end{aligned} \tag{3.65}$$

$$\begin{aligned}
\mathrm{E}_z = {} & \frac{P\rho_w}{4\pi}\cos\phi\int_0^\infty\frac{k^2 J_1(kr)}{1+\mathrm{R}_\mathrm{L}^\mathrm{TM}\mathrm{e}^{-2\beta_0 h}}\Big[\pm\mathrm{e}^{-\beta_0|z-z'|} \\
& -\mathrm{R}_\mathrm{L}^\mathrm{TM}\mathrm{e}^{-\beta_0|z+z'|}-\mathrm{e}^{\beta_0|z+z'-2h|}\pm\mathrm{R}_\mathrm{L}^\mathrm{TM}\mathrm{e}^{\beta_0(|z-z'|-2h)}\Big]\mathrm{d}k
\end{aligned} \tag{3.66}$$

– For the magnetic field:

$$\begin{aligned} B_r = \frac{P\mu_0}{4\pi}\sin\phi\int_0^\infty \Bigg[& \left(\pm\frac{J_1(kr)}{r\left(1+R_L^{TM}e^{-2\beta_0 h}\right)}+\frac{kJ_0(kr)-J_1(kr)/r}{1-R_A^{TE}R_L^{TE}e^{-2\beta_0 h}}\right)e^{-\beta_0|z-z'|} \\ & +\left(\frac{J_1(kr)}{r\left(1+R_L^{TM}e^{-2\beta_0 h}\right)}R_L^{TM}-\frac{kJ_0(kr)-J_1(kr)/r}{1-R_A^{TE}R_L^{TE}e^{-2\beta_0 h}}R_L^{TE}\right)e^{-\beta_0|z-z'|} \\ & +\left(\frac{J_1(kr)}{r\left(1+R_L^{TM}e^{-2\beta_0 h}\right)}+\frac{kJ_0(kr)-J_1(kr)/r}{1-R_A^{TE}R_L^{TE}e^{-2\beta_0 h}}R_A^{TE}\right) \\ & \times e^{\beta_0|z+z'-2h|}\pm\left(\frac{J_1(kr)}{r\left(1+R_L^{TM}e^{-2\beta_0 h}\right)}R_L^{TM}-\frac{kJ_0(kr)-J_1(kr)/r}{1-R_A^{TE}R_L^{TE}e^{-2\beta_0 h}}R_A^{TE}R_L^{TE}\right)e^{\beta_0(|z-z'|-2h)}\Bigg] \\ & \times dk \end{aligned} \tag{3.67}$$

$$\begin{aligned} B_\phi = \frac{P\mu_0}{4\pi}\cos\phi\int_0^\infty \Bigg[& \left(\pm\frac{kJ_0(kr)-J_1(kr)/r}{1+R_L^{TM}e^{-2\beta_0 h}}+\frac{J_1(kr)}{r\left(1-R_A^{TE}R_L^{TE}e^{-2\beta_0 h}\right)}\right)e^{-\beta_0|z-z'|} \\ & +\left(\frac{kJ_0(kr)-J_1(kr)/r}{1+R_L^{TM}e^{-2\beta_0 h}}R_L^{TM}-\frac{J_1(kr)}{r\left(1-R_A^{TE}R_L^{TE}e^{-2\beta_0 h}\right)}R_L^{TE}\right)e^{-\beta_0|z-z'|} \\ & +\left(\frac{kJ_0(kr)-J_1(kr)/r}{1+R_L^{TM}e^{-2\beta_0 h}}+\frac{J_1(kr)}{r\left(1-R_A^{TE}R_L^{TE}e^{-2\beta_0 h}\right)}R_A^{TE}\right) \\ & \times e^{\beta_0|z+z'-2h|}\pm\left(\frac{kJ_0(kr)-J_1(kr)/r}{1+R_L^{TM}e^{-2\beta_0 h}}R_L^{TM}-\frac{J_1(kr)}{r\left(1-R_A^{TE}R_L^{TE}e^{-2\beta_0 h}\right)}R_A^{TE}R_L^{TE}\right)e^{\beta_0(|z-z'|-2h)}\Bigg] \\ & \times dk \end{aligned} \tag{3.68}$$

$$\begin{aligned} B_z = \frac{P\mu_0}{4\pi}\sin\phi\int_0^\infty & \frac{k^2J_1(kr)}{\beta_0\left(1-R_L^{TE}R_A^{TE}e^{-2\beta_0 h}\right)}\Big[e^{-\beta_0|z-z'|} \\ & +R_L^{TE}e^{-\beta_0|z+z'|}+R_A^{TE}e^{\beta_0|z+z'-2h|}+R_A^{TE}R_L^{TE}e^{\beta_0(|z-z'|-2h)}\Big]dk \end{aligned} \tag{3.69}$$

where R_L^{TE} and R_L^{TM} are the coefficients that define the interactions of transverse TE and TM modes with the bottom of the sea, and R_A^{TE} those with air

- J_0 and J_1 are respectively the Bessel functions of order 0 and 1
- β_0 according to the wave number k is equal to:

$$\beta_0 = \sqrt{k^2 - \frac{i\omega\,\mu_0}{\rho_w}} \tag{3.70}$$

The previous equations expressing the amplitudes of the various components of the electric and magnetic fields according to the transverse modes can be used to establish the orders of magnitude (see details below). In deep water (the best case), we find then fields of the order of 10^{-12} V/A.m^2. In contrast, in shallow water, because of the presence of the surface (air), and its important contribution to the phenomenon of propagation, the field is $10^{-11.5}$ V/A.m^2 (Andreis and MacGregor 2008).

Details of these field calculations: they can be found in different articles cited in the bibliography. The complexity of these developments require high mathematical knowledge. They have high sensitivity if we have a lot of additional geological data that allow the inversion process. In Chap. 5 (Interpretation) we preferred to describe in detail the easiest methods using simplified analytical resolutions (forward problem) with some approximations, probably less accurate but, in our opinion, more demonstrative.

5.3 *Measurements: Evaluations of E and B Fields*

Considering on one hand the measures of the values of the electric field E and magnetic field B in two horizontal directions x, y perpendicular to each another, and the Laplace equation on the other hand:

$$\nabla^2 \mathrm{E} = \frac{\partial \mathrm{E}_x}{\partial x} + \frac{\partial \mathrm{E}_y}{\partial y} + \frac{\partial \mathrm{E}_z}{\partial z} = 0 \tag{3.71}$$

an equation that shows that the electric field is conservative (no charges accumulation), and that it derives from a scalar potential (gradient), then it becomes:

$$\frac{\partial \mathrm{E}_x}{\partial x} + \frac{\partial \mathrm{E}_y}{\partial y} = -\frac{\partial \mathrm{E}_z}{\partial z} \tag{3.72}$$

and:

$$\frac{\partial \mathrm{B}_y}{\partial x} - \frac{\partial \mathrm{B}_x}{\partial y} \propto \mathrm{E}_z \tag{3.73}$$

By solving these equations, we now get:

- For the horizontal transverse electric fields (x and y components):

$$\frac{\partial \mathrm{E}_x}{\partial x} + \frac{\partial \mathrm{E}_y}{\partial y} = -P\frac{\rho_{\mathrm{w}}}{4\pi}\cos\phi \int_0^\infty -\frac{\beta_0 k^2 J_1(kr)}{1+\mathrm{R}_\mathrm{L}^{\mathrm{TM}}\mathrm{e}^{-2\beta_0 h}} \left[-\mathrm{e}^{-\beta_0|z-z'|} + \mathrm{R}_\mathrm{L}^{\mathrm{TM}}\mathrm{e}^{-\beta_0|z+z'|} - \mathrm{e}^{-\beta_0(|z+z'|-2h)} + \mathrm{R}_\mathrm{L}^{\mathrm{TM}}\mathrm{e}^{\beta_0(|z+z'|-2h)}\right]\mathrm{d}k \tag{3.74}$$

- For the corresponding magnetic fields (x and y components):

$$\frac{\partial B_y}{\partial x} - \frac{\partial B_x}{\partial y} = -\mu_0 \frac{P\rho_w}{4\pi} \cos\phi \int_0^{\infty} - \frac{k^2 J_1(kr)}{1 + R_L^{TM} e^{-2\beta_0 h}}$$

$$\left[\mp e^{-\beta_0|z-z'|} - R_L^{TM} e^{-\beta_0|z+z'|} - e^{-\beta_0(|z+z'|-2h)} \mp R_L^{TM} e^{\beta_0(|z-z'|-2h)}\right] dk \quad (3.75)$$

The data can be then expressed (cf. Fig. 3.24) as complex numbers (real and imaginary parts) or in polar form, i.e., amplitude (vector) and phase (angle). These can be simultaneously measured by one device only, such as a synchronous detector (see Chap. 4, Sect. 5.4.2).

5.4 Elliptic Representation of the Fields

In practice we rarely record the pure modes of radial and azimuthal components of the horizontal transverse fields. For this reason, it is convenient for the interpretation of data, or to determine their orientation after that (Key and Lockwood 2010), to calculate these values and geometrically represent them.

If we consider, for example, two measures of amplitude $E_{1,2}$ and of phases $\varphi_{1,2}$, taken in perpendicular directions, such that we have (polar form) $E_1 e^{i\varphi 1}$ and $E_2 e^{i\varphi 2}$, it is possible to build from these values a polarization ellipse.

The latter is defined by the values of its small and large axes (p_{min} and p_{max}) and by the angle α_E, which makes, for example, one of these axes (ex: p_{max}) with an arbitrary direction (ex: *East*). We obtain in this case the following relations:

$$\begin{cases} p_{max} = \left|E_1 e^{i(\varphi 1 - \varphi 2)} \sin\alpha_E + E_2 \cos\alpha_E\right| \\ p_{min} = \left|E_1 e^{i(\varphi 1 - \varphi 2)} \cos\alpha_E + E_2 \sin\alpha_E\right| \\ \tan 2\alpha_E = \dfrac{2E_2E_1 \cos\,(\varphi_1 - \varphi_2)}{E_2^2 - E_1^2} \end{cases} \quad (3.76)$$

From this ellipse, oriented and declined with its geometric characteristics, we can imagine all sorts of graphs allowing us to visualize the different effects (Fig. 3.33).

We can also calculate, except for the scale factor ε, the ratio of the area of the ellipse to the one of a circle of unit radius $(p_{min}\ p_{max})/4\varepsilon^2$ or even use the advantages of vector analysis.

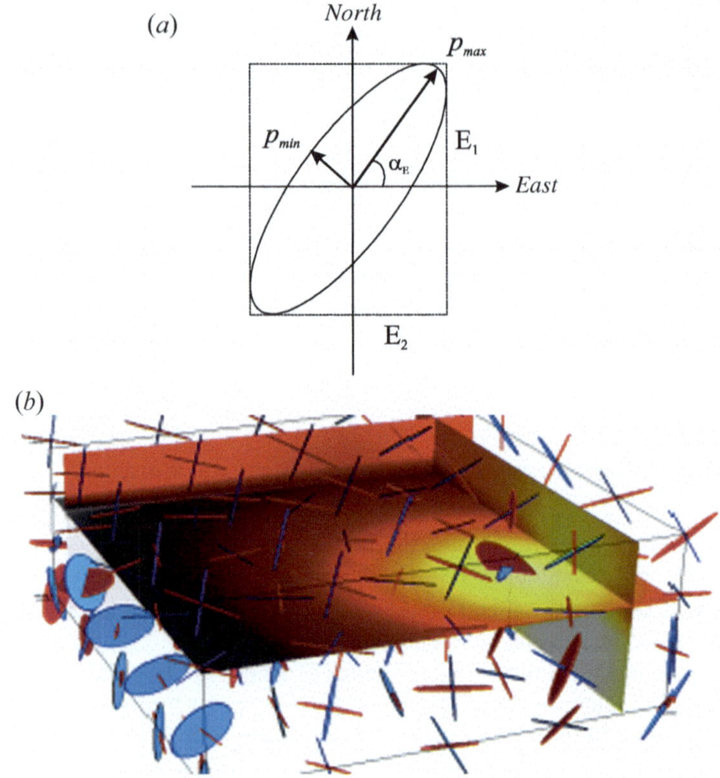

Fig. 3.33 Polarization ellipse of the transverse components of the horizontal fields $E_{1,\ 2}$ in a *North*/*East* axis system (*a*) and above field polarization around a conductivity anomaly (*b*) (According to Key and Lockwood 2010)

5.5 *Amplitude Units of the Electric Fields*

The values of these measured electric fields E representing a potential gradient are expressed in V/m or more precisely in $V/m/\sqrt{Hz}$ when considering the frequency. It is also possible to overcome the various operational conditions, to perform a normalization of the measurements.

To normalize the field values in relation to the source (emission dipole in general) whose power can vary from one transmitter to another, it is preferable to involve the dipole moment of the transmitting antenna (commonly known *P* in some publications), which corresponds to the product of the current flow I_a (intensity) by the antenna length *L*. The standard voltage is then equal to:

$$V_{[VA^{-1}m^{-2}]} = \frac{E}{P} = \frac{E_{[Vm^{-1}]}}{(I_a \times L)_{[Am]}} \tag{3.77}$$

a value that can be indifferently transferred from one piece of equipment to the other.

5.6 *Measurement, Evaluation of the Background Noise EM and the S/N Ratio*

The background noise can occupy a spectrum in the usable frequency band and be induced by a large number of anthropogenic and natural factors related to marine and geological activity (Kaushalendra Mangal Bhatt 2011). Its evolution over time and space is characterized by features that can be exploited by means of signal processing both temporally and spatially.

5.6.1 Temporally

The EM background noise is characterized by a voltage that varies unpredictably over time. This random function of time can be studied, taking into account its temporal development (deterministic approach) or by the laws of probability that govern it (ergodic noise).[33] Noise can be entirely or partially eliminated by analog or digital filtering (*hardware*: rejection filters → ad hoc analyzer) and/or data processing (*software*: sampling → statistical treatment).

For example, if we consider that for a sinusoidal signal of the form:

$$S(t) = A\cos\omega t \tag{3.78}$$

and for a noise level:

$$b(t) = a\cos\omega t + b\sin(\omega t) \tag{3.79}$$

the resulting signal is then of the form:

$$R(t) = \Gamma\cos(\omega t + \varphi_C) \tag{3.80}$$

a and b are known (deterministic variables) or unknown (independent random variables) functions of time that change over time according to the laws of chance (Gauss law) (Fig. 3.34).

The decomposition of the signal can then be obtained by a suitable measurement line composed, for example, of:

[33]When the study of one of its parameters by either method produces the same result, the function is called ergodic.

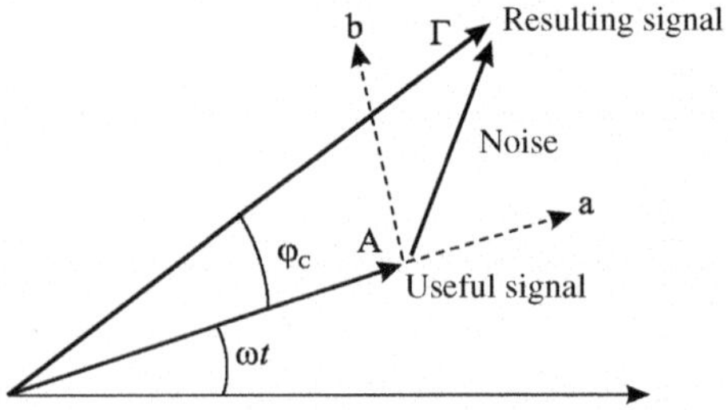

Fig. 3.34 Decomposition in the Fresnel plane of the resulting signal, sum of the useful signals and the noise

- A quadratic detector, which will have its output given by the squared modulus of the resulting signal (Γ^2)
- A linear sensor that will give the module of the resulting signal (Γ)
- A demodulator with a reference in phase with the useful signal (($\Gamma\cos\varphi_C$)

5.6.2 Spatially

In space, the background noise can vary from one place to another but is mainly dependent on the water depth (Cox et al. 1968). It is due:

- In coastal, littoral and harbor zones, mostly to human activities (traffic and maritime works), or more rarely to waves (Warburton and Caminiti 1964)
- In ocean areas, to natural water movements (currents, tides, etc.), to rapid changes in the earth's magnetic field, to the movement of earth currents, and to some causes as diverse and varied as earthquakes, waves gravity,[34] etc.

In the deep sea, at several thousand meters of water depth, we estimate by measurements the noise at several $\mathrm{pV/m/\sqrt{Hz}}$. On average, the background noise on the ocean floor is appreciated by service providers at 10^{-16} V/A.m^2 à 1 Hz, i.e., to at least 1 pV/m.

Then we obtain, in the best cases, an S/N ratio of about 1000 (or 1 nV/m ÷ 1 pV/m). On shoals, ambient noise may exceed $10\ \mathrm{nV/m/\sqrt{Hz}}$ with more important participation of surface waves (wind). In these most unfavorable conditions, the S/N ratio is therefore significantly reduced (Fig. 3.35).

If at medium to large water depths, the noise level does not affect the field measurements, in contrast to shallower depths, it is of the order of magnitude of the desired signals (cf. Fig. 3.26). This therefore gives the method, in spite of the fact of having a higher emission power, an S/N ratio even lower when the water layer is thin and when the presence of the air/sea interface increases the amplitude of the interference signals.

[34]Generally these latter correspond to epiphenomena restricted in time.

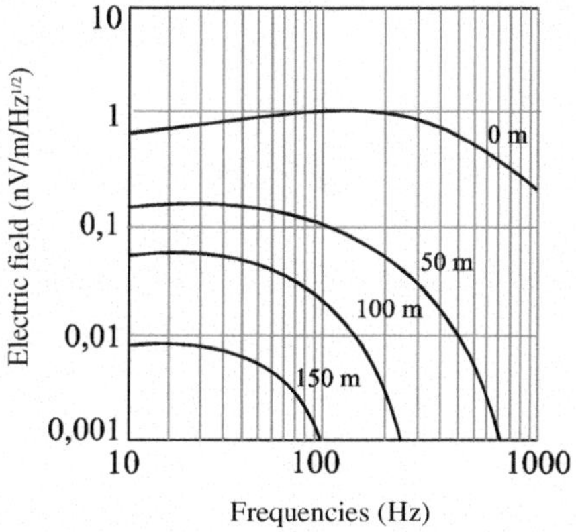

Fig. 3.35 Background noise at different frequencies (10–1000 Hz) in the first 150 m of water depth (According to Cox et al. 1968)

6 Definition of Acquisition Systems and Operational Procedures

Once the order of magnitude is determined and the background noise is measured, evaluated and quantified, a detection strategy must be implemented. It must also take into account operational constraints, which are severe in this type of investigation in a hostile environment.

6.1 Injections of Energy in a Geological Medium

As seen in the previous chapter, there are several ways to transmit electromagnetic energy in a material medium.

6.1.1 Types of Sources

For SBL techniques, depending on the type of aim, and magnetic sources aside, it is possible to select two of them:

- One without electrical contact with the seawater, which is to induce a magnetic field using an isolated turn (closed current loop) or a coil (several turns), in which circulates an alternating electric current
- The other with contact (open current loop). which is to directly inject the current through one or more immersed electrodes.

In both cases, the means of transmission of energy (induced currents or galvanic currents) are positioned slightly above the seabed so that the maximum power can penetrate the underlying grounds (cf. Fig. 3.36). Generally, the first method, called inductive, applies in geophysics with relatively high frequencies in order to penetrate any resistive layer (compact sand for example) as shown in surface

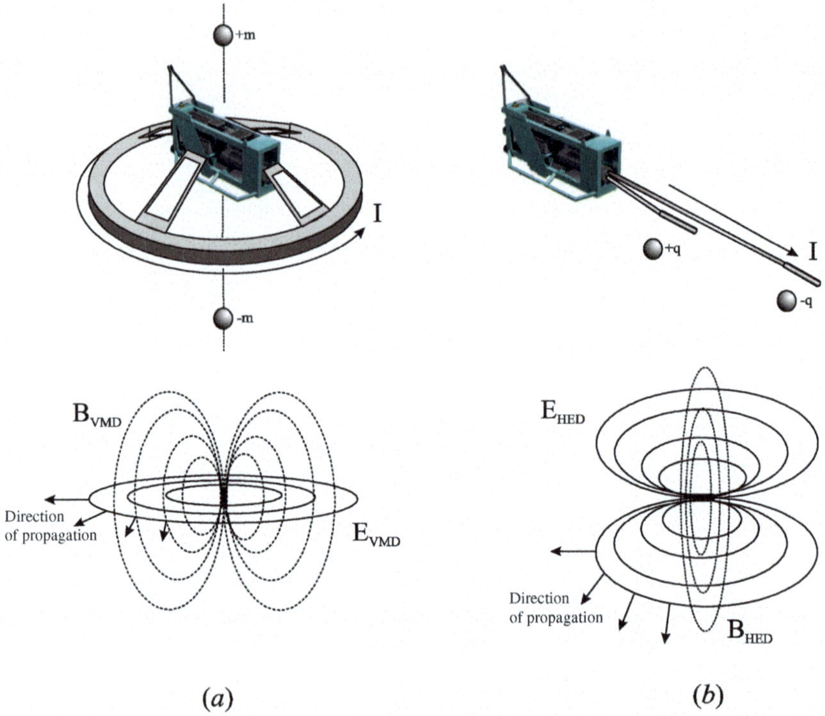

Fig. 3.36 Types of source with EM power. (*a*) Inductive source: circular coil in the horizontal plane corresponding to a *vertical magnetic dipole* or VMD (±m), which, for the moment, has not yet been used to our knowledge (A recently patented device (MacGregor et al. 2008) that creates horizontal fields and thus discriminates through the differentiation of the galvanic/inductive effects the anomalies attributed to hydrocarbons from the other heterogeneities of the seabed. A related device was proposed in the 1970s (Duroux 1974)). (*b*) Galvanic source with electrodes: *horizontal electric dipole* or HED (±q) currently in service. The galvanic source, placed in a horizontal plane, provides a sense of wave propagation in both the horizontal and the vertical plane, in contrast to the purely inductive source, which ensures that only in the horizontal plane

prospecting.[35] This type of source essentially generates some magnetic field and seems for the moment, except for certain geological configurations, not suitable for underwater detection of hydrocarbons. However, devices proposals are being studied for use in association with controlled sources (mCSEM) or telluric sources (mMT). This would have the advantage of being able to differentiate the galvanic and inductive effects, the latter only propagating in the horizontal plan (MacGregor et al. 2008).

The second method, called galvanic, uses an electrical current of very low frequency, which flows even better when the ground has good conductivity

[35]In marine exploration this could be the case above basaltic horizons or salt domes, diapirs, gas hydrates, etc. In the 1970s, very low frequency devices were proposed (Duroux 1974).

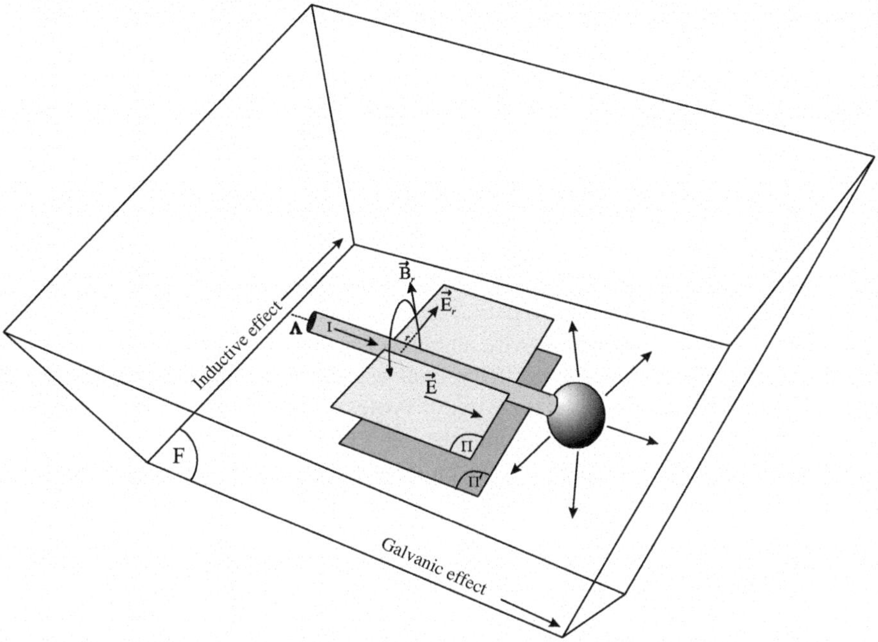

Fig. 3.37 Galvanic electric field $\vec{E}$ and induced electric field $\vec{E}_r$ in a horizontal plane, caused by the circulation of an electric current in an antenna (insulated cable) placed in a conductive medium and escaping through one of its ends

(limiting losses by the Joule effect at the injection electrode). This type of source is particularly adapted to receive galvanic dominant responses, but also inductive responses. Its energy also propagates vertically and horizontally. For now, the galvanic source with horizontal transmission (linear antennas) is the only one to be practically used in the context of oil exploration.

Exxon is now studying the possibility of using sources with vertical antennas (Fielding and Lu 2009).

6.1.2 Topology of the Transmitting Antenna

In EM seabed logging as it currently stands, the emission dipole is immersed in a relatively homogeneous and isotropic medium (seawater). This has the immediate effect of a total and uniform dissipation of energy in all space, followed by the coexistence in the crossed media of the inductive and galvanic phenomena (Walker and West 1992). These result in the presence of two types of energy transfer, more or less dependent on the conductive nature of the traversed media, i.e., of the sedimentary underlying series.

Metrologically, according to the law of induction on currents (Ampere's theorem), the use of an antenna where a current I flows, which escapes through one of its ends immersed in a conductive medium (seawater), creates in its proximity both a field of parallel force (galvanic field $\vec{E}$) and a field of force induced by B, perpendicular to its direction Λ (induced field $\vec{E}_r$). The value and the topology of the associated magnetic field are located in the Appendix A3.2.

If the transmitting antenna (cf. Fig. 3.37) is placed in a horizontal plan parallel to that occupied by the seabed (plan *F*), the two electric fields $\vec{E}$ and $\vec{E}_r$, of which the first is sensitive to the possible presence of a resistant layer and the other is not, can then be measured in the horizontal plane Π (containing the antenna) or in any other plane Π' parallel to the latter, for example placed above the seafloor (plan *F*).

To further minimize the problems of interpretation, it is important to have an as homogeneous as possible electric field.

At this time, purely inductive wire antennas are not used. The reader interested in the question especially for its theoretical aspects (calculations of the related fields) may consult with interest the work of Professor Bannister on the propagation of electromagnetic waves at sea (Bannister 1980) and that of Webster, the oldest, which reminds us of theoretical studies on submarine cables (Webster 1897).

6.2 Sensitivity and Topologies of Fields Measurements

The mCSEM is set in an exploration context of the type dipole–dipole where transmitter T and receiver R can be arranged in various layouts or topologies. Configurations *in line* (T/R aligned) and *broadside* (T/R parallel) are not sensitive to the same geological features. For example in the presence of a resistive layer (e.g., an HC reservoir), the field amplitude measurements are more sensitive in the *in-line* configuration (cf. Fig. 3.37) than in the *broadside* configuration (cf. Chap. 2, Sect. 2.5.5). However, the phase measurements of these same fields have reverse sensitivity (cf. Fig. 3.38).

The sensitivity also depends on the choice of the frequency, which results from a trade-off. At a high frequency, the effect of the oil reservoir on the measured field is important and may not be detected above the noise level (great attenuation). At low frequency, the S/N ratio is very good, but the fields are unable to resolve small scale structures (inferior to the wave length signal). In these conditions it results in bad resolution. In practice the frequency bandwidth is therefore narrow (0.01–10 Hz) and multiple frequency acquisition is needed.

6.3 Electrical Fields Measurements

The weakness of these electric fields, in both intensity and direction, requires instruments of high sensitivity with multiple acquisition channels corresponding to exhaustive information outlets. The sensors are thus composed of at least two antennas capable of measuring the intensities and phases of the transverse

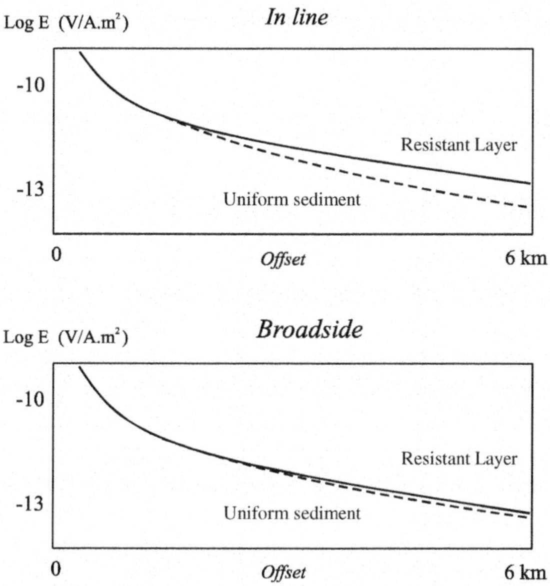

Fig. 3.38 Sensitivities to resistive layers in the configurations *in line* and *broadside*

components TE_x and TE_y (rectangular coordinates) of the electric field E in the horizontal plane. For each of the antennas, placed perpendicular to each other, the energy collection is performed:

- Either by biaxial measurements of a potential difference across two unpolarizable electrodes distant from a certain length even greater when the expected field is weak
- Or, more recently, by triaxial measurements of the current density, directly dependent on the surface collection of the electrical signal

In both devices, it is essential that the meter only samples minimum energy (electromotive force or current depending on the type of electrometer used)[36] so as not to locally disturb by its presence the distribution of the force field nearby, which would then have the disadvantage of generating fake events, making the interpretation obsolete.

6.4 Magnetic Field Measurements

For some applications, such as processing and separation of up/down waves (cf. Sect. 3.13), the acquisition devices may be complemented by magnetic sensors (triaxial magnetometers) to measure changes in the transverse components of the magnetic field.

[36]See Chap. 4, devoted to instrumentation.

6.5 Recordings, Calibrations, Standardization and Presentation of Results

Electric and magnetic fields are recorded according to the distance (offset) between the measuring devices (electrometers) and the source. The adopted unit of measure is V/A.m^2 (cf. Sect. 5.5). Before that, measurements are calibrated and standardized to take account both the variations of intensity of the source (not constant in time) and the differences in power due to the offset over all the receivers.

In situ calibration[37] corresponds to two successive stages prior to and after the run (source displacement). One or two reference R_{ref} sensors are available for that, installed at the end or edge of the profile, which continuously record the variations of the electric and/or magnetic fields throughout the survey.

Measurements standardization is an important mathematical phase of the acquisition processing, which allows us to reset the profiles among them (receiver by receiver) by an arithmetic operation that compensates for differences in calibration especially performed at the surface. This usually includes a displacement toward the origin (a shift toward the R_{ref}) and a rotation for the final alignment (Figs. 3.39, 3.40, 3.41 and 3.42).

6.6 Errors, Accuracy and Measurement Uncertainty

The interpretation by data inversion (cf. Chap. 5) requires that the instruments deliver information with maximum precision,[38] .i.e., in other words, by drastically reducing the effects of errors. This can only be done very often at the cost of decreasing the resolving power and very strict quality control (Fig. 3.43).

[37]Not to be confused with electrometer calibration, which is done on the surface ship before immersion (cf. Sect. 6.7).

[38]***Accuracy*** is evaluating the absence of errors e (e = measured value – actual value). When the latter are present, the equation becomes $Q_e = f(P) + e$, $e = Q - Q_e$

There are two kinds of errors: time-independent (static) errors and time-dependent (dynamic) errors, which respectively are in the form $e_S = g(Q)$, $e = g(Q, t)$

The coherent noise is, for example, a case of dynamic error. However, there are random variations related to the environment that can only be treated by statistical methods. The arithmetic average of a finite number N of measurements is then:

$$\overline{Q} = \frac{1}{N}\sum_{i=1}^{N} Q_i$$

The errors affecting the accuracy mainly come from the sensors, the processing and the location of the instruments (T/R). Errors can also be classified into systematic, random or accidental errors and mathematically defined. More simply, an absolute error is defined as the difference between the measurement and the true value, and a relative error as the ratio of the absolute error to the true or measured value.

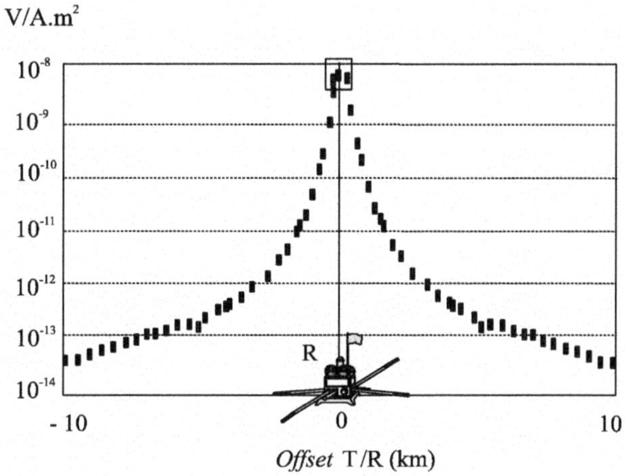

Fig. 3.39 Variation of the amplitudes of the electric field at the top of an anomaly, for a given receiver R as a function of the distance from the source (*offset*)

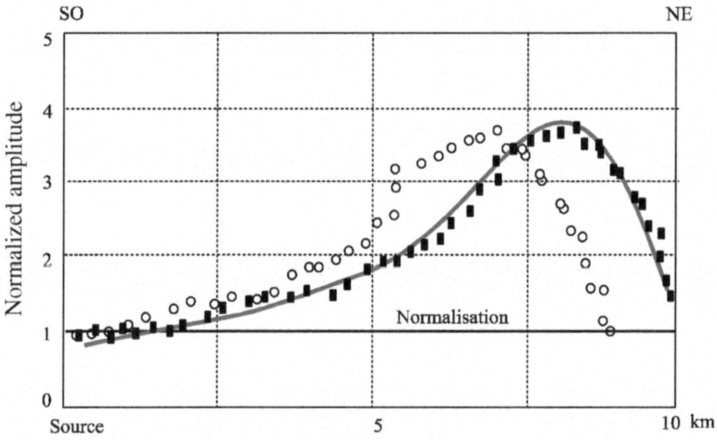

Fig. 3.40 Values of normalized amplitudes (electric field ■, magnetic fields ○) for a given sensor compared to a reference sensor in the function of the distance to the source (offset) and theoretical curve (in *gray*)

One of the errors introduced in SBL, which may limit an operation or even make the data unusable, is the uncertainty inherent in the location of the equipment and especially of the source, which moves continuously with more or less accuracy above the bottom.

For example, if we assume an uncertainty of 10 m in the depth ratio (δ/z_S) for an investigation carried out at a frequency of 1 Hz and for marine sediment resistivity of 1 Ω.m, it produces an error of about 2 %. At low bathymetries (z_S), when the skin depth (δ) is more important, then the error may reach 4 %.

In addition, measuring instruments must also possess intrinsic qualities that are defined by their accuracy, precision, repeatability, etc. (cf. Chap. 4, Sect. 2.4).

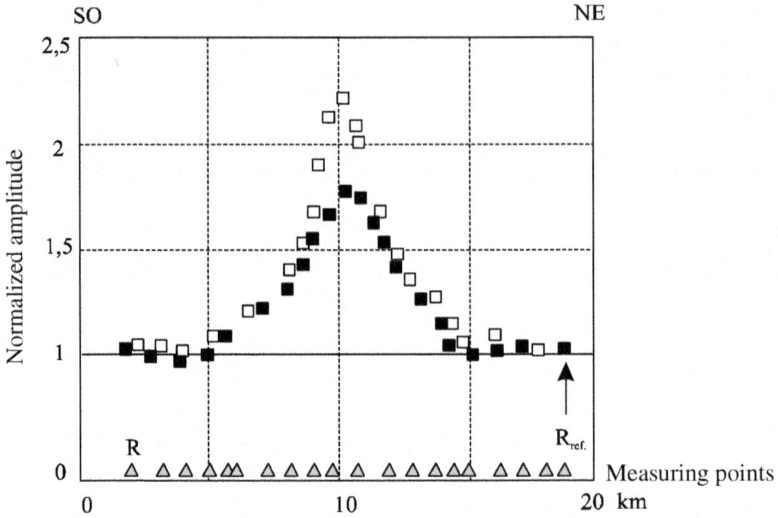

Fig. 3.41 Values of the normalized and calibrated amplitudes of the electric ■ and magnetic fields □ corresponding to the different measuring points R (receivers ▲)

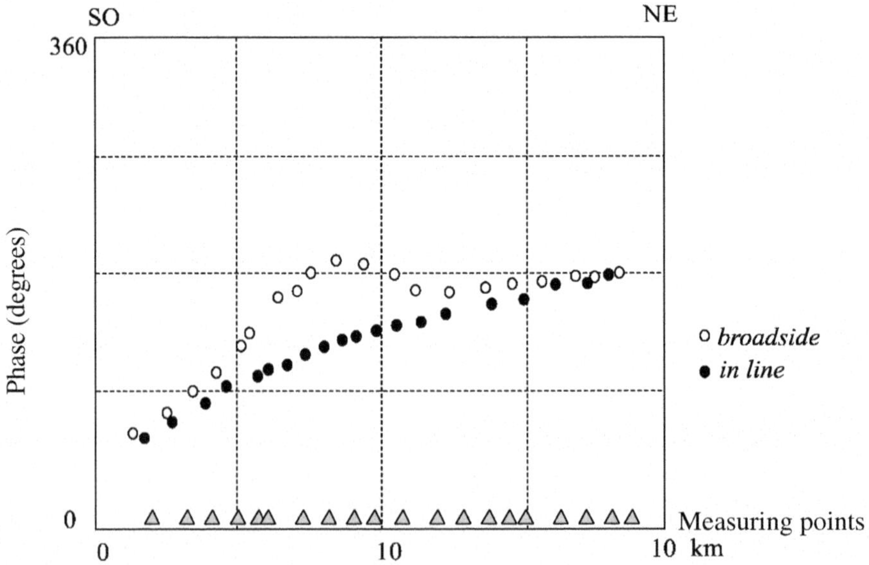

Fig. 3.42 Phase values (0–360°) in the *in-line* and *broadside* positions for different measuring points

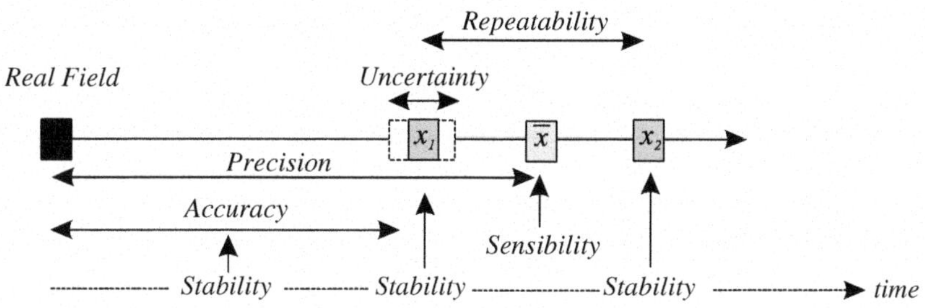

Fig. 3.43 Main qualities of field measuring instruments and their interdependence

6.7 Measurement Quality Control

Quality control is monitored at all stages of the acquisition and processing of data. In operation, this control is provided by a number of dispositions and devices, the main ones being:

- At the beginning of the survey, before immersion, electrometers are checked and calibrated (surface calibration) in an instrumented tank (a tank filled with water) or in an equivalent device located in the cargo hold.
- During the survey, the acquisition is controlled in situ by a reference electrometer (in situ calibration) whose data will provide a posteriori the normalization operation of all measures (cf. Fig. 3.41)
- At the end of the survey, the data are checked and, if necessary, the electrometers are calibrated a second time for corrections, especially those concerning instrumental drifts (including electronic).

During all phases of processing, data are continuously analyzed and verified by tests of convergence and certified at every stage of the acquisition process (sorting) up to the final interpretation phase. For now, there is no standard quality specific to SBL as may exist in other domains such as downhole well logging or marine seismics.

6.8 Optimal Conditions of Detection in EM SBL

The decision to operate in a future field is not acquired if optimal conditions are not met. The latter concern in particular:

- The water depth
- The depth of the reservoir
- The level of the electromagnetic noise, which affects the S/N ratio
- The contrast of resistivity (sediment/reservoir)
- The transverse resistivity
- The volume of the target

Of all these factors, the depth of the deposit is by far the most important parameter to take into account in the estimation of optimal conditions of detection. It is generally estimated through seismic data acquired during the previous phase of exploration. The other parameters are either readily available as already known or easily identifiable (background noise), or more difficult to understand, as for instance those concerning the specific resistivity of the grounds if a logging campaign has not been done in advance.

6.8.1 Index Feasibility: NAR

Using some of the previously explained data, when they can be considered reliable, as the automatic models were developed for their interpretation (including resolution of the forward problem), we can say that the operation is possible or not.

If we have a lot of information, we can also evaluate the chances of success with more or less accuracy on a given geological prospect. For this we establish a feasibility index (Stefatos et al. 2009). It corresponds then to the reduced ratio of the difference between the normalized amplitudes of the value of the transverse component on x of the fields E, normal and anomalous in the presence or absence of a deposit. In reference to Fig. 3.44, the NAR (normalized amplitude response) is then simply given by the formula:

$$\mathrm{NAR} = \frac{\left|\mathrm{E}_x^{\mathrm{res}} - \mathrm{E}_x^{\mathrm{bck}}\right|}{\mathrm{E}_x^{\mathrm{bck}}} \tag{3.81}$$

Not all prospects are therefore considered suitable for this type of investigation. However, we can consider that the percentage of inappropriate sites or situations, compared to the hundreds of research studies conducted to date, remains relatively low and estimated to be of the order of about 15 %.

More precisely, if we stick to operations in recent years, we can say that the survey can be attempted in three quarters of cases.

Experience has shown, for example with a less than 15 % NAR, it actually becomes difficult to differentiate a subsurface anomaly with its lateral and vertical effects from a deeper anomaly due to a hydrocarbon reservoir. The percentage of 15 % therefore seems to be the lower and significant limit today not to exceed to successfully operate a prospect in good conditions (Heshammer et al. 2010) (Fig. 3.45).

6.8.2 Need to Detect and Map the Pipelines

When the survey is carried out in the context of a valorization or an extension of an mature oil field or a field during operation, the network of present pipelines will interfere in the measurements (Sainson 2008, 2009). In this case, it is essential to

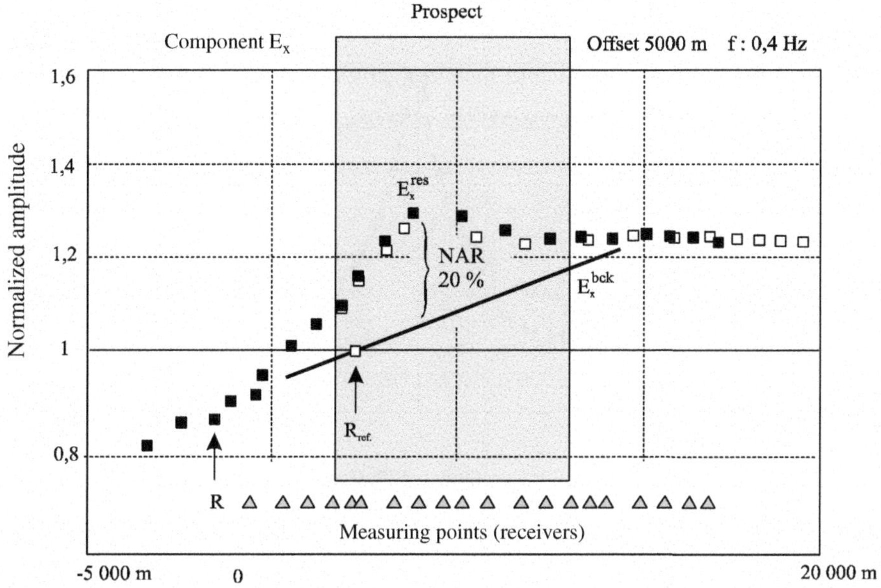

Fig. 3.44 Example of obtaining an NAR in the Barents Sea: 20 % in the prospect (According to Heshammer et al. 2010). A standardized amplitude of 1 indicates that the selected receiver R has the same standard amplitude as that of the reference receiver R_{ref}. An amplitude of 1.5 shows that the selected receiver R has a normalized amplitude 50 % greater than that of the reference receiver R_{ref}

Fig. 3.45 Feasibility index (%) on a set of 43 studies on west African offshore basins (According to Stefatos et al. 2009)

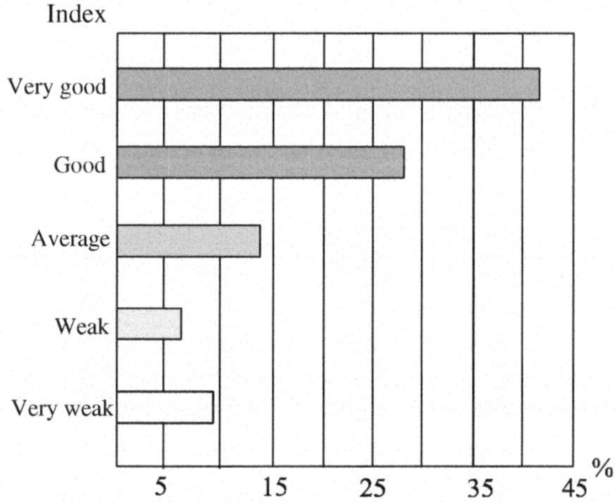

have precise mapping of underwater lines. It is not usually very easy to get, but there are internal means of auscultation of these pipes as intelligent pigs (cf. Chap. 6, Sect. 5.2), and even external means (sonar, ROV, AUV, instrumented fish, etc.) that immediately allow us to obtain relatively satisfactory results (Sainson 2007).

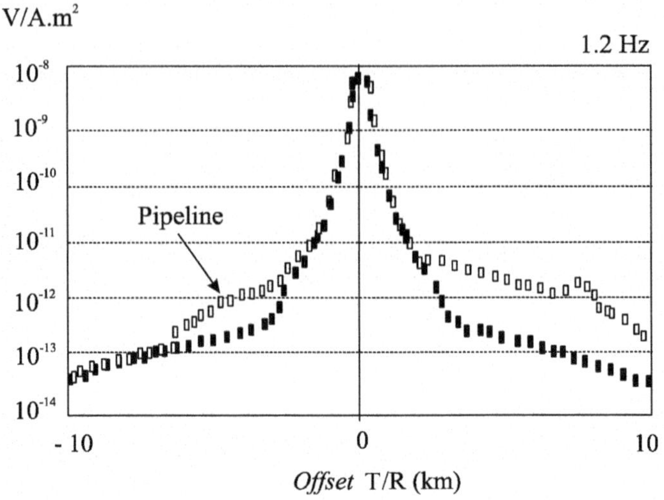

Fig. 3.46 Effect of a pipeline □ on field measurements on Frigg concession (North Sea) (According to Price et al. 2010)

We can note that no disturbance due to cathodic protection (DC) has a priori been observed so far (Fig. 3.46).[39]

Wells may also interfere. However, their vertical structure (completion and metal casing) should theoretically have lesser and more localized impacts on the data. Their trajectory can also be mapped by downhole directional type logging tools (Kong et al. 2009; Sainson 2010).

6.9 *Measurement Conditions Favorable in Electromagnetotelluric Detection*

As a general rule, the mEMT method (or mMT) uses a source of natural currents on a spectrum at very low frequencies, relatively wide, and where each band thereby defines the depth of investigation (cf. Sect. 3.3).

EM controlled source seabed logging has relatively favorable local conductivity conditions for depth investigation of a few hundred meters at most (low resistivity and high contrast). It is quite otherwise with mMT, which must consider investigating at much greater depths (several thousand meters) where the resistivity of the surrounding grounds then greatly increases with depth (less and less important presence of infiltration of saline water), thereby decreasing the contrast with the reservoir.

[39]The anodes only debit when there is a loss of electrical insulation; steel is then in contact with seawater (Sainson 2007).

6.9.1 Need to a Priori Identify Geological Criteria

In this very particular case of research, rather unfavorably, Archie's formula (cf. Eq. 3.9) allows us, relatively speaking, to predict to what extent the geoelectric conditions are most suitable for detection. More specifically it can be said that:

- For the reservoir, these conditions will be all the more interesting when its thickness and porosity and the temperature and salinity of its formation water are high.
- For the sedimentary series above and under the reservoir, these conditions will be and when their lateral facies and power variations are regular and progressive.

It is difficult to set orders of magnitude, since these parameters must be very closely examined relative to each other, all the more exactly as in these depths tectonics may also play a significant role. Clearly the possession, for the explored area, of seismic surveys (for structural features) and electric logs (for the electromagnetic properties of the layers) will be undeniable advantages to raise all indeterminacies.

We would like to remark that the depth of the reservoir does not directly intervene in theory. However, as the resolution power decreases with it, the targets should be all the more important than they are deep, which is valid for virtually all known potential geophysical methods (e.g., gravity).

6.9.2 Need for Seabed Measures

At the sea surface, the conductive water of conductivity σ is in perpetual movement (swell, waves, etc.). The local current density then depends not only on $\vec{E}$ (case of a conductor at rest: $\vec{J} = \sigma\vec{E}$) but also on the speed of movement $\vec{v}$ of the water mass in the terrestrial magnetic field $\vec{B}^T$. In this new case (of a moving conductor) the resulting electric field $\vec{E}'$ is then equivalent to:

$$\vec{E}' = \vec{E} + \vec{v} \wedge \vec{B}^T \tag{3.82}$$

The current density naturally becomes:

$$\vec{J} = \sigma\left(\vec{E} + \vec{v} \wedge \vec{B}_z^T\right) \tag{3.83}$$

For example, considering a water movement speed of 0.1 m/s, and a value of the vertical component of the terrestrial magnetic field $\vec{B}_z^T$ of 3.10^{-4} Wb/m^2 at this place (Cox et al. 1968), we obtain in the considered water column an additional field (Lorentz effect) of about 3 μV/m, i.e., of about the same order of magnitude as that coming from terrestrial signals in the same band ($\approx 10^{-2}$ Hz; mMT). At several

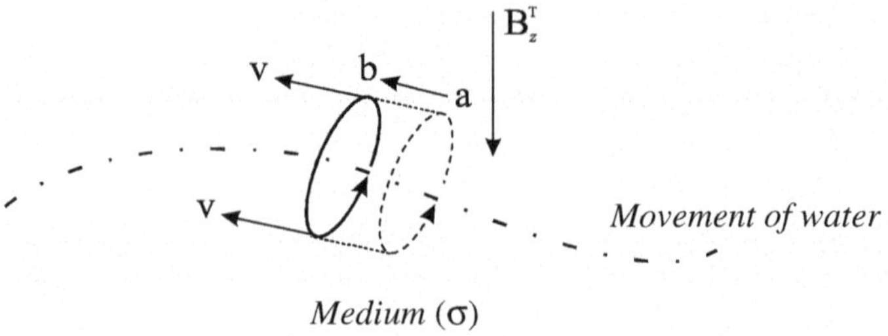

Fig. 3.47 The displacement (a → b) of the conductive mass of water in the earth's magnetic field induces an extra field: $\vec{v} \wedge \vec{B}_z^T$

Hz, generated by the water flow and the turbulence in the mCSEM bandwidth, the additional field is not more than a 1 μV/m (Fig. 3.47).

Seabed ocean currents (terrestrial tides, the Gulf Stream, etc.) can affect the measurements (Filloux 1987; Cox et al. 1978.). However, their longer and clearly identified periods can be eliminated by appropriate treatments (mechanical or electronic filters).

Furthermore, if it is also admitted that the pressure and temperature variations between the surface and the seabed change the chemical potential of water and we express it by a gradient $-\vec{\nabla}\zeta$, the current density is also assigned an additional term (Landau and Lifchitz 1969) proportional to this gradient, such as:

$$\vec{J} = \sigma\left(\vec{E} - a\vec{\nabla}\zeta\right) \tag{3.84}$$

the latter then being all the more lower than it is deep (a).

In order to eliminate interference voltages, it is desirable, in any event, to carry out field measurements on the seabed where marine movements and temperatures are minimal, the chemical potentials lower and the sensors immobilized by their anchorage. This also has the advantage of limiting the voltages induced by the movement of the arms supporting the sampling electrodes. Under these conditions, the equations (Cf. Eqs. 3.82 and 3.83) reduce to Ohm's law applied to a conductor at rest, ie., $\vec{J} = \sigma\vec{E}$

We would like to remark that in mMT prospecting theory shows that MT relations deduced from Maxwell equations, for a certain depth of the subsoil, are independent of those on the electromotive forces of the cut flow induced in the sea (Cagniard and Morat 1966).

6.9.3 Need to Take the Water Depth into Account

Apart from attenuation phenomena inherent to the intrinsic diffusion of waves in the water (see Chap. 4, Sect. 4.5.4), it is theoretically shown that the values of the electromagnetic fields are sensitive to the presence of the seabed in an even more important way when the water layer is thin (see Sect. 3.6.2) and the conductivity of the underlying ground is high.

Furthermore, it has been shown experimentally[40] that there is a significant decrease in these fields values with a concomitant increase in the water depth and the proximity of the bottom.

In these circumstances we accept that it is necessary for a correct interpretation of the measurements to imperatively make a reduction of the raw marine data[41].

6.9.4 Need to Control Anthropogenic Activity

Measurements can also be affected by electromagnetic interferences due to the presence of the survey vessels (ferromagnetic hull) and to their movement in the conducting water and the local geomagnetic field.

Those interferences:

- Of a static nature correspond to galvanic, electrochemical and electrokinetic currents (polarization cells, corrosion currents, stray currents, cathodic protection, etc.)[42]
- Of a magneto- and electrodynamic nature are equal to the eddy currents

They can also come from electric currents in the onboard equipment (engines, machinery, lighting, generators, cables, etc.), which circulate and escape the boat.

A few hundred meters away from the ship, the amplitudes of the electric and magnetic interference fields are respectively of the order of 1 μV/m and 10 pT on a frequency band of a few Hz.

These signals are easily locatable by identifying their frequencies a priori (location, identification and in situ measurements) and estimating their specific signature a posteriori[43] (Schermber 2006; Guibert 2009) (Fig. 3.48).

These anthropogenic noises finally need to be taken into account for all structures and industrial activities present on the prospect (harbor area, military and civilian maritime traffic, oil platform, subsea infrastructure, pipeline, etc.).

SBL measurements should therefore be made at several hundred meters at least from these amenities where these interference fields should absolutely be taken into account in filtering and correcting devices of measuring systems or data interpretation (cf. Chap. 6, Sect. 5.2).

---

[40]Campaign led by the LETI (Commissariat for Atomic Energy) in the Mediterranean in July 1991.

[41]Additional measurements simultaneously realized on land.

[42]The means of propulsion of the ship, the shaft and propeller especially, produce (by their more or less regular rotation in sea water) alternating currents that are variable on the fundamental frequencies and their harmonics (hash of the lines of force static → modulation of the currents → BF variable fields).

[43]After determining the characteristic frequencies (captures and measures), electromagnetic signatures are obtained by modeling (solving Maxwell's equations).

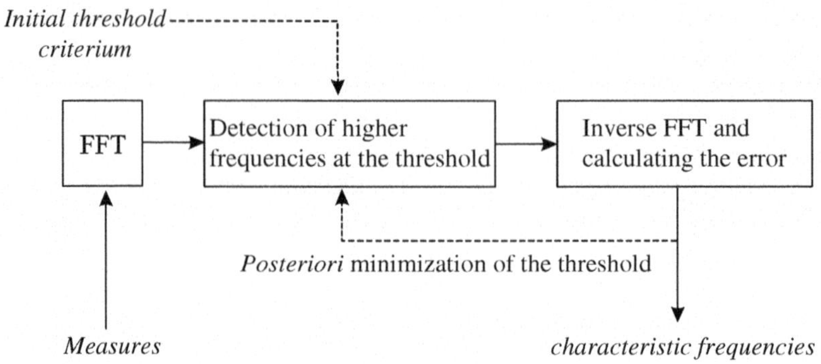

Fig. 3.48 Identification device (algorithm) of the characteristic frequencies of the propulsion engines processed by FFT (*fast Fourier transform*: numerical variation of the classical Fourier transform, and today the most widely used mean to quickly process the signals See, for example, the book by J. March et al.: *Signal Processing for Geologists and Geophysicists* (2004). According to Schermber 2006)

6.9.5 Need to Correct the "Coastal Effect"

The *coastal effect* was discovered in the nineteenth century.[44] It is the result of, among other things, the increase in the electrical conductivity of the grounds located along the coasts, now focusing the flow of telluric currents around the continents, and thus changing the spatial distribution of electric and magnetic fields in these special places (see Chap. 4, Fig. 3.15).

6.9.6 Need to Establish a Developed Transfer Function

From what has just been said, we easily understand that for a correct interpretation of telluric measurements, it is then necessary to use a transfer function much more complex than that recommended in mCSEM. It must take into account more precisely and in real time:

- The local effects such as electrical conductivity and the movement of the seawater
- The effects of the bathymetry (water depth) and the nature of the subsoil (resistivity)
- The regional effects such as the coastal effect

To establish these essential corrections, it is therefore necessary to have additional devices to evaluate these parameters (resistivity and current meters, a depth

[44]Effect demonstrated for the first time in 1879 by British scientists of the Post Office Telegraph Services (Mathias et al. 1924) and seriously studied from the late 1950s (Parkinson 1959, 1962; Coquelle and Mosnier 1969; Filloux 1967; Cox and Filloux 1974; Larsen 1975; Le Mouel and Menvielle 1976; Mosnier 1977).

sounder, etc.), to systematically perform reductions of marine data (as terrestrial variometers) and to own the associated interpretation software taking into account all these parameters.

6.10 *Sequence of an Operation of EM Seabed Logging*

To obtain good results (with respect to quality assurance), an SBL operation must absolutely follow a certain number of time steps. We briefly review some of them.

6.10.1 Preliminary Studies

The operation must be imperatively defined at the beginning with the client (preliminary studies), who sets objectives and delivers all necessary documentation regarding the prospect to establish the NAR. This information is also essential to prepare the mission and subsequently establish the process of inversion of the electromagnetic data that will be used for geological interpretation. In the case of an exploration of a new maritime area (block), additional operations (including seismics) are programmed. From the available elements, operators analyze the different possible geological scenarios (Fig. 3.49).

6.10.2 Survey Preparation

Except in special cases, the preparation of the SBL operation is the unique responsibility of the service provider who performs the survey. He is technically and legally responsible. He must implement all necessary means for the proper performance of the acquisition operations (logistics, mapping, positioning, security, etc.). This preparation must be all the more careful when the means are important (high tonnage ship, specialized teams, etc.) and beforehand requires all necessary permissions to cross territorial areas.

6.10.3 Data Acquisition

Aside from the techniques, the data acquisition must fulfill a certain number of requirements and meet certain criteria.

We can mention:

- *Quality control*, which relies on a quality approach to measuring instruments and the ethics of the personnel; as stated by some authors, "no bias should be introduced either by the data acquisition companies, or by oil companies … " (Theys 2006)

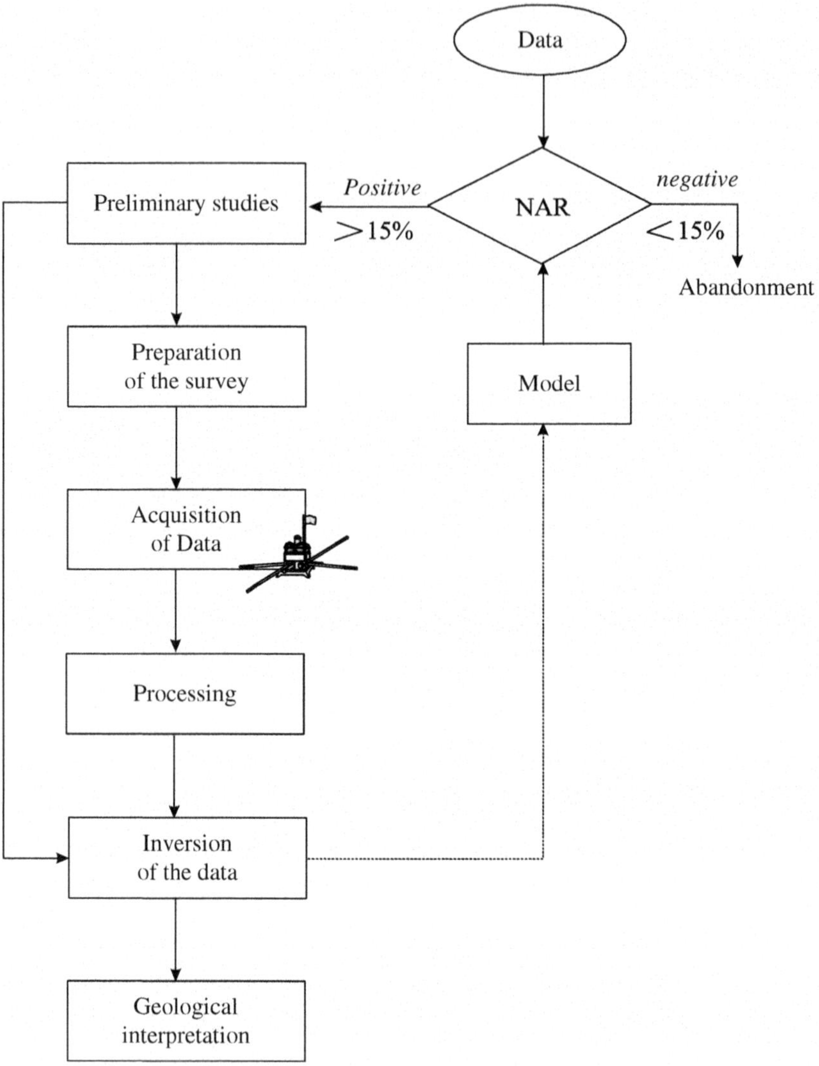

Fig. 3.49 Summary of the different phases of an SBL operation and of the preliminary studies to the final geological interpretation

- *Standards*, such as HSE standards (health, safety and environment) and good practice guides that are built over the operations and experience
- *Certification*, which involves notions such as standards, expertise, and independent external audit offices

6.10.4 Processing

Processing is a technical step that allows us to condition and process data to prepare modeling and inversion. The processing is done on the data analysis and formatting to make them homogeneous, coherent, and assimilable to mathematical models and to form 2D and 3D simulations in particular.

6.10.5 Simulation: Modeling and Inversion

Simulation, the phase that includes modeling (forward problem) and inversion made from the collected data, is an essential step in the interpretation of the data. This step requires extremely important calculation means (supercomputer, parallel computation, etc.) and highly qualified personnel (geophysicists). The quality and quantity of the external information are practically the guarantee of the quality of inversion[45] and of the success of the geological interpretation operation that follows.

6.10.6 Geological Interpretation

Geological interpretation is the final phase of the exploration process. Its results and conclusions correspond to the final document requested by the client. It should conclude with a strategic decision: to abandon, continue the geophysical investigations, or finally drill. This decision is generally that of the orders' providers, i.e., the geologist and the future (oil) producer under the guidance of the service company that has operated.

7 Conclusion

Metrology has set, according to the local geoelectric environment and the objectives to be achieved (geological target), the system performances to develop and the adjustments to be made. Furthermore, it states the means to implement to avoid unwanted signals. Theoretical calculations also show that the expected fields are very small (a billion times smaller than a small watch battery) but also—which is a nice surprise—that the seabed electromagnetic noise is also very weak, leading therefore to a signal-to-noise ratio that is very favorable for good detection. It now remains to develop low noise electronics (below the ambient noise) ensuring the entire measuring device sufficient resolution to address the interpretation phase with maximum precision and chance of success (Figs. 3.50 and 3.51).

[45]EM seabed logging has not been considered practically in its principle without the help of sophisticated methods of interpretation, particularly those involving data inversion methods that have emerged in practice in the last decade. Without the latter, which require the provision of additional external information (including seismic data), the measures could not take all their senses.

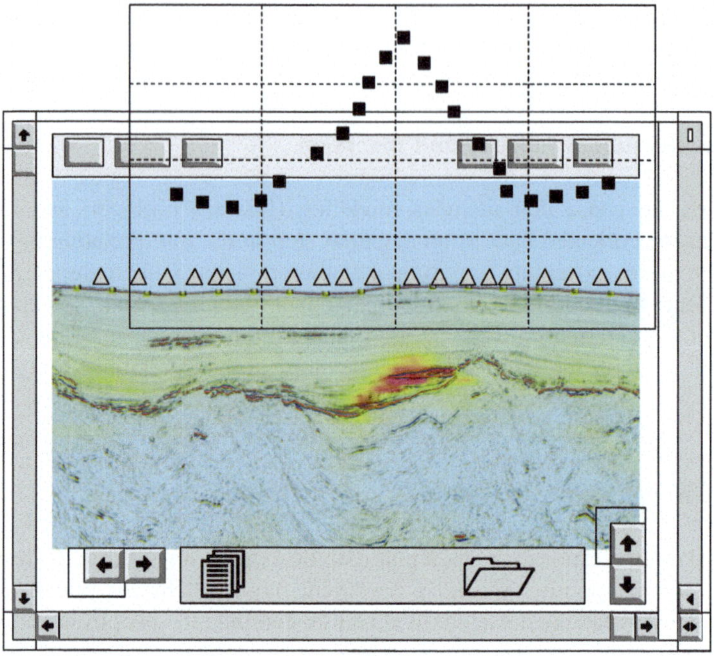

Fig. 3.50 Interpretation by inversion of electromagnetic data (SBL) on a seismic profile (vertical plan: section) (According to Johansen et al. 2008)

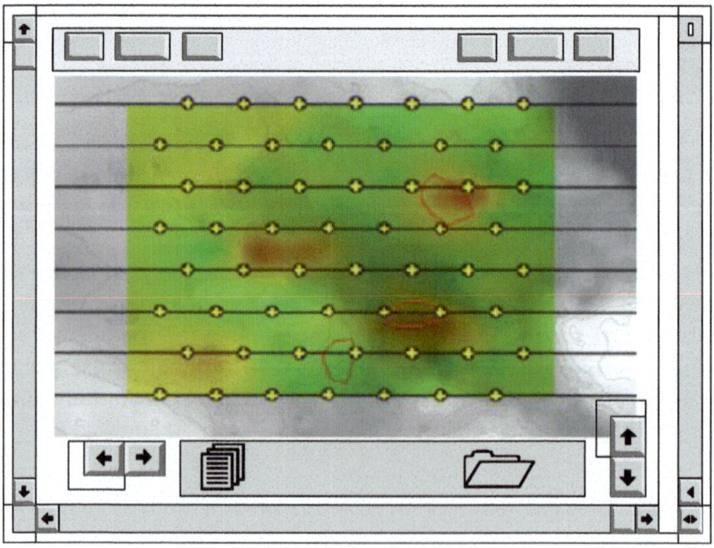

Fig. 3.51 Interpretation by inversion of electromagnetic data (SBL) on a set of profiles (horizontal plane). The crosses indicate the location of the measurements (According to Ellingsrud et al. 2004)

References

Abrikossov I, Goutman I (1986) Géologie du pétrole. Généralités, prospection, exploitation. Ed. Mir, pp 111–114

Admundsen L et al (2006) Decomposition of electromagnetic field into up-going and down-going components. Geophysics 71(5):G211–G223

Ambronn R (1928) Elements of geophysics as applied to explorations for minerals, oil and gas. Ed. McGraw-Hill, New York, p 177

Andreis D, MacGregor L (2008) Controlled source electromagnetic sounding in shallow water. Princip Appl Geophys J 73:21–32

Archie GE (1942) The electrical resistivity log as an aid in determining some reservoir characteristics. Petrol Trans AIME 146:54–62

Bahr DE, Hutton J, Syvitski J, Pratson L (2001) Exponential approximations to compacted sediment porosity profiles. Comput Geosci 27:691–700

Bannister PR (1980) Quasi-static electromagnetic fields. Scientific and engineering studies. Ed. Naval Underwater Systems Center. Newport Laboratory

Bannister PR (1987) Simplified expressions for the electromagnetic fields of elevated, surface, or buried dipole antennas. Compiled 87, vol 1, Scientific and Engineering Studies, Naval Underwater Systems Center, New-London, Connecticut

Becker K (1985) Large scale electrical resistivity and bulck porosity of the ocean crust, deep sea drilling project hole 504B. Costa Rica rift Init Rept DSDP 83:419–427

Becquerel M (1834). Traité expérimental de l'électricité et du magnétisme et de leurs rapports avec les phénomènes naturels. Ed. Firmin Didot frères

Becquerel M, Becquerel E (1847) Elément de physique terrestre et de météorologie. Ed. Firmin Didot frères, pp 655–662

Benderitter Y, Dupis A (1969) Prospection géophysique par la méthode magnéto-tellurique, vol 4. Ed. Gauthier-Villars, Paris, pp 251–273

Bitterlich W, Wöbking H (1972) Geoelektronik. Angewandte Elektronik in der Geophysik, Geologie, Prospektion, Montanistik und Ingenieurgeologie. Ed. Springer, Berlin, pp 115–117

Born M, Wolf E (1964) Principles of optics. Electromagnetic theory of propagation, interference and diffraction of light. Ed. Pergamon Press, Oxford, 808 p

Bruck R (1841) Electricité ou magnétisme du globe terrestre. Ed Imprimerie de Delevingne and Cllewaert Tome 1:41–52

Cagniard L, Morat P (1966). Extension de la prospection magnéto-tellurique à l'exploration offshore

Chave AD (2009). On the electromagnetic field produced by marine frequency domain controlled sources. Geophys J Int:29 p

Chave AD, Cox CS (1982) Controlled electromagnetic sources for measuring electrical conductivity beneath the oceans. 1. Forward problem and model study. J Geophys Res 87 (87):5327–5338

Chave AD, Luther DS (1990). Low frequency, motionally induced electromagnetic fields in the ocean. J Geophys Res 95(c5):7185–7200

Chen J, Alembaugh D (2009) Three methods for mitigating air waves in shallow water marine CSEM data. SEG Houston. International exposition and annual meeting, pp 785–789

Chouteau M, Giroux B (2006) Géophysique appliquée. Méthodes électriques. Notes de cours. Ecole Polytechnique, Montréal, p 11

Constable S (2010) Ten years of marine CSEM for hydrocarbon exploration. Geophysics 75 (5):75A67–75A81

Constable S et al (2008) Mapping offshore sedimentary structure using electromagnetic methods terrain effects in marine magnetotelluric data. Geophys J Int. doi:10.1111/j.1365-246X.2008.03975.x

Coquelle J, Mosnier J (1969) Etude de l'effet « bord de mer » dans la région de Quiberon., Mémoire ENS

Cox CS, Filloux JH (1974) Two dimensional numerical models of the California electromagnetic coastal anomaly. J Geomag Geoelectr 26:257–267
Cox CS et al (1968) Electromagnetic studies of ocean currents and electrical conductivity below the ocean floor, vol 4, part I. Ed. Willey Interscience, pp 637–693
Cox CS et al (1978) Electromagnetic fluctuations induced by wind waves on the deep-seafloor. J Geophys Res 86-Cl:431–442
d'Arnaud Gerkens JC (1989) Foundation of exploration geophysics. Method in geochemistry and geophysics, vol 25. Ed. Elsevier, Amsterdam, pp 541–559
Duroux J (1974) Procédé et appareil de prospection en mer par mesure de champ électromagnétiques. Brevet français no. 74 26465
Edwards N (2005). Marine controlled source electromagnetics: principles, methodologies, future commercial applications. Survey in Geophysics, pp 675–700
Eidesmo T, Ellingrud S, MacGregor L, Constable MC, Sinha S, Johansen FN (2002) Sea bed logging (SBL), a new method for remote and direct identification of hydrocarbon filled layers in deepwater areas. First Break 20.3 March, pp 144–152
Ellingsrud S et al (2004) Electromagnetic methods and apparatus for determining the content of subterranean reservoirs. US patent no. 6696839, Feb 24
Ellis DV, Singer JM (2007) Well logging for earth scientists. Ed. Springer, p 45 et pp 222–223
England and Wales high course (2009) Decision of justice between Schlumberger Holding Ltd and Electromagnetic Geoservices As (EMGS), p 5
Evjen HM (1938) Depth factors and resolving power of electrical measurements. Geophysics 3 (1–4):78–95
Fielding BJ, Lu X (2009) System and method for towing sub-sea vertical antenna. US patent no. 7541996, June 2
Filloux JH (1967) Oceanic electric currents, geomagnetic variations and the deep electrical conductivity structure of the ocean-continent transition of central California. PhD thesis, University of California, San Diego
Filloux JH (1987) Instrumentation and experimental methods for oceanic studies. Geomagnetism, vol 1. Ed. Jacob, pp 143–248
Fofonoff N, Millard RC (1983) Algorithms for computation of fundamental properties of seawater. UNESCO Tech Pap Mat Sci 44, 53 pp
Grand FS, West GF (1965) Interpretation theory in applied geophysics. McGraw-Hill, New York
Guegen Y (1997) Introduction à la physique des roches. Ed. Hermann, Paris
Guérin R (2007) Profondeur d'investigation en imagerie de résistivité électrique. 8ème colloque GEOFCAN, Bondy, France, 25–26 September, pp 27–30
Guibert A (2009) Diagnostic de corrosion et prédiction de signature électromagnétique de structures sous-marines sous protection cathodique. Doctoral thesis, Université de Grenoble, 218 p
Hamon BV (1958) The effect of pressure on the electrical conductivity of the seawater. J Mar Res 16:83–86
Harrington RF (1961) Time harmonic electromagnetic field. McGraw-Hill, New York, 480 p
Heaviside O (1899). Note on the electrical waves in seawater. Electromagnetic theory, pp 536–537. Ed. E. Benn
Heshammer J, Fanavoll S, Stefatos A, Danielsen J E and Boulaenko M (2010) CSEM performance in light of well results. The Leading Edge, January, pp 258–264
Horne RA, Frystinger CR (1963) The effect of pressure on the electric conductivity of the seawater. J Geophys Res 68:1967–1973
Ingmanson DE, Wallace WJ (1989) Oceanography: an introduction, 4th edn. Ed. Wadsworth Publishing, Belmont
Jackson JD (1998) Classical electrodynamics. Ed. Wiley, New York
Jiracek GR (1990) Near surface and topographic distortions in electromagnetic induction. Survey Geophys 11:162–203
Johansen et al (2008) How EM survey analysis validates current technology, processing and interpretation methodology. First Break 26, June, pp 83–88

Kaushalendra Mangal Bhatt (2011) Motion induced noise in marine electromagnetic data. Thèse
Key K (2009) 1D inversion of multicomponent, multifrequency marine CSEM data: methodology and synthetic studies for resolving thin resistive layers. Geophysics 74(2):F9–F20
Key K, Lockwood A (2010) Determining the orientation of marine CSEM receivers using orthogonal procrustes rotation analysis. Geophysics 75(3)
Kiyoshi B, Nobukazu S (2002) A new technique for incorporation of seafloor topography in electromagnetic modelling. Geophys J Int 150:392–402
Knudsen M (1901) Hydrografische, Tabellenth edn. Friedrichsen, Hamburg, 63 p
Kong JA (1986) Electromagnetic wave theory. Wiley, New York
Kong FN et al (2009) Casing effects in the sea-to-borehole electromagnetic method. Geophysics 14(5):F77–F87
Kraichman MB (1976) Handbook of electromagnetic propagation in conducting media. Ed. Navy Department Bureau of Ships, Appendix A.4
Kunetz G (1966) Principles of direct current resistivity prospecting. Ed. Bebruder Borntraeger, Berlin, p 9
Landau L, Lifchitz E (1969) Electrodynamique des milieux continus. Ed. Mir, Moscou, pp 150 and 352
Larsen JC (1975) Low frequency (0.1–6.0 cpd) electromagnetic study of deep mantle electrical conductivity beneath the Hawaiian Islands. Geophys J R Astr Soc 43:17–46
Le Mouel JL, Menvielle M (1976) Geomagnetic variation anomalies and deflection of telluric currents. Geophys J R Astron Soc 68:575–587
Li Y, Constable S (2007) Marine control source: 2D marine controlled source electromagnetic modelling. Part 2: effect of bathymetry. Geophysics 72(2):63–71
Li YG, Constable S (2010) Transient electromagnetic in shallow water: insights from 1D modelling. Chin J Geophys 53:737–742
Lodge OJ (1892) Modern views of electricity. Nature series. Ed. Macmillan, London, p 94
Loseth LO (2007) Modeling of controlled source electromagnetic data. PhD thesis, Université norvégienne, pp 167–168
Loseth LO et al (2006) Low frequency electromagnetic fields in applied geophysics: waves or diffusion? Geophysics 71(4):W29–W40
MacGillivray PR, Oldenburg DW (1990) Methods for calculating Fréchet derivatives and sensitivities for non-linear inverse problem: a comparative study. Geophys Prospect 38(5):403–416
MacGregor L, Andreis D, Tompkins M (2007) Electromagnetic surveying for resistive or conductive bodies. US patent no. 2007/0288211, December 13
MacGregor L et al (2008) Electromagnetic surveying for hydrocarbon reservoirs. US patent no. 7,337,064, Feb 26
Meyer JJ et al (1969) Handbook of ocean and underwater engineering. Ed. McGraw-Hill, New York, p 3.38
Mittet R (2008) Normalized amplitude ratios for frequency domain CSEM in very shallow water. First Break 26:47–54
Mosnier J (1977) Le sondage magnétique différentiel. Courrier du CNRS, no. 26, October, pp 6–14
Nover G (2005) Electrical properties of crustal and mantle Rocks : a review of laboratory measurements and their explanation. Survey in geophysics, vol 26. Ed. Springer, pp 593–651
Ohm GS (1827) Die galvanische Kette, mathematisch bearbeitet. Ed. TH Riemann, Berlin, 246 p
Parkhomenko EI (1967) Electrical properties of rocks. Ed. Plenum Press, New York, 314 p
Parkinson WD (1959) Direction of rapid geomagnetic fluctuations. Geophys J R Astron Soc 2:1–14
Parkinson WD (1962) The influence of continents and oceans on geomagetic variations. Geophys J R Astron Soc 4:441–449
Pavlov D et al (2009). Method for phase and amplitude correction in controlled source electromagnetic survey data. US patent no. 2009/0133870. May 28
Pirson JP (1950) Element of oil reservoir engineering. Ed. McGraw-Hill, New York, p 58

Price A, Mikkelsen G, Halmilton M (2010) 3D CSEM over Frigg: dealing with cultural noise. SEG Denver, annual meeting, pp 670–674

Razafindratsima S et al (2006) Influence de la teneur en eau sur les propriétés électriques complexes des matériaux argileux. Laboratoire IRD/UMPC

Robert M (1959) Géologie des pétroles. Principes et applications. Ed. Gauthier-Villars, pp 87–90

Rosten T, Admundsen L (2008) Electromagnetic wave field analysis. UK patent no. 2412739

Rosten T et al (2005) Electromagnetic seabed logging: a proven tool for direct hydrocarbon identification. OTC paper 17483, Houston, 5 p

Rubin Y, Hubbard S (2006) Hydrogeophysics. Water science and technology library. Ed. Springer, Dordrecht, p 196

Sainson S (2007). Inspection en ligne des pipelines. Principes et méthodes. Ed. Tec et Doc, Lavoisier, pp 231–242 and pp 73–74

Sainson S (2008) EM response to conductive structures immerged in conductive media. Confidential report

Sainson S (2009) EM subsea pipelines effect on mCSEM survey. Confidential report

Sainson S (2010) Les diagraphies de corrosion. Acquisition et interprétation des données. Ed. Tec et Doc, Lavoisier, pp 77–81

Schermber JL (2006) Identification et caractérisation de sources électromagnétiques. Application à la discrétion des moteurs de propulsion navale. Doctoral thesis, Université de Grenoble, 229 p

Seguin MK (1971) La géophysique et les propriétés physiques des roches. Ed. Les presses de l'université de Laval, Québec, p 130

Serra O (2004) Well logging, vol 1 Data acquisitions and applications. Ed. Technip, 688 p

Skilling HH (1942) Fundamentals of electric waves. Ed. Wiley, pp 118–122

Stefatos et al (2009) Marine CSEM technology performances in hydrocarbon exploration: limitations or opportunities. First Break 27, May, pp 65–71

Stratton JA (1961) Electromagnetic theory. Ed McGraw-Hill, p 314 et p 374

Sunde ED (1948) Earth conduction effects in transmission systems. Ed. D Van Nostrand, New York, p 20

Theys P (2006) Utilisations des données pétrophysiques en temps réel et pendant toute la vie d'un champ: critères de qualité et d'éthique. Communication SAID (SPWLA), Schlumberger, Clamart, 5 April

Thomsen, Menache (1954) Instructions pratiques sur la détermination de a salinité de l'eau de mer par la méthode de Mohr-Knudsen. Bulletin de l'institut océanographique, no. 1047, 30 July

Tompkins MJ, Weaver R, MacGregor LM (2004) Sensitivity to hydrocarbon targets using marine active source EM sounding: diffusive EM imaging methods. EAGE. Ed. OHM

Von Aulock W (1953) The electromagnetic field of an infinite cable in seawater. Bureau of Ships Minesweeping Branch technical report, no. 106

Walker PW, West GF (1992) Parametric estimators for current excitation on a thin plate. Geophysics 57:766–773

Warburton F, Caminiti R (1964) Champ magnétique induit par les vagues de la mer. Bull Am Phys Soc 2(3):347

Webster AG (1897) The theory of electricity and magnetism being lectures on mathematical physics. Ed Macmillan, London, pp 534–540

Weiss CJ (2007) The fallacy of the shallow water problem in marine CSEM exploration. Geophysics 72(6):A93–A97

Whitchead CS (1897) The effect of seawater on induction telegraphy. Physical Society

Chapter 4
Instrumentation and Equipment

The difference between theory and practice is that in theory there is no difference between theory and practice, but in practice, there is one.

(Jan Van de Snepscheut)

Abstract This chapter describes in detail (electrical, electronic, mechanical and signal processing characteristics) the various measuring instruments used in prospecting, i.e., the electrical current sources and field receivers. The transmitters are mobile dipoles of very low frequency able to deliver currents of several hundred amperes under the water, or even telluric sources caused by magnetic storms (solar wind) or atmospheric storms (Schumann resonance). Seabed fixed receivers are vector electrometers and magnetometers able to record the horizontal components of electric and magnetic fields in a bandwidth covering the frequencies of 0.01 to 10 Hz. The accuracies of these low noise instruments (several $\mathrm{nV}/\sqrt{\mathrm{Hz}}$) are respectively about one $\mu\mathrm{V}/\sqrt{\mathrm{Hz}}/\mathrm{m}$ and one nT. Generally the signal-to-background noise ratio is very favorable to the detection and can reach a factor of 1000. Finally, we briefly mention the operational means and procedures with the different equipment and a few elements of signal processing used during the seabed logging survey.

Keywords Electrometer • Magnetometer • Extremely low frequencies

Preamble

Please note that the descriptions and technical developments that follow only apply to fields measuring instruments, whether artificial (controlled source) or natural (telluric source). The equipment used for DC investigations is the same

S. Sainson, *Electromagnetic Seabed Logging*, DOI 10.1007/978-3-319-45355-2_4

except for a few details such as onshore materials and therefore are not described here.[1]

1 Introduction

The instrumentation physically corresponds to the practical implementation of the metrology developed in the previous chapter. It describes the equipment and acquisition materials (hardware) including power sources (transmitters), the measuring instruments themselves (receivers) and operational resources currently used in the offshore industry or in the phase of being so in the near future (vessels).

2 Measurement Requirements: Quality Criteria for the Instruments

Electrometric and magnetometric investigation, on one hand, requires us to measure electrical and magnetic fields of very low amplitude and, on the other hand, to assess the background noise, which should not exceed the EM field in absolute values. In these very unfavorable conditions for acquisition, the meter must meet extremely strict instrumental imperatives, which will directly influence its design, performance and in situ conditions of use. These will finally give it an acceptable (or not) quality criterion.

2.1 First Imperative: Power Consumption and Electromagnetic Discretion

Like all measuring precision devices, the field sensor must not interfere with the magnitudes that must be considered. In the specific case of the electrometric measures, this overriding criterion is even more true when the device is itself immersed in the measuring medium.

This requirement will then require that:

- Internally to the instrument, the amount of energy required for its operation is negligible compared to that which may be taken, i.e., the one indirectly coming from the information source (subsoils)

[1]These instruments, such as *SYSCAL pro Deep Marine*, are generally marine versions of onshore resistivity meters (waterproof and treated against marine corrosion according to international standards). The reader may refer to conventional prospecting works (Koefoed 1979 and more recent works) and to the instructions of the manufacturers (Iris instruments, Scintrex, etc.) for more information on these materials.

– Externally, the surrounding environment is not disturbed by the forms of the instrument and its electromagnetic properties, or at least the electromagnetic disturbance resulting from its presence is offset by adapted means

In general, we do not control the energy source, whether artificial (displacement, power, variable amount due to remoteness, etc.) or natural (random telluric currents). The measuring instrument must therefore be characterized by a near-zero energy levy—in other words, its intrinsic consumption should be as low as possible or even negligible in relation to the quantities to be measured.

In addition, the system must be designed so that its presence is, from the electromagnetic point of view, as "transparent" as possible with respect to the local environment (seawater), both on the static and dynamic plan, adopting for that a strategy either a priori of electromagnetic discretion or a posteriori of correction of the faults. In most cases, to enhance the efficiency of the system, these two complementary techniques are allowed and practiced, the second one also being involved in the measurement process itself.

2.2 Second Imperative: Correctness

The second imperative, no less important, concerns the output indication of the measuring instrument, which must only depend on the input variable. This means that the instrument should neither add nor subtract parasitic quantities of the same species but of different origin (usually at random). This is the case, for example, of the amplification of the signals, which generates noise and drift, for example when the latter are relatively slow. It is then imperative to control these phenomena by adapted signal processing and if necessary by means of specific compensation. This correctness[2] requirement is common to all precision measuring devices but is especially important for those performing measurements (voltage or amperage) of very low amplitudes in disturbed environments (in the presence of ambient noise).

2.3 Third Imperative: Fidelity

The instrument shall be true, i.e., it must not distort or alter the applied input signals. Otherwise, the distortions must be detected, quantified and corrected by suitable passive or active processes.

[2]That is the tendency of the device to be closer to the actual value of the measured events. Errors affecting the *correctness* come from stable defects of the tool, poor or inadequate *calibrations*, a bad choice of *scales of corrections*, etc.

2.4 Additional Qualities

To these three major requirements, you can add the qualities of:

- *Linearity*, which represents the constant of the ratio of the output signal to the input signal, depending on the characteristics of the input signal (in particular the amplitude, frequency, and phase)
- *Sensitivity*,[3] which describes the ability of the instrument to measure small signals or highlight low variations more or less significant on average (resolving power)
- *Accuracy*, which represents the evaluation of the absence of errors from the sensor (cf. Chap. 3, Sect. 6.6)

which are, after all, requirements common to all quality measuring precision devices.[4]

2.5 Signal-to-Noise Ratio

In the measurement and recording of small amounts of energy (electric fields), the noise (N) is a determining factor in the validity of the measure (S). It can be of different origin, natural or artificial, internal or external to the evaluation apparatus. In all cases it is essential to identify and to minimize it to the maximum extent in order to obtain the highest possible signal-to-noise ratio (S/N) (Fig. 4.1).

This aim is achieved on one hand by taking measurements in an environment as quiet as possible (the bottom of the sea, away from sources of anthropogenic noise) and on the other hand by processing the signals with suitable means (low noise electronic and *homodyne detection* for example) (Table 4.1).

[3]The *sensitivity* is the ratio of the variations INPUT/OUTPUT.

[4]The metrological quality, which can be considered as the ability of the instrument to give consistent results for the requested requirements, also depends on many factors such as:

- The *uncertainty*, which represents the amplitude of the field of measurements in which the true value is located with a given probability (degree of dispersion)
- The *repeatability*, which is an in situ estimation of the *accuracy*. It is the difference between two distinct measurements of the same event, made under the same conditions and with the same sensor
- The *stability*, which is the ability of the meter to give the same results in the same conditions
- The *reproducibility*, which is the ability to differentiate two measures with the same method under the same conditions with different equipment and possibly different personnel, etc.

These intrinsic qualities can be evaluated theoretically for some of them and practically for others (*calibration*, *control section*, etc.). They can be related to *reliability studies* (fault tree or FT, mean time between failure or MTBF, etc.).

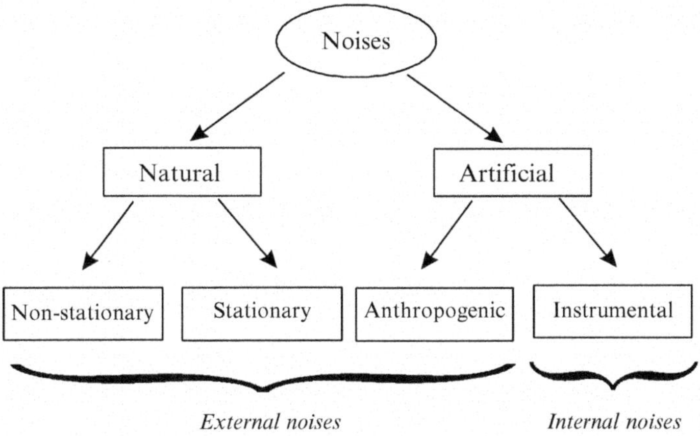

Fig. 4.1 Nature of the different noises that can damage measures with significant errors. Moreover, these noises can be consistent or not, and therefore should be treated in a differentiated manner (According to Khesin et al. 1996)

Table 4.1 Summary of metrological data (magnitude of electric field values) that define the sensitivity of measuring equipment

Electrical field at 10 km mCSEM (*injection 1500 A, antenna 200 m length*)		
EM noise (*telluric field*)	Seabed (*direct field*)	Geological formations (*refracted field*)
≈ several pV/m	≈15 nV/m	≈ several nV/m

See Chap. 3

3 History of Measuring Equipment

Measuring equipment has been developed for different frequency bands corresponding to various applications ranging from telecommunications to detection and underwater surveillance (Bruxelle 1995), via, of course, earth physics and applied geophysics.

One can roughly define three LF (*low frequency*) bands, the first of which, corresponding to the lowest frequencies, can be used in the context of SBL (Fig. 4.2).[5]

The first submersible equipment[6] measuring electromagnetic fields more specifically allocated to electric fields[7] was constructed in the 1960s. The aim was to

[5]Longer times correspond to magnetic surveys used in global geodynamics (very deep layers).

[6]Profiling in a continuous regime used, from the 1930s onward, wireline equipment (electrical coring), slightly modified, whose characteristic cables (named flute) was their tightness. The measuring equipment consisted of a potentiometer located on land or in a boat (see Introduction, photographs, Figs. 1.7, 1.8 and 1.9).

[7]At the same time, magnetic variometers of more or less long periods were built and tested (see Sect. 5.7).

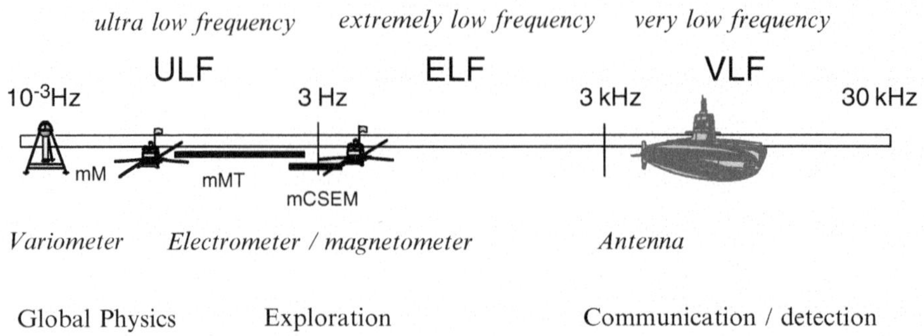

Fig. 4.2 Names of the frequency bands and associated spectra. Only the ULF bands (mMT) and ELF bands (mCSEM) are used in marine geophysical investigations. The VLF band, also reserved for subsurface underwater communications, is however used in onshore prospecting surveying (mining geophysics)

highlight the *Schumann resonances* (see Sect. 4.5.1) in the ELF band and more confidentially in the VLF band (background noise measurement) for underwater detection (Bleil 1964), for marine alert or signaling systems, for submarine command and control communication with the study of the seafloor effect on signal propagation (Wait 1972; Burrows 1978; Aldridge 2001; Merill 2002; Lanzogarta 2012) included in US Navy programs (Advanced Research Projects Agency and Office of Naval Research): *Sanguine* (1958–1967), *Seafarer* (Surface ELF Antenna for Addressing Remotely Employed Receivers) in 1975, *Shelf* (Super Hard ELF), *Pisces* (Pacific Intertie Strategic Communications ELF System) and finally the ELF project (1981–1989).

In the 1970s other devices appeared, this time in the ULF (*ultralow frequency*) band in the context of basic research on the study of the terrestrial globe and, more specifically, physical oceanography (Filloux 1973a, 1977, 1978a, b, 1985; Cox et al. 1978; Young and Cox 1981; Webb et al. 1985; Hekinian 2014).

These voltage measurement materials were designed for evaluation of natural EM fields induced by large oceanic phenomena (ocean currents, tides, movements of seawater, etc.), determination of the influence of the bottom or the seaside (*seaside effect*) or even structural study of the midocean ridge, island arcs, etc.—basic geological objects in understanding the theory of plate tectonics (Hallan 1976; Cox et al. 1973).

At this time, the following were developed:

- *Long base horizontal* devices (Filloux 1967; 1973a, b; Cox et al. 1971) or *vertical devices* (Harvey 1974) designed around potential difference measures between two sampling points, sometimes distant by several kilometers (sensitivity of about 0.1 mV/m)
- *Short base* devices (Mangelsdorf 1968; Filloux 1974), more compact (from 100 to a few meters) designed around electromechanical elements (sensitivity of about 10 nV/m)

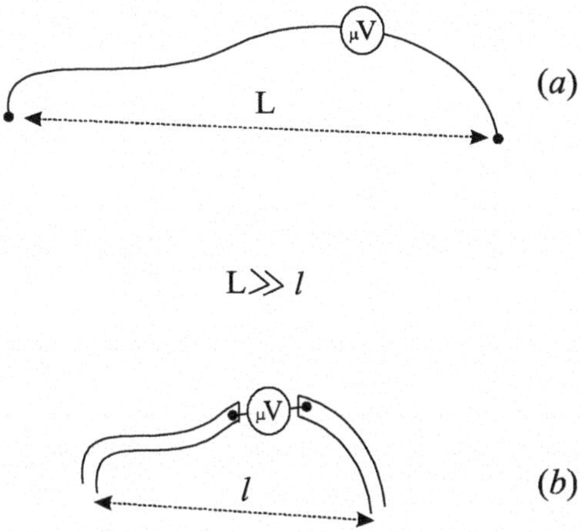

Fig. 4.3 The two principles adopted for electric field measurement at sea: (*a*) a *long base* device composed of electrodes spaced by several kilometers (L) connected by isolated and tight electrical cables to a microvoltmeter (µV) with high input impedance; (*b*) a *short base* device composed of pipes of a few meters (l), made of insulating material (glass or PVC), each open at one end and channeling seawater up to measuring points, weakening the effects of the movements of the latter

These were designed to compensate for the drift caused by polarization changes induced by phenomena other than telluric ones (induction), and were based on measurements of potentials or rather of potential differences between two distant points (Fig. 4.3).

To remove the instrument drifts and the electrode and electronic noise limiting the sensitivity of the devices, electromechanical devices were early conceived, developed and used early on (Mangelsdorf 1968).

For example, the rotary valve system electrically and alternately connects both electrodes to the microvoltmeter, then doubling the useful signal (voltage) by modulating it to the beat frequency (opening/closing valve or intake valve cycle).

Under these conditions, we obtain for two values of the potential V_1 and V_2 corresponding to two opposite positions 1 and 2 of the electromechanical valve (cf. Figs. 4.4, 4.5 and 4.6):

- A value of the amplitude of the electric field (potential reported to the distance l) equal to:

$$|E| = \frac{1}{2l}(V_1 - V_2) \tag{4.1}$$

- A noise level independent of l, with an amplitude equal to:

$$V_{Noise} = \frac{1}{2}(V_1 + V_2) \tag{4.2}$$

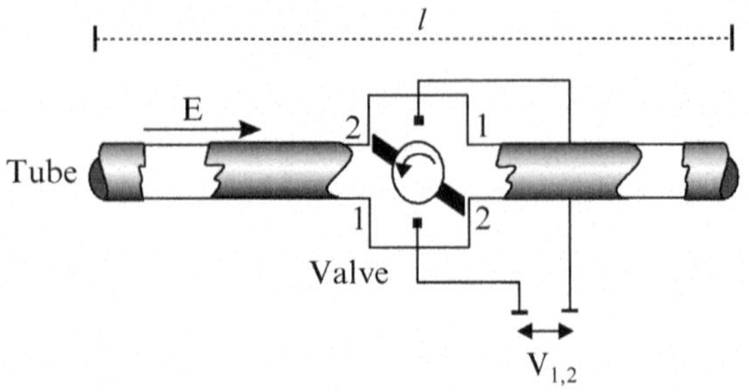

Fig. 4.4 Schematic diagram of an electromechanical device or *salt bridge chopper* (cf. Plate 4.1a). This apparatus eliminates, among other things, the noise of the electrode by mechanical modulation (alternating closures/openings 1, 2 due to an obturation mechanical valve acting as an electrical switch with two positions, 1 and 2) at a rate of a few inversions per minute, thereby limiting instrumental drifts over long periods. The field thereby cut out (modulated) is then rebuilt (remodulated) electronically. The sensitivity of such an instrument is 10 nV/m for a measuring line 6 m in length (According to Filloux 1974)

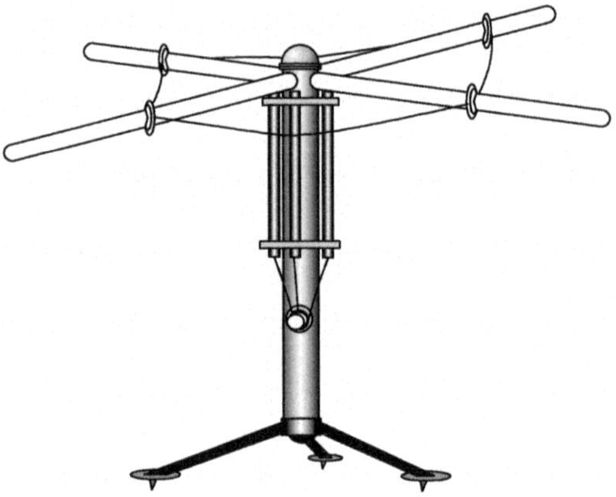

Fig. 4.5 Two-way seabed electrometer of the *salt bridge type* with tubes and Ag-AgCl electrodes (not shown) mounted on a tripod, developed in the 1970s (According to Filloux 1985 and Chave et al. 1991)

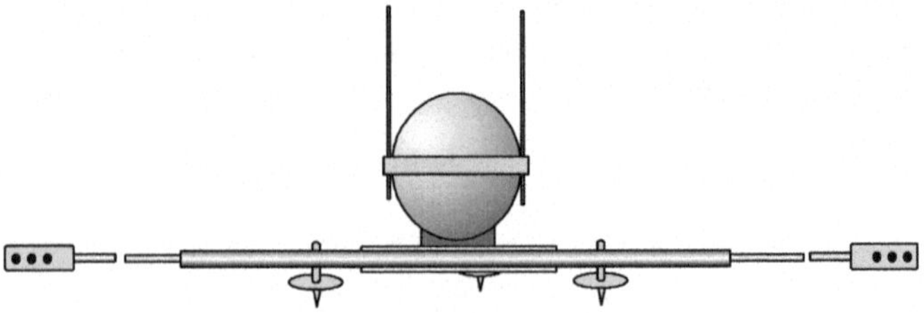

Fig. 4.6 Seabed electrometer, a device for measuring potential difference, made from unpolarizable electrodes (cf. Fig. 4.23b) separated by 9 m and with low noise amplification (According to Cox et al. 1981)

It was then in the 1980s, to deal with the important advances in underwater acoustic discretion (military atomic submarines) and with the magnetic detection limits in force at that time (Mosnier 1986; Latour and Toniazzi 1990) that investigations into electromagnetic noise measurements were conducted and equipment was built (Constable and Snrka 2007).

More specifically, in the field of applied geophysics, previous experimental and theoretical studies had proved the effectiveness of *transverse electric field* measures for the controlled source systems in the remote detection and localization of low dimension resistivity anomalies (Sainson 1982).

The same research led, sometime later, to the development of a new type of material (a current density electrometer), compact and more sensitive than the potential difference electrometer, thus allowing its use, among other things, in mining, underground storage, geothermal applications, and oil wells on small diameter logging tools (wire line), and related devices for underwater exploration (mMT) and submarine detection (Sainson 1984; Mosnier 1984; Rakotosoa 1989; Bruxelle 1995, 1997).

It was in the 1990s that the first electrometers (for mMT) appeared in their current configurations (Constable et al. 1998a, b), built in series by Statoil and then by EMGS (for mCSEM). It must be remembered that at that time several university teams were offering similar prototypes were not necessarily patented or brought to public attention.

The 2000s, which saw the rapid business development of SBL techniques and more particularly mCSEM, allowed the realization of relatively compact equipment, thereby fixing the data acquisition technology.

The generic system generally includes:

- A mobile transmitter (submersible fish) composed of a horizontal emission dipole (collinear with the surface of the sea floor) of a length of about 200 m towed by a boat at a speed of 1 to 2 knots (1 m/s) and moving about 100 m at most above the seafloor
- Tuned field receptors (a few tens), laid on the seabed and equidistant one from each other, which, at the passage of the transmitter, measure the variations in the depth of the EM induced field

all forming, with the processing and data interpretation techniques, an original concept of geophysical exploration.

4 The Transmitter

Historically, the first transmitters delivering EM energy (considered as a dipole source) were proposed in the late 1970s (Cox 1981). Presently, transmitters in their principle appear only in the form of a *horizontal electric dipole* (HED). This configuration is the only way of offering excitation in the TE and TM modes (see Chap. 2, Sect. 2.7.4.3). Specifically, it corresponds to a linear antenna placed in a horizontal plan (see Chap. 2, Sect. 2.5.4). Tests of a *vertical magnetic dipole* (a loop or horizontal coil where the moment is given by the number of turns) are being studied but are not yet offered to the service.[8] However, this type of dipole (VMD) can only provide excitations in the TM mode.

The emitting source is immersible and is towed by the surface vessel (at a speed of ≈ 1.5 knots) through a watertight electrohauling cable, which also provides the electrical power needed to operate (power and control). It consists briefly of:

- A controllable power transmitter
- A wire antenna
- Carrier wave telemetry
- A device for navigation and positioning the current injection means

The power transmitter and related equipment are embedded in a cradle fitted with gripping and pulling elements of a tow-fish type.[9] The equipment must have slightly negative buoyancy, which allows the whole vehicle to immerse itself easily and remain at a constant depth during the traction at speeds of a few knots on average.

Moreover, the linear antenna must have almost neutral buoyancy to move as horizontally as possible behind the fish.

Some systems are equipped, on the antenna end, either with an automatic dynamic balancer device (active control), enabling the entire line to maintain a straight path and a flat attitude, or simply with a floating anchor (passive control).[10]

The elevation of the fish and antenna as the bathymetry are continuously recorded using echo sounders placed at the head of the fish and the end of the antenna. The position of the fish can also be controlled by means of a surface buoy equipped with a GPS (global positioning system) receiver (Fig. 4.7).

4.1 Power Source

A major challenge is to bring the power in the fish which cannot actually store it. The transmitter power is then supplied directly by the connecting cable (250 kW eligible), which also allows the fish to be pulled. The cable brings to it a low amperage current under a high voltage delivered by an atmospheric generator group or by the ship's own resources (onboard alternator). At this level (the fish), a electrical converter then allows the current intensity to be raised by lowering the voltage to a value of several hundred amperes at working frequencies between 0.05 and 5 Hz.[11]

---

[8]Similar methods have been proposed in the past (see Chap. 1, Sect. 6.5).

[9]In the future, the source may be contained in an AUV, which would have the advantage of removing the source from the surface means and could operate under any weather or bad oceanic conditions, or even under the ice.

[10]A parachute shaped and centrally holed flexible piece.

[11]The emission is on a spectrum that lies below the band ELF (extremely low frequency) reserved for underwater communications (military submarines), which extend to the ranges VLF (very low frequency) and LF (low frequency).

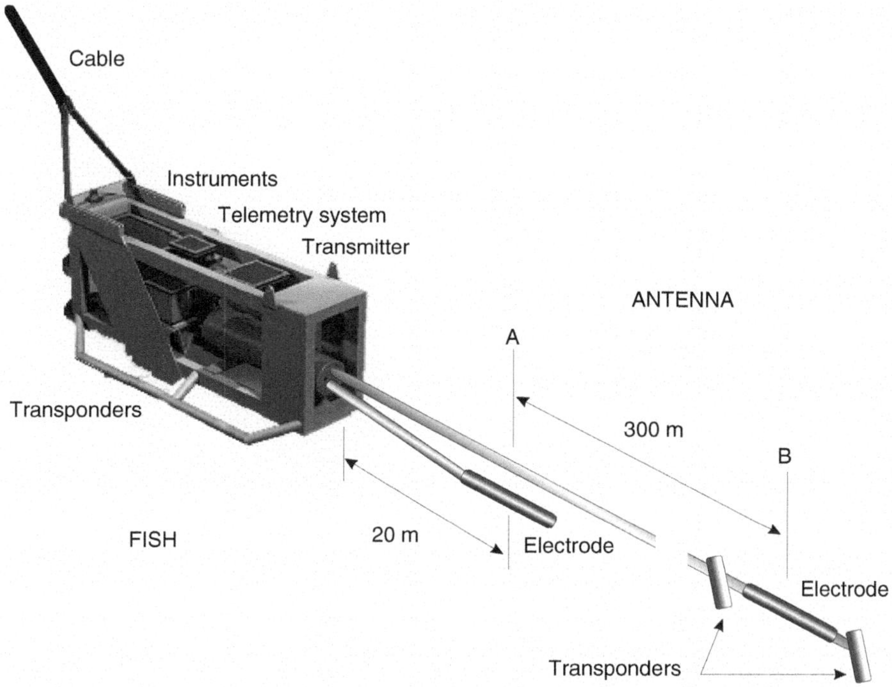

Fig. 4.7 3D view of an underwater electromagnetic source of the HED type composed of a power transmitter housed in a streamlined fish and its wire antenna equipped at both ends with current emission electrodes (cf. Plate 4.2c). The entire transmission device is equipped with acoustic transponders on the antenna and with an echo sounder (3.5 kHz) on the fish, for its location in space and in relation to the background. The immersed transmitter can be completed by a relay transponder, an acoustic altimeter, an inertial navigation system, a sound velocity device, which increases the accuracy of the acoustic travel time to range conversion, and a bathy-thermograph to correct the local acoustic measurements. Some of these instruments can be installed behind the source (cf. Plate 4.3d)

The power emitted by the presently most efficient sources is about 50 kW, which gives, for example 10 km away, in seawater, some values of electric and magnetic fields respectively equal to about 1 nV/m and 0.001 nT. Information about the quality control of the emission (waveform) and its steering are sent to the boat via the electromechanical cable by high–transfer speed telemetry.

Recollection about the notion of power: during a very short time interval dt, energy is equivalent to the product UIdt where U and I are the voltage and the current intensity at that moment. The energy involved for a longer period T (power) corresponds simply to the integral of this product, such as:

$$W = \int_0^T UI\,dt \tag{4.3}$$

Considering the alternative type, or more precisely in a sinusoidal regime, the *effective power* is then given according to the effective values (rms)[12] of U and I and their phase difference φ such as:

$$W_{eff} = \frac{U_{max}}{\sqrt{2}} \cdot \frac{I_{max}}{\sqrt{2}} \cdot \cos \varphi \qquad (4.4)$$

where cos φ corresponds to the *power factor*.

4.2 Power Electronics

The conditioned signal at the surface (on the boat) is sent to limit power losses at a frequency of 60 Hz under high voltage (1800–3000 V)[13] and low amperage (5 A) in the traction electric cable (several hundred meters' length depending of the survey's depth).

In the fish, the voltage is then lowered and the amperage raised using a electrical transformer, which also acts to match the impedance with the seawater (cf. Fig. 4.8). The injection frequency (electrode output) is operated by a bridge power diode frequency divider or by power thyristors or power transistors (insulated gate bipolar transistor: IGBT), thus lowering the frequency of 60 Hz to a few hertz (1 Hz on average) or below.

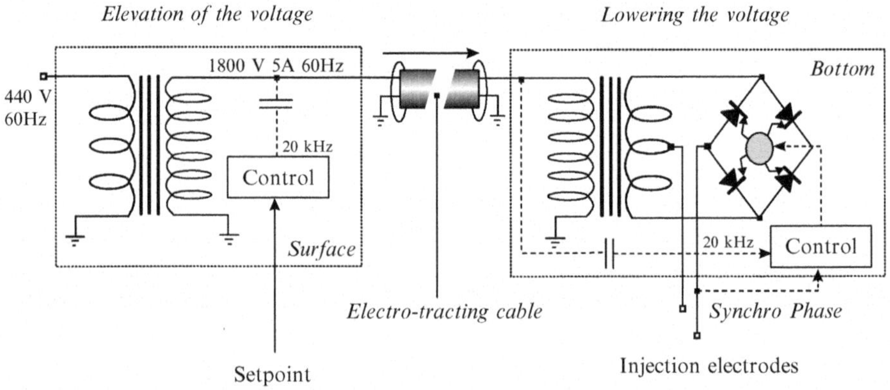

Fig. 4.8 Schematic diagram of the power electronics (surface and *bottom*) of the current emission device. The power is conveyed by an electrohauling cable of about 100 m (According to Chave et al. 1991)

[12]Root mean square, or RMS, i.e., literally the *square root of the mean square*.

[13]It is a fundamental principle of the transmission of electrical energy (e.g., high power line).

The steering of the whole system (waveform, quality control, frequency, power, regulator, etc.) is supplied from the surface via an adapted control system (CS) whose instructions are sent at a frequency of 20 kHz on the carrier low frequency wave delivering the power (60 Hz). The CS is synchronized to GPS time using IEEE 1588 (Precision Time Protocol).

For more details on the embedded systems, especially for those who use more complex devices (signal elaboration), the reader may also refer to the technical literature (see the bibliography at the end of the volume) and to various patents on the subject (Sinha et al. 1990; Constable 2013; EMGS 2015).

4.3 Type of Generated Wave

Generally the transmitted wave is of a sinusoidal or squared shape and continuous during time, but may be of a more complex shape for more sophisticated processing in reception (increase of the signal-to-noise ratio, harmonic elimination, etc.). In this case, it is no more possible to directly use the transformer (cf. Fig. 4.8).

The company WesternGeco (Leedert et al. 2009), for example, has adopted wave types of a *sweep* shape or made of a superposition of several frequencies, and uses analog generators (BF oscillator with power amplifier).

In terms of energy, it is known that it is preferable to use square-shaped waves (cf. Fig. 4.9), due to their high dynamics caused by a sudden rise (vertical slope of the curve). The average power of this type of wave is about two times greater than that delivered by a sinusoidal wave (see Chap. 2, A2.1).

Mathematically, it has been shown that a square wave is the sum of a sine waveform, composed of a wave of the same frequency (ω) supplemented by a series of odd harmonics with n terms of decreasing amplitude such that the expression of the latter is equivalent to:

$$\frac{4}{\pi}\left[\sin(\omega t) + \frac{\sin(3\omega t)}{3} + \frac{\sin(5\omega t)}{5} + \frac{\sin(7\omega t)}{7} + \frac{\sin(9\omega t)}{9} + \cdots + n\right] \quad (4.5)$$

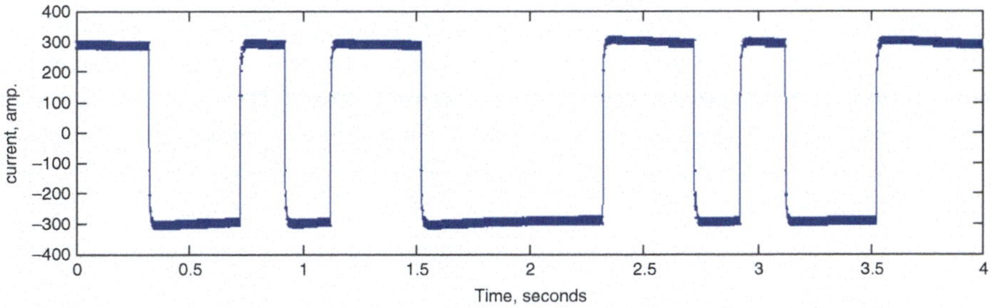

Fig. 4.9 Form of a transmission signal (modulated square wave) of 600 A peak to peak over a period of 4 s (×2)

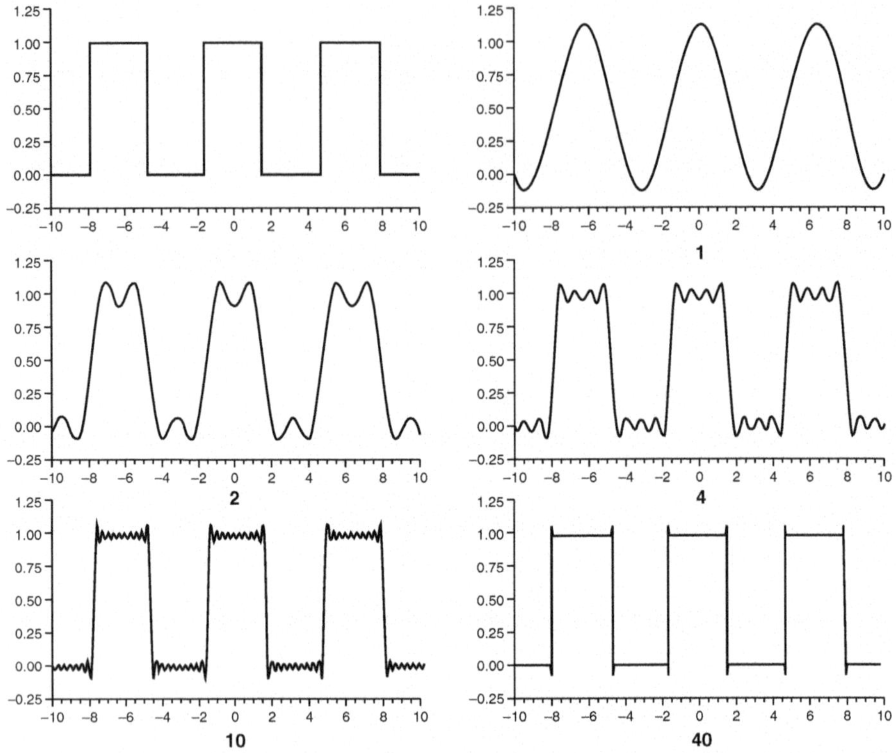

Fig. 4.10 Construction of a square wave from a rectangular function made of the sum of sinusoidal functions at 1, 2, 4, 10 and 40 terms

Conversely, the square signal can be decomposed into sine waves by Fourier transforms (cf. Fig. 4.10).

Practically, a square wave may be obtained electronically either by the clipping of a sinusoidal signal, or by a plurality of frequency dividers and a summator (analog methods), or even numerically using, for example, a programmed sequence of n sinusoidal functions (microprocessor).[14]

Furthermore, given that the current amplitude is changing inversely proportionally to the frequency, and that the attenuation increases with the latter, some authors have proposed to sequence the signal (switching) every 100 ms, for example (cf. Fig. 4.9). According to operational data, the signal power can then be optimized by a more sophisticated device obtained by the *Monte Carlo method* (Mittet and Shaug-Pettersen 2007).

[14]With this arrangement, it is more difficult to obtain high powers.

4.4 Transmitting Antenna

The transmitting antenna is composed of two strands of composite, insulating materials, each fitted at its end with an electrically conductive massive metallic electrode, thereby forming an electric dipole. These are generally spaced at several hundred meters (200–300 m in average).

The purpose of the transmitting antenna is to transfer the maximum energy to the medium in which it is located. For this, its characteristics must meet a number of electrical criteria that must be in close relationship with the concerned environment, i.e., the seawater. The energy is mainly transmitted to the water by conduction, at the injection electrodes,[15] which assumes perfect insulation of the connecting cables.

However, some of this energy could be dissipated by the *Joule effect* (σE^2) as heat if precautions (including thermal insulation) are not taken. Moreover, the immersion of the antenna in water, a very good conductor of heat, and moreover at a few degrees Celsius, allows the latter, proportionately, to quickly evacuate all its calories and therefore cool quasi-instantaneously.

4.4.1 Electrode Contact Resistance

At the considered frequencies (LF), the electrical characteristics of the antenna, resistance, inductance and impedance can be regarded with good approximation as independent of the frequency (see Appendix A4).

The sea contact resistance R_c is an ohmic resistance that is by far the most important compared to the radiation resistance $R_r = 2\pi l_e^2/3\varepsilon c\lambda^2$ (Fleury and Mathieu 1958). For a conductive electrode (electronic conductivity), the contact resistance is calculated as in DC,[16] i.e., essentially depending on the geometric characteristics of the electrode and on the resistivity of the seawater surrounding ρ_W. For a cylindrical electrode, such as those currently used (whose axis is coincident with that of the strands of the wire antenna), it is given in ohms by the following equation (Sunde 1949):

$$l_e \gg d_e \quad \rightarrow \quad R_{c[\Omega]} = \frac{\rho_W}{2\pi l_e}\left[\ln\frac{2l_e}{d_e} - 1\right] \qquad (4.6)$$

where l_e and d_e are respectively the length and the diameter of the current emission electrode.

From Eq. (4.6), the length of the emitting part therefore has more influence on the resistance than its diameter. In practice, it is essential that the current injection

---

[15] We have seen (Appendix A3.2) that most of the energy is transmitted through this way.

[16] At the considered frequencies (LF), the resistance in the marine environment cannot be considered as an impedance. The capacity aspect is negligible in this case (see Appendix).

electrodes have contact resistances as low as possible, to dissipate in consequence the minimum energy at their level (reduction of heating due to the Joule effect and electrochemical effect) and so that maximum electrical power can be then quickly transferred in the medium (water). In the best case, they can reach 0.01 Ω.

In total, the contact resistance of the antenna, all electrodes together (×2), varies between 0.10 Ω and 1 Ω, values under which it is physically difficult to go.

4.4.2 Inductance of the Antenna

The inductance of the antenna L_a, expressed in micro-henrys, is the translation of the inductive effect of the electrical cable between the two electrodes, caused by the movement of the variable electrical current along it.[17] This electrical characteristic here corresponds to a magnetic field created across a surface (flux). Its expression (Rosa 1908), depending on the length L of the antenna and the area a of its section, is given by the following formula:

$$L_{a[\mu H]} = 0,2L\left[\ln\frac{4L}{a} - \frac{3}{4}\right] \quad (4.7)$$

A typical linear antenna of several hundred meters in length therefore has an average inductance of approximately 500 μH, which is balanced by a capacitance of about 800 μF across the transmitter output.

4.4.3 Impedance of the Antenna

The impedance Z_a of the antenna is defined as the ratio of the voltage U_a on the current I_a circulating therein and is expressed in ohms, as is the contact resistance. It somehow determines the energy transfer that occurs between itself and the elements constituting the input of power W (driving force of the generator). It must adapt to them and be as low as possible so that the maximum energy can be transmitted to the transmitting electrodes and then to the marine environment.[18] It is equal to:

$$Z_{a[\Omega]} = \frac{U_{a[V]}}{I_{a[A]}} = \frac{W}{I_a^2} \quad (4.8)$$

[17]It may be recalled here that the alternating current tends to flow in the periphery of the cable (*skin*) and all the more easily when the frequency is high.

[18]Ideally it is desirable to have perfect impedance matching between the different elements and emission bodies. The transformer can then automatically play that role especially when the electrical characteristics vary over time.

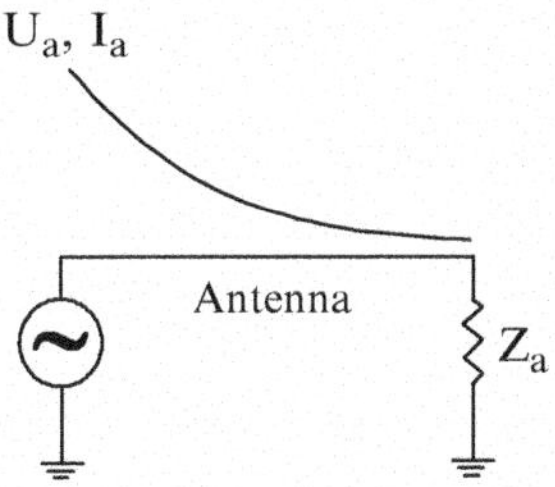

Fig. 4.11 Electrical schema equivalent to the transmitting antenna, and shape of the evolution of the voltage and the current (*top*), which decreases with distance from the generator

For example, for a power W of 100 kVA (1000 A, 100 V rms), the impedance of the antenna is approximately (100 ÷ 1000), i.e., 0.1 Ω (Fig. 4.11).

4.4.4 Dipole Moment of the Antenna

The dipole moment *P* of the antenna represents convenient data. It is determined as the product of the current intensity I_a by the length *L* of the antenna ($I_a \times L$) and is given in A.m. The dipole moment corresponds to a compromise between the length of the antenna and the intensity emitted. A too-long antenna is difficult to implement under the seabed. On the other hand, the dipole approximation cannot be used and there is a loss of resolution. In another way, increasing the intensity also has drawbacks; doubling of the intensity amplitude requires four times the transmitted power. In all cases, because of various losses in the cable and the different components, the transmitted power is twice the dissipated power in the seawater. The dipole moment is, depending on the available power W (either according to the formula 4.8: $W = I_a^2 Z_a$), equal to:

$$P_{[A.m]} = I_a L = \sqrt{W \frac{M_e}{\rho_e d_e}} \tag{4.9}$$

where M_e is the total mass of the electrode, and ρ_e and d_e are respectively the electrical resistivity and the density of the metal forming the electrode.

In this way, instrumentally, the expression (4.9) on one hand allows us to somehow standardize the emission characteristics of any filamentary antenna of any amperage and any length, and on the other hand to be able to perform within a certain limit objective comparisons between different devices with different components and qualities. The dipole moment of a conventional antenna is approximately 10^5 A.m. Currently some proposed devices can be 20 times more powerful, now using a 7 kA source (Roth et al. 2013; EMGS 2015).

4.4.5 Initial Conditions of the Antenna

The antenna, measuring several hundred meters long, is not rigid.[19] It will therefore deform during the survey operation itself. In DC investigation, if the cable is well insulated, the current will flow between the two electrodes and the field distribution will only depend on their arrangement and spacing. In contrast, in a harmonic regime, the insulated cable will participate along its entire length in the propagation, then behaving like an infinite number of magnetic dipoles whose directions will be confused with those of the small deformations caused by different movements of the line during the *run*.

This phenomenon, particularly noticeable at high frequency, disappears as we use wavelengths λ sufficiently large compared to the length L of the transmitting antenna. At the frequencies ordinary used (<1 Hz), for a propagation speed of the electromagnetic waves of 1600 m/s in seawater (see Chap. 3, Sect. 4.1), this corresponds to an antenna length of about $(1600 \div 8)$, i.e., 200 m.

$$L << \lambda \quad \rightarrow \quad L << 1600\,\text{m} \tag{4.10}$$

These lengths are also compatible with the objectives of detection, resolution, definition, etc. (cf. Chap. 3) (Table 3.2).

4.4.6 Radiation Pattern of the Transmitting Antenna (Wire)

We consider (Loseth 2007; Loseth et al. 2006) a horizontal wire antenna whose length L is small compared to the wavelength (far field criterion), in which circulates a harmonic current $I_a\ (\omega)$, of density:

$$\vec{J}_0 = I_a L\, \boldsymbol{\delta}(r)\, \hat{z} \tag{4.11}$$

where $\boldsymbol{\delta}$ is the Dirac function (cf. Eq. 2.8, Chap. 2).

Table 4.2 Summary table of emission antenna characteristics (According to EMGS)

Characteristics[a]	Values (mksA)
Length	200–300 m
Diameter	0.02–0.04 m
Contact resistance	0.1–1 Ω
Impedance	0.1 Ω
Inductance	300–500 μH
Dipolar moment	10^5 A.m
Frequency	0.5–5 Hz
Power	50–100 kVA

[a]These characteristics may vary from one material to another and are given here only as an indication (see, for example, EMGS source specifications 2015)

[19]A feature difficult to obtain for such length.

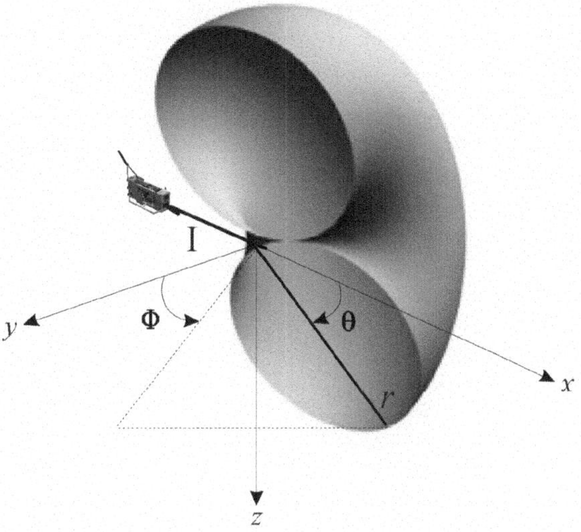

Fig. 4.12 Radiation pattern in the far field (e.g., in open water) of the electric field for an antenna (electric dipole) placed on the axis Ox. The lobes are symmetrical with regard to the axis of the antenna and to its normal. The angles θ and Φ respectively fit into the planes x, z and x, y

The distributions of the electric and magnetic lines of the EM field in the frequency domain, at a distance r from the source (where $r >> L$) in relation to the direction $\hat{z}$, in spherical coordinates (dependence only in r and θ) according to Fig. 4.12, then respectively follow the forms:

$$\vec{E}\,(\vec{r},\omega) = \frac{ik\widetilde{\zeta}I_a L}{4\pi r}e^{ikr}\left[-\left(1-\frac{1}{ikr}+\frac{1}{(ikr)^2}\right)\hat{\theta}\,\sin\theta + \left(\frac{1}{ikr}-\frac{1}{(ikr)^2}\right)2\hat{r}\,\cos\theta\right]$$

$$\vec{H}\,(\vec{r},\omega) = \frac{ikI_a L}{4\pi r}e^{ikr}\left(1-\frac{1}{ikr}\right)\,(-\hat{\Phi}\,\sin\theta) \tag{4.12}$$

with k being the number of waves equal to:

$$k = \sqrt{\omega^2\mu_0\varepsilon - i\omega\mu_0\sigma}$$

and $\widetilde{\zeta}$ representing the complex impedance[20] and defined as follows:

$$\widetilde{\zeta} = \sqrt{\frac{\mu_0}{\widetilde{\varepsilon}}} = \sqrt{\frac{\mu_0}{\left(\widetilde{\varepsilon}+\frac{i\sigma}{\omega}\right)}} \tag{4.13}$$

where μ_0 and σ are respectively the magnetic permeability and the electrical conductivity of the medium, ω the angular frequency ($2\pi f$), and $\widetilde{\varepsilon}$ is the complex permittivity.

[20]The impedance is defined here as a function of angular frequency (ω).

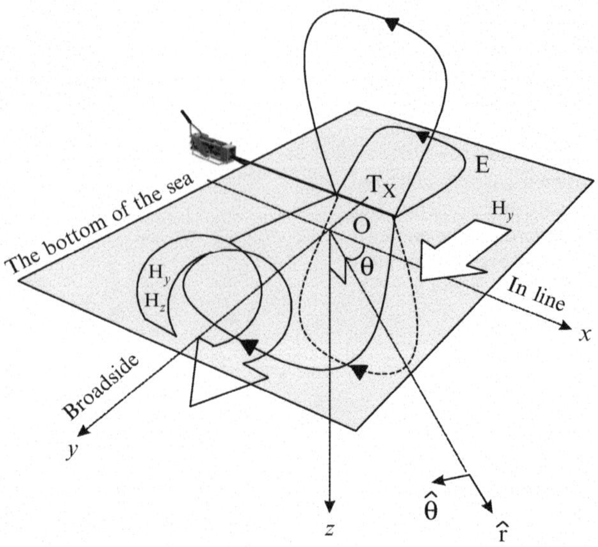

Fig. 4.13 Topology of the electric $\vec{E}$ and magnetic $\vec{H}$ fields in the *in-line* configuration (along *x*) and *broadside* (along *y*) for a transmitter (Tx antenna disposed in the direction O*x*) placed above the bottom of the sea

For the considered frequencies (≈1 Hz), as the wavelength in seawater is very much greater than the length of the antenna (1600 m >> 200 m at 1 Hz), the emissive source is assimilated into an EM dipole whose radiation pattern is shown in Fig. 4.12.

In such conditions of electromagnetic radiation, assuming a spherical distribution around a point electrode (see Chap. 3, A.3.2), the distributions of electric $\vec{E}$ and magnetic $\vec{H}$ fields follow the laws of Maxwell (cf. Fig. 4.13).

4.4.7 Characteristics of Circular Antennas

Circular antennas, composed of a single coil or a winding, are not being used at the moment. We therefore give here as a guide their electrical characteristics (Wait 1957; Kraichman 1962; Galejs 1965).

When they radiate in a medium conductor of electricity, the external resistance R_{ex} and reactance X, which define these arrangements, are given by the following expressions:

$$R_{ex} = \frac{8}{3\sigma_w a}\left(\frac{a}{\delta}\right)^4 \tag{4.14}$$

and:

$$X = X_0 - \frac{2\pi}{3\sigma_w a}\left(\frac{a}{\delta}\right)^5 \tag{4.15}$$

where σ_w, a and δ respectively are the conductivity of the seawater, the radius of the coil and the skin depth, and X_0 is the reactance in an empty space.

For the antenna consisting of *N* contiguous turns (coil), the terms R_{ex} and X are then multiplied by N^2.

We would like to remark that today, to reach increasingly important depths of investigation, and given the more and more severe operational constraints (the noise level of the receptors), it is necessary in these circumstances to increase the power of the source. It goes without saying that the latter cannot be increased to infinity. In this spirit, it is necessary to optimize the efficiency of the antenna in its context. This is achieved mainly by efforts to improve electrical characteristics by reducing losses by the Joule effect and by strengthening the isolation, using previously, for example, optimization studies (Noorhana et al. 2012) developed from numerical models and simulations (finite integration method or FIM).

4.5 Telluric Sources

Telluric methods (mMT and mET) use natural electromagnetic sources, externally and/or internally to the marine environment. These come partly from celestial space (the sun) and partly from the sea (hydrodynamic phenomena).

As for the external sources, the electromagnetic energy comes from the interaction in the atmospheric plasma between ionized particles from the sun (solar wind) and the earth's magnetic field (Bleil 1964; King, Newman et al. 1967; Jacobs et al. 1987; Nickolaenko 2013).

As for the internal sources, the energy comes from much more complex phenomena related to the marine environment and in particular to the movement of water masses in the regional geomagnetic and telluric fields (Parkinson 1983) (Fig. 4.14).

4.5.1 Sources External to the Environment: Geomagnetic Effect and Schumann Resonance

In addition to its intraterrestrial fluctuations (geological scale), the magnetic field of the earth also undergoes extraterrestrial variations over shorter periods:

- Distant, such as those from the sun (solar wind and *magnetic storms*),[21]
- Closer, such as those from the *magnetosphere*
- Even closer, such as those from resonance phenomena in the cavity formed by the Earth and the ionosphere (*Schumann resonance*)

[21]Flow of solar plasma from the *coronal mass ejection* (CME) of ionized gas escaping from the sun at speeds of more than 2000 km/s according to cycles of intense activity and more or less long quiet periods (of about 11 years).

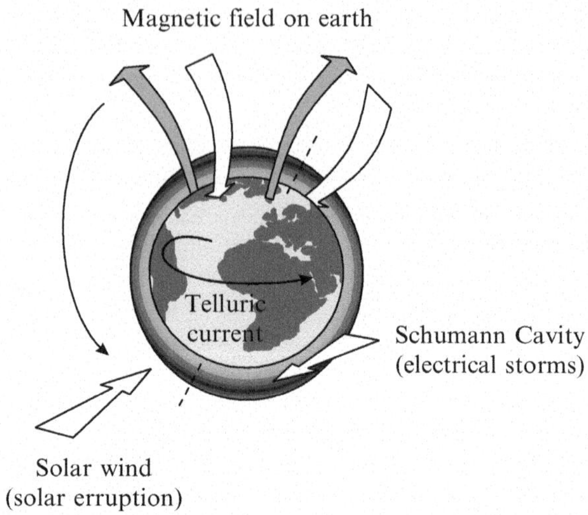

Fig. 4.14 Simplified diagram of the origins of the telluric currents flowing on the surface of the globe both onshore and offshore

The amplitude duration and timing of these disturbances, expressed in γ,[22] are highly variable:

- 50 γ (mean latitude) to 100 γ (geomagnetic equator) for periods of 24, 12, 8 and 6 h for diurnal fluctuations[23]
- 150 to 200 γ during magnetic storms covering periods from 1 week to 1 hour (10^{-6} to 10^{-4} Hz)
- 1 γ or less for periods of less than 1 minute (0.002–5 Hz)

External magnetic fields generated in the ionosphere and magnetosphere vary both temporally and energetically with high amplitudes. In electrical conductors of large dimensions, whether industrial (*pipelines*) or natural (oceans), these fields will induce (Faraday's law) currents and electric fields $\vec{e}$ such as:

$$\frac{\partial \vec{b}}{\partial t} = -\vec{\nabla} \wedge \vec{e} \tag{4.16}$$

where $\vec{b}$ then is the sum of the primary and secondary induced magnetic fields in the conductor. Considering this time that the electric field induced in the conductor follows Ohm's law:

[22] *Geomagnetic usual unit*: 1 γ is equal to 10^{-9} Tesla (T) or 1 nanoTesla (nT). 1 T is equal to 1 Wb/m^2. The earth's field in France has a total amplitude close to 50,000 γ and its horizontal component is about 20,000 γ.

[23] These variations have a pseudoperiodicity of 26 days, corresponding to the rotation period of the sun, and are limited to the illuminated face of the earth.

$$\vec{j} = \sigma \vec{e} \tag{4.17}$$

and that in turn it generates an induced magnetic field such that:

$$\mu_0 \vec{j} = \vec{\nabla} \wedge \vec{b}_{ind} \tag{4.18}$$

we can then deduce, using the above three equations, that:

$$\frac{\partial \vec{b}}{\partial t} = -\vec{\nabla} \wedge \frac{1}{\mu_0 \sigma} \vec{\nabla} \wedge \vec{b}_{ind} \tag{4.19}$$

which is the equation representative of the temporal variations of the magnetic field induced in the seas.

Several authors (Jones 1983; Kuvshinov and Olsen 2005) calculated the magnetic field induced in the oceans from data acquired during *magnetic storms*. The spatial distribution of these fields across the terrestrial globe clearly shows a concentration of lines of force around the continents (*seaside effect*),[24] thus corroborating the precise measurements from 1960 (Mosnier 1967).

Finally, other researchers, to clarify these calculations, took more specifically into account the hydrodynamic aspects such as water movements in these special places (Chave and Luther 1990) (Fig. 4.15).

Fig. 4.15 Map of the vertical component of the magnetic field induced in the oceans by the magnetic storm *Bastille Day* on July 15, 2000 (at 9:30 p.m.) (According to Kuvshinov and Olsen 2005)

[24]An effect demonstrated for the first time in 1879 by British scientists from the Post Office Telegraph Services (Mathias et al. 1924).

As an external source, we can also add, among others, that concerning the trapping of the electromagnetic perturbations caused by electrical discharges of *atmospheric storms* (lightning, 20–50 kA) in the cavity formed by the surface of the earth and the ionosphere (Schumann 1952; Balser and Wagner 1962; Surkov and Hayakawa 2014).[25] With respect to the EM waves, the cavity acts as a *spherical wave guide* whose fundamental resonance frequency f_1 and the harmonics f_n related to the dimensions of the earth are equal to:

$$\begin{aligned} f_{1[\mathrm{Hz}]} &\approx \frac{v(\sigma)}{2\pi r_{\mathrm{T}}} \\ f_{n[\mathrm{Hz}]} &= f_1\sqrt{n(n+1)} \end{aligned} \tag{4.20}$$

with $n = 2,3,.....$ and where r_{T} and $v(\sigma)$ are respectively the earth's radius (≈6378 km) and the speed of propagation of the electromagnetic waves depending on the local electrical conductivity σ.

If we now consider that the speed of propagation of the electromagnetic waves in the atmosphere is approximately equal to that of light c in a vacuum (≈300,000 km/s), the fundamental frequency is then established around 7.5 Hz. This is partly true firstly because of the latitude (cf. Tables 4.3a and 4.3b), and secondly because of the anisotropic nature of the electric conductivity σ of the ionosphere, whose values σ vary as a function of the altitude h_a.

These variations then follow an exponential law and can be expressed by the equation:

$$\sigma(h) = \sigma(h_0)\exp[\beta(h_a - h_0)] \tag{4.21}$$

where

- h_0 is equal to 60 km altitude[26]
- For diurnal periods, $\sigma(h_0)$ is equal to 4.63×10^{-8} mho/m and β is equal to 0.308/km
- For nighttime, $\sigma(h_0)$ is equal to 4.5×10^{-10} mho/m and β is equal to 0.44/km.

Table 4.3a Day and night variations of the velocity (v/c) function of the latitude at the frequency 7.5 Hz. In the *Schumann cavity*, electromagnetic waves travel in June at a speed of about 75 % of the speed of light in a vacuum

Latitude (degree)	v/c (diurnal)	v/c (nocturnal)
90	0.726	0.743
45	0.745	0.764
10	0.771	0.769
0	0.782	0.776

[25]The phenomenon was discovered theoretically in 1948 (Schumann 1948) and proved experimentally 5 years later (Schumann and Koenig 1954). For a more theoretical approach, the interested reader may refer to the work of Professor Wait (Wait 1996).

[26]Beyond that distance, the atmosphere acts as a filter. Below it, the long length electromagnetic waves are reflected.

Table 4.3b Schumann resonance frequencies: observed values of fundamental and pseudoharmonics by different authors and in different places of world

Frequencies	f_1 (Hz)	f_2 (Hz)	f_3 (Hz)	f_4 (Hz)
Balser and Wagner (*USA*)	7.8	14.1	20.3	26.4
Benoit and Houri (*Tunisie*)	–	14.4	20.8	26.8
Chapman and Jones (*UK*)	8.0	14.1	20.0	26.0
Stefant (*France*)	7.85	14.2	19.9	26.25
Rycroft (*UK*)	7.8	14.1	20.0	26.0

According to Selzer (1972)

More precisely, two oceanographic campaigns, carried out in the 1960s, highlighted *Schumann resonances* on the seabed:

- One in the first layer of 300 m of water, with the use of the diving saucer of Captain Jacques-Yves Cousteau (Soderberg 1966, 1969)
- The other at depths reaching 5000 m, with the French bathyscaphe *Archimède* (Selzer 1966, 1968; Launay et al. 1964)

4.5.2 Sources Internal to the Environment: Magnetoelectrodynamic Effect

Internal sources concern more or less large movements of the hydrosphere, i.e., those of the oceans and seas (Brahtz et al. 1968). The movement of these water masses, important and electrically conductive at different levels and scales (ocean currents, tides, waves, etc.), in the local or extended geomagnetic field induces electric fields that in turn give rise to secondary magnetic fields that will oppose the primary field (laws of Faraday and Ampere). Electromagnetically these related hydrodynamic phenomena were the first to be studied in detail (see Chap. 1, Sect. 6.3).

The complex set of the temporal variations of these magnetic fields, which interact, is at the origin of the current flow and the creation of induced electric fields in the sea. Their amplitude then depends on the direction of the transmitted wave and on the attenuation due to seawater.

4.5.3 Transmitted Wave Direction

We can show (Porstendorfer 1960) that the impact of the electromagnetic radiation from the airspace on a surface is of the form:

$$\sin \alpha_1 = \sin \alpha_0 \frac{1}{c\sqrt{2\sigma_w T}} \tag{4.22}$$

where α_0 is the angle of incidence of the ray (air) and α_1 is the angle of the refracted ray in seawater at the interface of these two components (cf. Fig. 4.16).

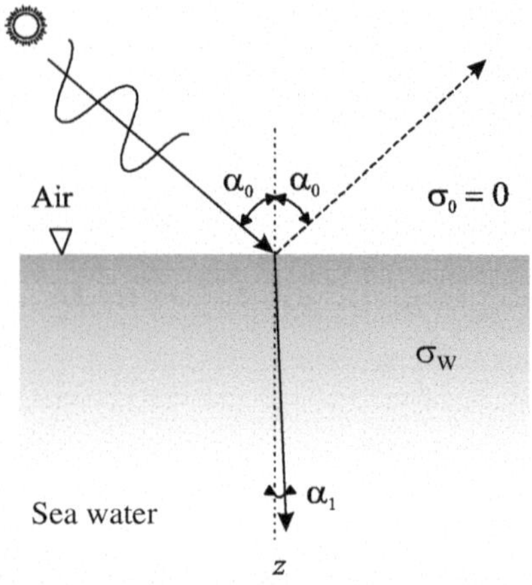

Fig. 4.16 Reflection and refraction of electromagnetic waves from solar activity at the surface of the ocean

For an EM wave arriving at the surface of the ocean (0.2 Ω.m) at a velocity c of 300,000 km/s (air) for a period T of 10 s we obtain, for example:

$$\sin\alpha_1 = \sin\alpha_0 \frac{1}{5\times 10^5} \rightarrow 0 \tag{4.23}$$

The refracted radiation, whatever its impact on the surface of the sea, then penetrates vertically into the water (90°).

4.5.4 Electromagnetic Field Attenuations

Considering now that the electromagnetic waves are planes and that they penetrate vertically, i.e., perpendicularly to the separating surfaces (air/water and then water/sediment), it is then possible by solving the *equation of Helmholtz* to obtain the expressions of the magnetic (y component) and electrical (x component) fields, depending on the depth (z component) and the height of water h, such as:

$$\begin{cases} B_y(z) = B_0 \dfrac{(1+q)\,e^{-\gamma(z-h)} + (1-q)\,e^{+\gamma(z-h)}}{(1+q)\,e^{+\gamma h} + (1-q)\,e^{-\gamma h}} \\ E_x(z) = E_0 \dfrac{(1+q)\,e^{-\gamma(z-h)} - (1-q)\,e^{+\gamma(z-h)}}{(1+q)\,e^{+\gamma h} - (1-q)\,e^{-\gamma h}} \end{cases} \tag{4.24}$$

with:

$$\gamma = (1-i)\sqrt{\pi f \mu_0 \sigma_W} = \frac{1-i}{\delta} \quad \text{and} \quad q = \sqrt{\frac{\sigma_W}{\sigma_S}} \tag{4.25}$$

where δ is the skin depth (see Chap. 3, Sect. 3.3.1) and q is the square root of the ratio of the electrical conductivity of seawater σ_w on one of the marine sediments σ_s.

Letting h tend to infinity, we find the law in $e^{-z/\delta}$ that characterizes the attenuation of plane waves in conducting media (Figs. 4.17 and 4.18).

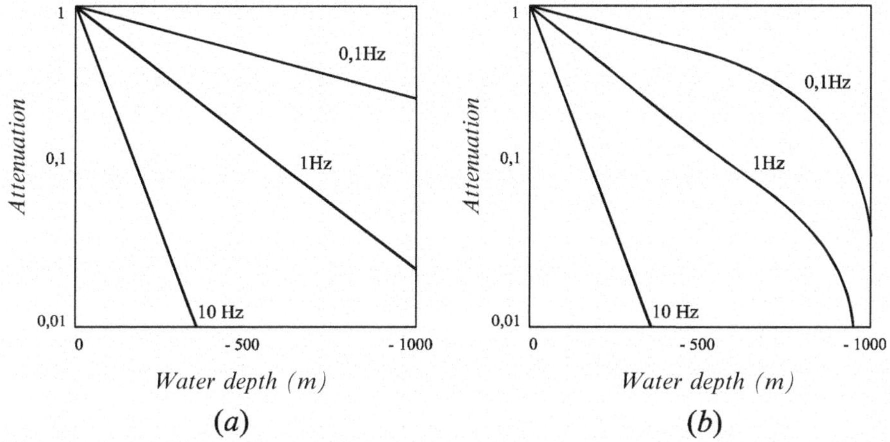

Fig. 4.17 Attenuation of the geomagnetic field in open water at infinite depth (*a*) and on a shoal of 1000 m (*b*). Below 1 Hz the relation is not linear

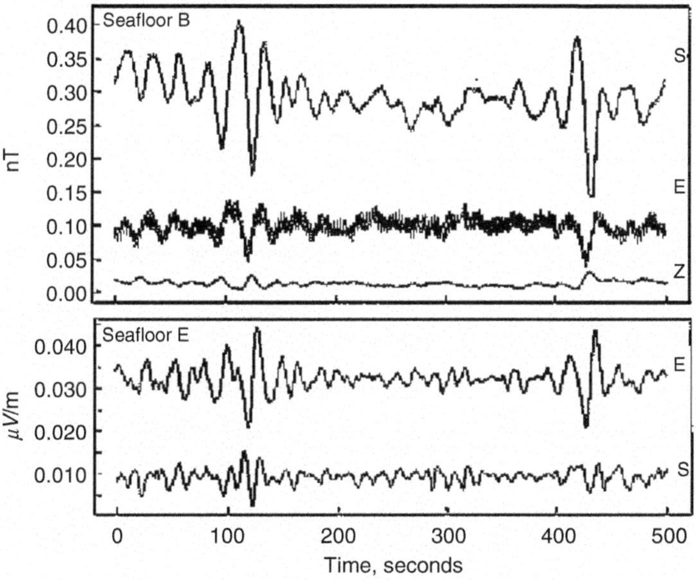

Fig. 4.18 Shapes of east–west and north–south variations of the telluric fields (magnetic field in nT and electric field in μV/m) (According to Constable et al. 1998a, b)

5 Receivers

The appreciation of the flow of an electric current through a medium can be realized in different ways. In geophysics it is quantified by the evaluation of an electric and/or magnetic field. The measurement is usually indirectly done either by potential measurements or more precisely by potential difference measurements with the use of a microvoltmeter, or by measurement of flux or of induced magnetic fields using, for example, a magnetometer. These last measures are relatively marginal and confined to special prospecting applications. In an electrically conductive medium, such as seawater, the electric fields can also be evaluated by measurements of current density.

At these auscultation frequencies (ELF), the receivers then sample the geoelectric environment every 0.04 s approximately.

The transmitted energy is dissipated in the different environments via different paths that are more or less long and fast depending on the reflections and refractions at the different discontinuities (air/water/sediment/reservoir).

In the majority of cases, SBL uses electrometers as signal receivers (that is, in the vocabulary of electricians) devices (here, field sensors) that supposedly do not partly consume the energy to be measured. If it is relatively easy to assess potential or high amplitude current (electrostatic and electrodynamic systems), it is not true for the measurement of very small amounts of electricity.

These electrometers have essentially to provide in a given direction some data from measurements of the amplitude and phase of the electric fields, corrected for the *offset* values and immediate environment (Fig. 4.19).

Corrections are made afterward, depending on the degree of uncertainty (Zach et al. 2009).

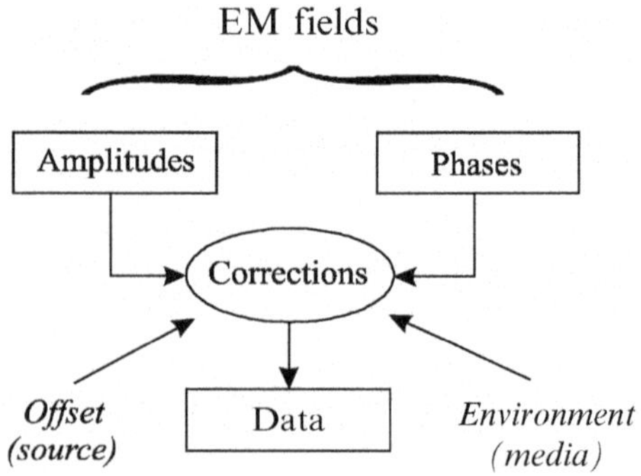

Fig. 4.19 Schematic diagram of the acquisition and field data corrections (amplitudes, phases) assigned to electrometers in a given direction

5.1 Types of Receivers

In an electrically conductive medium, on a purely electrokinetic plan, indirect field measurements must be made with equipment that creates as little disturbance as possible in the local field of forces (power lines I).

If we consider, for example a measuring voltage (V) between two distant points M and N (cf. Fig. 4.20), the capture of the current i in this case should be a minimum that is to say, the current through the device must be very small compared to I ($i \ll I$). However, if you decide to make an amperage measurement (A), then the instrument must possess in the absolute an internal resistance equivalent to that of the interrupted line of current such that the withdrawn current will be equivalent to the one flowing through the line of force ($I = I'$).

In general, instruments that meet these criteria (low sampling or low internal power consumption) are classified as *electrometers*. They can either measure the potential difference between two points then provide an output voltage, or measure the current flow (amperage). In both cases, the measured values are dependent on the dimensions of the apparatus, whose size and electrical characteristics (nature of the materials used) may themselves introduce local distortions in the measured field of force.

In the methods of field measurement and more particularly for seabed logging, the receivers are of the vector type (decomposition of the total field). The two or three components of the field, whose directions are generally perpendicular to each other, are then simultaneously measured on different channels.

The electrometric measures (electromagnetic fields) can then be seen from different principles, by measurement of a potential difference

- By current measurements
- By magnetic measurements

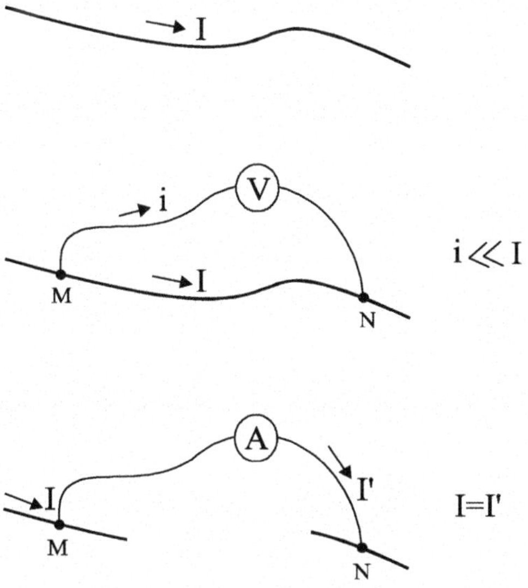

Fig. 4.20 Voltage sampling conditions at very high impedance (V) or current sampling conditions at very low impedance (A) in the apparatus so as not to (close to the sensor) disturb the local electric field

and in various devices and methods:

- By fixed *seabed sensors*
- By subsurface mobile *streamer cables*
- By downhole *logging tools*

these devices being currently in the experimental stage or developed.

5.2 *Potential Difference Vector Electrometers*

Potential difference electrometers (pd) are currently the only instruments to be industrially used and have so far proven their effectiveness in certain limits.

5.2.1 Principles and Technology of pd Sensors

We know that the evaluation of an average electric field (equivalent to a scalar potential gradient dV along an element of length dl as $\vec{E} = -\vec{\nabla}V$) may be commonly performed by measures of the potential difference V[27] between two collection points M and N ($\Delta V_{M,N}$) separated by a distance l (cf. Fig. 4.21) as:

$$E = -\frac{dV}{dl} = \frac{V_M - V_N}{l} \tag{4.26}$$

Usual receivers consist of a sealed box (1/2 m^3) enclosing all the electronic components of data acquisition (amplification cards, memory, power supply, etc.). In space, the electric field vector can be measured on several components in the directions of a right trihedron, for example. This allows us, among other things, to distinguish side effects and state the direction.

Under these conditions, the antennas supporting the collection points are positioned perpendicularly to each other and fixed in pairs on the housing (cf. Fig. 4.22). They consist of two insulating flexible arms in composite fiber full or formed of polypropylene tubes each about 5 m length, which support at their ends the

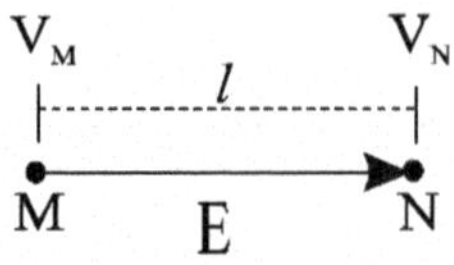

Fig. 4.21 The value of the electric field E is defined by the potential difference $\Delta V_{M,N}$ between two points M and N, or between two small equipotential surfaces, separated by a distance l

[27] In a given direction, the average field is not determined by the derivative of the potential, but by a potential difference between two points.

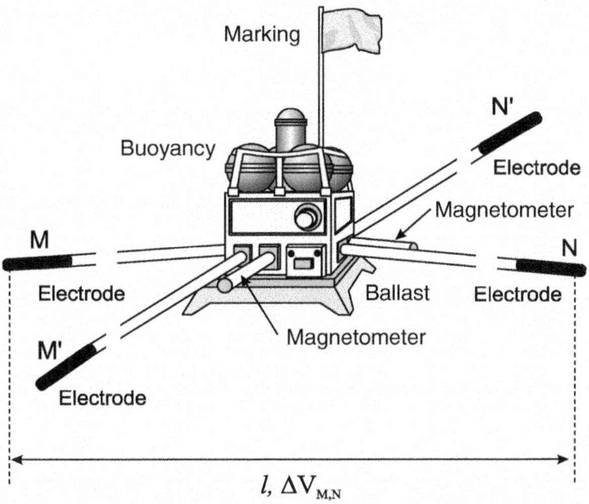

Fig. 4.22 SBL receiver comprising in its center the acquisition tight case (amplification, processing and storage of data, MN and M′N′ antennas and buoyancy spheres at the *top*). Such a device has dimensions (electrode to electrode) of 10–20 m for the models currently in service. Magnetic field sensors placed parallel to the antennas can be added to the electrometric device (cf. Plate 4.2). The tracking and reporting of the sensor at the surface is provided by a flag (day) and/or a glow torch of the Xenon type (*night*) and sometimes by a VHF device for localization in bad weather (fog)

conductive collection electrodes (MN and M′N′). All the strands forming the antenna thus form an electrical dipole (TE mode) of approximately 10 m in length.

The anchoring of the apparatus on the seabed is provided by a degradable concrete ballast (soluble in seawater and chemically inert), fixed under the waterproof container, which at the end of the mission automatically separates (via a released mechanism piloted by an acoustic transponder trigger) from the acquisition device to let it come to the surface. The solid ballast (about 200 kg) is designed (weight, shape, adhesion, etc.) to ensure the instrument has maximum stability at the bottom of the sea.

To ensure the hydrostatic equilibrium of the set (box and arms), syntactic foam spheres are arranged on the top of the box to adjust the buoyancy and ensure their rise when the ballast is dropped. The foam is composed of 10–200 μm diameter glass microspheres placed in a plastic resin matrix. The density is about 0.55 times that of seawater (600 $kg \cdot m^{-3}$).

The device can also be completed in a third way corresponding to the measurement of the vertical field and possibly by two or three directional magnetometers (TM mode) placed in the same directions, but corresponding to the transverse fields.

In a conductive medium such as seawater, electric fields, the result of a voltage drop caused by a rapid dissipation of energy, are necessarily of very low intensities. The measurements are then limited by the spacing l between the two electrodes

(10 m in general for the existing industrial equipment). Thus, if the use of an instrumentation amplifier with high input impedance eliminates contact resistances (electrodes/fluid), making them negligible compared to the one of the measurement system, it nevertheless takes into account only a small portion of the signal. This therefore gives the measuring equipment a very poor performance for maximum space.

One solution would be then to increase the distance l between the two electrodes, but the electric field then could no longer be considered as locally uniform and the congestion would become a disability (see Sect. 5.2.9). Moreover, in most cases, measurements are also limited by the background noise of the apparatus (white noise having a voltage with at least an amplitude of $10\ \mathrm{nV}/\sqrt{\mathrm{Hz}}$).

5.2.2 Sampling Electrodes

To eliminate the polarization phenomena caused by seawater on metals[28] (contact between a liquid phase containing chlorides and a degradable and metallic solid phase), we preferably use unpolarizable type electrodes[29] or metal/metal-ion electrodes (cf. Fig. 4.23). At the moment, they are built around electrochemical couples such as Ag/AgCl (silver metal Ag bathing in its saturated silver salt solution Ag^+ + Cl^-),[30] thereby forming a half-cell potential (Fig. 4.24).

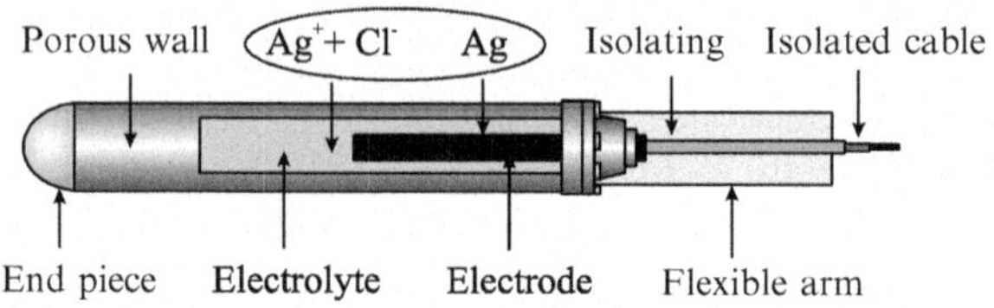

Fig. 4.23 Details, in longitudinal section, of a potential sampling impolarizable electrode located at the ends of each arm in contact with seawater. The reversible reaction becomes $Ag + Cl^- \leftrightarrow AgCl + e^-$

[28]The contact between two different phases (liquid for seawater, solid for the electrodes) causes electrochemical phenomena at the interface, inducing electromotive parasitic forces. The electrodes of M/MCl (M for metal) type were preferentially chosen because of the presence of high concentrations of chloride ions (Cl^-) in seawater. The electrodes of Ag/AgCl type have an annual drift of a few millivolts per year. We could also use the combination $Pb/PbCl_2$. This one indeed has a lower drift of the order of 0.2 mV/month and a starting polarization of less than 0.2 mV with a thermal drift of about 200 μV/°C (Petiau 2000). These electrodes are used on MT NOMADE™ electrometers from Ifremer.

[29]The invention of the unpolarizable electrode or nonpolarizing electrode is due to the Italian Professor Matteucci for his studies on telegraphy (Matteucci 1862) and was more specifically applied to the exploration of ore deposits in 1882 by Barus in the USA (Barus 1882) and then in 1912 by Conrad Schlumberger for the method of *spontaneous polarization* or the PS method (Schlumberger 1913).

[30]The mechanism of unpolarizable electrodes is based on *Nernst's formulation* (see Ives and Janz 1961).

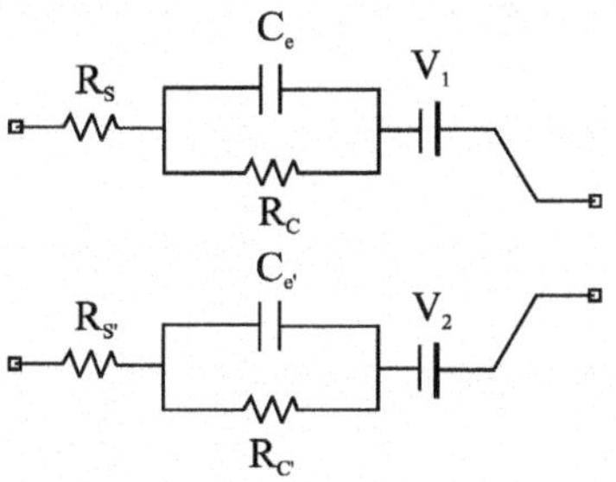

Fig. 4.24 Equivalent electrical circuit of the impedance characteristic of an impolarizable electrode system of the EMGS type. R_s, R_c and C_e are respectively the resistance of the electrolyte, the charge transfer and the capacitance (According to Havsgard et al. 2011)

The potential measuring electrodes are characterized on the electrokinetic plan by their contact resistance with regard to seawater and by their electrochemical noise. These electrodes are carried either on solid strands (fiberglass, EMGS type), connectors then being deported outside (visible cables), or at the end of hollow strands in polyvinyl chloride or polypropylene containing the connecting cables.

The potential of the unpolarizable electrodes comes from the *Nernst* equation such that the resultant voltage is equal to:

$$u = u^0_{AgCl} - \frac{K_m T}{F} \mathrm{Log}(\alpha_{Cl^-}) \tag{4.27}$$

where u^0_{AgCl}, K_m, T and F are respectively the normal potential of the electrode (i.e., 0.22 V at 25 °C), the perfect gas constant (i.e., 8.314 J.mol^{-1}.K^{-1}), the absolute temperature, Faraday's constant (96,485 C.mole^{-1}), and $\alpha_{Cl}{}^{-}$ the activity of chloride ions (Cl^-).

The variations in temperature and salinity (see Chap. 3, Sect. 2.1.1) introduce continuous interference voltages, which are equal for the Ag/AgCl electrodes to 350 μV/°C and to 500 μV/%$_{salinity}$ (Drever and Sanford 1970).

After a week of storage, the DC potential between the two electrodes is below 0.01 mV. To reduce these voltages, the electrodes are always matched in pairs together with adapted systems, so as to minimize the polarization potential difference.

5.2.3 Contact Resistance of the Electrodes

The contact resistance between the "solid" metal of the electrode and the "liquid" electrolyte (seawater) results from a problem of electrochemistry, which is more or less complex to deal with. It is easier then to study the phenomenon experimentally than theoretically. The results are given for the Ag/AgCl electrodes in Fig. 4.25.

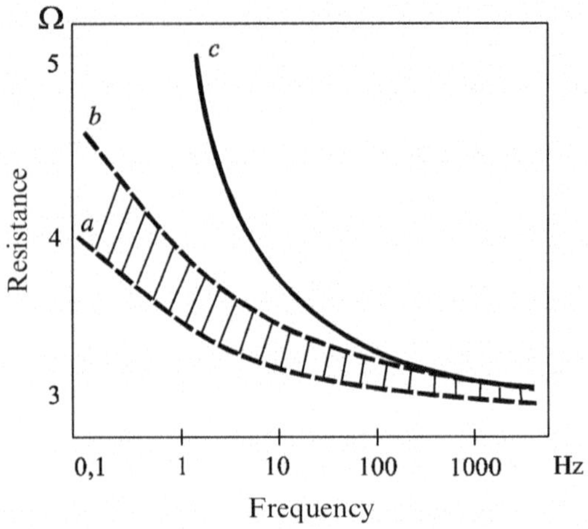

Fig. 4.25 Graph of the ohmic drop for an Ag/AgCl electrode of 5 cm diameter and 15 cm length according to the frequency, bathing (*a*, *b*) or not (*c*) in a seawater-type electrolyte (According to Cox et al. 1981)

When collecting the potential, the ohmic drop caused by these electrochemical reactions is low (a few ohms) compared to the very high input impedance of the sensors' instrumentation amplifiers (several megohms). Furthermore, the experiments show that the contact resistance decreases as the frequency increases, corresponding to a decrease in polarization phenomena (cf. Fig. 4.25).

5.2.4 Electrode Noise

Tested in a laboratory the noise electrode (Ag/AgCl couple) in the range of 0.02 to 20 Hz is below 2 nV rms, which corresponds to the noise of the electronics (Cox et al. 1978) and to the Johnson or thermal noise (see Eq. 4.27) (Fig. 4.26).

Today, the most effective electrodes have a noise level of about 4.5 pV/m (Havsgard et al. 2011). However, it seems difficult to go below the threshold of $1\,\mathrm{nv/m/\sqrt{Hz}}$, knowing that the best amplifier chains have a noise reduced at the input to about $0.5\,\mathrm{nv/m/\sqrt{Hz}}$ (cf. Sect. 5.2.6).

5.2.5 Measurement Electronics

The measurement electronics have been imperatively developed around low noise components and more specifically operational amplifier types TL61 or TL081 (BiFET), for example, to which are added active and passive filtering devices for detection and correction of the environmental effects.

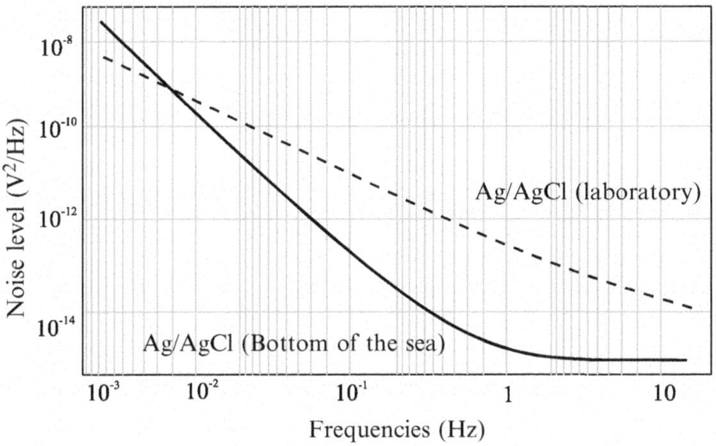

Fig. 4.26 Energetic spectral comparison of the level of noise of the electrode on the seabed and in a laboratory (According to Constable et al. 1998a, b)

Signal Amplification and Detection

The electrodes ($\times 2$) of each channel (M, N) and (M′, N′) are connected in the circuit input to a low noise differential amplifier of gain 1000, for example,[31] followed by an RC filter (stop offset voltages generated by the floating input amplifier), a chain of selective low pass amplifiers (high frequency filter) and an integrator to eliminate residual voltages. The amplitude and phase are then detected by *ad hoc* processes as synchronous detection (an example will be seen later; see Sect. 5.4.2), and then corrected from the errors of the offset and orientation of the sensors (cf. Fig. 4.31). To fully reduce the drift and give the detection electronics a resolution of 1 μV we particularly use low noise preamplifiers stabilized by a *chopper* (Fig. 4.27).[32]

Other measuring devices have been proposed (cf. Figs. 4.28 and 4.29), having for example as a first preamplification stage a midpoint transformer also stabilized

[31] The advantage of the *differential amplifier* lies in the fact that its inputs are perfectly matched thanks to the symmetrical arrangement. This has the effect of almost completely compensating for drifts of low periods. This device also has the advantage of eliminating similar parasitic voltages simultaneously affecting the two inputs of the amplifier stage. At this point one can choose differential amplifiers with floating inputs and incorporated gain whose output voltage relative to the mass of the electronic system is directly proportional to the difference of the input voltages. Current technologies allow us to produce this kind of amplifier. These amplifiers also have programmable gains up to 1000, which allow us to raise tension very quickly without significantly affecting the signal by the instrumental noises that would occasion a conventional amplifier chain. For example, one can use a *precison instrumentation amplifier*, such as the Analog Devices AD624.

[32] Electromechanical or electronic switching allowing us to modulate the signal to a higher frequency and to restore it at the end of amplification by *synchronous demodulation* (Schwartz 1959). This eliminates then more or less long term drifts, characterizing, among others, electrochemical noises.

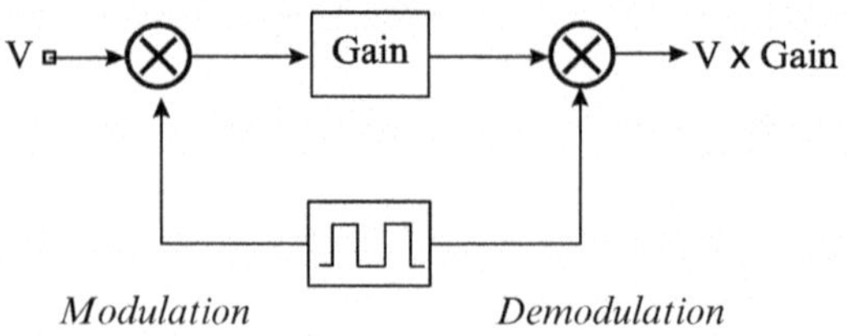

Fig. 4.27 Principle of *chopper* preamplification. To eliminate, among other things, slow drifts, the signal is synchronously modulated and demodulated upstream and downstream from the amplifier (gain). We also use double inverters instead of multipliers

Fig. 4.28 Electronic block diagram (one path) of an instrumentation amplifier for field receivers measuring potential difference, supplying in the output of the circuit the amplitude and the phase of the electric field

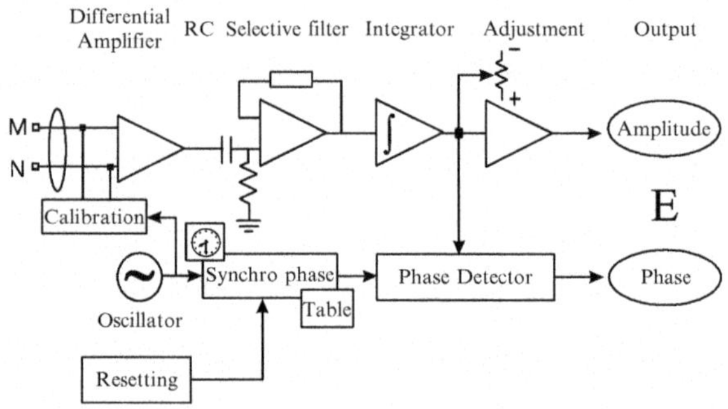

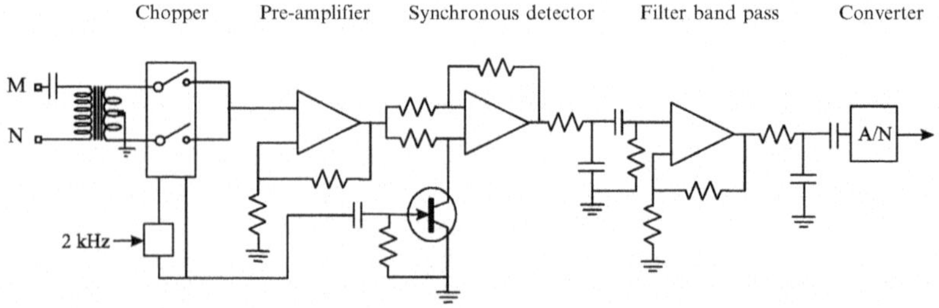

Fig. 4.29 Electronic diagram (one path) of an instrumentation amplifier, stabilized by a chopper at the secondary winding of the transformer, with a transistor synchronous demodulator (2 kHz clock) and isolation, amplifier and buffering transformer. Application to the field receivers measuring potential difference (According to Cox et al. 1981)

by a chopper (Cox et al. 1981) or using a Deaton electric field chopper amplifier, which uses a MOSFET bridge (Webb et al. 1985).

For passive detection systems (telluric source and marine magnetotelluric survey) the reader may refer for example to Professor Constable's patent (Constable et al. 1998b).

5.2.6 Noise Amplification, Noise Level and Background Noise

Related to a resistance of 4 Ω, the average value of the resistance of the electrode (cf. Fig. 4.26), the amplification chain noise voltage e_{noise} is a function of the noise voltage of the amplifiers $\approx 0.6\ \text{nV}/\sqrt{\text{Hz}}$ and of the thermal noise (cf. Sect. 5.3.3.4) $\approx 0.3\ \text{nV}/\sqrt{\text{Hz}}$ (Northrop 1989) such as:

$$e_{\text{noise}} = \sqrt{0.6^2 - 0.3^2} \approx 0.6\ \text{nV}/\sqrt{\text{Hz}} \tag{4.28}$$

The noise level is accordingly proportional to the noise amplification voltage and to the bandwidth Δf such that:

$$v_{\text{noise}} = \sqrt{e^2_{\text{noise}} \cdot \Delta f} \tag{4.29}$$

For example, if during a period of 10 s, Δf is 0.1 Hz, with a voltage e_{noise} of about 1 nV, then we will have a noise voltage v_{noise} in output amplification equal to 3×10^{-10} V rms. On the electric field, noise is then reported to the distance l separating the measuring electrodes, i.e., v_{noise}/l or 3×10^{-11} V/m for an antenna of 10 m length.

On the sea floor, considering the power emission, the noise η depending on the moment ($I_a L$) of the transmitter antenna is equal to:

$$\eta = \frac{v_{\text{noise}}}{I_a L l} = \frac{\sqrt{e^2_{\text{noise}} \cdot \Delta f}}{I_a L l} \tag{4.30}$$

e.g., for an injection current of about 500 amps, this corresponds for a 200 m length antenna to a noise value of 3×10^{-16} V/A.m^2 at 0.1 Hz.

In fact, in commercial operations, the background noise in the best case is slightly higher (10^{-15} V/A.m^2 in average over all frequencies studied), even much higher in shallow waters near the coast where the frequency band is more spread out (Fig. 4.30).

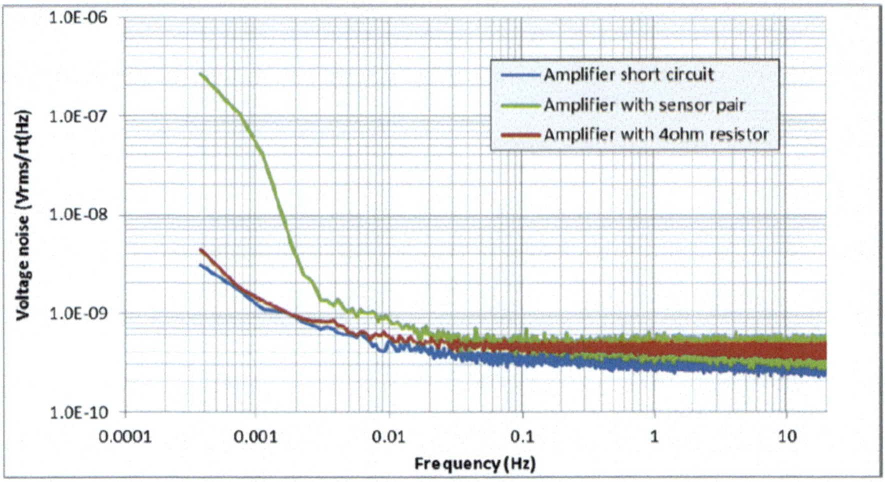

Fig. 4.30 Example of noise recording for a pair of electrodes and a 4 Ω input resistance (According to Havsgard et al. 2011)

Phase Corrector (Offset and Orientation Effects)

During acquisition, it is necessary to correct errors on the phase values introduced by the offset and the orientation unknown by the electrometers on the seabed as was pointed out above (see Chap. 2 Sect. 4.2.2).

More generally, when, for one reason or another, the clocks of the energy transmitter and the receivers are not perfectly synchronized, it is possible to accurately reposition them during the passage of the source above the electrometers. We know that at this precise moment (at the top) the signals are in phase and the intensities are maximum.

We can then automatically perform on the detected intensity peaks a phase correction from the measured fields ϕ_{Ex} in the *in-line* configuration with the help of a numerical table ϕ_{TAB} pre-established from a synthetic model taking into account the frequency ω, the electric conductivity of seawater σ_w, the length L of the antenna of the transmitter, and the offset from the center of the antenna r_0.

At this exact moment, the delay times $\Delta\tau$ are thus equivalent to the time difference between these two fields (measured and calculated) such that we have:

$$\Delta\tau = \frac{\phi_{Ex}(\omega) - \phi_{TAB}(\omega, L, r_0, \sigma_w)}{\omega} \tag{4.31}$$

or, more precisely, by performing the test in several frequencies, such as to obtain the average of the $\Delta\tau$ (Mittet et al. 2007):

$$\overline{\Delta\tau} = \frac{1}{n_\omega}\sum_\omega \frac{\phi_{Ex}(\omega) - \phi_{TAB}(\omega, L, r_0, \sigma_w)}{\omega} \tag{4.32}$$

The estimation of these differences by one of these means allows us to adjust the values of the erroneous phase introduced by the relative orientation of the receivers relative to the emission source.

Digital Electronics

The outputs of the overall measurement system (preamplification, detection, amplification, processing, etc.) are then directed to a 24-bit analog/digital converter (ADC) (cf. Sect. 5.9.2).

Sampling is performed within a range of frequencies from 30 to 1000 Hz.[33] Data are then digitally processed using, for example, an embedded microcontroller. Moreover, this has the possibility of optionally controlling the arm deployment actuators for measurements and releasing the ballast for the ascent of the receivers to the surface. Data are then temporarily stored in RAM (random access memory) to minimize the power consumption and the noise during acquisition, before being permanently stored on ROM (read-only memory)—mass storage such as flash cards (capacities up to 64 GB) or mini–hard disk, completed by an Ethernet interface, allows data recovery without opening the instrument. To control and synchronize all the data acquisition, the instrument must have onboard precise electronic clocks to keep time. These clocks piloted by a quartz oscillator must operate at a constant temperature. They are placed in a thermostatically controlled box or can be regulated with a calibrated microcomputing system with the use of a conversion table for temperature/frequency.

Auxiliary Devices

The acquisition device itself is complemented by:

- A magnetic or electronic sensor–type direction compass (north direction),
- An attitude sensor for the antenna orientation (tiltmeter, inclinometer)
- A pressure sensor for bathymetry and depth (depth meter)
- A communication transmitter (acoustic transponder)

and, where appropriate, location sensors (acoustic beacons).

When electrometers lack these systems and more particularly the orientation system, some authors (Key and Lockwood 2010) propose to use a posteriori the parameters of the ellipses of polarization of the electric fields for this task (see Chap. 3, Sect. 5.4) (Fig. 4.31).

---

[33]The *sampling frequency* is here dependent on the speed of movement of the transmitter and on the depth of immersion of the measuring devices, and partly sets the *longitudinal resolution*.

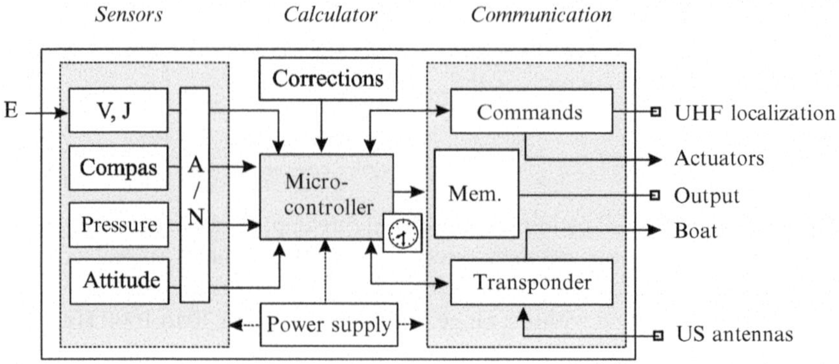

Fig. 4.31 Diagram of an entire electrometric data acquisition system including the annex sensors and the calculation, control and communication means

5.2.7 Consumption and Packaging

In general, the electronic acquisition part consumes less than 500 mW. The power supply is composed of lithium batteries, the most successful today (about 500 Wh/kg, with very low internal impedance, etc.) providing autonomy for several days or, if appropriate, several months. All these electronic components and related sensors are enclosed in a waterproof, unmagnetic case made of plastic, able to withstand the immersion pressure (at least 150 bars). For example, the potential energy of a standard sphere (17 in.) at 4000 m (operational maximal pressure) is 1.7 MJ (see note 5, Chap. 3). The collapse strength at depth, as a function of the mechanical and geometric characteristics of the packaging, is given by different analytical formulas (Dehart 1969; Sainson 2007).

We would like to remark that today we no longer use metals for the construction of cases, which prevents corrosion and galvanic effects (electrical currents).

The electrical wire connections from the field sensors (minimum $\times$ 4) pass through watertight connections (for example, Jupiter waterproof plugs collected on an outer housing fixed to the case).

5.2.8 Intrinsic Characteristics of Potential Difference Electrometers

Electrometers offer, depending on the manufacturer, almost the same features and performances. A large number of elements are in fact common to the different sensors (Constable 2013) (Table 4.4).

Table 4.4 Summary of the main characteristics of seabed pd electrometers

Characteristics	Values
Number of paths (electrical field)	2 or 3
Converter (ADC)	24 bits
ADC noise level	10^{-13} V^2/Hz at 0.01 Hz
Consumption	450 mW (3 channels)
Maximum sampling	1 Hz
Bandwidth	10 s–1 Hz
Data storage capacity	1–2 Gb
Energy, power	Lithium batteries
Navigation system	Edgetech (manufacturer)
Noise (voltage)	10^{-18} V^2/Hz at 1 Hz
Noise (field), 10 m length antenna	3×10^{-16} V/A.m^2 at 1 Hz
Weight in air	300 lb
Immersed weight	30 lb
Dimensions	10 m overall
Maximal immersion depth	4000 m

[a]These specifications may vary from one manufacturer to another

5.2.9 Disadvantages of Electric Field Measurements by Potential Difference

If we desire, for one reason or another, to increase the sensitivity of the measuring apparatus, to obtain a greater depth of investigation for instance or better resolution, it is then necessary to increase the distance l separating the potential measuring electrodes ($-\mathrm{dV} = \mathrm{e.d}l$) in proportions that cannot withstand the operational constraints (too great a congestion of the materials), especially now in the context of industrial operations. Moreover, this constraint also includes side effects such as those caused by the induction phenomena related to displacement or due to the movements of the arms carrying the two electrodes (antennas).

Indeed, at the point of measurement, the potential can be finally expressed in the form of a balance of fields and electromotive forces whose integration into the closed circuit of the MNOM arrangement formed by the electrodes MN and the measuring apparatus O (cf. Fig. 4.32*c*), is equal to:

$$\oint_{\mathrm{MNOM}} \left(-\vec{\nabla}\mathrm{V} + \vec{\mathrm{v}} \wedge \vec{\mathrm{b}} - \frac{\vec{\mathrm{j}}}{\sigma_{\mathrm{w}}} - \frac{\partial \vec{\mathrm{a}}}{\partial t} \right) . \, \mathrm{d}\vec{l} = \vec{0} \tag{4.33}$$

where the order of the four terms in parentheses corresponds to the order of their contributions to the medium of conductivity σ_w of:

- The electrostatic field where V is the scalar electric potential (*Coulomb force*)
- The field induced by the movement $\vec{v}$ (velocity) of the arms in the local magnetic field $\vec{b}$ (*Lorentz force*)

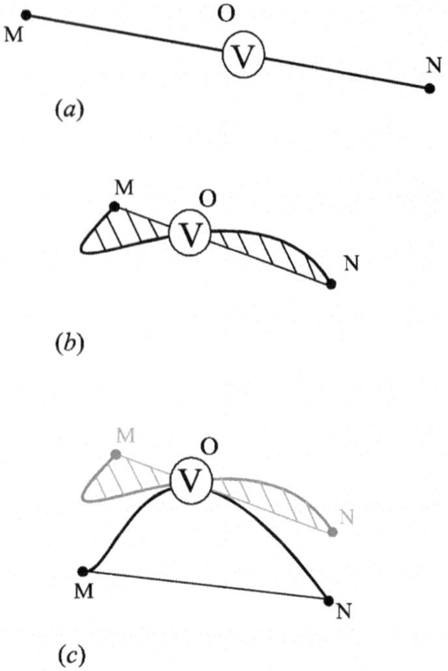

Fig. 4.32 Favorable case for measuring pd (*a*) and unfavorable case: arm or cable not extended (*b*) or moving (*c*)

- The resistance of the medium to electric currents (*ohmic resistive force*)
- The field induced by temporal variations of the magnetic field (*eddy current*)

Only the first two terms, concerning *Coulomb* and *Laplace forces*, are defined in the same way over the entire periphery of the MNOM circuit. The following two terms are different according to the path, either directly electrically closed on the seawater (MN path, cf. Fig. 4.32*a*) or passing through the receiver O (MON path, cf. Fig. 4.32*b*).

Then, using *Stokes' theorem*:

$$\oint_{\text{outline}} \vec{u} \,.\, d\vec{l} = \oiint_{\text{surface}} (\vec{n}.\vec{\nabla} \wedge \vec{u}) \,.\, dS \tag{4.34}$$

and the vector identity:

$$\vec{\nabla} \wedge \vec{\nabla} p = \vec{0} \qquad \forall p \in \Re \tag{4.35}$$

we may treat then the integrals of the gradient and the *eddy currents* such that we have:

$$\int_{\text{arms}} \frac{\vec{j}}{\sigma_w} . \, d\vec{l} = \int_{\text{arms}} \left(\vec{v}_a \wedge \vec{b}\right) . \, d\vec{l} + \int_{MN} \left(\vec{v} \wedge \vec{b}\right) . \, d\vec{l} - \int_{MN} \frac{\vec{j}}{\sigma_w} . \, d\vec{l} \tag{4.36}$$

where $\vec{v}_a$ is the relative velocity of the arm, the antenna or the cable naturally stirred into the water or towed by a boat (*streamer*).

If the measuring apparatus (microvoltmeter in this case) has infinite resistance, the left term of the above equation is then the voltage U at the terminals of the arm, i.e., ultimately the local electric field e_a considering the tense and straight arm. We have then in this preferential position (maximum field):

$$U = \int_{arms} \vec{e}_a . d\vec{l} = \int_M^N \left[\frac{\vec{j}}{\sigma_w} - (\vec{v} - \vec{v}_a) \wedge \overleftarrow{b} \right] . d\vec{l} \tag{4.37}$$

However, if the arm or the cable is not tense, or if it has curves or even (the worst case scenario) if it moves in the local magnetic field (thus initiating flux variations in $-\partial b/\partial t$ in the nonzero surface defined by the arm and the straight line joining the electrodes), the collected voltage and the resulting electric fields are then no longer representative of the electrical phenomena we wish to study (cf. Fig. 4.32).

We can therefore easily realize that in this seabed environment—hostile and inaccessible with fluctuating magnetic and hydrodynamic conditions[34]—instruments with a base several tens of meters or even hundreds of meters long, which could then clearly have higher sensitivity, cannot be used without the support of a rigid structure (frame) of large dimension.[35] Provided one can also assess the position of these sampling lines in relation to the underwater topography, such arrangements may not be compatible with industrial constraints and especially those of laying and raising these materials offshore.

For these practical reasons, another method is now proposed. Based on a different physical principle, this original technique presents an ultracompact device whose overall dimensions are then totally unrelated to the wired devices currently recommended and even less to those with a long base used until now in earth physics, bringing then to the measuring equipment perfect rigidity.

5.3 Current Density Vector Electrometers

Current density vector electrometers (active detection), first developed for mining and hydrogeological research (Sainson 1982, 1984), then for passive detection of underwater signatures (Mosnier 1984; Rakotosoa 1989; Bruxelle 1997),[36] are now

[34]It is the same for the towed equipment.

[35]This is what happens in measurement of the vertical component of the electric field (see Fig. 4.40 and Plate 4.3a).

[36]In 1969 the US Navy filed a patent application for a primitive detection system using current measurements (Pittman and Stanford 1972). We do not know if this system has been used. However, all components, hardware and signal processing of the detection device are present (cf. Plate 4.1b).

going to be introduced on the market of seabed logging (Babour et al. 2008, 2009; Besson et al. 2009a, b, 2010).[37]

Indeed, they represent very substantial improvements in performance especially by their dimensions (about 1 m versus 10 m and more for the traditional electrometers) and by their sensitivity, thus responding to the problem raised by pd electrometers.

5.3.1 Principle and Technology of Current Density Sensors

To overcome the disadvantages of the potential difference electrometers (poor performance, large dimensions, low resolution and definition, etc.), and more particularly to remove pd interference due to some relative movements of the antenna (see Sect. 5.2.9), one of the recommended solutions is, *knowing the resistivity* ρ of the propagation medium (seawater), to evaluate this time not an electric field, strictly speaking, but an average current density $\vec{J}$. This latter represents at a given point a vector with the direction of flow, the way of displacement and the amount of electricity passing through a unit area taken perpendicularly to the direction of the current.

The value of the field $\vec{E}$ is this time no longer dependent on the distance between the sampling electrodes but on their surface *S*. Thus, under Ohm's law ($\vec{E} = \rho\vec{J}$), we have:

$$E = \rho \frac{dI}{ds} = \rho \frac{I}{S} \tag{4.38}$$

If we now compare the previous potential difference measurement device (cf. Eq. 4.26) between two points M and N, integrated in a *single dimension* (length *l*) such that:

$$V_M - V_N = \int_l \vec{E}.d\vec{l} \tag{4.39}$$

with the current measurement, this time integrated in *two dimensions*, that is, more precisely on a surface (*S*) such that:

$$I = \iint_S \vec{J}.d\vec{s} \tag{4.40}$$

[37]Incidentally, I would like to thank very briefly the Schlumberger company, its subsidiary WesternGeco and especially its engineers, for, after 30 years, funding my research works and those of my colleagues, of interest… I hope therefore that the rigor and honesty that usually drive scientists will allow us to finally restore, in all objectivity, the scientific contributions of each.

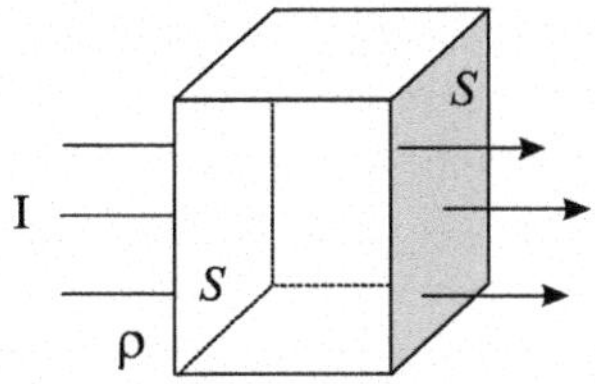

Fig. 4.33 Principle of evaluation of the electric field from a measurement of current density dI/d*S* expressed in A/m^2 and a measurement of resistivity ρ. The current I (A) is collected from the conductive surfaces *S* (m^2) arranged here on both sides of a more or less resistant prism immersed in a conductive medium

we can easily see that this surface, for much lower dimensions (at least ten times lower), thus presents a yield ($I = |\vec{J}| \cdot S$) significantly higher compared to the potential measurement related to the length (see Eq. 4.39) (Fig. 4.33).

In this case, virtually all of the signal energy is used, especially when current measurements I are made through an instrument with very low input impedance (a current transformer, for example). In these special conditions, the signal-to-noise ratio (S/N) is then greatly improved. On the other hand, Ohm's law represents an equality that is valid not only in magnitude but also in direction (electric field and current density are vector quantities), which makes this technique connecting these two vectors more directive and more sensitive than those using measurements of electric fields by the potential difference.

In its basic configuration, the current density electrometer (CDE) consists of two thin metal plates (thickness here plays no role), of surface *S*, parallel faces plated on an insulating support of any shape (cylindrical or parallelepipedal, even prismatic). These are connected to the primary winding of a toroidal transformer, with the current flowing in it measured by the secondary winding (cf. Fig. 4.34).

With such a device (cf. Figs. 4.34*a*, *b*), we can easily realize that the impedance Z_o of the measuring circuit (contact resistance of the electrodes and impedance of the transformer included) is well below that provided by the insulator (between the plates and occupying the place of the slice of the displaced fluid). This causes an immediate distortion of the force field around the sensor itself.[38] For a correct and meaningful measurement of the ambient electrical field and to solve this problem, a specific correction and calibration are therefore necessary (Sainson 1984).

Corrective Systems

The correction that aims to take into account in the measurement the distortion of the field lines due to the sensor consists in this case of restoring the electrokinetic balance by appropriate means. This correction can be considered in three forms:

[38]The idea to channel the current through the sensor (ferromagnetic core) was included in the development of a magnetic sensor (Baxendale 1989).

- The instrumental method (*hardware*): by matching in the direction considered, the impedance Z_m of the measuring circuit (cf. Fig. 4.34) with an impedance Z_ρ equal to the value of the resistance of the removed water layer of resistivity ρ, by appropriate electronic methods (i.e., variable resistance)
- The computing method (*software*): by calculating the impedance Z_ρ representative of the close environment and of the operational conditions (for example,

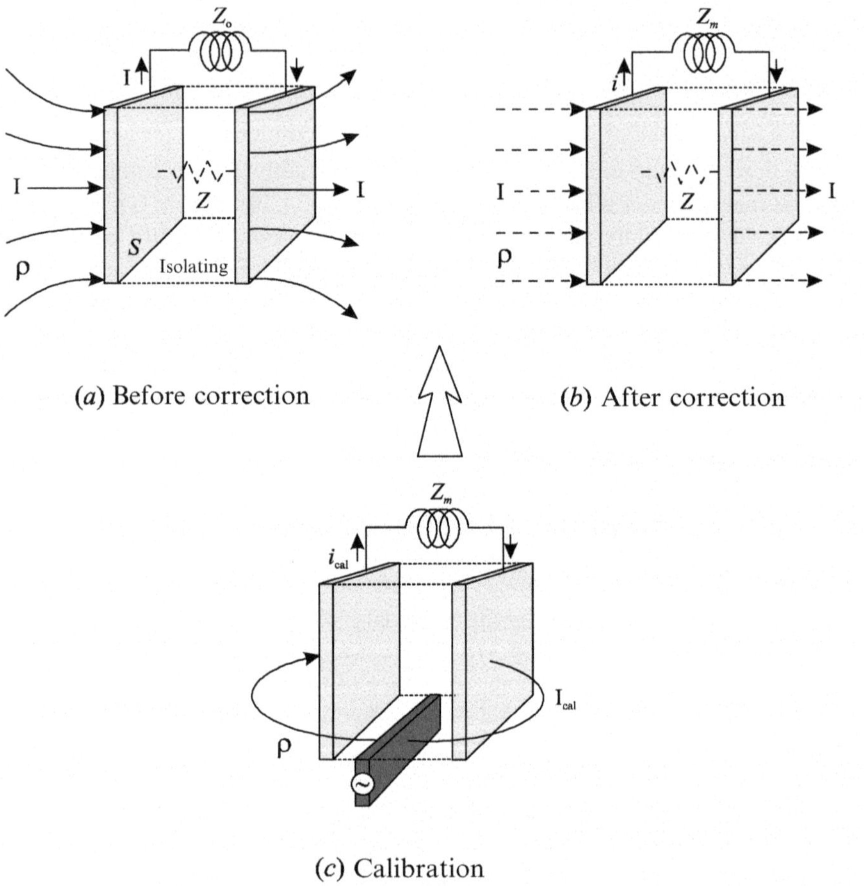

Fig. 4.34 Principle of detection of the current density electrometer (CDE) (According to Sainson 1984). (*a*) CDE before correction: the impedance Z (insulator + leaks) is larger than the impedance of the measuring circuit Z_o. The current then passes completely therein, thereby changing in its proximity the distribution of the field lines, thereby causing a *capture effect* (not desirable). (*b*) CDE after correction: the impedance of the measuring circuit is equal to the impedance of the fluid volume displaced by the sensor ($Z_m = Z$). This therefore behaves as if a current of the same intensity I were crossing through it, and the current *i* taken by the measuring circuit is about one million times smaller than I. In this case the distribution of the field lines is no more physically disturbed by the instrumentation and the anomalous field measurements after calibration can then be performed properly. (*c*) CDE calibration phase: temporary injection of a constant tuned current, calibrated in intensity (I_{cal}) near the sensor electrodes allowing calibration of the sensor

those provided in the calibration phase) or by evaluating a coefficient of capture (in comparison with theoretical values) or even by simulating the EM field with more or less sophisticated mathematical models[39]
- The mixed method using the transfer function of the sensor, which eliminates the value of the seawater conductivity (Baicry 2015a, b)

Calibration Devices

The calibration of the CDE is performed by injecting near the sensor an electric current that is homogeneous and calibrated in intensity. Received values are then compared to theoretical values, taking into account the topology of the device and the geoelectric parameters of the surrounding environment obtained by specific processing and acquisition means.

Sensitivity

The sensitivity of such an arrangement depends on several instrumental factors (work frequency, resistance noise, S/N ratio of the amplifiers, etc.) and environmental factors (EM background noise, contact resistances of the electrodes varying with the nature of the electrolyte, etc.).

For the developed instruments (active area of a few square centimeters to several square decimeters) and in good measuring conditions, sensitivity ranges from a few hundred $\mathrm{pV/m/\sqrt{Hz}}$ to a few $\mathrm{nV/m/\sqrt{Hz}}$ beyond 1 Hz.[40] This has the advantage, for example, to increase the detection accuracy by increasing in frequency. Moreover, with such an instrument, packed on itself, adverse effects on the phase due to the gap (shifts) on the different channels E_x, E_y, E_z are largely minimized or even completely removed.

Overall Dimensions

These receivers, which can be used with controlled sources, today occupy, all components combined, a volume of at most a cubic meter for a resolution at least a hundred times greater than present detectors (Fig. 4.35).

They also have the advantage, thanks to their small dimensions, to limit or even eliminate the *amplitude offset* (see Chap. 2, Sect. 4.2.1). We can also predict in the

[39]We can use, for example, numerical resolution methods such as the *finite element method* (adapted to the nonlinear medium) or the boundary integral method, under the *Neumann boundary* conditions for the insulating part of the electrometer and the *Dirichlet boundary* conditions for the conductive portion of the electrodes. In 3D, this method has the advantage of taking into account the volume current density.

[40]Below 1 Hz the noise level increases, particularly when the detector is used in mMT (see below).

Fig. 4.35 Tight prototype of a current density and controlled source electrometer: transverse fields T_x and T_y. The photograph shows two pairs of surface measurement electrodes matched with their preamplification system: toroidal transformer, differential amplifier, filters, synchronous detector, etc.. . . (According to Sainson 1982)

future with system miniaturization even smaller dimensions comparable to those of 3D seabed seismic sensors.

A major advantage of these compact vector electrometers is to have also an element for measuring the vertical component of the electric field (cf. Figs. 4.37a and 4.37b) of very small dimensions (1 m or less) compared to those recommended for conventional electrometers (10 m height at least, requiring in addition a very bulky superstructure). This saves space (cf. Fig. 4.45a) and drastically reduces the effects of topography (see Chap. 3, Sect. 3.6.2).

Directivity and Focus

Having nondeformable collecting surfaces, more or less wide, perfectly parallel to each other, gives the measuring instrument, in spite of very small dimensions, much better directivity than that provided by pd wired electrometers, less rigid and with more changing dimensions. The current detection device also allows us to bring perfect orthogonality to the evaluation of the transverse components (Fig. 4.36).

Depending on the nature of the studied fields, it is possible, to improve performance, to place the sensor in mechanical focusing devices such as manifolds (insulating tubes) or current concentrators (tronconical pipes).

5.3.2 Measurement Electronics for Active Sensors (Controlled Source)

The electronic circuit is similar to that developed for potential difference measurements. The potential points of the previous device are then replaced by two surface electrodes connected to the primary winding of a toroidal transformer ad hoc (with a specific impedance). The terminals of the secondary winding are connected to a

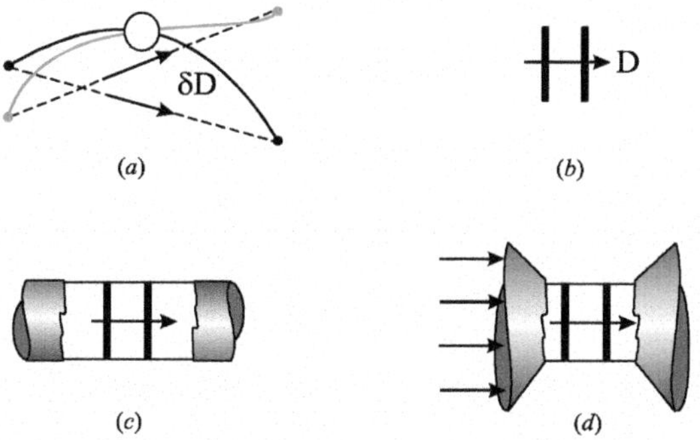

Fig. 4.36 Comparison of the directivities D of pd wired electrometers (*a*) and current density electrometers (*b*). Even for small angles (small changes in the direction of the electric field) the variations of directivity δD may be important. To increase the directivity, it is also possible to place the sensor in a collector (*c*) or a current concentrator (*d*)

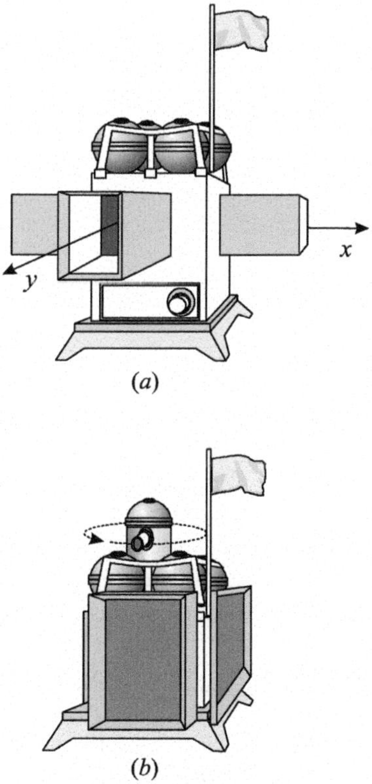

Fig. 4.37a Current detection electrometers of the CNRS type for mMT measures: (*a*) with two horizontal channels with an equivalent impedance system with a current collector focusing device; (*b*) with two horizontal channels (*rectangular* capture surface) with a rotary calibration device (According to Rakotosoa 1989)

differential amplifier with floating inputs whose midpoint is connected to the mass circuit (point 0 V of the electrical DC power supply +15 V 0–15 V). The detection is performed by a phase detector, the different principles of which are further explained in Sect. 5.4.2. The system is then completed by an independent device

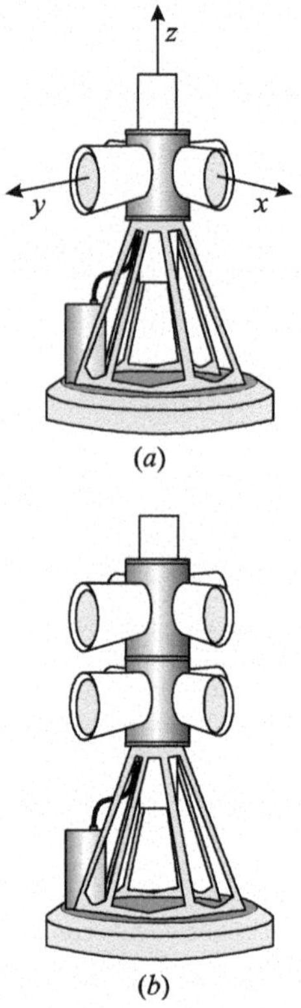

Fig. 4.37b Current density static electrometers for submarine detection, of the GESMA type (Atlantic Submarine Studies Group), also usable in mMT (cf. Plate 4.3): (*a*) with three channels (two horizontal and one vertical components); (*b*) with five channels, multiband ULF and ELF (four horizontal and one vertical components)

to evaluate the resistivity of the medium (seawater), the most suitable method (Ivanoff 1975) being the Wenner probe (cf. Figs. 4.38 and 4.39).

5.3.3 Measurement Electronics for Passive Sensors (Telluric Source)

Telluric field measuring requires much more sophisticated instrumentation for detection and analog signal processing, because of random variations in these EM fields and the very wide bandwidth of the interesting signals. As frequencies or periods, of any shape (nonsinusoidal), are not known in advance and are always drowned in background noise, which is often important, it is a real instrumental challenge to obtain an acceptable signal-to-noise ratio[41] (Reich and Swerling 1953).

[41]This is generally the case for all passive detection methods and methods for listening to EM natural noises where a signal of very small amplitude is often drowned in high background noise.

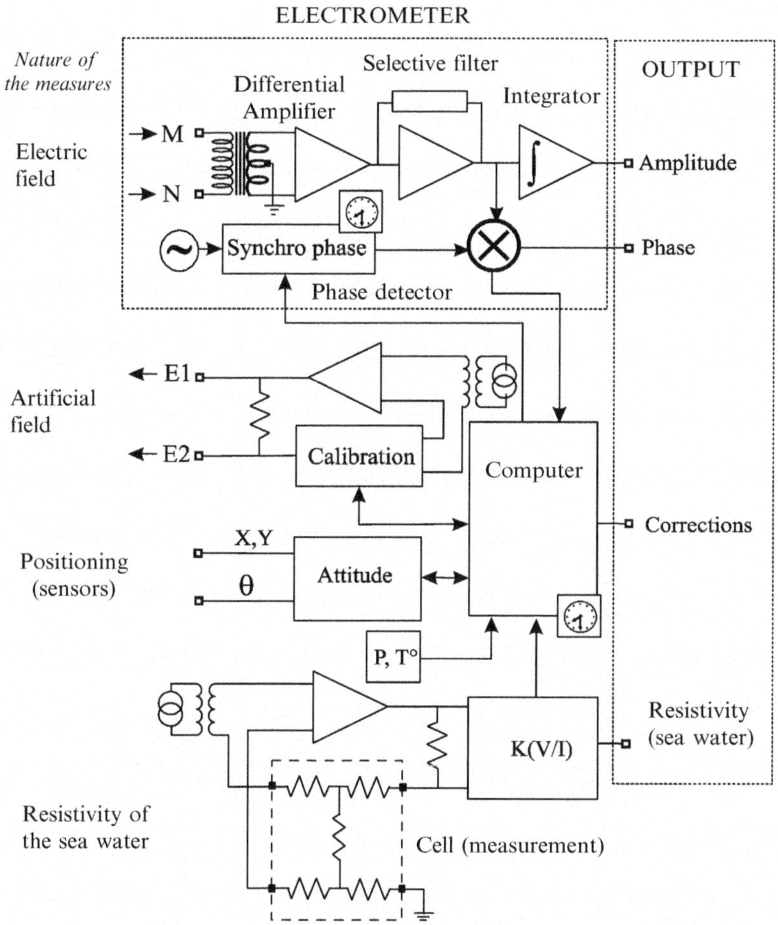

Fig. 4.38 Simplified electronic diagram of current density active receivers provided with attitude correction and intensity calibration systems, and a probe for measurement of water resistivity

The measurement system, the first of its kind (Rakotosoa 1989), briefly comprises an amplification device, which is more or less broadband (*a*) supplemented by a spectrum analyzer (*b*) followed by an analog computer with quadratic average (*c*). The analog circuit input is supplemented by a calibration device (*d*) analyzing the performances of the detection chain during acquisition (cf. Fig. 4.40).

Broadband Amplifier

As for selective systems of controlled source equipment, the broadband amplifier comprises, in this order:

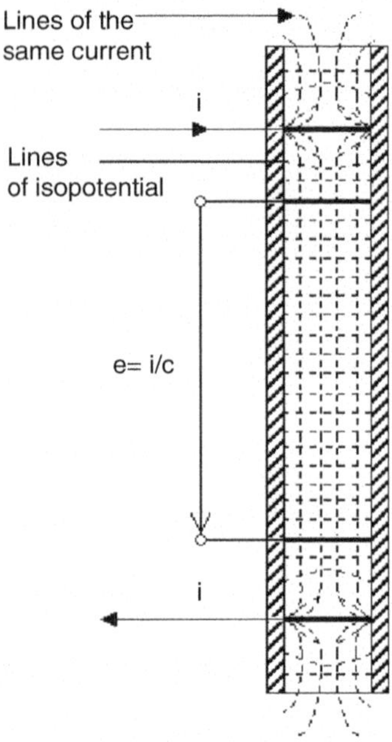

Fig. 4.39 Wenner-type conductivity probe with four guideline electrodes. Seawater is in contact at the ends of a PVC tube

- A toroidal transformer into which is inserted a feedback loop, which transfers the current in opposition to the phase with the input current, allowing us to obtain at each moment a voltage output proportional to the input current
- A servo instrumentation preamplifier comprising a differential amplifier whose input parity is provided by the midpoint of the secondary winding of the transformer (connected to a mass circuit), and whose gain must be large enough to ensure maximum efficiency of the flux feedback device
- An output amplifier ensuring an adequate level to attack the spectrum analyzer

The presence of the feedback loop in the servo circuit allows us to continuously adapt the impedance of the transformer secondary winding to that of the primary winding.

Spectrum Analyzer

The output of the broadband amplifier cannot be operated effectively without knowledge of the frequency content of the capture signal and its temporal variations. The addition of a spectral analysis device thus allows us to get the energy for each frequency contained in the spectrum. Analogically, this is achieved by multiplying the random signal (telluric source) by a reference signal of known frequency established in the bandwidth of the acquisition system. A low pass filter is then used to eliminate the side-band beats to only keep those of the main band.

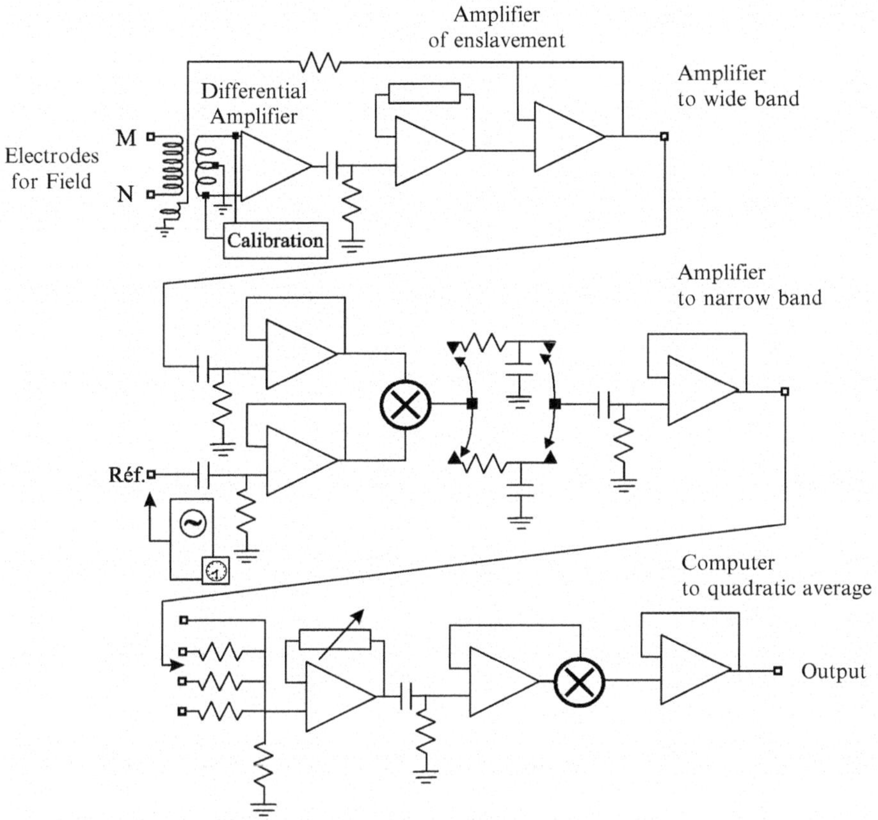

Fig. 4.40 Electronic diagram (one path) of an amplification system with feedback flow and spectral (analog) analysis in current density measurement passive receivers (mMT) (According to Rakotosoa 1989)

The spectrum analyzer comprises:

- A narrow band amplifier, consisting of an electronic multiplier ⊗ (AD534 from Analog Devices, for example) whose inputs are those of the source signal (measurement) and those of the frequency reference. The outputs are connected to low pass passive filters which then limit the width of the band. The signal is then sent to a high pass filter with a cutoff frequency of 0.1 Hz
- A fixed gain amplifier (100 for example) followed by an impedance adapter (follower)

The reference synthetic signal must be established by appropriate means (analog or digital) on a continuous spectrum, stored in mass memory, for example. In this case, it is necessary to use a digital/analog converter.

Quadratic Average Analog Calculator

The narrow band amplifier only determines the relative amplitude of any frequency in a signal. To obtain the energy peak corresponding to each frequency it is necessary to square the amplitude and integrate its value over an RC cell.

Schematically, the quadratic average calculator is composed of:

- A stage of variable gain amplification, followed by a high pass filter and a follower
- An analog multiplier
- An integrator, which eliminates rapid variations in the signal

Calibration of the Amplification and Detection Device

Calibration is performed by substituting the capture device (electrodes/transformer) with a discrete resistor, for which noise can be a priori calculated (see below) or evaluated in real time in local conditions. Knowing the noise of this resistance, we can then estimate the overall noise of the amplifier chain (amplifiers, switches, modulator, etc.) (Fig. 4.41).

The noise of such resistance is usually *thermal noise*, also called *Johnson* noise (Johnson 1928). It is particularly due to the thermal and constant agitation of electrons in electrical conductors and was quantified by *Nyquist* (Nyquist 1928; Goldman 1948).

Resistance R, at the temperature T (°K) in the absence of any applied voltage, is subjected to spontaneous displacement of electrons, which move in all directions. On average, the algebraic electron velocity is zero and the average current is zero too; however, the average of the velocity square is not zero. Across the resistance

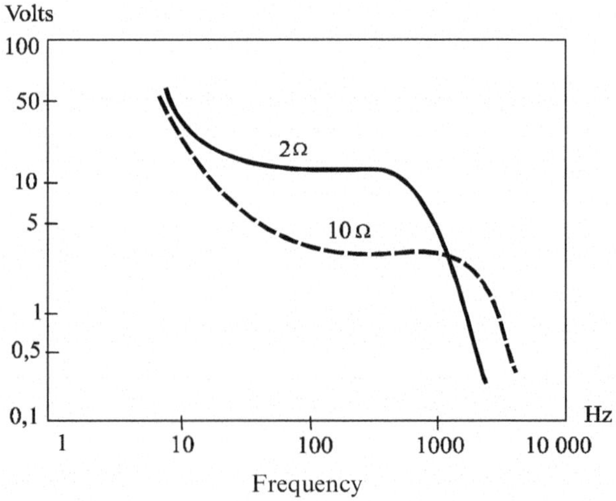

Fig. 4.41 Noise resistance curves from an amplification device for calibration of the acquisition system (According to Rakotosoa 1989)

therefore remains a noise voltage of instantaneous value U (Kittel and Kroemer 1995). The quadratic mean value of the noise (efficient noise electromotive force) is the root of the mean square of the voltage, i.e.:

$$U_{eff} = \sqrt{\overline{U}^2} \tag{4.41}$$

For a frequency band of width Δf, the noise voltage at the terminals of the resistance in thermal equilibrium is therefore equal to:

$$\overline{U}^2 = 4k_B TR\Delta f \tag{4.42}$$

where k_B is *Boltzmann*'s constant ($1{,}38.10^{-23}$ J/K).

By taking into account all the semiconductors, a more accurate formulation would be to use the spectral density $p(f)$ derived from *Planck*'s law about the blackbody radiation such as:

$$\overline{U}^2 = 4k_B TR\Delta f \, p(f) \tag{4.43}$$

with:

$$p(f) = \frac{h_P f}{k_B T} \frac{1}{e^{h_P f / k_B T} - 1} \tag{4.44}$$

where h_P is *Planck's* constant ($6.63.10^{-34}$ J.s).

The relation $p(f) = 1$ (in the radiofrequency range) corresponds to a constant spectral density, hence the designation of white noise by analogy with white light, which contains all of the light spectrum.

5.3.4 Sampling Electrodes: Impedance, Noise and Polarization

In the context of current density measurements, the sampling electrodes play a capital role all the more important when the origin of the signals is natural and when the capture surface is extended (cf. Fig. 4.33). On purely electrokinetic and electrodynamic terms the current sampling electrodes may be characterized by their contact impedance (kinetic data) and their noise (dynamic data).

These two characteristics vary differently over time, more or less randomly, so that it is necessary to know them instantly in order to evaluate them precisely.

Impedance and noise variations actually depend on many physicochemical parameters intrinsic to the electrodes (type of metal), to the immersion medium (composition of the electrolyte) and to the ambient thermodynamic conditions (temperature, pressure, presence of dirt, pH, etc.). These variations are more exactly due to the electrochemical interface reactions (ion exchange) between the electrodes and seawater, such as those that cause corrosion (Sainson 2010).

Contact Resistance of the Electrodes

The contact resistance of the electrodes, their internal resistance with the impedance of the transformer, determines the distribution of the current lines near the receiver. If the internal resistance of the electrode and of the processor remains relatively constant, it is not the same for the contact resistance, which will vary depending on local electrochemical and thermodynamic conditions.

The complexity of the involved phenomena makes theoretical calculations difficult to conduct and indeed unrepresentative of the phenomena. For this, it is therefore preferred to use experimentation with protocols agreeing on many specific measurement conditions.

In these assembly conditions (cf. Fig. 4.42), the impedance is simply given by the formula:

$$Z_{[\Omega]} = R_{[\Omega]}\frac{V_E}{V_R} \quad (4.45)$$

where V_E and V_R are measured in volts respectively at the terminals of the voltmeter and the resistance R (Fig. 4.43).

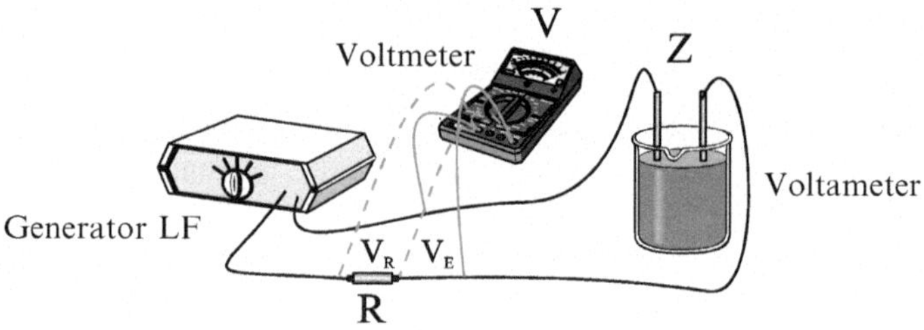

Fig. 4.42 Experimental protocol for evaluating contact resistances (impedance bench) composed of a low frequency generator (LF), a voltmeter where two electrodes dip in an electrolyte (salted water at 15 g/l) with resistance (R) and a voltmeter (V) for high input impedance (According to Rakotosoa 1989)

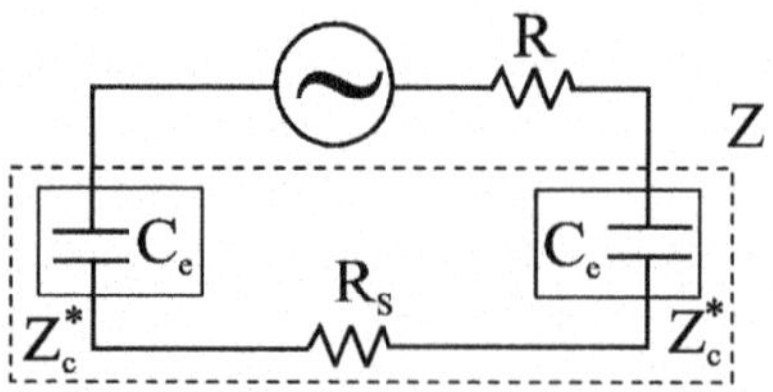

Fig. 4.43 Equivalent electrical diagram of the impedance measurement bench (cf. Fig. 4.42) where, more precisely, R_s is the resistance of the electrolyte volume and Z_c^* is the contact impedance at each electrode

For metals such as lead, nickel, stainless steel, chromium or silver, the impedance decreases with the frequency (from 500 Ω to 70 Ω). However, for steel, the impedance is the lowest (50 Ω) and remains constant regardless of the frequency used.

Electrode Noise

The sensitivity of the apparatus is also limited by the noise of the capture elements (electrodes), which are the site of important electrochemical reactions, at their contact surface with seawater.

To determine this specific noise we can make a measurement of the global impedance of the voltmeter at a given frequency through the assembly described above (cf. Fig. 4.42), or more preferably we may use a spectrum analyzer for which the internal resistance would have previously been evaluated (Rakotosoa 1989).[42]

Compared with other metals such as chromium, lead, and silver, steel has the lowest noise (5 dB at 1 Hz) to 1 dB at 50 Hz.

Polarization of the Electrodes

These different electrochemical interface phenomena (resistance and noise) provide additional against-electromotive force at each sampling current electrode, thus creating an imbalance between the potentials of the two electrodes, or polarization, the causes of which can be due, among other things:

- Either, for distinct electrodes but of the same kind, to differential diffusion of metal ions in seawater (differential degrees of oxidation)
- Or, for electrodes of different composition, to an electronic imbalance (more or less electronegative metals, roughly easily yielding electrons to the aqueous medium)
- Or even, for the same electrode, to a difference in local ionic concentrations of the present electrolyte (*Nernst* potential) (Fig. 4.44)

[42]For more information on noise measurement protocols, the reader may refer to the thesis of Mr. Urbain Rakotosoa UMPC (Rakotosoa 1989).

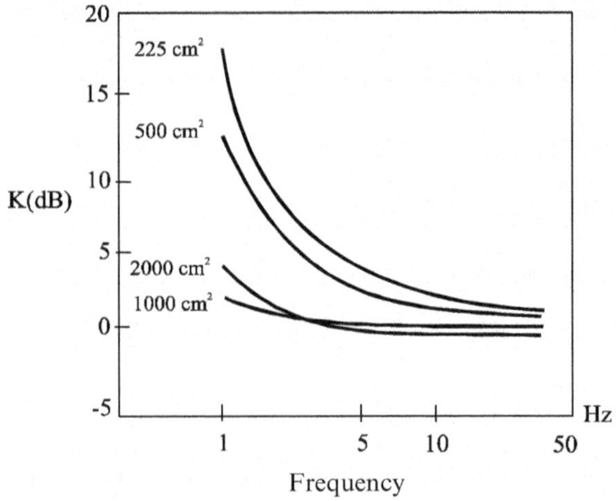

Fig. 4.44 Electrode noise for different stainless steel surfaces (225–2000 cm^2) as a function of frequency (1–50 Hz) (According to Rakotosoa 1989)

5.4 Streamer Electrometers with Horizontal Current Injection

Seabed electrometers, as they exist today, are bulky (cf. Fig. 4.45a and 4.45b) and heavy to implement, and should be arranged so as to cover the entire area of the prospect. They have therefore a significant operational cost, at this time higher than that of seismic reflection. It is for these reasons that research is now conducted to set these instruments on streamers pulled by the surface vessel (Scholl and Edwards 2007; Babour et al. 2008; Davidson 2009; Tenghamn et al. 2010; Swidinsky 2011) or to integrate them on solid seismic streamers enjoying then a common acquisition (Mattson and Anderson 2010).

This technology, currently employed by the company Petroleum Geo-Services (PGS 2010; Mattson et al. 2011; Lindqvist 2012; Mattsson et al. 2013; McKay et al. 2014; 2015) for exploration in shallow water (control of air waves), if it proved its effectiveness, would have as a benefit coverage of larger areas and could be integrated into seismic equipment forming a complete measurement system while, of course, minimizing acquisition costs, which logically could be halved[43] (Fig. 4.46).

The very small size of the streamers (a few centimeters in diameter at most) do not allow us to use the technology of potential difference electrometers, which are too large. In contrast the current density electrometers can easily be integrated on rigid cables of the small diameter streamer type, used, for example, in marine seismic acquisition (cf. Fig. 4.47).

We can therefore at this stage imagine all sorts of arrangements that may increase the collecting surface. The electromagnetic source can be placed at a

[43]Moreover, these systems may have common telemetry devices, treatment, etc. (similar transfer frequencies).

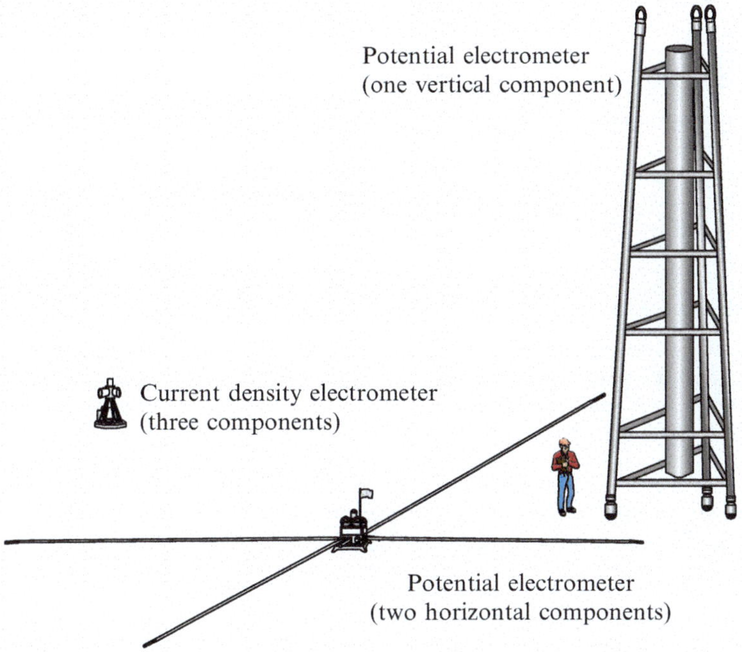

Fig. 4.45a In the *center*, dimensions of the receivers with horizontal components (TE_x, TE_y), currently in service, and *right*, vertical component (TE_z) electrometer embedded in a high rigid structure (rigid tripod stabilized by dead weights for each leg). With such dimensions, an increase in the sensitivity cannot be reasonably considered anymore. *Left*, current density receiver approximately 100 times more precise than its homologs on the right

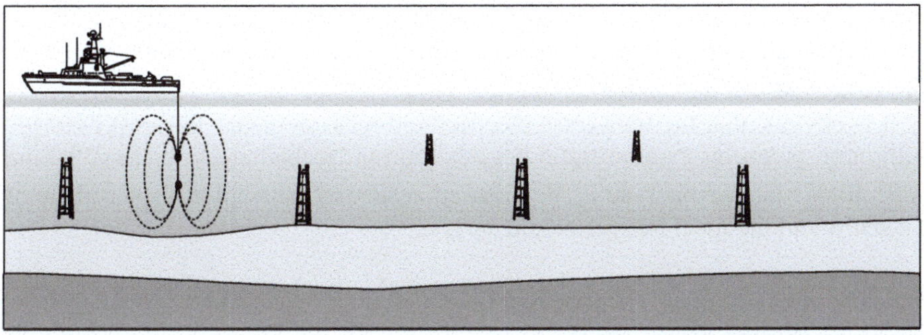

Fig. 4.45b Use of sensors measuring vertical component R_z (cf. Fig. 3.45a and Plate 4.3a) with a vertical stationary pulsing source T_z (2500 A) in the time domain (transient VED). Service proposed by the Petromarker company to increase the horizontal resolution. The challenge is in the measurement of signals E_z smaller than they are in horizontal configurations (Holten et al. 2009; Cuevas and Alumbaugh 2011)

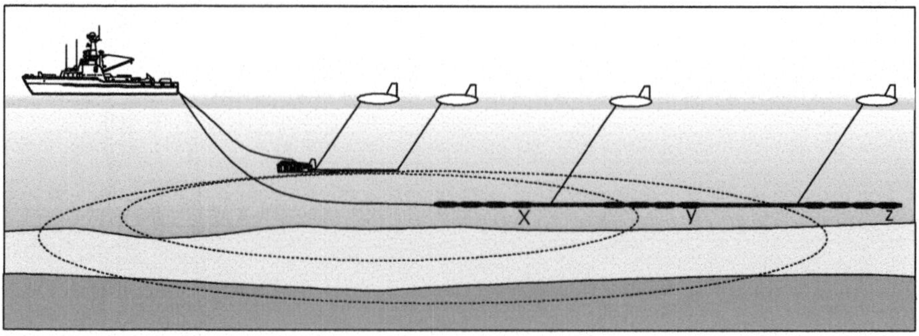

Fig. 4.46 Acquisition device of the streamer type. The source and electrometers X, Y, Z are connected with the same bottom line. The positioning is performed by acoustic transponders attached to the cable ends. The device may also be used in transient mode (time domain) for acquisition in shallow sea (According to PGS 2010)

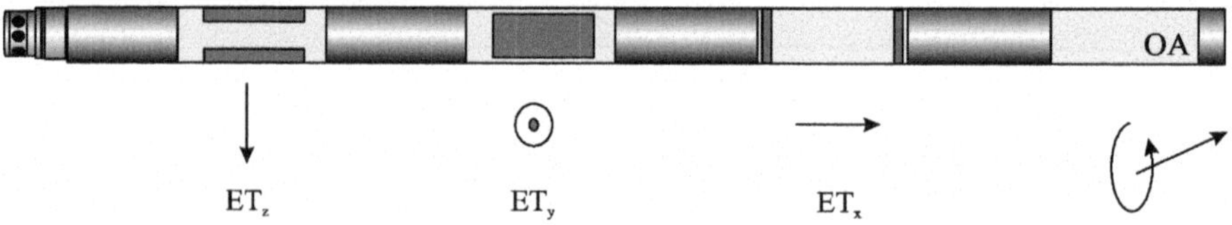

Fig. 4.47 Measuring current density vectorial electrometer, with three channels (TE_x, TE_y, TE_z), arranged, for example, on a *solid streamer*. The acquisition is completed by an orientation and attitude module (OA) attached to the three sensors. The set is mechanically fixed on a rod or a rigid bar

distance, either laterally or in depth above the streamer, or collinearly to it. It can also be placed on either side of the receiving device, as it was done in the early detection attempts (Sainson 1984).

It is obvious that the transverse electric field measurements $\vec{E}_T$ must be made with more precision than on static devices (current seabed electrometers), taking imperatively into account the $\vec{E}_m$ field variations due to the relative displacement of the sensors in the local magnetic field such as:

$$\vec{E}_T = \vec{E} + \vec{E}_m \tag{4.46}$$

which can be written as a function of the instantaneous speed $\vec{v}_r$ and the magnetic induction $\vec{B}_l$:

$$\vec{E}_T = \vec{E} - \vec{v}_r \wedge \vec{B}_l \tag{4.47}$$

a problem that is solved by an additional acquisition (measures $\vec{v}_r$ and $\vec{B}_l$) and by adapted electronic processing of correction and compensation.[44]

5.4.1 Acquisition Electronics

The fact that the electromagnetic source and field sensors are somehow combined can greatly simplify the signal processing. Indeed, theoretically the simultaneous movement of the source and the streamer (at the same speed) almost allows data processing in real time. In practical terms, wired connections can be established, then greatly simplifying communication systems, consequently removing all wireless telemetry problems that are quite common on current devices.

5.4.2 An Example of Detection: Synchronous Detection

The detection of phase variations in the reception can be solved simply by using, for example, a *synchronous detection* device derived among others from *homodyne detection* techniques[45] developed in the 1960s (Rowe 1965; Bennett and Davey 1965). We can give here an example of coherent detection with a simple electronic circuit (*lock-in* detection) used successfully, for example, for acquisition on a streamer (Sainson 1984).

Synchronous detection is the very type of simple and elegant implementation that allows us to detect and follow simultaneously, analogically (without delay), the variations in amplitude and the phase movement of one element of voltage or current compared with another, drowned in high background noise.

The principle of this basic device, which combines speed (real time), accuracy and wide dynamics, consists of the electronic multiplication of the received signal (clean enough) by a reference signal generally taken directly from the transmitting source. Thus the method only brings out the phase variations as a function of the amplitudes V_R.

In terms of processing only (theory), if we consider a received signal of sinusoidal amplitude V_R as $y_l = V_R \cos \omega t$ and a reference signal (also called a replica or copy), also sinusoidal, of amplitude V_r taken on the injection of current of the emitter $y_2 = V_r\ (\cos \omega t + \varphi)$, and if a multiplication of the two signals (analog multiplier, for example) is carried out, then we get:

[44]In the case where the *streamer* is immersed at the surface, the speed of the water movements should also be taken into account (see Chap. 3, Sect. 6.9.2).

[45]Processing by amplitude demodulation.

$$\begin{aligned} y_1 y_2 &= V_R V_r \cos \omega t \ \cos \ (\omega t + \varphi) \\ &= V_R V_r \cos \omega t \ (\cos \omega t \cos \varphi - \ \sin \omega t \sin \varphi) \\ &= V_R V_r \cos^2 \omega t \ \cos \varphi - V_R V_r \cos \omega t \ \sin \omega t \ \sin \varphi \end{aligned} \tag{4.48}$$

as:

$$\cos^2 \omega t = \frac{1}{2} (\cos 2\omega t + 1) \qquad \text{and} \qquad \cos \omega t \ \sin \omega t = \frac{1}{2} \sin 2\omega t \tag{4.49}$$

from which, finally:

$$\begin{aligned} y_1 y_2 &= \frac{V_R V_r}{2} [(\cos 2\omega t \ \cos \varphi + \ \cos \varphi) - (\sin 2\omega t \ \sin \varphi)] \\ &= \frac{V_R V_r}{2} [\cos \ (2\omega t + \varphi) + \ \cos \varphi] \end{aligned} \tag{4.50}$$

This expression corresponds to the superposition of an AC component $V_R V_r /2 \cos (2\omega t + \varphi)$ and a DC component $V_R V_r / 2 \cos \varphi$. The low pass filter placed at the end of the circuit only restores the continuous term, which then directly depends on the amplitude of the input signals and their phase shift with the source signal. In other words, the output average voltage is proportional to the amplitude of the input signal and varies as the sine of the phase difference. The signal-to-noise ratio improvement means here reduction of the bandwidth, which is limited in this particular case to the cutoff frequency of the low pass filter.

From a purely electronic point (cf. Fig. 4.48), it is often necessary to introduce in the multiplier input a reference signal phase shifter to specifically refine the settings phase (catching up or calibration). It also helps to decouple at this level the measurement and reference signals (excitation) by introducing if necessary a connecting optical coupler (opto-isolator) to suppress, among other things, parasitic return currents[46] (induced currents), which are important in this type of measure.

To improve the stability of the system, we can also replace the sinusoidal signal by a square signal, or substitute the electronic multiplier with a single or, even better, a double automatic electronic inverter driven by the frequency of the reference signal (cf. Fig. 4.49).

[46]Prevents return current loops between transmitters and receivers, and galvanic couplings.

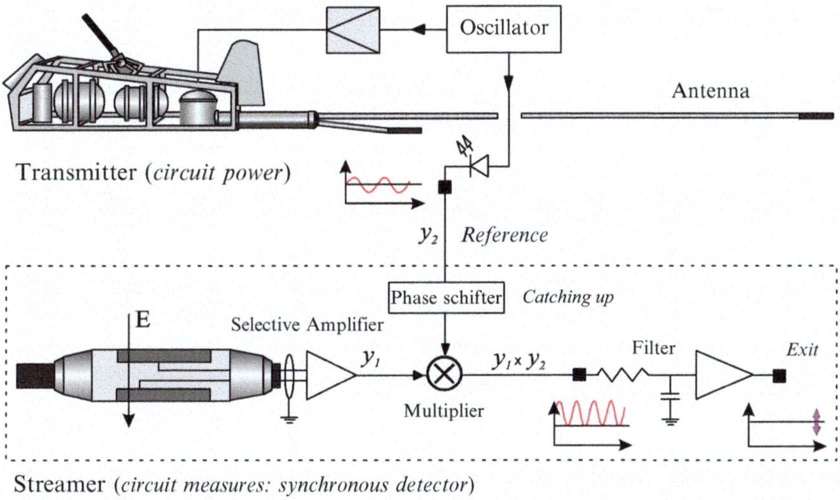

Fig. 4.48 Simplified electronic diagram of an acquisition electrometric device of the EM streamer type. The detection can be done using a synchronous detection assigned to each channel. The correction devices, such as the induction corrector, are not shown in the diagram

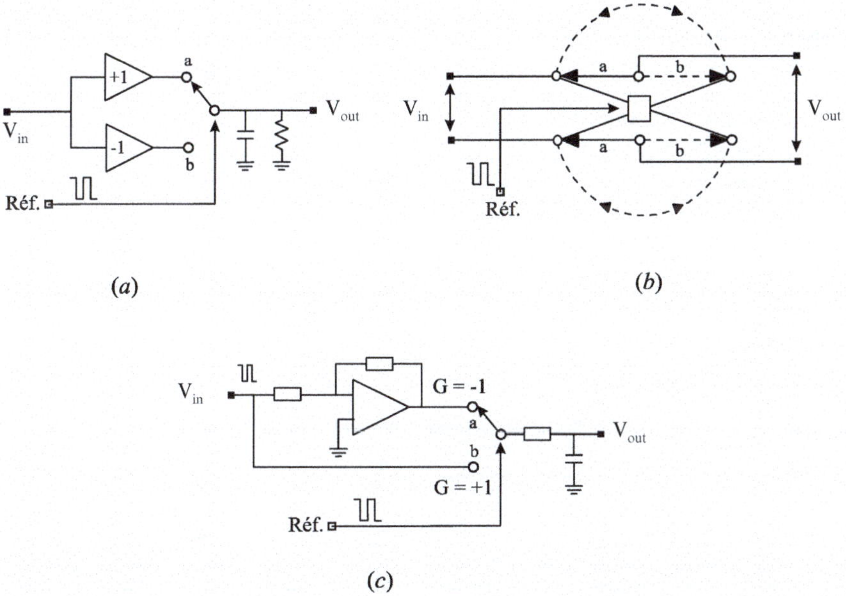

Fig. 4.49 Two types of synchronous detection, with single (*a*) and double (*b*) reverser and analog phase detection (*c*) for rectangular signals, controlled by the reference signal (According to Sainson 1982, 1984)

5.5 Streamer Electrometer with Vertical Injection

To increase the accuracy, or even the depth investigation, some authors (Alumbaugh et al. 2010) today advocate the use of vertical devices (sources and receivers) working in TM mode (see Chap. 2, Sect. 2.7.4.3) (Fig. 4.50).

In this particular mode of action, the current density electrometer described above is indicated to reduce dimensions and increase operational maneuverability but also to eliminate parasitic fields caused by the distortion of the imposing structure that carries the sampling electrodes. In this sense the magnetometric method also appears to be more suited to this type of investigation, if however corrective systems are expressly set up (motion correction).

5.6 Downhole Electrometers

The technology of the current density electrometers, because of their small size and better directionality compared to those of pd electrometers, allows an integration of the field sensors, in particular on the transverse components, on *logging tools* or wire line probes of very small diameters (2 inches).

We can therefore assume that their use downhole, especially in exploration drilling (*wildcat*), would extend the electromagnetic investigations at the level even in the immediate periphery of the oil reservoir. In these favorable conditions, disregarding the influence of joint casings forming the production column (Kong et al. 2009), this technology could provide a very specific statement of:

- The extent of the deposit
- The state of fracturing
- The illumination of the lower portion of the traps
- The evaluation of the reservoir, etc.

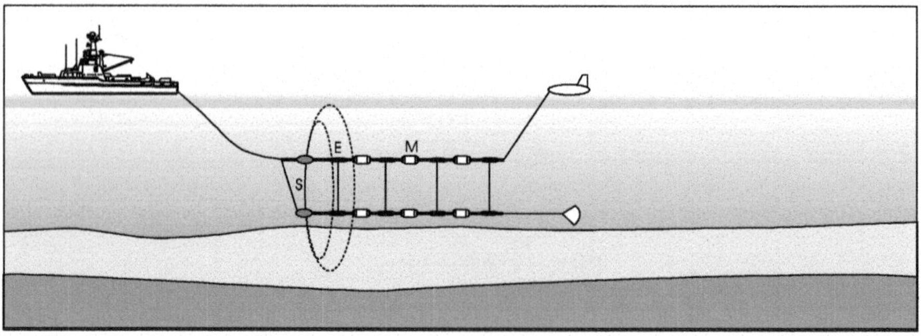

Fig. 4.50 Acquisition device of the *vertical streamer* type. The source S is also vertical (VED). Measurements are made in TM mode by electrometers E (*vertical* component) and magnetometers M. The device can also be used in transient mode (According to Alumbaugh et al. 2010)

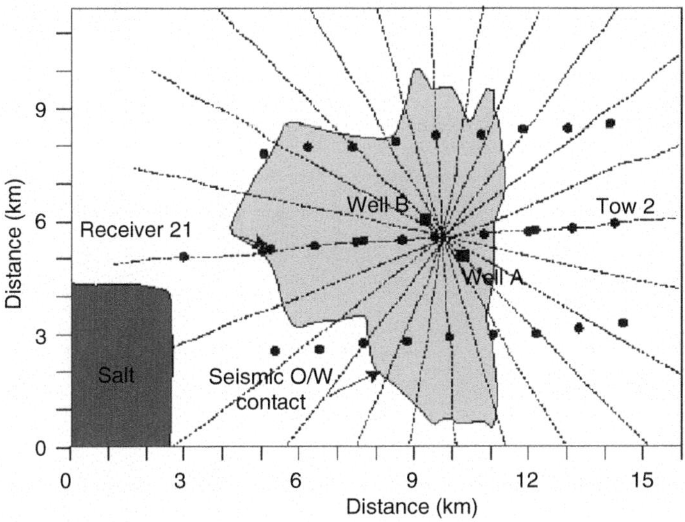

Fig. 4.51 Prospect topology that can be valorized by downhole investigations (According to Constable and Snrka 2007)

For example, we can consider, like what is done in borehole seismics (VSP,[47] *walk-away*, *cross-well*, etc.) measurements at different depths (sensors) step by step and a different source offset using the same power transmitter (mCSEM) (Figs. 4.51 and 4.52).

We can also imagine all sorts of combinations, including seabed and downhole sensors for the control of the evolution (*time-lapse technique*) and the *monitoring* of the oil deposits leading to 4D imaging (dimensions and time), or logging probe equipment with a triaxial magnetometer for downhole MT measuring or for down/surface differential measurements (cf. Fig. 4.53).

5.7 Vector Magnetometers and Variometers

Electrometers are usually (not always) accompanied by magnetic field sensors. In certain applications the latter are used alone or independently (TM mode).

The first submersible magnetometers were proton or *nuclear resonance* magnetometers of the Varian type developed after the Second World War (Vacquier 1972), and *magnetic resonance* or *optical pumping* magnetometers[48] tested in the

[47] Vertical seismic profile.

[48] Magnetic resonance equipment in *rubidium vapor* (Bloom 1962), *potassium vapor* developed at ENS Paris in the late 1960s (Mosnier 1967) or *cesium vapor*. The principle of these magnetometers is based on optical pumping and the *Zeeman effect*. Electrons in the atoms of a gas follow different orbits representing highter or lower energy levels. In an external magnetic field, these energy levels are split into energy sublevels or *Zeeman states*. The energy differences of the states are proportional to the strength of the magnetic field. Generally, the sensing component consists of an element vapor lamp, a filter to select one spectral line, a circular polarizer, a vapor cell and a photocell.

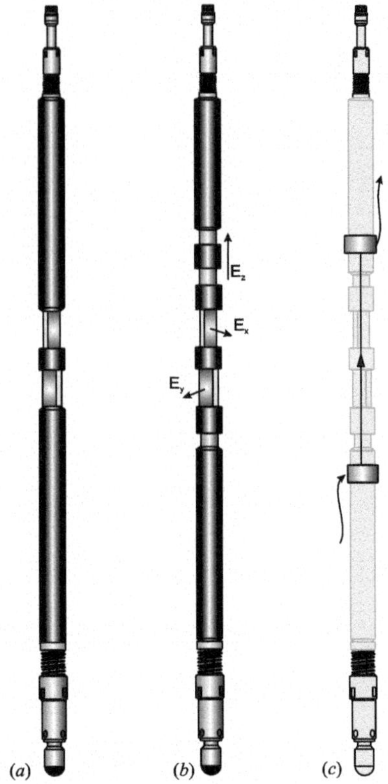

Fig. 4.52 Downhole electrometric sonde (wire line logging tool): (*a*) with two horizontal channels (components $E_{x,y}$); (*b*) with three channels, (horizontal components $E_{x,y}$ and vertical E_z); (*c*) with guard electrodes. The diameter of these probes can be a few centimeters, which allows them to be introduced easily in exploratory drillings (*wildcats*) (According to Sainson 1984)

1960s (Lokken 1964), directly arising from the research of Professors A. Kastler and J. Mosnier (ENS Paris).

However, these devices (scalar magnetometers), despite their high sensitivity (0.01 γ), can only measure the total field (modulus of the magnetic field) and not its directions. This means in other words that when the field vector changes under the action of an interference field, only the component parallel to the latter is accessible for measurement[49] (Mosnier 1967).

This type of magnetometer can, however, be advantageously used to calibrate scalar magnetometers and thus provide absolute measurements.[50]

For these reasons, other devices have been designed for variational measures of horizontal and vertical components, such as vector magnetometers and *rapid run variometers*, locked in direction.

[49]The normal components only change the field at the second order.

[50]Recently CNES and ESA have used on their observation satellites (Swarm constellation) vector magnetometers calibrated by a *pumping helium 4* scalar magnetometer (operation of the Zeeman effect) capable of measuring absolute magnitudes (without drift and bias) of the three components of the magnetic field (provider: CEA-LETI Grenoble).

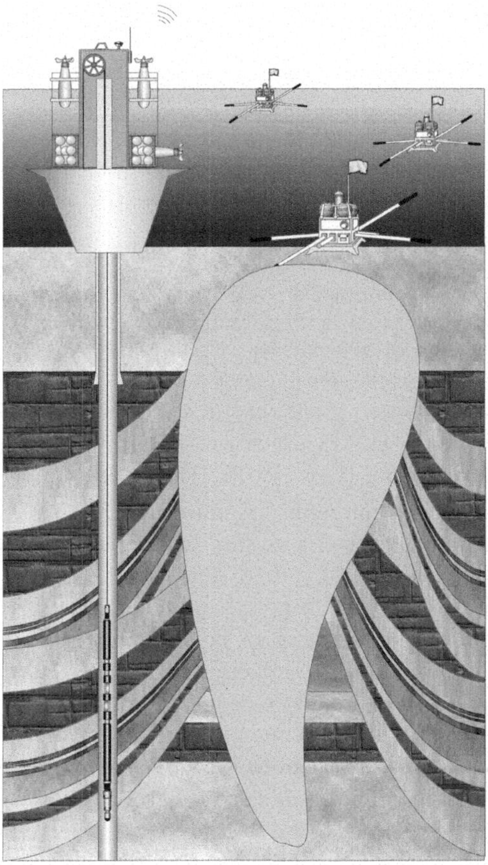

Fig. 4.53 mMT differential sounding using a downhole sonde E, H (set up by an automatic submersible logging unit of the Nadia type, for example) and seabed sensors (magnetometer and electrometer). This type of research could eventually provide an interesting perspective on complex structures such as those accompanying the salt domes and especially their lower part, using, for example, the vertical gradient technique (involving simultaneous EM downhole and ocean bottom recordings)

5.7.1 Induction Magnetometer (Search Coil)

Induction magnetometers or *search coil* magnetometers are composed of a coil of small diameter surrounding a core of high magnetic permeability. This coil, placed in the direction of the magnetic field fluctuations, provides at its terminals an electromotive force proportional to induction $\vec{B}$, and equal, depending on the number of windings n and the area of the coil A, to: $nA\partial_t(\vec{a}.\vec{B})$, where $\vec{a}$ is the unit vector parallel to the axis. Of very high sensitivity (about $0.1\,\text{pT}/\sqrt{\text{Hz}}$ at 1 Hz), these magnetometers have the advantage of being very compact (from a few centimeters to 10 cm) function of the frequency and consume very little power but they are limited in their bandwidth (Kaufman and Keller 1981; Chave and Jones 2012; Coillot et al. 2014).

5.7.2 Fluxgate Magnetometers

Magnetometers using saturated magnetic flux are vector magnetometers that measure either the three components of the magnetic field or the two horizontal

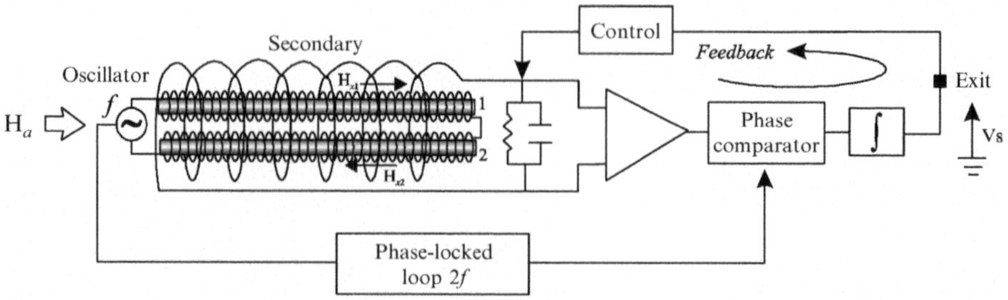

Fig. 4.54 Principle diagram of a *fluxgate* magnetometer with a phase lock. The operation of this sensor is based on the properties of the hysteresis cycle of the μ-metal. The output voltage Vs changes are proportional to changes in the field H_a

components, preferably in the ELF band. They are of the *flux-gate* type (e.g., Scintrex MFM3, Geoscan or Bartington) incorporated in the 1970s into seafloor instruments (Law 1978). Their accuracy generally approaches 1 nT, and they possess a minimum of noise amplitude of 4 $pT/\sqrt{Hz}$ at 1 Hz (Dlamini 2010) with a $1/f$ power spectrum associated with Barkhausen noise (core material).

Very generally, these sensors are composed of two primary windings (1 and 2 in Fig. 4.54) each disposed around a core with high magnetic permeability (μ-metal, permalloy, ferrite, etc.), coiled in opposite directions and placed side by side. The amplitudes of the induction fields $H_{a1,2}$ generated in these coils (oscillator beating at 660 Hz) are opposed for the fundamental and third harmonic and in the same direction for harmonic 2. The latter, which then corresponds to the variations of the local magnetic field in the direction of a (H_a), is received by a secondary winding common to both coils and then processed with a system of amplification (gain 10^5) and detection (phase locking loop + phase comparator). In this case, the instrument behaves as a phasemeter, of which the phase variations directly depend on the amplitude variations H_a.

Posing now:

$$\vec{B} = \vec{H} + 4\pi\vec{M} \tag{4.51}$$

where M is the magnetization intensity,[51] we obtain for each coiled bar 1 and 2:

$$\begin{cases} |B_1| = |H_1 + 4\pi M_1| - H_a \\ |B_2| = |H_2 + 4\pi M_2| - H_a \end{cases} \tag{4.52}$$

so that we find in the secondary winding a flux:

[51]The intensity of the magnetizing field M and the magnetic field H are related by the relation of proportionality: $M = \chi_m H$, where χ_m is the magnetic susceptibility of the material, i.e., its ability to magnetize, and is equal to: $\chi_m = \mu_r - 1$.

$$\phi = (B_1 - B_2) .f \tag{4.53}$$

or if coils 1 and 2 are matched:

$$\phi = 2H_a .f \tag{4.54}$$

These cylindrical magnetometers (one direction) are 50 cm long and a few centimeters in diameter,[52] and perfectly fit on the equipment bearing the electrometers (cf. Fig. 4.22). To withstand pressures and limit the noise due to deformations on housings, the coils can be immersed in mineral oil.[53]

5.7.3 Enslavement Field Variometers

These variometers, as torsion fiber magnetometers, were initially developed to study the *declination* of the earth's magnetic field (Filloux 1967, 1970; Hutton 1976). Their measurements were extended later to the two horizontal components with a simultaneous recording north/south (H type) and east/west (D type) and to their transient variations over time in the ULF (Mosnier 1970a, b; Mosnier and Yvetot 1972) (Fig. 4.55, Table 4.5).

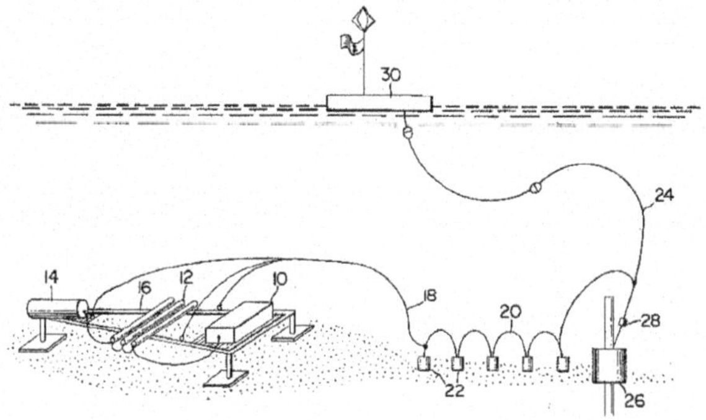

Fig. 4.55 First submarine magnetometer with field enslavement (According to Filloux 1967)

Table 4.5 Values of the stability of directive Mosnier variometers in the ULF band

Stability	Periods
0.05 nT	5 s to 10 min
0.2 nT	1 h to a few hours
0.5 nT	1 day to a few days

[52]The reader will find in the literature all the information necessary for the understanding of these instruments, widely used in the fields of submarine detection and geophysics (Delcourt 1990).

[53]These magnetometers are briefly described in numerous applied geophysics works. The reader will find, however, a comprehensive description of these materials in the work of Dr. Geyger (Geyger 1964).

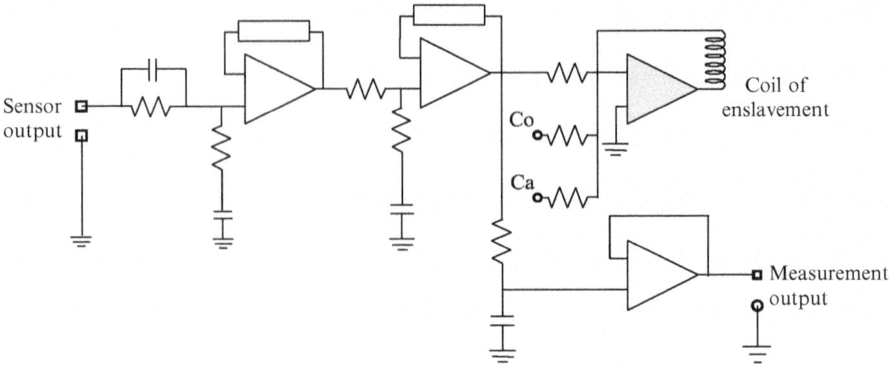

Fig. 4.56 Details of the electronics of the capacitive variometer with field enslavement (According to Yvetot 1980)

Fig. 4.57 Declinometer design in the 1970s with servo Helmholtz coil control. These variometers were developed in the 1990s and are still in use today (Ecole Normale Supérieure of Paris, Yvetot 1980)

The main element of these devices is usually a moving cylindrical magnet, sensitive to changes in the magnetic field, embedded in a nonmagnetic light-alloy frame (aluminum), hanging by a thread without torsion (Mosnier and Yvetot 1977; Yvetot 1980). This frame forms with four metal plates C_1, C_2, C_3, C_4, electrically connected in pairs, two capacitive groups C and C′ (cf. Fig. 4.57) inserted in an electronic circuit (cf. Fig. 4.56).

At equilibrium, when the capacities are equal ($C = C'$), the magnet is perfectly aligned along the axis of the system. Under the action of an external magnetic field, the rotation of the magnet causes a change in the two groups C and C′ (Fig. 4.57).

To increase sensitivity, the entire device is set in a capacitive divider so that when group C increases and group C′ decreases (rotation), an output voltage appears in phase with the reference signal (*synchronous detector*), and in the opposite phase in the opposite case (cf. Figs. 4.58 and 4.59).

Amplified, the output voltage is then sent into the *Helmholtz coils* so that the newly created field then opposes the field that produced it. This feedback field then rebalances the mobile equipment with a time constant fast enough and lower than

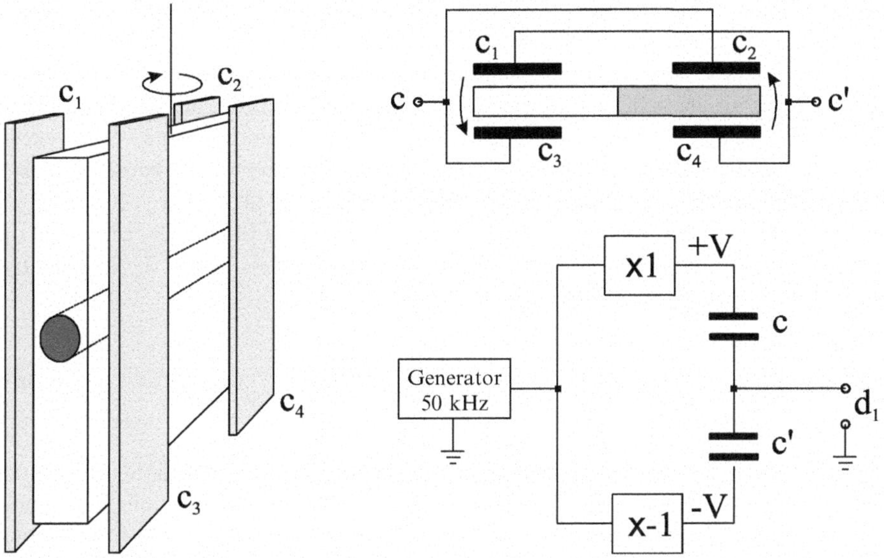

Fig. 4.58 Mechanical details and electrical diagrams (power) of a capacitive sensor with field enslavement (According to Mosnier and Yvetot 1977)

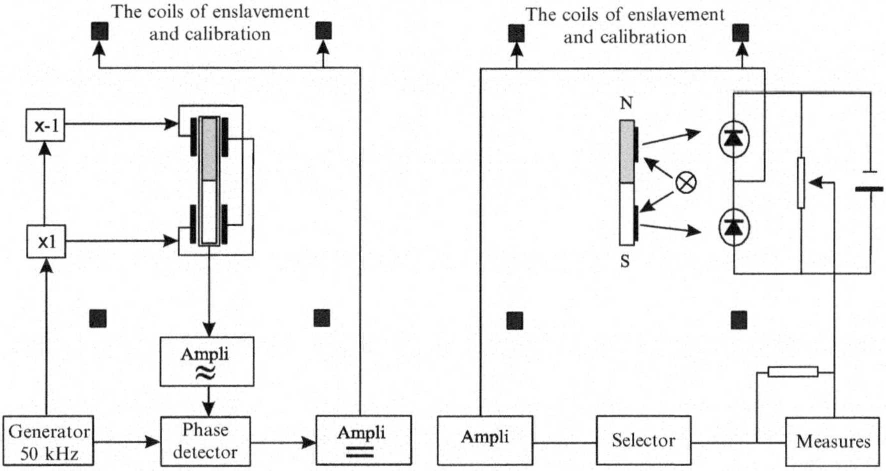

Fig. 4.59 Principle schematic diagrams of (*left*) a capacitive variometer with field enslavement (Helmholtz coils with calibration and compensation devices) and (*right*) a variometer with optical detection by photodiodes (According to Yvetot 1980 and to Porstendorfer 1975)

the local field variations. The intensity of the magnetic *feedback field* B is given by the formula:

$$B_{[\gamma]} = 899\frac{nI_{c[A]}}{R_{[\Omega]}} \tag{4.55}$$

where n and R are respectively the number of turns of wire and the resistance of the coil, and I_c is the electric current flowing therein.

On the seabed, where we are unable to precisely position sensors, the magnet is controlled in an artificial director magnetic field issued, for example, by a set of Helmholtz coils.[54] It makes any angle with one of the components of the earth magnetic field. When using two sensors H and D, the director fields of these two components are then mutually perpendicular (Mosnier 1974).

Under these conditions, the variations of H and D, as a function of the angle of rotation of the magnet α and of the amplitudes of the signals d_1 and d_2 supplied by the two variometers, are equal to:

$$\begin{cases} dH = d_1 \cos\alpha + d_2 \sin\alpha \\ dD = d_1 \sin\alpha - d_2 \cos\alpha \end{cases} \tag{4.56}$$

On submarine developed equipment (measures of the magnetic field horizontal components), the variometers D (declination) and H (with an additional auxiliary coil) are mounted in gimbals (two degrees of freedom), allowing them to equilibrate in the horizontal plane when placed freely on the seabed (Durand and Mosnier 1977). These instruments of greater dimensions than the fluxgate magnetometers are particularly suitable for the measurement of variations in telluric fields in the ULF band. They have a precision of over 0.1 γ in 1 hour.

In the future, it is expected that the equipment at the developer's disposal may be more sensitive magnetic sensors showing noise levels (sub-pT/Hz$^{1/2}$ at 1 Hz) comparable to those of SQUID magnetometers (operating at very low temperatures)[55] and based on magnetoresistances such as:

- Anisotropic magnetoresistance (AMR)
- Giant magnetoresistance (GMR)
- Giant magnetoimpedance (GMI)

5.8 *Magnetotelluric Receptors*

To investigate the deep layers of the subsoil (more than 1000 m), we use passive sensors (EM listening) in the ULF frequency range. They are usually composed of two telluric lines for measurements of the horizontal electric perpendicular field components (E_x and E_y) and a vector magnetometer for the measurement of the three magnetic field components (H_x, H_y and sometimes H_z). According to the

[54]Circular coils placed on either side of the conductive amagnetic clevis carrying the magnet, in which an electric current flows.

[55]SQUID: superconducting quantum interference device. We can also imagine the use of the radiofrequency SQUID working at high temperature and developed in the late 1990s (Fagaly 2006).

observed frequencies, these sensors are either search coils, fluxgate magnetometers and electromechanical variometers for longer periods.

We can note that fluxgate magnetometers have higher noise than search coils (×40) at periods shorter than 500 s but can still be used to collect seafloor magnetotelluric data in the 40–100 s bandwidth (Constable 2013).

In these devices, the onboard computer then estimates for different frequencies (sounding) the Cagniard impedances Z (see Chap. 2, Eq. 2.63) such that:

$$\begin{cases} E_x = Z_{x,y} H_y \\ E_y = Z_{y,x} H_x \end{cases} \tag{4.57}$$

representative of the distribution of conductivities of deeper structures.

One can also use only the three components of the magnetic field by forming:

$$H_z = X_{zx} H_x + X_{zy} H_y \tag{4.58}$$

and using the record of in-phase (real part: Re) and quadrature (imaginary part: Im) components such that:

$$\begin{cases} X_{zx} = \dfrac{H_z}{H_x} = \underbrace{\mathrm{Re}(X_{zx})}_{\text{Phase } 0^\circ} + i \underbrace{\mathrm{Im}(X_{zx})}_{\text{Phase } 90^\circ} \\ X_{zy} = \dfrac{H_z}{H_y} = \underbrace{\mathrm{Re}(X_{zy})}_{\text{Phase } 0^\circ} + i \underbrace{\mathrm{Im}(X_{zy})}_{\text{Phase } 90^\circ} \end{cases} \tag{4.59}$$

Such integrated instruments using this method are described, for example, in papers concerning material manufactured by Flinders University of South Australia (Heinson et al. 2000) or in French patent N° 2,896,044 (D'Eu et al. 2006). Magnetometers that equip INSU[56] Magellan stations (the same patent) are triaxial fluxgate sensors type TFM100G2 from the firm Billingsley Magnetics, and the potential plugs are electrodes Pb/$PbCl_2$ type Petiau/CRG Garchy[57] made by the company SDEC France (cf. Plate 4.3c). These measurement stations are equipped, among others, by an analog biaxial inclinometer (Geomechanics, Model 900) for reference in the horizontal plane (°*x*, °*y*).

Today virtually all seabed sensors, including those used with controlled sources, are of this type and can be interchangeably used in mMT or mCSEM (Fig. 4.60, Table 4.6).

[56]*Institut national des sciences de l'univers* (National Institute of Sciences of the Universe): the French state agency responsible for coordinating research activities in the fields of earth sciences, oceans and space.

[57]Geophysical Research Centre of Garchy of CNRS, based in the Nièvre state (France).

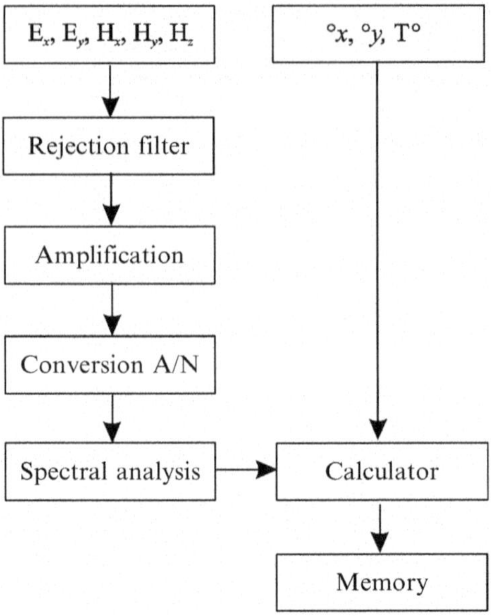

Fig. 4.60 Principle schematic diagram of mMT station acquisition. The detected and filtered spectrum, corrected from inclination values, is then analyzed and frequencies are discriminated. For each frequency, field values are extracted and translated into impedance by the computer and finally stored (flash memory). These data are either teletransmitted to the surface via a US/VHS (submerged buoy, floating antenna) communication device or discharged through a wired connection as the stations return to the ship. The entire device is powered by lithium batteries

Table 4.6 Summary of the performances of the Magellan mMT stations

Probe	Range of measures *(precision, drift)*		Acquisition
Analog vectorial magnetometer	±60,000 nT on 3 axis 0.1 nT, ±0.6 nT/°C	±60,000 nT on 3 axis 0.1 nTcc	1 point/min
Analog inclinometer	±25° on 2 axis 0.01° over the whole range of T°	±10° on 2 axis 1″ (0.016°)	1 point/h
Electrometer	±70 μV/m on 2 axis ± 10 nV/m integrated on 6 m	±70 μV/m on 2 axis ± 10 nV/m integrated on 6 m	1 point/min

According to CNRS/INSU

5.9 Digital Processing of the Reception Signals: Generalities

At these low frequencies the digital signal processing is relatively simple and easy to implement.

5.9.1 Acquisition or Sampling Frequency

The theory of sampled systems or sampling uses very different mathematical techniques, which are regularly updated in numerous publications. The reader will find an excellent introduction to it in the basic book by Tschauner (1963).

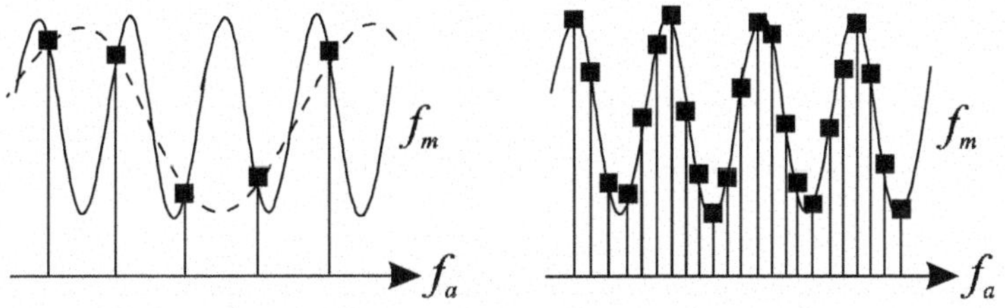

Fig. 4.61 Acquisition frequency (*dashed line*) and sample (marked with the symbol ■) too low (*left*) for the adopted sampling to be representative of the signal to be recorded (in *bold*). The signal in this case is considered to be subsampled. On the *right*, the sampling is correct

The acquisition frequency is given as a function of the numerical resolution of the acquisition system and the desired resolution.[58] It somehow defines the sampling of the analog signal and the amount of information to be stored, and must be of course higher than the working frequency (≈1 Hz) (Fig. 4.61).

Mathematically, the sampling process is equivalent to multiplying the analog signal $x(t)$ by a Dirac pulse sequence $\boldsymbol{\delta}(t)$ called a *Dirac comb* (or *Shah function*) such that:

$$x_e(t) = x(t) \cdot \boldsymbol{\delta}(t) \tag{4.60}$$

The resulting function $x_e(t)$ then corresponds to a sequence of pulses modulated in amplitude by the signal $x(t)$. To respect the waveform and its characteristics, it is essential that the pulses are close enough. To follow this criterion, we must then rely on a sampling f_a minimum twice the frequency[59] of the signal $x(t)$ (*Nyquist-Shannon* sampling criterion).

$$f_{a[\mathrm{Hz}]} \geq 2f_m \tag{4.61}$$

For example, for an application with a maximum frequency f_m of 1 Hz to be measured, the sampling frequency f_a will be at least 2 Hz.

5.9.2 Analog-to-Digital Converter

Today, data from detection systems must be converted into a language directly readable by microprocessors. The analog measurements from the sensors are therefore translated into a series of numbers, which are easier to process and

[58] It is related to the travel speed v of the source (1.5–2 knots) and given by v/f_a. For example, at a speed of 1 m/s, and an acquisition at 1 Hz, the resolution will be 1 m. For a typical survey, during towing, data are collected at a spacing of 10–20 ms.

[59] Note that frequencies above half the sampling frequency (or *Nyquist* frequency: $f_N = f_a/2$) generally introduce spectral overlapping, called *aliasing*. To remove this effect, before the sampling operation, we place a low pass analog filter or *anti-aliasing filter* whose cutoff frequency is then the highest frequency to be recorded.

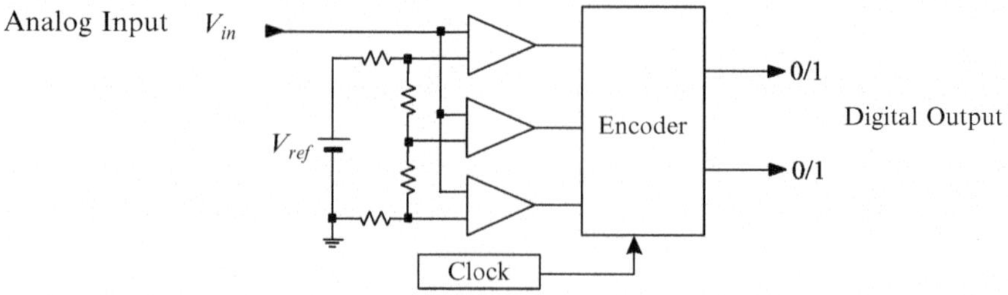

Fig. 4.62 Example of an A/D 2-bit flash converter (voltages are referred to according to the mass). At the output of the comparator, the voltage can take up to four values (0, 0, 0), (1, 0, 0), (1, 1, 0) or (1, 1, 1). In binary code (encoder output), they are converted in the present case into words of 2 bits (00, 01, 10, or 11)

store. This is the analog-to-digital converter or ADC,[60] which is responsible for this operation. Generally, this is done after sampling.

Instrumentally, the ADC behaves as a voltage (or current) comparator. For example, in Fig. 4.62, the input voltage V_{in} is thus compared to three reference voltages developed from a single voltage V_{ref} and three dividers (resistances). Then, thanks to a binary encoder (a combinational circuit with logic gates), these signals can thus have the values 0 or 1, forming at the output words of 2 bits, proportional to the input voltages. For negative voltages a second symmetrical converter is required in addition. In these situations, the increase in the numerical resolution is made with an increase in the comparisons, i.e., the number of bits (number of cycles: 2^n).

The voltage resolution of such an A/D converter is equal to the ratio of differences between the high and low reference voltages in input on the binary number 2^n.

$$R^{A/D}_{[V]} = \frac{(V_h - V_l)}{2^n} \tag{4.62}$$

Today we almost exclusively use 24-bit converters at sample rates of up to 1000 Hz (125 Hz per channel)

5.9.3 Digital Resolution of Converters

The resolution of an analog/digital converter corresponds to the value (dimensionless) of the change in input voltage that causes a change of one unit in the output number N (with $N = 2^{n-1}\ a_{n-1} + 2^{n-2}\ a_{n-2} + \ldots + 2^1\ a_1 + 2^0\ a_0$).

$$R^N = \frac{1}{2^n} \quad \text{or} \quad R^N_{[\%]} = \frac{1}{2^n} \times 10^2 \tag{4.63}$$

[60]There are several types of ADC, such as converters with weighted resistances, double-ramp converters, successive approximation converters, sigma/delta converters, or even-flash converters.

The conversion is all the more accurate when the digital resolution of the converter is small. The resolution can also be expressed as a percentage. For example, for an ADC of 8 bits the digital resolution will be $1/2^8 \times 100$ or 0.39 %.

5.9.4 Dynamics

The concept of dynamics covers several aspects of the measure. This occurs both in the analysis of the analog signals at reception and in the conversion and digital encoding.[61]

Dynamics of the Analog System (Amplifiers, Filters, etc.)

The measurement of analog signals poses the problem of the amplification of their amplitude, which may have very different values. We appraise the system's ability to measure this diversity of scale by its dynamics, which theoretically represent the absolute value of the ratio of the amplitudes of the largest to the smallest signal. Practically, the dynamics that represent this variation ratio are expressed in decibels (dB) and, in this case, they correspond to 20 times the decimal logarithm of this quotient such that:

$$\mathrm{D}^{\mathrm{A}}_{[\mathrm{dB}]} = 20 \log\left|\frac{A_{\max}}{A_{\min}}\right| \tag{4.64}$$

For example, for values after amplification ranging from 1.5 to 7.5 V, the dynamics of the system will be about 14 dB.

Dynamics of the Digital System (Converter)

The dynamics of the A/D converter are also given in dB as a function of the number of bits n by the expression:

$$\mathrm{D}^{\mathrm{A/D}}_{[\mathrm{dB}]} = 20 \log 2^n \tag{4.65}$$

For example, a 16-bit converter has a dynamic range of about 96 dB (or $20 \log 2^{16}$).

[61]For further information, the reader can refer to books concerning works on measurement systems (Paratte and Robert 1996).

5.10 Data Processing

In a seabed logging operation (mCSEM), depending on the offset source, the useful signals generally arrive in a delayed manner at the receptors.

When the source is close to the sensors, the direct waves arrive first, while when it is removed, the surface waves arrive first. Under these conditions, wave identification is necessary.

The first step of the treatment is that of signal identification:

- It has been seen that the direct waves and surface waves are *down waves*, while the waves from the reservoir are *upgoing waves*.
- We have also seen that, in a conventional survey, the useful waves, despite a greater propagation speed (HC reservoir), usually arrive later.

We can therefore logically discriminate these waves (down and up) by specific means both in the frequency domain and in the time domain (see Chap. 3, Sect. 3.13).

The deterministic aspect of data sorting (e.g., rejection filters) can be furthermore improved by an ad hoc statistical treatment and possibly *a measure of entropy* (quantitative aspect).[62]

We can also qualitatively evaluate the addition (duplicates, ghosts, false detections, etc.) and/or simply the loss of information by a probabilistic treatment *(probability of detection* or POD),[63] which will be adapted to the detectors (quality and environment), their relative position, the media crossed, the coverage and the resolution of the survey, such that for an EM detecting device (quadratic detector Q, linear detector L and demodulator D) we have for each element and for probability $P\{\Gamma\}$:

$$\begin{cases} \mathrm{POD_Q} = P\{\,\Gamma^2\rangle\ \mathrm{k}\} \\ \mathrm{POD_L} = P\{\,\Gamma\rangle\ \mathrm{q}\} \\ \mathrm{POD_D} = P\{\,\Gamma\cos\varphi_\mathrm{C}\rangle\ \mathrm{K}\} \end{cases} \tag{4.66}$$

where k, q and K then represent the thresholds of each of the processing elements.

[62]This in some way evaluates the mess in the data entry system. The *entropy* can be defined as a function proportional to the logarithm of the probability of a sequence of average data corresponding to a message, a sequence or a signature. The main interest of entropy is its additivity property. Indeed, when two signatures are composed, their probabilities are multiplied and their entropies are added. As, on other hand, amounts of information are also added, it is clear that the entropy is a convenient measure of the detectable amount of information and, more specifically, that which is necessary to describe its location and complexity.

[63]The notion of POD was introduced in 1947 following the work, during the Second World War, on radar (Marcum 1947; Woodward 1953). These studies were designed to determine the probability of an event on the basis of hypothesis tests run from an objective criterion whose result was estimated by the joint probability of two types of errors with an antagonistic behavior (ignorance or appearance upon the detection of an event). Therefore, these works have been amplified as covering all areas of detection at large (Grigorakis 1997). In the oil sector, POD is mainly used for the processing of data for *pipeline inspection* by instrumented pigs (Sainson 2010).

In practical terms, the series of measurements for each electrometer are segmented in sequences of the same duration and the data are distributed according to a Gaussian function normalized in weight and width. For each time interval we then apply a discrete Fourier transform or DFT,[64] which finally provides the amplitude and phase ready to be calibrated and corrected.

For the study of telluric answers (mMT), the signal processing is more complex to carry out. Apart from analog techniques (see Sect. 5.3.3), and those of harmonic analysis of evaluation of the spectral density power P_D for a signal time function $f(t)$:

$$P_D = \frac{1}{2T}\left| \int_{-T}^{T} f(t)\, e^{-i\omega t} dt \right|^2 \quad \text{when } T \to \infty \tag{4.67}$$

decomposition methods based on convolution[65] or autocorrelation[66] were proposed in the 1960s to supplement the difficulties of previous methods (Madden 1964).

In the analytical domain, the convolution product is the result of a mathematical operation (noted by *), which allows us to compose between them two functions or two signals x and h (depending on time) through an integration such as:

$$y(t) = x(t) * h(t) = \int_{-\infty}^{\infty} x(\tau) \,.\, h(t-\tau)\, d\tau = \int_{-\infty}^{\infty} x(t-\tau) \,.\, h(\tau)\, d\tau \tag{4.68}$$

In fact this operation is more convenient to use and easier to calculate in its digital form. Then the signal is no more defined at any instant t but on a sample k for a given period corresponding to the signal sampling frequency f_e.

Then we obtain:

$$\begin{gathered} y[k] = x[k] * h[k] = \sum_{l=0}^{\infty} x[l] \,.\, h[k-l] = \sum_{l=0}^{\infty} x[k-l] \,.\, h[l] \\ y[k] = y[t_k] \qquad t_k = k/f_e \qquad y[k\langle 0] = 0 \end{gathered} \tag{4.69}$$

[64] Discrete spectral representation of the sampled signal (delimited time window).

[65] We must remember that this process is legitimate only when the stationary character is present to a degree sufficient for the desired accuracy. This remark is essential to understand the limitations of these methods.

[66] The *autocorrelation* developed in the late 1940s (Bode and Shannon 1950) is used as a function of time $f(t)$ to measure the statistical dependence of a value $f(t + \tau)$ in relation to an initial value $f(t)$. Then the autocorrelation maintains the frequency information but loses the phase information. It is a kind of similarity indicator between two fairly similar samples of the same signal, separated by t, and is constructed by convolving them together. If the random function $f(t)$ is known up to time t, the possibilities of variation are then distributed according to a beam of more or less divergent curves.

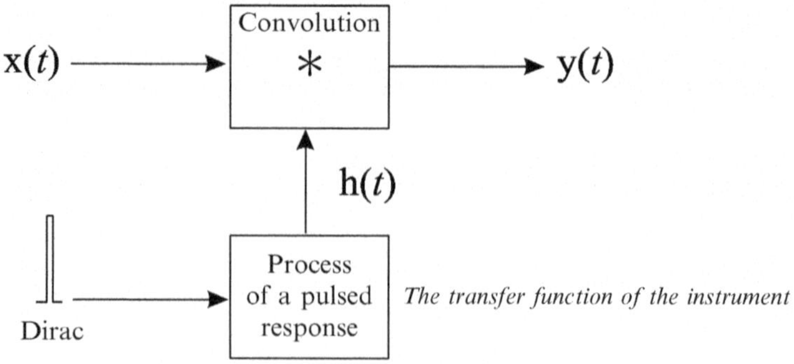

Fig. 4.63 Schematic diagram of an operation, or convolution product, where the transfer function of the instrument is previously determined by the injection of a Dirac (very well identified function)

The signal x[] represents the signal to be processed, y[] the currently processed signal, the product of the convolution, and h[] by definition the impulse response of the system (Fig. 4.63).[67]

The direct application of this formulation, however, results in a large number of calculations, which can be greatly reduced through an interesting mathematical property of the convolution product (Jennison 1961), i.e., the Fourier transform (FT) or its "digital equivalent", the fast Fourier transform (FFT).[68]

6 Operational Means

Operational capabilities include all equipment for the implementation of the acquisition hardware. These means are heavy and comparable to those of marine seismic surveys and are collected on a special vessel.

6.1 Surface Vessels and Operations

The size and number of electrometers require the use of seagoing vessels of high tonnage. The latter must be equipped with all air navigation instruments (GPS) as well as submarine instruments (acoustic transponders). Today there are ships exclusively dedicated to this activity, taking into account all the operational

[67]For example, to the input of an amplifier, an infinitely short pulse characterized in this case by a Dirac can only give a signal spread in time (called an impulse response). The purpose of the *convolution operation*, which mathematically corresponds to a product, then consists of determining the shape of the output signal (of any sort and deformed by the meter), knowing in advance the impulse response of the system (assumed to be linear).

[68]The reader will find additional information in the technical literature (Smith and Smith 1996).

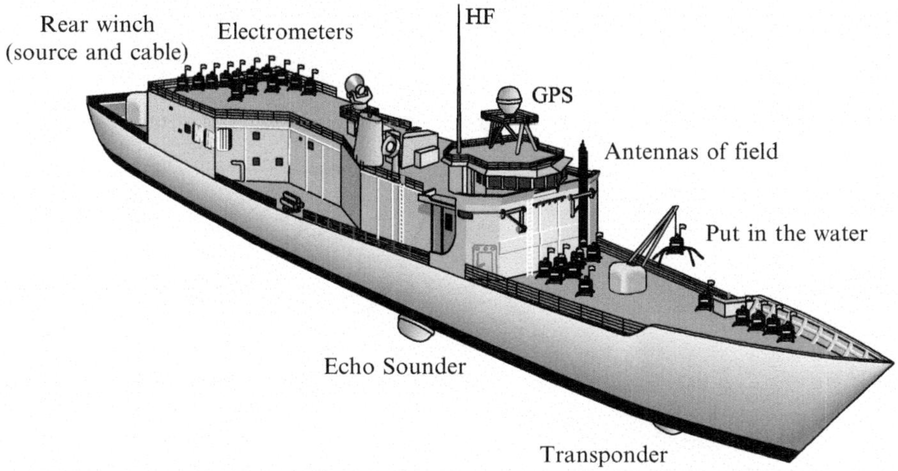

Fig. 4.64 Electromagnetic exploration vessel capable of carrying many tens of field sensors. The antenna strands (eight per sensor) are independently stored on the deck and assembled just before the launch (cf. Plate 4.2)

constraints (see Plate 4.2). These vessels can carry up to 200 EM receivers and multiple power sources.

Electromagnetically, the prospecting vessel must have an electromagnetic signature as low as possible or otherwise clearly identifiable (see Chap. 3, Sect. 6.9.4) (Fig. 4.64).

6.2 Navigation and Bathymetry Instruments

The interpretation of data requires a thorough knowledge of the precise position of the various components of the acquisition chain on one hand and of the topology of the measurement locations on the other hand.

During the exploration campaign, the field transmitter and the receivers are located and identified through acoustic transponders connected to the ship by ultrasonic telemetry. These position data can also be confirmed by electromagnetic field measurements in the time domain.

The height of the fish containing the power source is estimated from an onboard echo sounder and the dynamic behavior of the transmitting antenna (pitch, roll and yaw) is analyzed from attitude sensors attached to both ends (inclinometers, accelerometers, gyros, etc.).

As for the surface vessel, it is equipped with a conventional echo sounder (3.5 kHz), or with side-scan sonar or even with synthetic aperture sonar (SAS) to record variations in seafloor relief on the entire route, which are identified by the GPS devices and the MRU (motion reference unit) onboard.

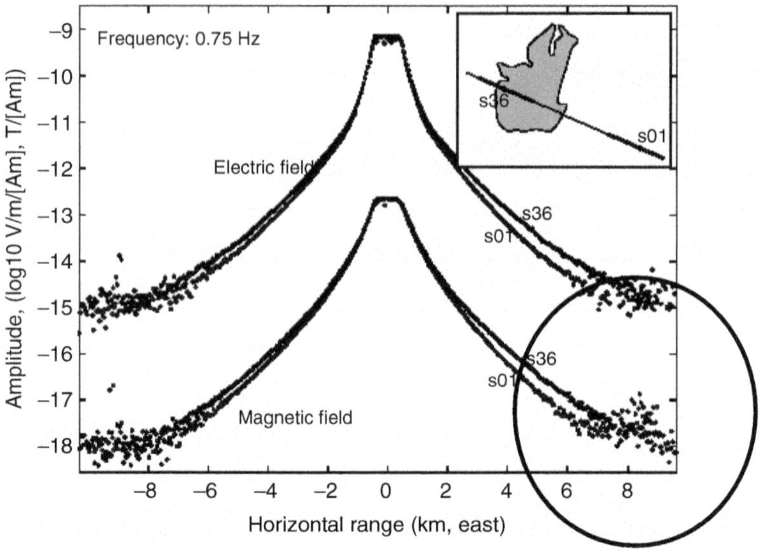

Fig. 4.65 Dispersion measurements beyond a certain offset (6 km) (According to Constable 2010)

In practice (offset 3000 m), due to oceanic and meteorological conditions, or to problems of systems calibration, the positioning error can reach 50 m. Beyond 6000 m the acquisition data show a significant dispersion that can question the validity of the measures (cf. Fig. 4.65).

Positioning the receivers is done using an immersed ultrasonic (12 kHz) communication system, short base transponder type (short base line or sbl) or ultrashort base (ultrashort base line or usbl) whose different elements (weights, buoys, floats, etc.) have been after immersion previously positioned and localized from the surface (Sainson 2007) by a radio link VHF (very high frequency) via the boat (GPS)[69] and capable of working up to ranges of 10 km and water depths of 6000 m. The pitch, roll and heading of the vessel need to be continuously monitored using a motion reference unit (MRU) (Figs. 4.66 and 4.67).

6.3 Procedures for Implementation

The arrival in the area is carried out via GPS tracking and the immersion of the autobuoying elements (acoustic transponders) with verification of their efficiency (acoustic and microwave links). After verification and calibration, electrometers are then immersed (negative buoyancy → descent) at intervals as regular as possible, depending on the chosen acquisition geometry. Then the actual data acquisition

[69]This chain is similar to all radio localization oceanographic devices.

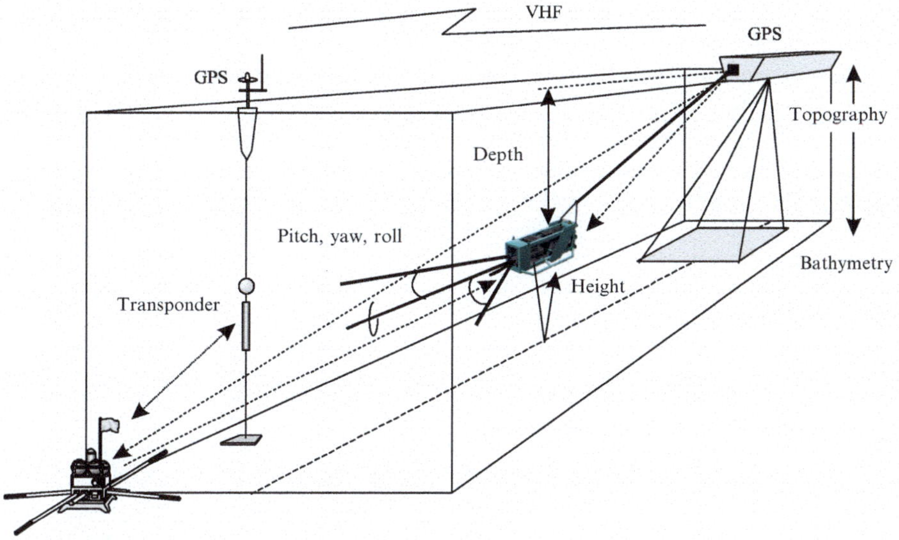

Fig. 4.66 Echolocation and radio-positioning devices that locate the mobile source and receivers according to the submarine topography and bathymetry

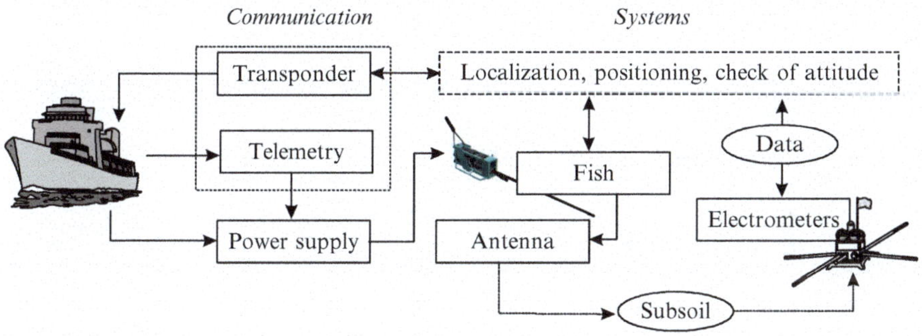

Fig. 4.67 Diagram of the system including communication, attitude control and fish location devices

operation is performed firstly by measurements of background noises, then by successive passages of the source above the field sensors according to the predetermined profiles. Quality control of the measures ensures the end of the survey.

Electrometers are then relieved (positive buoyancy → ascent) and recovered at the surface. The weights that remain at the bottom are soluble, thus ensuring almost zero ecological impact.

All these operations must be carried out under meteorological ocean conditions as good as possible so as not to damage or lose the equipment. Moreover, it is possible and even recommended to perform SBL in conjunction with high resolution seismic acquisition (a combined job).

All data, including the values of electric and magnetic fields, their positions, measures relating to the intervention conditions (attitude, resistivity of seawater, etc.) are discharged from the acquisition systems on onboard computers to apply the quality control and to be then automatically processed (*quick look interpretation*) (Fig. 4.68).

6.4 Related Measures

In some cases, additional oceanographic local measurements are needed. These are made during the survey and are intended:

- Firstly to assess the *static parameters* of seawater (pressure, temperature, electrical conductivity) to make corrections on field measurements (raw data)
- Secondly to estimate the *dynamic parameters* of the medium, speed and direction of the ocean current in particular, to evaluate the nature, origin and amplitude of the present electrodynamic noise.

For it we use a vertical submerged buoy at a few meters above the bottom of the sea, containing oceanographic measurement instruments (current meter, thermometer, pressure sensor and electrical conductivity probes).

6.5 Configurations of the Survey

Like conventional seismic prospecting the mEM survey can be geometrically configured according to the objectives (recognition, detailed mEM, etc.). These specific arrangements are usually predetermined at the beginning of the study by computer simulations (modeling by solving the forward problem, for example).

After this practically essential design phase we then place the survey receptors at positions as accurate as possible to obtain a profile (1D), mapping (2D) or imaging (3D). We do the same to choose the directions of movement of the EM transmitter in strict accordance with the location of the downhole sensors. We define then a research coverage that is more or less exhaustive, adapted to the problem and then taking into account:

- The packing density of the measuring instruments
- The resolution defined in advance
- The assumed or estimated depth of the target
- The measurement accuracy (presence of errors)
- The oceanic conditions, including the electromagnetic background noise, etc.

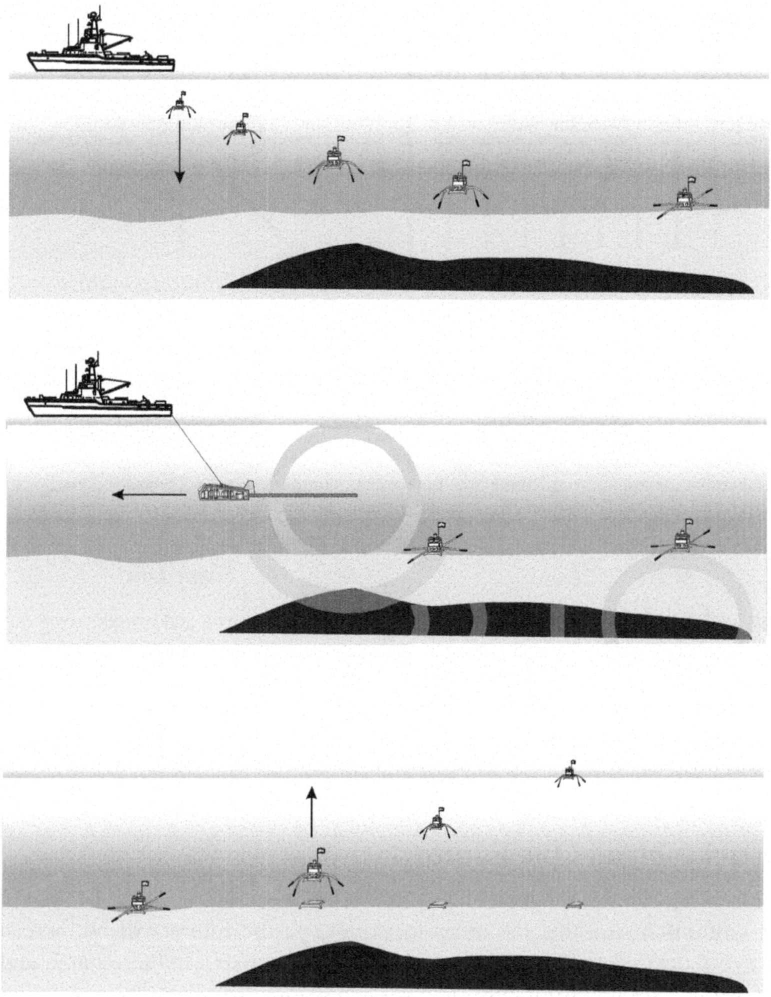

Fig. 4.68 Various stages of the commissioning of the measuring device. Field electrometers are launched from the surface (boat) with negative buoyancy (ballast), which allows them to go down. After the acquisition phase on the seabed, they are recovered at the surface through automatic discharge of the ballast, which then gives them positive buoyancy (rise)

This schematically gives exploratory configurations in a grid, star, cross or even just in line (cf. Fig. 4.69), giving them specific operational times and costs.

The configuration survey depends on the local geology defined by the seismic survey, on the economical objectives and on the interpretative means that will subsequently be implemented. The interpretation approach involves several steps. The first step is to start with one profile (1D forward modeling and inversion) or several profiles (multiple-azimuth) to determine the resistivity layers and the

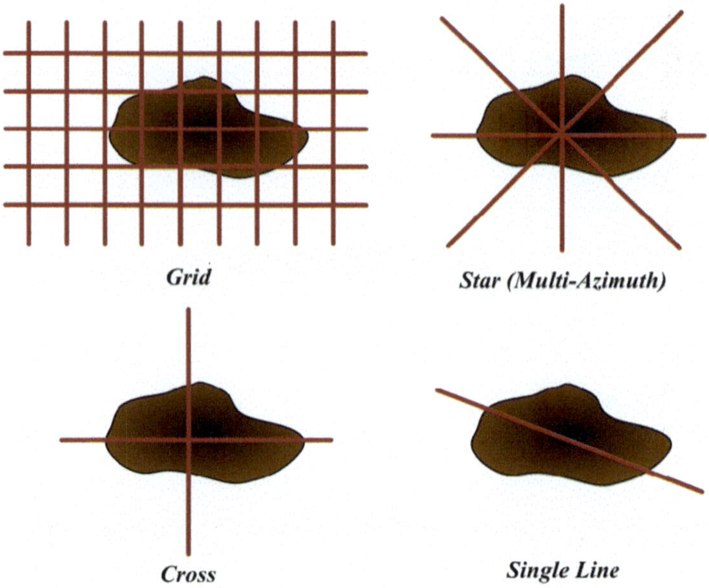

Fig. 4.69 Different survey configurations offering different services and results

anisotropic degree (cross-traces). The second step is to produce 2D and 3D models (pseudosections or volumes, etc.).

7 Conclusion and the Future

It goes without saying that the areas to explore in the future will be increasingly hostile (very deep-sea areas and arctic zones in particular). In these harsh and very difficult to access environments (high immersion, presence of sea ice, ocean currents, etc.) where the penetration of vehicles and instrumentation is risky and uncertain, automatic "intelligent" systems must imperatively take over.

Today, the instrumentation is moving toward the conception of integrated sensors providing measures of all six components of the electromagnetic field, using, to reduce the instrumental noise, an ASIC electronic conditioner (CMOS 0.35 μm). These small dimension sensors are based on the principles of the current density electrometers, on the search coil (cubic induction design) for the magnetometric part (Boda et al. 2014) or on the optical pumping broadband magnetometers (Baicry 2015a, b) (Fig. 4.70).

Ideally, one might think that streamer type methods, whose coupling transmitter/subsoil/receiver is invariant (transmitter and receivers are combined) and therefore easier to use, will probably be more reliable and efficient.

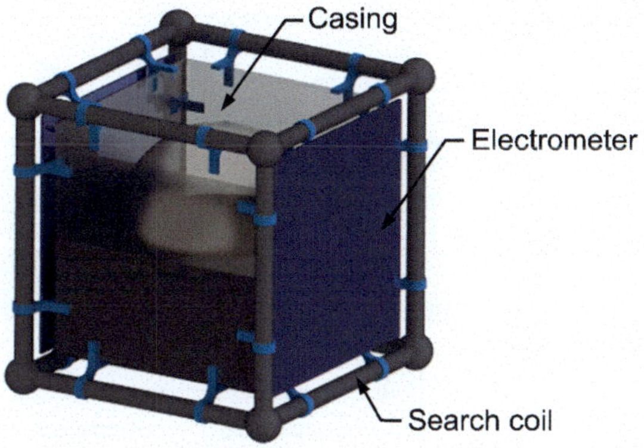

Fig. 4.70 Concept of an mEM miniature induction sensor or OBEM (Ocean botton Electromagnetic) receiver of a few square decimeters for simultaneous measurement of the six components of an electromagnetic field (accuracies: 0.5 μV/m/$Hz^{1/2}$, 0.7 pT/$Hz^{1/2}$): simple applications or in a phase array rather reserved for deep-sea mining (Boda et al. 2014)

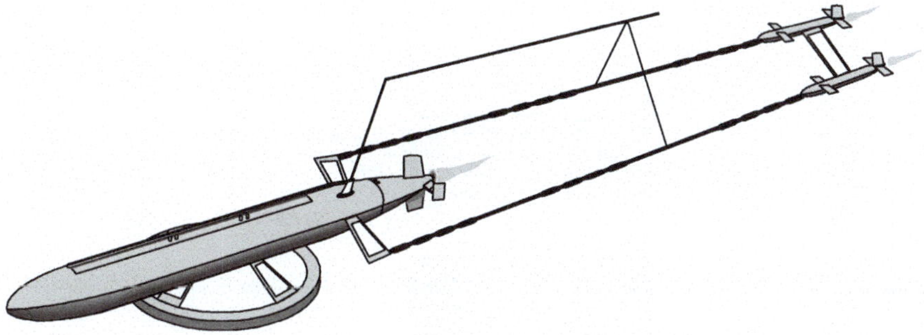

Fig. 4.71 Example (artist's view) of a compact submersible EM system of the future (AUV/streamer) with "double galvanic/inductive transmission" (vertical magnetic dipoles down and horizontal electrical dipoles up) to automatically operate under the ice or in very deep water. The craft propulsion and power supply for the electromagnetic sources could be provided, for example, by fuel cells perfectly adapted in an anaerobic medium and already in operation on some automatic submersibles. To limit noise, measurements can be done in the time domain (transient) as the engine has stopped

We can perfectly imagine today the use of submersible drones of the AUV (autonomous underwater vehicle) type,[70] but with larger dimensions using underwater recognition LIDAR, dragging behind them antennas/streamers, and also more specifically the use, for example, of multiple sources[71] or of rotating fields for more precise identification of the characteristics of the subsoil geological layers (Fig. 4.71).

---

[70]Underwater drones are already used for the monitoring of offshore installations such as wellheads and pipelines. For now, these robots have a mission of supervisory and long distance observation unlike ROVs (wire guided) dedicated to local tasks.

[71]In this case, the resultant field is the sum of the fields generated by the different sources (*superposition principle*).

a

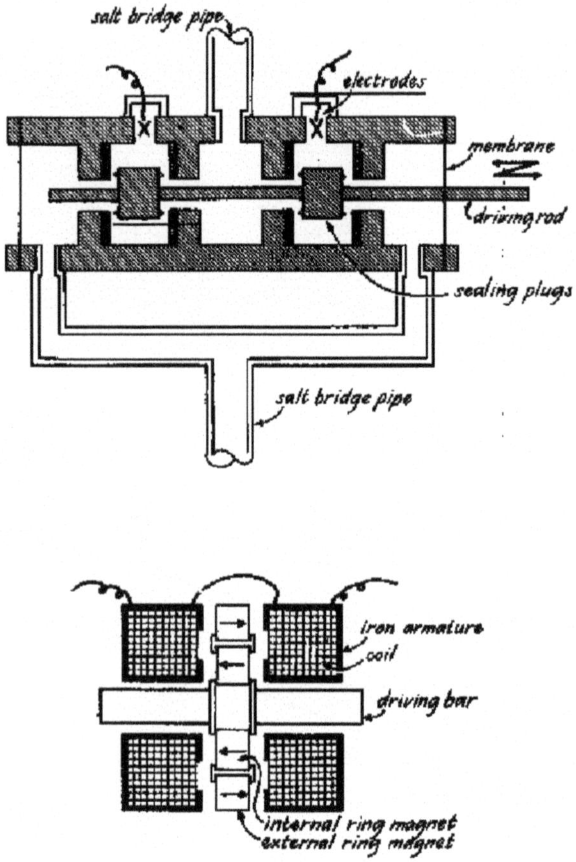

b

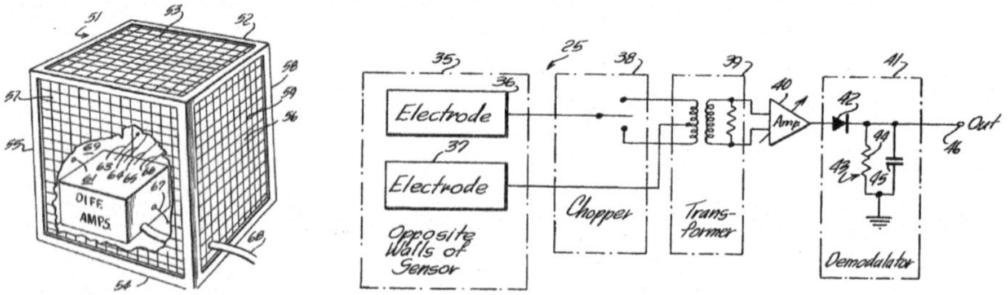

Plate 4.1 (**a**) Details of one of the very first electromechanical systems for the measurement of the electrical low noise field, called *Salt bridge chopper* (According Filloux 1973). (**b**) Details of the first current measurement device, proposed by the US Navy for underwater object detection (According to Pittman and Stanford 1972)

Plate 4.2 (**a**) Galatea: company EMGS vessel built by the BOA shipyards. (**b**) Standart horizontal electric fields sensor (2 electrometers with 2 magnetometers). (**c**) Electromagnetic power source (transmiter) before immersion

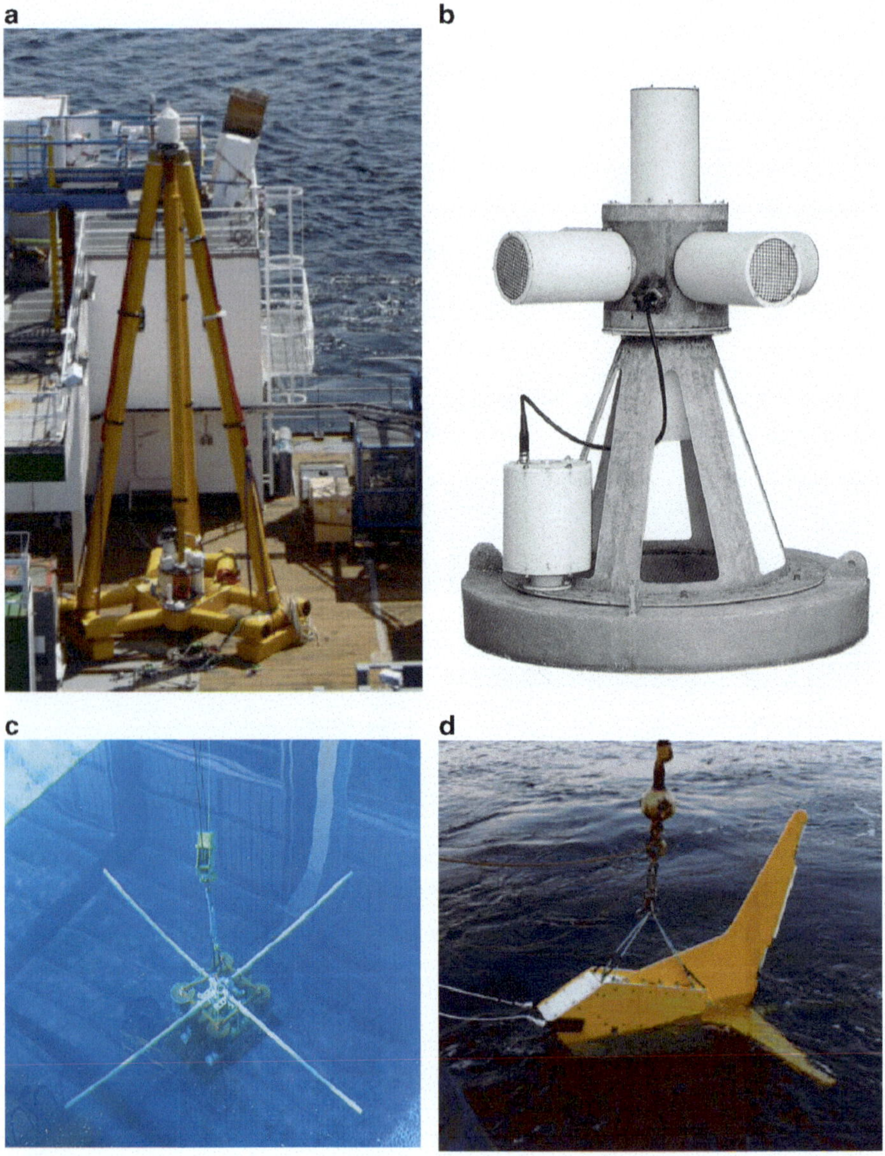

Plate 4.3 (**a**) Vertical electric field sensor by Petromarker (2007). (**b**) Current density electric fields sensor CNRS/GESMA type (1990). (**c**) Magellan submarine station (CNRS/INSU, 2006). (**d**) Submersible sensor towed behind the source (Scripps Institution of Oceanography, 2009)

References

Aldridge B (2001) ELF history. Extreme low frequency communication. Pacific Life Research Center. PLRC-941005B, 6 p

Alumbaugh DI et al (2010) Surveying using vertical electromagnetic sources that are towed along with survey receivers. US patent no. 2010/0013485 Jan. 21

Babour K et al (2008) Methods of electromagnetic logging using a current focusing receiver. Patent PCT WO 2008/121772

Babour K et al (2009) Sensor cable for electromagnetic surveying. Patent PCT WO 2009/0295394

Baicry M (2015a) Performance improvement of a current based marine electrometer. In: Marelec conference, Philadelphia, USA, 18 June

Baicry M (2015b) Etude d'un magnéto-électromètre marin: conception, dimensionnement optimisé et réalisation d'un prototype. PhD thesis, Grenoble University, 11 Dec, 148 p

Balser M, Wagner C (1962) Observations of earth–ionosphere cavity resonances. Nature 638:641

Barus CW (1882) US geological survey, mon. 3, pp 309–367 and 400–404

Baxendale JE (1989) Underwater electric field sensor. UK patent no. 8726908

Bennett WR, Davey JR (1965) Data transmission. Ed McGraw-Hill, New York, pp 136–137

Besson C et al (2009a) Systems and methods for calibrating an electromagnetic receiver. Patent PCT WO 2009/006077

Besson C et al (2009b) Methods for electromagnetic measurements and correction of non-ideal receiver responses. Patent PCT WO 2009/0001985

Besson C et al (2010) Methods for electromagnetic measurements and correction of non-ideal receiver responses. Patent PCT WO 2009/0001985

Bleil DF (1964) Natural electromagnetic phenomena below 30 kc/s. Ed. Plenum Press, New York, 470 p

Bloom AL (1962) Applied optics. p 61

Boda M, Coillot C, Sainson S (2014) Electromagnetic sensor for SMS geophysical survey. Concours mondial de l'innovation 2030, Paris

Bode HW, Shannon CE (1950) Linear least square smoothing and prediction theory. Bell Telephone Laboratories, technical publications, monograph B-1732. Ed. The Bell System Technical Journal, 9 p

Brahtz JP et al (1968) Ocean engineering. Physical and hydrodynamic factors. Ed. Wiley, Chichester, p 204

Bruxelle JY (1995) Analyse de la propagation électromagnétique en milieu marin et méthode de localisation spatiale d'une source dipolaire. Doctoral thesis, Université de Lille

Bruxelle JY (1997) Electric field measurement at sea. In: Ocean 97 MTS/IEEE conference proceedings, vol 1, pp 547–550

Burrows ML (1978) ELF communications antennas. Ed. Peter Peregrinus, Stevenage, 245 p

Chave AD, Jones AG (2012) The magnetotelluric method. Theory and practice. Ed. Cambridge University Press, Cambridge, pp 437–440

Chave AD, Luther DS (1990) Low frequency, motionally induced electromagnetic fields in the ocean. J Geophys Res 95, n c5:7185–7200

Chave AD, Constable SC, Edwards RN (1991) Electrical exploration methods for the sea floor. Electromagnetic methods in applied geophysics, vol 2, Chap. 12. Ed. Society of Exploration Geophysicists, Tulsa, pp 931–966

Coillot C, Moutoussamy J, Boda M, Leroy P (2014) New ferromagnetic core shapes for induction sensors. J Sensors Sensor Syst

Constable S (2010) Ten years of marine CSEM for hydrocarbon exploration. Geophysics 75 (5):75A67–75A81

Constable S (2013) Instrumentation for marine magnetotelluric and controlled source electromagnetic sounding. Geophys Prospect 61(Suppl 1):505–532

Constable SC, Snrka LJ (2007) An introduction to marine controlled source electromagnetic methods for hydrocarbon exploration. Geophysics 72(2). Ed. Society of Exploration Geophysicists, Tulsa.

Constable S et al (1998a) Marine magnetotellurics for petroleum exploration. Part 1: sea-floor equipment system. Geophysics 63(3):816–825
Constable SC et al (1998b) Seafloor magneto-telluric system and method for oil exploration. US patent no. 5770945
Cox C (1981) Electromagnetic active source sounding near the East Pacific Rise. Geophys Res Lett 8:1043–1046
Cox C et al (1971) Electromagnetic studies of ocean currents and electrical below the ocean floor. The sea, vol 4 part 1. Ed. Wiley Inter-Science, pp 637–693
Cox C et al (1973) Plate tectonics and geomagnetic reversal. Ed. Freeman, San Francisco, 702 p
Cox C et al (1978) Electromagnetic fluctuations induced by wind waves on the deep sea floor. J Geophys Res 83:431–442
Cox C et al (1981) An active source EM method for the sea floor. Scripps Institution of Oceanography, San Diego, CA, Technical report
Cuevas NH, Alumbaugh D (2011) Near source response of a resistive layer to a horizontal or vertical electric dipole excitiation. Geophysics 76(6):F353–F371
Davidson PA (2009) Cable type electromagnetic receiver system for subsurface exploration. US patent no. 7602191
Dehart RC (1969) External pressure structures. In: Handbook of ocean and underwater engineering, pp 9.3–9.20
Delcourt JJ (1990) Magnétisme terrestre. Masson, Paris
D'Eu JF, Cairns G, Tarits P, Jegen Kulscar M, Dubreuil A (2006) Dispositif de mesure géophysique pour l'exploration des ressources naturelles du sol en domaine aquatique. Brevet français no. 2896044
Dlamini VE (2010) Electromagnetic wave sensor for seabed logging using the fluxgate technology. Universiti teknologi Petronas, Perak
Drever RG, Sanford TB (1970) A free fall electromagnetic current meter. Instrumentation. Proceeding of the IERE conference on electronic engineering in ocean technology, pp 353–370
Durand P, Mosnier J (1977) Magnétomètre sous-marin pour l'étude des composantes horizontales du champ transitoire. Annales de géophysique, Tome 33, fac.4, pp 519–526
EMGS (2015) EMGS transmitter system Self Xpress, 2 p
Fagaly RL (2006) Superconducting quantum interference device instruments and applications. Rev Sci Instrum 77:100101. doi:10.1063/1.2354545
Filloux JH (1967) An ocean bottom D component magnetometer. Geophysics 32(6)
Filloux JH (1970) Magnetometer utilizing a mirror and a magnet the latter being movable relative to said mirror to a connect and disconnect position. US patent no. 3508142, 21 April
Filloux JH (1973a) Oceanic electric currents, geomagnetic variations, and the deep electrical conductivity of the ocean-continent transition of central California. PhD thesis, University of California, San Diego
Filloux JH (1973b) Techniques and instrumentation for study of natural electromagnetic induction at sea. Phys Earth Planet Inter 7:323–328
Filloux JH (1974) Electrical field recording on the seafloor with short span instruments. J Geomag Geoelec 26:269–279
Filloux JH (1977) Ocean floor magneto-telluric sounding over North Central Pacific. Nature 269:297–301
Filloux JH (1978a) Observation of very low frequency electromagnetic signals in the ocean. Advances in earth and planetary sciences, vol 9. Ed. U. Schmucker, pp 1–12
Filloux J H (1978b) North Pacific magneto-telluric experiments. Advances in earth and planetary sciences, vol 9. Ed. U. Schmucker, pp 33–43
Filloux JH et al (1985) The Tasman project of sea floor magnetotelluric exploration. In: 4th ASEG conference, pp 221–224
Fleury P, Mathieu JP (1958) Courants alternatifs ondes hertziennes. Ed. M Eyrolles, Paris, pp 250–251

Galejs J (1965) Admittance of insulated loop antennas in a dissipative medium. IEEE Trans. Antennas Propagation, vol Ap.3 no. 2, pp 229–235

Geyger WA (1964) Non linear magnetic control devices. Basic principles, characteristics and applications. Ed. McGraw-Hill, London, pp 328–378

Goldman S (1948) Frequency analysis, modulation and noise. Ed. McGraw-Hill, New York, pp 381–403

Grigorakis A (1997) Application of detection theory to the measurement of the minimum detectable signal for a sinusoid in gaussian noise displayed on a lofargram. DSTO-TR-0568. Aeronautical and Maritime Research Laboratory, Melbourne, 48 p

Hallan A (1976) Une révolution dans les sciences de la terre: de la dérive des continents à la tectonique des plaques. Ed. du Seuil, Paris

Harvey RR (1974) Derivation of oceanic water motions from measurement of the vertical electric field. J Geophys Res 79(30):4512–4516

Havsgard GB et al (2011) Low noise Ag/AgCl electric field sensor system for marine CSEM and MT applications. Geoservices ASA

Heinson G, Constable S, White A (2000) Episodic melt transport at a mid-ocean ridge inferred from magnetotelluric sounding. Geophys Res Lett 27:2317–2320

Hekinian R (2014) Sea floor exploration. Springer, New York, 293 p

Holten T, Flekkoy EG, Singer B, Blixt EM, Hanssen A, Maloy KJ (2009) Vertical source, vertical receiver electromagnetic technique for offshore hydrocarbon exploration. First Break 27:89–93

Hutton VRS (1976) The electrical conductivity of the earth and planets. Rep Prog Phys 39 (6):513–515

Ivanoff A (1975) Introduction à l'océanographie. Propriétés deseaux de mer, 2 vol. Ed. Vuibert, Paris

Ives DJ, Janz GJ (1961) Reference electrodes: theory and practice. Academic, New York, 651 p

Jacobs JA et al (1987–1991) Geomagnetism, 4 vols. Ed. Academic Press

Jennison RC (1961) Fourier transforms and convolution for the experimentalist. Pergamon, New York

Johnson JB (1928) Thermal agitation of electricity in conductors. Phys Rev 32:97–109

Jones AG (1983) The problem of current channelling: a critical review. Geophys Surv 6:79–122

Kaufman AA, Keller GV (1981) The magnetotelluric sounding method. Elsevier, Amsterdam, 596 p

Key K, Lockwood A (2010) Determining the orientation of marine CSEM receivers using orthogonal procrustes rotation analysis. Geophysics 75(3)

Khesin BE et al (1996) Interpretation of geophysical fields in complicated environment. Kluwer Academic, Dordercht, 368 p

King JW, Newman WS et al (1967) Solar-terrestrial physics. Ed. Academic, New York, pp 107–211

Kittel C, Kroemer H (1995) Thermal physics. Ed. Freeman, San Francisco, pp 98–102

Koefoed O (1979) Geosounding principles, 1: resistivity sounding measurements. Methods in geochemistry and geophysics. Ed. Elsevier, Amsterdam, pp 7–18

Kong FN et al (2009) Casing effects in the sea-to-borehole electromagnetic method. Geophysics 14(5):F77–F87

Kraichman MB (1962) Impedance of a circular loop in an infinite conductive medium. J Res Nat Bur Std D 66(4):499–503

Kuvshinov A, Olsen N (2005) Modelling the ocean effect of geomagnetic storms at ground and satellite altitude. Earth observation with CHAMP. Ed. Springer, Berlin, pp 353–358

Lanzagorta M (2012) Underwater communications. Synthesis lecture on communications. Morgan and Claypool Publishers, San Rafael, 115 p

Lanzagorta M (2013) Underwater communications. Synthesis lectures on communications. Ed. Morgan & Claypool Publishers, San Rafael, 130 p

Latour P, Toniazzi C (1990) Magnetic detection: detection for the future? In: Undersea defence technology conference proceedings, UDT 90, London, 7–9 Feb, pp 71–76
Launay et al (1964) ULF environment of the sea floor. In: Symposium of the ULF electromagnetic fields. Paper no. 13, Boulder, Colorado
Law LK (1978) An ocean bottom magnetometer: design and first deployment near the Explorer Ridge (abstract). Eos Trans AGU 59:235
Leedert et al (2009). Signal generator for electromagnetic that produces a signal having an analog continuous waveform. WesternGeco. US patent no. 2009/0243614, 1 October
Lindqvist UP (2012) Method and system for calibrating streamer electrodes in a marine electromagnetic survey system. US patent no. 8,198,899
Lokken JE (1964) Instrumentation for receiving electromagnetic noise below 3,000 cps. Natural electromagnetic phenomena below 30 kc/s. Ed. Plenum Press, New York, pp 373–416
Loseth LO (2007) Modelling of controlled source electromagnetic data. PhD thesis, Norwegian University of Science and Technologies
Loseth LO et al (2006) Low frequency electromagnetic fields in applied geophysics: waves or diffusion? Geophysics 71(4):W29–W40
Madden T (1964) Spectral, cross-spectral and bispectral analysis of low frequency electromagnetic data. Natural electromagnetic phenomena below 30 kc/s. Ed. Plenum Press, New York, pp 429–448
Mangelsdorf PC (1968) Gulf Stream transport measurements using the salt bridge GEK and Loran navigation. Trans Am Geophys Un 49:198
Marcum JI (1947) A statistical theory of target detection by Pulsed Radar. ASTIA document no AD 100287. Research memorandum, vol 754. Rand Corporation, Santa Monica, 418 p
Mathias E et al (1924) Traité d'électricité atmosphérique et tellurique. Ed. Presses universitaires de France, Paris, pp 468–469
Matteucci C (1862) Sur les courants électriques observés dans les fils télégraphiques. CRAS 55:264–266
Mattson J, Anderson C (2010) Toward the integration of electromagnetic and seismic survey. Offshore Magazine, May
Mattson J et al (2011) Marine resistivity streamer survey produces high-quality data in the North Sea. World Oil 232:25–40
Mattsson J, Englemark F, Anderson C (2013) Towed streamer EM: the challenges of sensitivity and anisotropy. First Break 31:155–159
McKay A, Bergh, KF, Bhuiyan AH (2014) Determining resistivity from towed streamer EM data using unconstrained inversion. Tie to well and discovery examples. 76*th* EAGE annual conference & exhibition, Extended abstracts
McKay A, Mattsson J, Du Z (2015) Towed streamer EM reliable recovery of sub-surface resistivity. First Break 33:75–85, April
Merill J (2002) A history of extremely low frequency (ELF) submarine radio communications. Ed. Strong Books, an imprint of Publishing Directions, 96 p
Mittet R, Shaug-Pettersen T (2007) Shaping optimal transmitter waveforms for marine CSEM surveys. SEG San Antonio annual meeting, pp 539–543
Mittet R et al (2007) On the orientation and absolute phase of marine CSEM receivers. Geophysics 72(4):145–155
Mosnier J (1967) Application des techniques de pompage optique à l'étude des gradients géométriques, cas particulier de l'effet bord de mer. Doctoral thesis, ENS de Paris, Laboratoire de physique, p 4
Mosnier J (1970a). Variomètre sensible pour l'étude pour l'étude de la déclinaison : étude théorique. Annales de la géophysique. Tome 26
Mosnier J (1970b) Les magnétomètres sensibles améliorent notre connaissance du champ magnétique terrestre. Gauthier Villars, pp 75–80
Mosnier J (1974) Dispositif et mesure du champ magnétique terrestre. Brevet français no. 74 15502

Mosnier J (1984) Device for measuring an electric field in a fluid conductor, and method using such a device. CNRS. Brevet français no. FR19840019577
Mosnier J (1986) La détection magnétique des navires. Aspect géophysique. Revue: L'armement. NS no. 4 September, pp 140–149
Mosnier J, Yvetot P (1972) Nouveau type de variomètre asservi en direction. Annales de géophysique. Tome 28
Mosnier J, Yvetot P (1977) Nouveau type de variomètre à asservissement de champ et capteur capacitif. Annales de géophysique, Tome 33, fac. 3, pp 391–396
Nickolaenko A (2013) Schumann resonance for tyros: essentials of global electromagnetic resonance in the earth–ionosphere cavity. Springer Japan, Tokyo
Noorhana Y, Nadeem N, Majid Niaz A, Muhammad K, Hasnah Mohd Z, Afza S (2012) Modeling of antenna for deep target hydrocarbon exploration. J Electromagn Anal Appl 4:30–41
Northrop RB (1989) Analog electronic circuits: analysis and applications. Ed. Addison-Wesley, Reading, pp 251–253
Nyquist H (1928) Thermal agitation of electricity charge in conductors. Phys Rev 32:110–116
Paratte PA, Robert P (1996) Systèmes de mesure. Traité d'électricité, vol 17. Presses polytechniques et universitaires romandes, Lausanne
Parkinson WD (1983) Introduction to geomagnetism. Ed. Scottish Academic Press Ltd, Edinburgh, p 6
Petiau G (2000) Second generation of lead-lead chloride electrodes for geophysical applications. Pure Appl Geophys 157:357–382
PGS (2010) A prototype electromagnetic streamer: latest advance. Tech link. vol 8, no. 8. Ed. PGS.
Pittman EP, Stanford RA (1972) Electric field sensor. US patent no. 3,641,427, Feb
Porstendorfer G (1960) Tellurik. Grundlagen, Messtechnik und neue Einsatzmoglichkeiten. Ed. Akademie Verlag, Berlin, pp 24–26
Porstendorfer G (1975) Principles of magneto-telluric prospecting. Geopublication associates. Ed. Gebruder Borntraeger, Berlin, p 97
Rakotosoa U (1989) Appareillage de mesure des très faibles champs électriques en milieu marin: application à la mise en évidence des signaux électromagnétiques induits dans la mer. Doctoral thesis, Université Paris 6, Pierre et Marie Curie, 98 p
Reich E et Swerling P (1953) The detection of a sine wave in Gaussian noise. J Appl Phys 24:289–296
Rosa EB (1908) The self and mutual inductance of linear conductors. Bulletin of the Bureau of Standards, Washington, DC
Roth F, Lie JE, Panzner M, Gabri PT (2013) Improved target imaging with a high-power deck-mounted CSEM source—a field example from the North Sea. In: 75th EAGE conference & exhibition incorporating SPE EUROPEC 2013, London, UK, 10–13 June
Rowe HE (1965) Signals and noise in communication systems. Ed. Van Nostrand, Princeton, p 78
Sainson S (1982) Etude des champs électriques transverses: électromètre à ddp et à densité de courant pour la détection d'anomalie de conductivité dans un milieu conducteur de l'électricité. Rapport Laboratoire de géophysique appliquée ENS/CNRS, 30 p
Sainson S (1984) Etude d'une méthode de détection électrique d'anomalies conductrices voisines d'un forage. Doctoral thesis in geophysics, Université d'Orléans, 177 p
Sainson S (2007) Inspection en ligne des pipelines : principes et méthodes. Ed. Lavoisier, Paris, 332 p
Sainson S (2010) Les diagraphies de corrosion. Ed. Tec et Doc Lavoisier, Paris, 547 p
Sainson S (2016) A current density electrometer to explore ore bodies and subsea hydrocarbons (forthcoming)
Schlumberger C (1913) Procédé d'investigation du sous-sol. Brevet français no. 457.661
Scholl C, Edwards RN (2007) Marine downhole to seafloor dipole–dipole electromagnetic methods and the resolution of resistive targets. Geophysics 72 p. WA 39
Schumann WO (1948) Elektrische wellen. Hanser, Munchen, 324 p

Schumann WO (1952) Über die Dämpfung der elektromagnetischen Eigenschwingnugen des Systems Erde/Luft/Ionosphäre. Zeitschrift und Naturforschung 7a:250–252
Schumann WO, Koenig H (1954) Uber die Beobachtung von Atmospherics bei geringsten Frequenzen. Die Naturwiss. Heft 8, Jg.41, pp 183–184
Schwartz M (1959) Information transmission, modulation and noise. A unified approach to communication systems. Ed. McGraw-Hill, New York, pp 160–167
Selzer E (1968) Mesures électromagnétiques effectuées en mer profonde _a l'aide du bathyscaphe Archime'de: campagne de Gre'ce. Ann Inst Oceanogr 46:19–28
Selzer E (1972) Variations rapides du champ magnétique terrestre. Handbuch der Physik. Geophysik III Teil IV. Ed. Springer, Berlin, p 327
Selzer E et al (1966) Compte rendu d'enregistrements électromagnétiques _a très basses fréquences fait en bathyscaphe en eaux profondes. Agard Conf Proc 20:595–606
Sinha MC, Patel PD, Unsworth MJ, Owen TRE, Mac-Cormack MRJ (1990) An active source EM sounding system for marine use. Marine Geophy Res 12:59–68
Smith MS, Smith WW (1996) Handbook of real-time fast Fourier transforms: algorithm. Ed. IEEE, 468 p
Soderberg EF (1966) Undersea ELF measurements of the horizontal E-field to depth of 300 meters. Agard conference proceedings no. 20, pp 453–470
Soderberg EF (1969) ELF noise in the sea at depth of 30 to 300 meters. J Geophys Res Space Phys 74:2376–2387
Sunde ED (1949) Earth conduction effects in transmission systems. Ed. Van Nostran, New York, p 71
Surkov V, Hayakawa M (2014) Ultra and extremely low frequency electromagnetic fields. Ed. Springer, Tokyo, pp 57–106
Swidinsky A (2011) Transient electromagnetic modelling and imaging of thin resistive structures: applications for gas hydrate assessment. PhD thesis, University of Toronto, p. 16 and 128
Tenghamn SRL et al (2010) Receiver streamer system and method for marine electromagnetic surveying. US patent no. 7834632, 16 Nov
Tschauner J (1963) Introduction à la théorie des systèmes échantillonnés. Dunod Orléans, Paris
Um ES (2005) On the physics of galvanic source electromagnetic geophysical methods for terrestrial and marine exploration. University of Wisconsin–Madison, Madison, 428 p
Vacquier V (1972) Geomagnetism in marine geology. Elsevier oceanography series, vol 6. Ed. Elsevier, Amsterdam, pp 28–30
Wait JR (1957) Insulated loop antenna immersed in a conductive medium. J Res Nat Bur Std 59 (2):133–137
Wait JR (1972) Projet Sanguine. Science 178:272
Wait JR (1996) Electromagnetic waves in stratified media. Chap. 6. Ed. IEEE Press
Webb SC, Constable SC, Cox CS, Deaton TK (1985) A seafloor field electric instrument. J Geomag Geoelectr 37:1115–1129
Woodward PM (1953) Probability and information theory, with Application to Radar. Ed. Pergamon Press, 128 p
Young DP, Cox C (1981) Electromagnetic active source sounding near the East Pacific Rise. Geophys Res Lett 8:1043–1046
Yvetot P (1980) Réalisation d'un variomètre horizontal à asservissement de champ et capteur capacitif. Mémoire de DESS. Université Paris VI, Paris
Zach JJ, Frenkel MA, Ostved-Ghazy AM, de Lugao P, Ridyard D (2009) Marine CSEM methods for 3D hydrocarbon field mapping and monitoring. Socieda de brasileira de geofisica. In: 11 th international congress of the Brazilian geophysical society, 6 p

Chapter 5
Interpretations and Modeling

Theory is when one knows everything but nothing works.
Practice is when everything works but nobody knows why.

(Albert Einstein)

Abstract This chapter describes the various methods of interpretation of electromagnetic data. After a succinct recollection of the approach to solve the forward problem determined from the Maxwell equations and Ohm's law, several methods of resolution are proposed. These involve either analytical 1D models considering some (quasistatic) approximations or numerical models for higher-order dimensions. Analytical equation resolutions have been favored for their educational value considering geological canonic models and relatively simple integration methods. Then we recall some data inversion techniques, which allow us to directly access specific resistivity values of the subsoil and therefore to detect the presence or absence of hydrocarbons. Finally we describe the analog models that concretely allow us to establish special detection devices or check some assumptions.

Keywords Forward problem • Data inversion • Maxwell equations • Ohm's law • Analytical model • Numerical model • Quasistatic approximation

Preamble

It is only very recently that geophysicists have been interested in the interpretation of models with resistive heterogeneities. Until the 2000s, apart from civil engineering activities (e.g., detection of cavities), electromagnetic prospecting, which mainly concerned mining and hydrogeological activities such as geothermal prospecting, primarily aimed to detect electrically conductive anomalies, usually placed in resistant rocks. It was the advent of electromagnetic seabed logging that

S. Sainson, *Electromagnetic Seabed Logging*, DOI 10.1007/978-3-319-45355-2_5

led to a renewal of interpretation methods with the inclusion of this particular typology.[1]

EM seabed logging mainly developed with the advent of automatic methods for data inversion, which mathematicians quickly seized upon. There are many, all more or less understandable to the neophyte. For this reason, without underestimating the importance of these critical operating techniques, which are one of the essential links in the chain of interpretation, it was deliberately decided to develop, in this chapter, analytical and numerical interpretative methods (resolution of the forward problem) used with inversion algorithms, then highlight the difficulties and limitations of these techniques of interpretation. Despite its length, this part therefore gives only a brief overview of this, however, important aspect of the interpretation of the data, for which the reader will find supplements in more general books and more specific items listed in the bibliography.

To illustrate the different interpretation concepts of mEM prospecting and to introduce the higher dimension models, 1D simple modelings are considered in this chapter. The 1D approximation provides considerable insight into the physics of mCSEM (*in-line* acquisition). 2D and 3D models are mentioned in some particular cases to access some important information such as target size, lateral resolution, discrimination, and use of *broadside* data. In terms of number of pages, this chapter is the most important. It propose an overview of different interpretation methods. To avoid overloading it and complicating the understanding, the details of the calculations have been postponed to the appendix relative to this chapter at the end of the volume.

1 Introduction

To explore marine subsoil heterogeneities by mCSEM or mMT, it is necessary to measure and afterward to build an image of it from the study of electric and/or magnetic fields on the seabed. Theoretical knowledge of the propagation phenomena is then a dominant element in the interpretative stage of the data, but, regarding the complexity of the involved geological structures, it can be naturally only reducing.

Data acquisition (collecting, signal processing, transmitting, recording, etc.) and geophysical mapping of the results are the first and last steps in a process that requires an equally fundamental and important intermediate phase, which is the interpretation. It is in this intermediate phase that the main difficulty of the electromagnetic methods—and more generally of any geophysical method—lies.

[1]As in previous chapters, we recall, throughout the presentation, a brief history of the interpretation methods, allowing us to follow the evolution of ideas and concepts that have led to current techniques.

Mapping in one or more dimensions is the final document allowing the future operator to place exploration and then production drillings. It must provide, from n acquisition data ξ_n, the distribution of the conductivity σ in the space x, y, z, in the form, according to the technique adopted, of a diagram of the conductivities as a function of the frequency ω (frequency domain) or even as a function of the time t (time domain),[2] such that, in summary, we have:

$$\xi_n \rightarrow f(\sigma, x, y, z, \omega, t) \tag{5.1}$$

In fact, as the subsoil is generally heterogeneous, techniques and measuring devices can only provide, often after calculations, integrated or apparent values, which do not precisely correspond to the real conductivity of the underlying grounds.[3] Then the purpose of interpretation is ultimately to go from the values of the *apparent conductivities* σ_a to the distribution of specific conductivities σ, characteristic of the different investigated geological strata.[4]

$$\sigma_a \rightarrow \langle\sigma\rangle_{x,y,z} \tag{5.2}$$

To achieve this objective, it is necessary to establish cause-and-effect relationships[5] between the structure of the subsoil, its components and the physical phenomena (propagation/diffusion) involved. The material connection between these elements is the instrumentation. This provides then, directly or indirectly, the potential values, the field values (electric and/or magnetic fields) and the current density values used then by the interpretive systems.

In certain circumstances—the rarest—these measurements allow direct access to the desired information on the studied system. This is called direct measurement. This is, for example, the case when it is possible to connect the measures by accurate deterministic physical laws. Under these conditions, the conductivity σ_a is associated with the field $\vec{E}$ or with the current density $\vec{J}$ measurements by an explicit mathematical operator ϑ such that:

$$\vec{E}, \vec{J} \xrightarrow{\vartheta} \sigma_a \tag{5.3}$$

[2]Scope of action of transient methods, for example.

[3]Only electric logs (wireline or MWD) are able to provide the "true" resistivity. This assumes, of course, that the geological layers are crossed by a drilling (borehole).

[4]This corresponds, as discussed later, to the resolution of the inverse problem.

[5]"To interpret is to establish relations of cause and effect between the structure of the subsoil and the highlighted physical phenomena, it is to deduce what are (or possibly what are not) the geological structures compatible with a specific anomaly. These relations of cause and effect between structure and anomaly are conceivable only in numerical form, in other words mathematics. Whether one likes it or not, the interpretation is only a mathematical problem "(Cagniard 1950).

In other situations—the most frequent, such as SBL—the researched parameters are not directly accessible for measurement. This is called indirect measurement. In these special circumstances, we can only have an implicit operator ϑ', to which it is necessary to add additional information (i) collected from additional experiments, past or simulated.

$$\vec{\mathrm{E}}, \vec{\mathrm{J}} \xrightarrow{(i) \rightarrow \vartheta'} \sigma_{\mathrm{a}} \tag{5.4}$$

Specifically, and in essence, the interpretation of data in the broad sense is based:

- A priori on mathematical models taking into account the physical laws that govern the EM propagation of energy in specific topological and physical contexts (modeling)
- A posteriori on numerical simulations coming from these mathematical models taking into account the acquisition data (geological interpretation)

The difficulty of these theoretical investigations often lies in the fact that they cannot provide a single solution. Indeed, field measurements, performed at the bottom of the sea, for example, generally correspond, in the subsoil, to different possibilities of distributions of electrical conductivities. For this purpose, to remove any uncertainty and ambiguity, and to move toward the most plausible solution, it is then necessary to add other information, external to the EM investigation itself, influencing near and far, and usable by the system, as those concerning, for example:

- The marine environment (near) of the sensors in the sea (orientation, positioning, physical nature of the medium, etc.)
- The geological environment (far) of the underlying grounds, given by direct (geological) or indirect (geophysical) studies[6]

One response to this delicate problem of interpretation is one that integrates in particular the principle of inversion data. So far, this solution, which uses multiple sources of information from various sources, seems to be the most suitable for the interpretation of mCSEM measurements whatever the relative complexity of the present geological environment is (Fig. 5.1).

Data inversion, more widely known as resolution of the inverse problem, involves, apart from EM acquisition measures (values of electric or/and magnetic fields), a relatively heterogeneous external source of additional data.

These can come from:

---

[6]The mathematical problem applies above all to geological data, i.e., to facts of observation that fix reality at a certain moment in time and in space. We can thus arrive at valid conclusions only if we take into account this natural reality, which can be obtained by sufficient knowledge of the local geology. The more numerous and reliable the geological data are, the less theoretical and long the mathematical work of the interpretation will be.

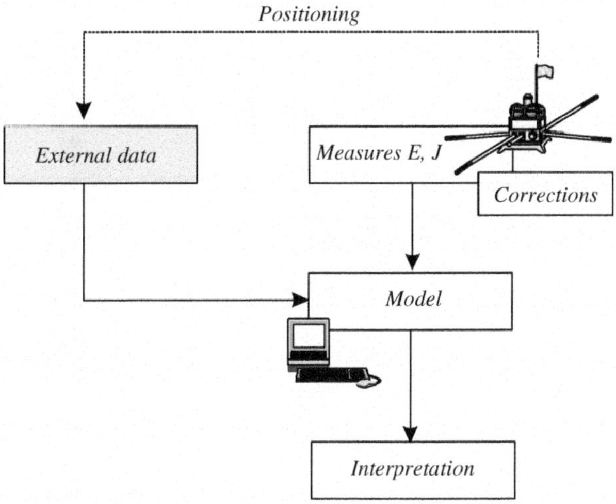

Fig. 5.1 Interpretation concept for SBL data introducing external data to form a coherent interpretation model

– Geophysical acquisition data, such as those indirectly derived from seismics (sections, profiles, maps, block diagrams, etc.), from gravimetry, magnetic surveys and/or geological data (facies, sedimentology, stratigraphy, petrophysics, tectonics, structural data, etc.), or directly from the subsoil (coring, samples of rocks, logs, cuttings, etc.).
– Data modeling, resulting from the resolution of forward problems (synthetic traces) adapted to the considered detection problematic

With all of these additional data, it is then possible by increasing the number of them and their qualities, to strive for uniqueness of the solution, but without achieving it (Fig. 5.2).

Advances in computing, and especially with large computers today, allow us to process a large amount of information in record time (Commer et al. 2008). However, they are dependent on acquisition means that are often limited either in the number of sensors, or in their measurement parameters. SBL equipment allows us to simultaneously acquire the quantitative, dimensional, temporal and frequency characteristics of the contributing electromagnetic fields. With all these elements, supplemented by as many annex data as possible, the inversion problematic can be accordingly approached with maximum certainty and also consistency, thanks today to the sophistication of the proposed interpretative models (Fig. 5.3).

If the processing of data acquisition, by simple resolution of the forward problem on its own, is now relegated to the background (because it is substantially solved for simple cases), it remains that this processing, in one form or another, remains the cornerstone of measurement inversion techniques. For this reason, the reader will not find here comprehensive development of several examples of analytical resolutions, showing the complexity of the calculations at this elementary level, despite the simplicity of the proposed models (geometric structures especially). Numerical

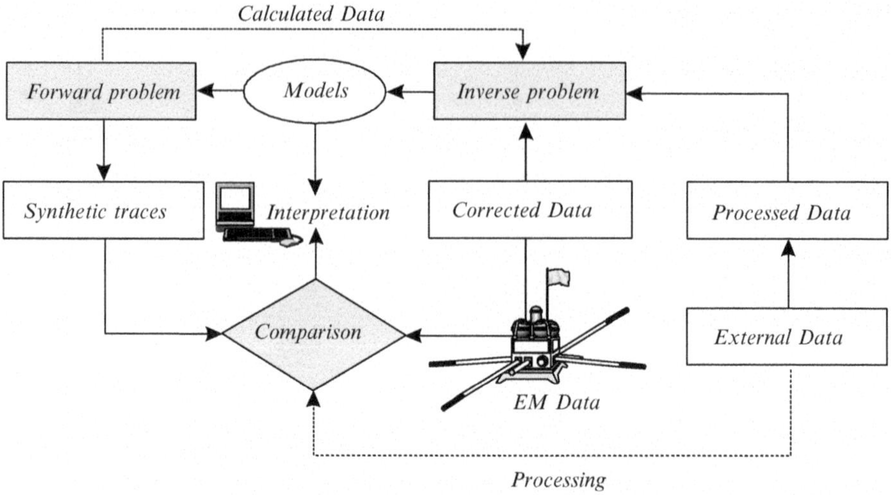

Fig. 5.2 Concept of data inversion for the interpretation of SBL measures implementing several sources of experimental data (seismics, well logging, etc.) and theoretical data (solutions of the forward problem) compatible with the model (processing)

Fig. 5.3 Super computer used for processing, modeling and interpretation of electromagnetic data. Computing power (parallel computing) can handle today a very large number of data especially when used in the inversion process. These computers, such as Aurora of EMGS, are currently able to perform more than 70 trillion calculations per second

models with more elaborate formalism and more eclectic mathematical techniques will only be recalled here for reference, referring the reader interested in these methods to specialized monographs, more comprehensive articles, or even specialist thesis works.

The calculation in the space of electromagnetic fields generated by dipoles immersed in seawater is most often done by introducing the *Hertz potential vector* and breaking the waves into two modes: TM (transverse magnetic) and TE (transverse electric). For example, and for a geological canonic model, EM answers in TE and TM modes are given in Chap. 3, Sect. 5.2, for the broadside configuration. The exact computing involves *Sommerfeld integrals* and requires the use of *Hankel transforms* to evaluate them—mathematical tools that are quite heavy to handle, such as numerical integrations with adaptive digital filtering or with specific algorithms, which perform automatic integration programs (Chave 1983, 2009). For this principal reason, for a first step (elementary or first course). I preferred to retain approximate formulations, as did J. T. Weaver and other authors (Weaver 1967; Kraichman, 1970; Habashy 1985; Bannister 1987a, b) (using direct analytical integrations) with certain simplifying assumptions considering mainly *quasistatic approximation*, with an EM transmitter and receiver considered as motionless (Galilean referential), and with an electromagnetic propagation that occurs in horizontal stratified media or in the presence of isometric conductivity anomalies for some simple examples of modeling.

For numerical methods, we have decided to give a simple overview and to explain some examples rather than expose a comprehensive coverage of all used methods. To researchers interested in studying these complex subjects, we particularly recommend reading the two books *Mathematical Methods in Electromagnetic Induction* (Weaver 1994) and *Electromagnetic Methods in Applied Geophysics* in two volumes (Nabighian 1987), which include applications to seabed measurements.

The following discussion describes, as simply as possible, for teaching purposes, the different modeling methods, included or not in the processes of solving inverse problems, now universally used. The reader will appreciate then, for example, the theory of electrical images, specifically developed for underwater surveying (tabular model) and which was also the first implemented method of onshore interpretation. The reader who knows the *Dirichlet problem* (i.e., to find a harmonic function verifying that the Laplacian is equal to zero with limit conditions) will also discover among other things the limitations of theoretical physics with the analytical solution of the Laplace equation in a relatively simple environment (sphere model) allowing us to understand the problem of pseudo-3D isometric interpretation and side detection.

Until now the computational complexity has allowed for these investigations only DC or low frequency current under certain simplifying conditions (quasistatic approximation). This limited on one hand the opportunities for investigation (resolutions), and secondly failed so far to include virtually all these methods in the concepts of inversion. However, due to the timeliness of the simulations they provide, these different resolution techniques can be advantageously used to:

- Quickly test the feasibility of new exploration concepts
- Establish orders of magnitude
- Optimize the parameters of power transmission by reducing noise
- Define and increase resolution
- Define the acquisition geometries between the transmitter and receiver
- Make comparisons between different methods of interpretation
- Reinforce the interpretation of more or less composite models

- Deal with simple cases of interpretation
- Perform quick-look interpretation (fast field computation during surveying at sea)
- Be used in conjunction with numerical models
- Develop embedded correction devices (galvanic effect)[7]

Numerical methods included in the inversion process are then briefly treated, taking then into account part of the geological complexity and more particularly the specific typology of the forms of the grounds and especially the deposits, which has always been lacking in previous methods.

2 Reminders and General Information on Methods of Interpretation

Detecting anomalous geological objects, i.e., discrete heterogeneities placed in a so-called normal environment,[8] poses from a mathematical point of view both the hardest and easiest problems. It is the most difficult problem when we attempt to characterize the nature itself of the geological anomaly, i.e., concerning its size, its geometry, its electrical properties and especially its location. It is the easiest problem when we simply establish only its presence. Thus we implicitly admit that something is more conductive or more insulating than its immediate environment.

In the technical literature, we do not moreover find a simple criterion to discern whether the foreign body is a better or worse conductor than the environment that surrounds it. At this level, we can ask, of course, if a quantitative approach is not the best solution to this complex problem of interpretation, especially when we have at the moment more and more powerful computers and increasingly numerous and substantial sources of information. It is largely this problem that actually answers the methods of inversion, used today in data interpretation of SBL.

The ultimate goal of data interpretation is to go from the measured field values (acquisition) to a probable distribution of the specific conductivities, or at least approaching it, and to finally draw a spatial mapping (laterally and in depth) that is as complete and realistic as possible (uniqueness of the solution).

Two options are then available to the geophysicist engineer in charge of the interpretation, to give a geological sense of the results:

- One, when the data are sufficient, consists of going back to the original characteristics, i.e., mathematically solving the *inverse problem*.

[7] It is easier for now to memorize a series of abacuses than to run numerical calculation codes on embarked units and more particularly on seabed sensors.

[8] See the definition in Appendix 2.5.

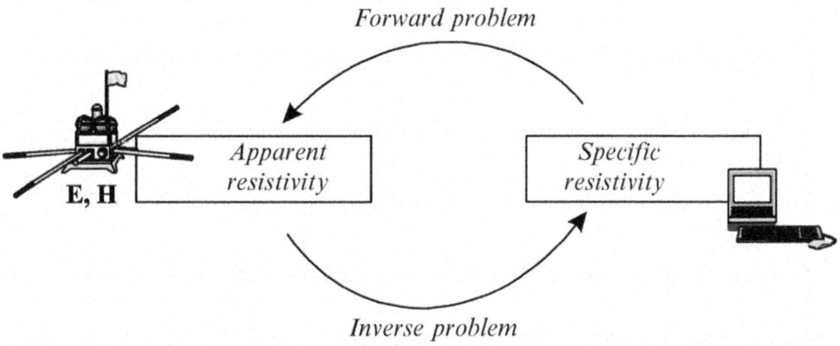

Fig. 5.4 From specific resistivity obtained using a synthetic model, propped up in depth, the forward problem provides the apparent resistivity then comparable to data calculated from measurements. Conversely, from the apparent resistivity obtained from the field measurements, the inverse problem goes back automatically to the specific resistivity values

– The other, when the data are partial, consists of directly comparing data (measures) to simple theoretical models developed from mostly analytical theoretical calculations (*forward problem*) (Fig. 5.4).

For lack of anything better, with adaptation of existing systems of acquisition, electromagnetic prospecting contented itself for a very long time with reducing hypotheses and simplifications to interpret these data by solving forward problems (comparison of data with the results of theoretical calculations). However, in recent years, advances in mathematics, numerical analysis, optimization algorithms, tests, etc.; better understanding of geological phenomena; and especially increases in the computing power of computers have allowed us to discuss, with much confidence and success, interpretation by the data inversion process (Fig. 5.5).

2.1 Resolution of the Forward Problem: Generalities

Solving the forward problem may have very different aspects from the realization of analog simulators[9] in the creation of computer programs derived from a variety of mathematical algorithms. These can be used either in "implicit" interpretation of data by comparing the model a priori with the data acquisition, or in processes of data inversion. Solving the forward problem is the basis for the development of models ranging from constructing models similar to reality (within the scale factor)[10] to analytical or numerical resolution of complex equations (partial derivative differential equations) (Fig. 5.6).

Forward modeling then consists of calculating the apparent conductivity σ_a for a distribution of specific conductivities σ corresponding to a factual geometric model

[9]The first to be put into practice and still used.

[10]Not always easy to get.

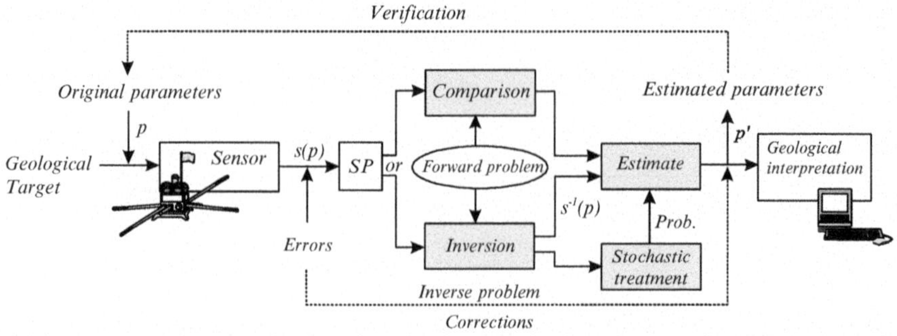

Fig. 5.5 Place of data interpretation in the electromagnetic acquisition chain. This occurs after the capture and treatment phases of signal processing (*SP*) and the development of one or more geometric models, intended to be representative of the target (deposit). (Before talking about the deposit we prefer to talk about the target.) On the interpretive level, solving the forward problem consists of formulating several model proposals issued by the operator to obtain synthetic results. The resolution of the inverse problem is a proposal by the machine of one or more models based on data derived directly from the acquisition (measures)

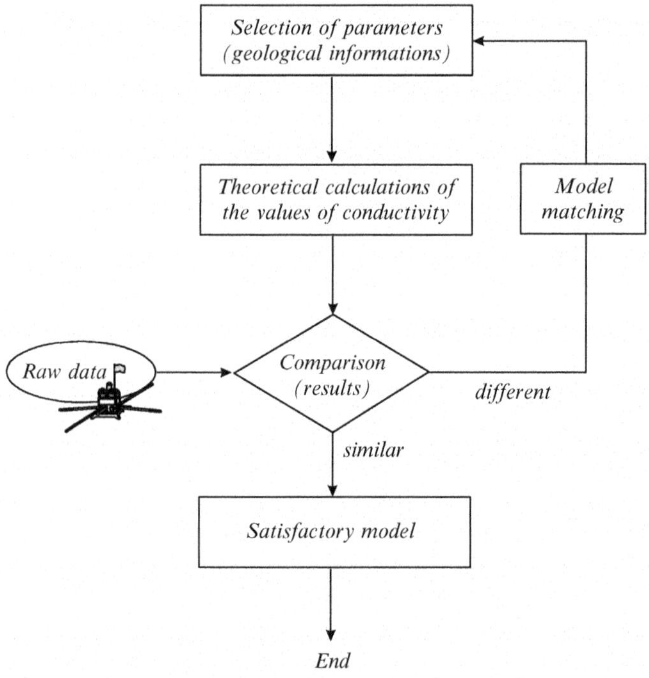

Fig. 5.6 Algorithm chart for resolution of the forward problem (*forward modeling*)

(synthetic and nonreal). This operation corresponds to the inverse operation in the equation (cf. Eq. 5.2), i.e.:

$$\langle\sigma\rangle_{x,y,z} \rightarrow \sigma_{\mathrm{a}} \tag{5.5}$$

2.1.1 Analog Models

These models have been used for a very long time for lack of finding mathematical equivalents (Schlumberger in 1911; Heiland 1931; Beaufort 1944; Boissonnas 1946; Janovskij and Molocnov 1959). Varied, they still allow a fast and pragmatic approach especially when it is a question of studying comparable interactive phenomena (an electrical/mechanical analogy for example) where homothetic ratios are possible (models) and also compatible with the frequency aspect (wavelength) of the method used (see Sect. 6). They are much used in feasibility studies of new prospecting methods to practically validate the concepts, simulate the answers and fix orders of magnitude. They often provide the link between the sensor part, which is related to or in intimate contact with the external environment (e.g., adaptation of impedance) and the elements to identify and locate (e.g., a geological anomaly). We find these models in several forms. From a rheostatic tank of several square meters of surface to models that are automated[11] or not (networks of resistances, capacities welded between them), these models today can also be in the form of codes of computer simulation that are more flexible and convenient for use.

2.1.2 Analytical Models

Analytical models are also very old and have been the basis of classical theoretical physics (Lamb 1924). Laplace (*Traité de mécanique céleste*, 1799, 1805) and Fourier (*Théorie analytique de la chaleur*, 1807, 1822), followed by Poisson and Lamé, were the creators of analytical integration methods in mathematical physics.

Present in all the major treatises of electromagnetism in the nineteenth century (Maxwell 1873; Heaviside 1899) and subsequently supplemented by numerous physicists (Jeans 1933; Brillouin 1938; Smythe 1939; Stratton 1941), analytical models often correspond to the exact resolution (integration) of partial differential equations allowing us to simulate the physical behavior of the studied processes, posing very precise initial conditions (on the source), limit conditions (on the environment) and boundary conditions (on the interfaces).

[11]In the 1950s, these models, locked in cabinets, were completely automated and controlled by a panel where it was possible to set different combinations with electromechanical switches.

They are of interest at all stages of the taking of information for geometric configurations, often very simple but sufficient to supply orders of magnitude.

These models are, for example, constructed from solving the well-known wave equation:

$$\nabla^2 \vec{\psi} - \frac{1}{c^2}\frac{\partial^2 \vec{\psi}}{\partial t^2} = 0 \tag{5.6}$$

which can be transcribed, among others, into rectangular coordinates[12]:

$$\nabla^2 \equiv \frac{\partial^2}{\partial x^2} + \frac{\partial^2}{\partial y^2} + \frac{\partial^2}{\partial z^2} \tag{5.7}$$

The function ψ, which obeys this equation (cf. Eq. 5.6) then describes a physical quantity that evolves in space and time:

$$\psi\left(\frac{n_x x + n_y y + n_z z}{c} - t\right) \tag{5.8}$$

where n_x, n_y, and n_z are numbers that satisfy the condition $n_x^2 + n_y^2 + n_z^2 = 1$ and are the projections on the axes of the unit vector normal to the wave front.

The waveform defined by ψ, which propagates at the speed c in the same environment, remains unchanged during propagation (assumption) (Fig. 5.7).

This equation can also be reduced to the Helmholtz equation when the velocity of propagation c is no longer taken into account (low frequencies) or to the Laplace equation when the frequency dependence (ω) is not necessary as at very low frequencies (quasistatic approximation) and in direct current (static approximation).

More generally, depending on the magnitude of frequency, we then obtain:

- At high frequencies (application to onshore geological RADAR, for example), the wave equation in the time domain where the terms of diffusion and propagation coexist:

$$\nabla^2 \vec{e} - \mu\sigma \frac{\partial \vec{e}}{\partial t} - \mu\varepsilon \frac{\partial^2 \vec{e}}{\partial t^2} = 0 \tag{5.9}$$

- At low frequencies (application to mCSEM and mMT, for example), the diffusion equation also called the *Helmholtz equation*[13]:

[12]As discussed later, cylindrical and spherical coordinates are also used.

[13]At the frequencies, the propagation term (second term) disappears.

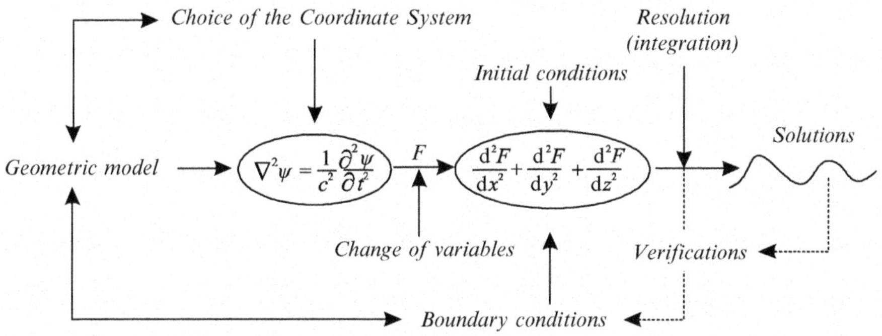

Fig. 5.7 Example of a conventional mathematical model obeying the waves propagation equation where c is the propagation speed of the waves. The geometric model is developed from a priori available geological data (structural, stratigraphic, etc.) and geophysical (seismic) data. This equation can be written in the form, after a change in variables, of a partial differential equation in different coordinate systems (Cartesian, cylindrical or spherical). In a rectangular coordinate system for example, the potential function $\psi(x, y, z, t)$ for a permanent harmonic regime becomes $F(x, y, z).e^{j\omega t}$ (with $\omega = 2\pi/T$ where T is the period of oscillation of the waves). The solutions to these equations, such as the Bessel equation, generally admit expressions as integrals (resulting from series development) with particularly special *Bessel functions* (also called cylinder functions), *Neumann, Hankel functions* (cf. special functions at the end of the volume) modified or not (and also linear combinations of these) that can be calculated from numerical tables or from their recurrence relation (development also in series)

$$\nabla^2 \vec{e} - \mu\sigma \frac{\partial \vec{e}}{\partial t} = 0 \tag{5.10}$$

– At very low frequencies (application to electrical subsurface sounding, for example), when the latter tend to zero ($\omega = 0$) and particularly for DC, the *Laplace equation*:

$$\nabla^2 \vec{E} = 0 \tag{5.11}$$

Or in the presence of a current source I, the *Poisson equation* (see Chap. 2, Eq. 2.8):

$$\nabla^2 V = -I\boldsymbol{\delta}(r - r_S) \tag{5.12}$$

These equations implicitly define, with the frequency, the sensitivity of the different techniques used in geophysical prospecting (see Chap. 2, Sect. 3.4).

Except in exceptional circumstances:

– A tabular model representative of a succession of layers of different thickness and resistivity, but limited in number and lateral extension
– Isometric models (sphere, disk, cylinder, plate) buried or submerged, etc.

where special techniques:

- Of power series development
- Of formulations derived, for example, from the geometric optics such as those of the electrical images (*Maxwell* or *Thomson* theory),
- Of resolution such as that proposed in his time by the physicist S. Stefanescu[14]
- or even integration,

were found, the analytical solution of these equations to calculate the apparent resistivity, in the state of mathematics for more complex geometries, remains unsolved for now.[15]

With the geology presenting, most of the time, objects of observation that are more or less complicated (anticlinals, domes, faults, etc.), where a side resolution is generally sought, new techniques of modeling, this time numerical, were proposed and finalized with success.

2.1.3 Numerical Models

Numerical models, which appeared significantly in the early 1980s, thanks to the increase in the computing power of computers and to improvements in the languages of scientific programming (Fortran and in the pass Basic particularly), have massively replaced other types of modeling. As with analytical models, they correspond to the mathematical solution of differential equations; however, they bring to it an unequaled flexibility necessary for the simulation of complex phenomena or geometries. In general, they are used when there is no analytical solution to the problem, which is the case for discrete anomalies and sometimes for geological environments with complicated topologies. They give approached calculations. These performances are obtained by an automatic discretization in simple elements of the propagation medium and by the resolution point by point of the equations in the knots of the meshing so developed, adapting itself then perfectly to the local limit conditions. Mention may be made, for example, of numerical methods using the techniques of resolution by:

[14]Weber's law allows us to connect the images theory developed by Hummel to that of S. Stefanescu (Lasfargues 1957).

[15]The integration of linear partial differential equations with a constant coefficient was built very naturally at the beginning (A. Cauchy) based on theorems of existence and uniqueness, then providing analytical integrals in the form of power series, where the coefficients were calculated by recurrence from the initial conditions and boundary conditions. According to their research, physicists realized pretty quickly that certain series of special functions (Bessel functions, Legendre polynomials, etc.) highlighted, better than the power series, the properties of integrals. However, these integrals such as the series converge only very slowly, which limits the calculation of apparent resistivity in practice.

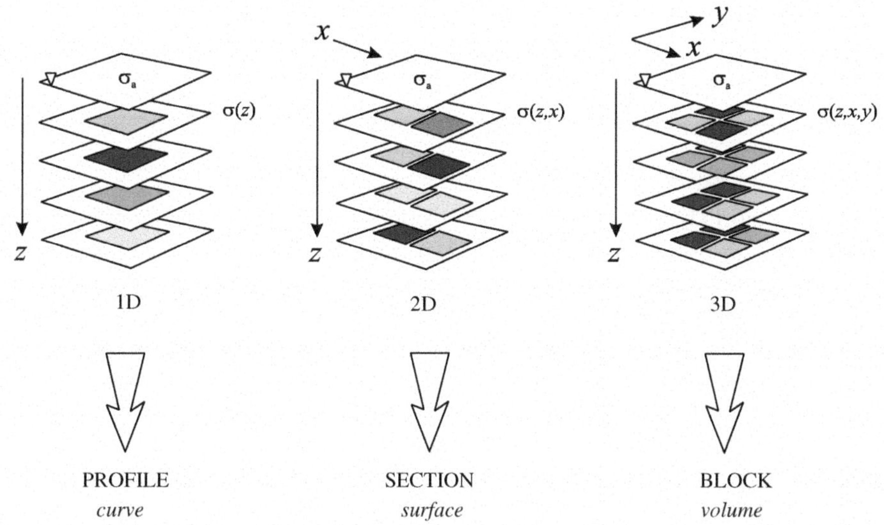

Fig. 5.8 1D, 2D and 3D models. At the measuring point ∇, the 1D model gives a distribution of conductivity in depth. The 1D model is shown at the various measuring points by a profile (curve) corresponding to a constant investigation depth. The 2D model is represented by a section or a surface where the different resistivity zones are bounded by closed curves (contour lines). The 3D model is distinguished by a relief visible from any point of view. When axes of symmetry can be identified, we may also form intermediate models of the 2.5D type imaged, for example, by a relief surface (cf. Fig. 5.10)

- The *finite elements* method (FEm)
- The *finite differences* method[16] (FDm)
- The *finite surface* or *volume integration* or *finite integration* technique (FIt)
- The *boundary elements* method (BEm)

These mathematical models are often organized in the form of automatic computer codes, which are more or less user friendly, according to their level of achievement (academic or commercial).

Unlike analytic codes, numerical codes require a long process of programming, and their convergence in extreme circumstances or those that leave the strict frame of their operation is sometimes difficult to obtain. The calculation time is much longer and the solutions are approximate and depend on the quantity and quality of data acquisition (Fig. 5.8).

By the 1980s, some researchers, such as A. D. Chave and C. S. Cox for mCSEM, and R. N. Edwards and S. J. Cheesman for t-mCSEM, had presented most successful models of interpretation for horizontal and vertical injections of current, giving then the values of specific fields in modal form (see Chap. 3, Sect. 5.2).

[16]It is necessary to know that these methods offer incomparable implementation flexibility with respect to the complexity of the considered media, but require on the other hand the monopolization of significant computational resources not always compatible with the data stream.

2.2 *Resolution of the Inverse Problem: Generalities*

The most attractive method that comes naturally to mind for the provision of geophysical maps (spatial distribution of conductivity reflecting the location of the oil site, for example) is that of determining the causes of the detected phenomenon from data collected by the measuring instruments. In other words, this is the step that consists of going back to the original information from the recorded signals or even rebuilding a model from measurements. That being said, the desired quantities are often not directly accessible to measurement. However, they may be connected to it by known physical laws. Practically, this approach is far more difficult to achieve.[17] Indeed, for lack of a sufficiently evolved mathematical tool, and of numerous measures, the inverse problem for a long time remained inaccessible to direct interpretation of data. However, some geophysicists have proposed, since the 1930s, inversion formulas for tabular structures within DC electrical soundings (Langer 1933, 1936; Slichter 1933) or for cylindrical structures adapted to the electrotelluric method (Baranov 1951). It was only in the years 1960–1970 that the technique became widespread (Backus and Gilbert 1967; Inman et al. 1973) through application, among others, to EM study of the deep conductivity of the earth mantle (Parker 1970; Bailey 1970).

In general the resolution of the inverse problem occurs around an operation of minimization of the differences between measured values and values calculated theoretically, and briefly passes through a phase of modeling and optimization which consists of:

- Solving the forward problem (theory), which primarily aims to calculate the response of a more or less precise physical model corresponding to the accepted geological reality (modeling)
- Repeating several times the sequence by adjustment, each time minimizing by ad hoc methods the differences between the various simulations (optimization)

all this in order to obtain from the apparent resistivities measured (data of acquisition):

- The calculated apparent resistivities (synthetic responses from the resolution of the forward problem)
- The final distribution of the estimated specific resistivities coming from the successive iteration processes and alternately completed by plausibility tests

Of course, this approach will be all the more effective in instances where the sources of information are numerous, such as those involving seismic acquisition data, which are very frequent in marine exploration and often such data are the only ones available.

[17]The reader will find in the works of Professors Parker (Parker 1994) and Scales (Scales et al. 2001) an excellent introduction to the resolution of the inverse problem in the field of geophysics. The works of Professor Tarantola, however, are more difficult to access (Tarantola 1981, 2005).

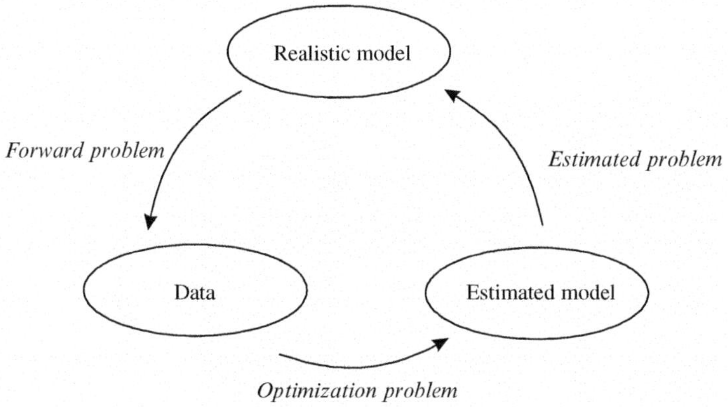

Fig. 5.9 Diagram summarizing the decomposition of the treatment of the reverse problem in models as part of SBL

Unlike the forward problem, the inverse problem is an unsolved problem, i.e., its solutions are not unique and have to aim toward a likely and acceptable convergence criterion, which then depends on many parameters often exogenous to the model. The process of the inverse problem can be summarized in the resolution of three problems, in the following order: forward, optimization and estimation (cf. Fig. 5.9) offering two models from the data acquired: one estimated, the other one realistic.

2.3 *Simulation: Generalities*

The simulations correspond to the interpretation phase itself through the use of an informatics code of automatic calculation or a mathematical model presented in the form of software modeling, commercial or not. It is at this level that the engineer can understand the geological parameters (structural, tectonic, geomorphological, etc.) and the lithological and petrophysical parameters, by the resolution (depending on the type of computer program used) of either:

- The forward problem, which, according to the parameters introduced in the machine, provides the synthetic responses representative of the phenomena influenced by the characteristics of the model, or
- The inverse problem,[18] which, based on the responses (experimental, synthetic or other field measurements), allows us to provide the characteristics of the subsoil.

[18]The resolution of the inverse problem must, according to the French mathematician J. Hadamard match to "*a well-posed problem*"—i.e., that its solution must imperatively respond to three separate criteria: to exist, be unique and stable with respect to measurement errors.

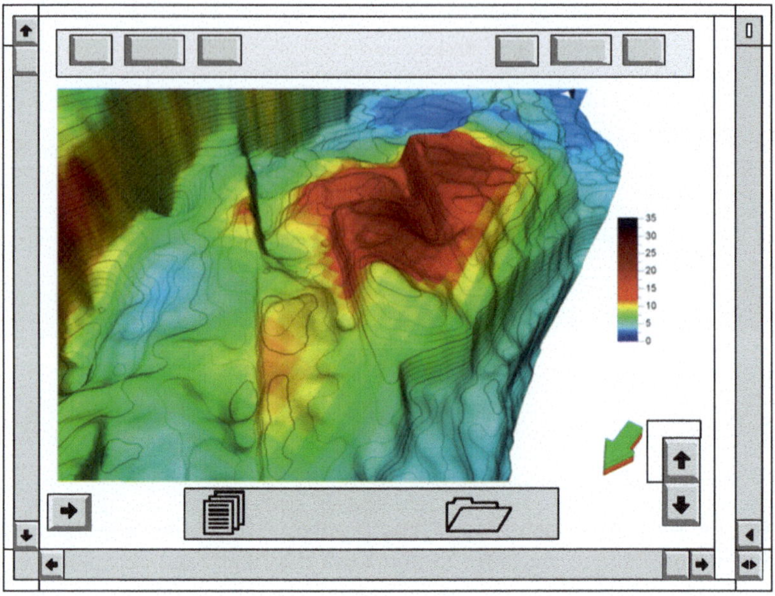

Fig. 5.10 Results of an interpretative 2.5D (surface) simulation. According to EMGS (2010)

Currently, for electrographic techniques, the second alternative is by far the most commonly used; the first method has been more rarely used so far.

In what follows, for educational purposes, several interpretative alternatives, developed over the years, are proposed, corresponding to more or less complex models (Fig. 5.10). They are based on very different mathematical algorithms, which will enable readers to appreciate the diversity of the reasoning and the limits of interpretation in each case, in a more or less circumscribed space (1D, 2D or 3D).

2.4 Choice of Interpretation Methods

To describe all the methods of automatic interpretation currently used would widely go beyond the frame of this work, especially since most of them are extremely complex for the novice.

We can say that the analytical methods, the first to be offered in applied geophysics, developed around 1D models corresponding in particular to a stratified subsoil, often limited to a few layers. Except in very rare cases, for relatively simple heterogeneity geometries such as spheres and cylinders (by agreeing on "symmetric" properties of the equations), 2D or 3D solutions were proposed.

As geology is generally much more complex than these models, new techniques of 2D and 3D calculations were developed. In the case of the numerical methods, which allow us to model more or less any shape provided, we possess enough information to faithfully propose a modeling. We touch here on the interface existing in the interpretation between the resolution of the forward and inverse problems, where the more or less thorough knowledge of the local geology is an asset to the proposal of plausible and realistic physical models.

In what follows, only demonstrations with limited developments are set to allow the reader to appreciate the diversity of the interpretation models and also to fix the limits.

3 Analytical Models

During the development of electrical prospecting, several theoretical approaches to interpretation have been proposed.

The very first mathematical models to be suggested were analytical models under *the reciprocity theorem* (Wenner 1916).[19] Chronologically, these were initially developed in Germany and the USA from the theory of electrical images (Hummel 1928; Ehrenbourg and Watson 1931; Roman 1934), then in France, with the Schlumberger school (Stefanescu, Maillet, Kostitzine, etc.), in the Soviet Union and in Italy. Most of these models were built from the resolution of appropriate differential equations.

In this case, the potential in each of the layers and for any resistivities and thicknesses, and the solution of these equations is then formally expressed either:

- In the form of defined integrals including Bessel functions, or
- In the form of Fourier series

Almost all of these models were designed in fact for surface prospecting involving an air/ground interface (insulator/conductor), and sometimes adapted in very rare cases to marine experiments at the interface of marine water/seabed (Litvinov 1941; Terekhin 1962; Patella and Schiavone 1974; Schiavone and Patella 1974).

This type of approach can be mathematically conceivable only for tabular representations of geological layers[20] (superposition of horizontal and infinite strata with different and especially well marked conductivities), therefore offering no side resolution, and only for steady-state currents. More complex models such as that of the analytical sphere have been solved, giving then the possibility of an azimuthal detection (→ co-lateral resolution).

At the end of the book, the reader will find in the Appendices A5 (1, 2, 3, 4, 5, and 6) different analytical models corresponding to various conceptual approaches, and two related automatic programs. These were chosen for their historical interest (the images method, for example) and especially for their instructional and teaching

[19]Cited by Wenner but not demonstrated by the author (see, more precisely, Appendix A2.2).

[20]With few exceptions (*mise à la masse*, spontaneous polarization), electrical methods have never been used for the detection of a local anomaly, this type of investigation being reserved for the field of electromagnetic methods. The reader will find details on the development of these techniques in general works on electrical prospecting (Keller and Frischknecht 1966).

qualities (Shen and Kong 1983). Through these examples, specifically designed for marine applications (air/water/seabed), we can understand, despite the simplicity of the modeled structures, the relative complexity of the calculations that were developed and the theoretical artifices cleverly used. However, if these models are too simple for the interpretations, sensu stricto, of complex geological sets, the ease in computer programming (discrete formulations or formulas), the convenience of use (quickness of execution, almost immediately, of the program) and the accuracy of their results can now make these models valuable tools, and useful:

- As part of the setting or definition of limits, as part of the comparison, verification or determination of orders of magnitude, as part of the setting of detectability thresholds, etc., for technical and interpretation studies of new submarine prospecting methods
- In the improvement of numerical algorithms, providing, for example, specific anomalous analytical responses (stratified environment, model of a cylinder or of a disk, etc.) in addition to the numerical model whose computation time can reach several tens of hours

For theoretical and practical reasons, these models were calculated for continuous currents but can also be used for the interpretation of investigations carried out at low and very low frequencies (quasistatic approximation),[21] especially when we consider that galvanic effects are predominant (cf. Chap. 2, Sect. 2.3.2).

3.1 Reminders

These very important reminders set, according to the needs (target, resolution, definition, etc.) and to the type of envisaged and practiced investigation (principles and methods of prospecting), the limitations of the theoretical models that were usable and practically exploitable. Moreover, with some approximations, some theoretical problems can be solved by the use of Ohm's law, which is more manageable than the Maxwell equations.

However, the reader will find, in some books, some solutions (1D) in the frequency and time domains (Kaufman and Keller 1983).

3.1.1 First Reminder: Quasistatic Approximation and Ohm's Law Applicable to Conductors

For conductors of electricity, at low as well as very low frequencies, the displacement currents associated with polarization phenomena are negligible regarding the

[21]DC marine models (calculations) are not restricted as in situ investigations (measurements) and may in some cases be useful for the interpretation of low frequency data.

conduction currents. On the theoretical level, if the temporal variations are neglected, inductive effects can be removed, which means considering the currents as stationary currents. In these special conditions, the derivative of the electric displacement vector $\vec{d}$ is negligible compared to the current density $\vec{j}$ such that:

$$\vec{\nabla} \wedge \vec{h} = \frac{\partial \vec{d}}{\partial t} + \vec{j} \rightarrow \vec{\nabla} \wedge \vec{h} = \vec{j} \tag{5.13}$$

We can then assume that the total current density is equal to the current density only relating to the conduction such that:

$$\vec{j} = \vec{j}_C \tag{5.14}$$

This equality actually characterizes the stationary approximation of Maxwell's equations and is based globally on the same expressions of potential as in the static regime.

It concerns in particular the propagation of electricity in seawater and marine sediments. Under this approximation, potentials, electric fields, etc. can then be obtained by the same expressions as those governing the steady states, i.e., in this case, direct currents or DC (currents that do not vary with time). The main constitutive law governing these currents is *Ohm's law* (see Chap. 2, Sect. 2.4.4):

$$\vec{J} = \sigma \vec{E} \tag{5.15}$$

This local and linear relation connects the electrical field vector $\vec{E}$ to the current density vector $\vec{J}$ by the electric conductivity σ of the medium (or its inverse, its resistivity ρ). In most cases, it is possible to involve the source term in the law (cf. Chap. 2, Eq. 2.4b).

This approximation, also known as low frequency, which tends to neglect the inductive effect of currents and to keep only the galvanic effect, allows us to approach in certain circumstances DC modeling in an analytical or numerical form, thus echoing the main principles of electrostatics (static approximation) as the laws governing the distribution of potentials (see Chap. 2, Sect. 2.3.1).

3.1.2 Second Reminder: Limitation of Quasistatic Approximation

The quasistatic approximation is meaningful only if the geometry of the evaluation device, i.e., the one that includes all the acquisition systems (transmitting and receiving), is invariable. By this we mean that the device must not be deformed in the space or that the relative position of the transmitter and receiver does not vary.[22] In other words, the electrical field varies much more slowly than the duration of its travel, i.e., in practice at a work frequency of 1 Hz, for example, the transmitter speed must be less than 1 m/s (about 2 knots, currently ship speed), which is not a problem of time when the trap is precisely located in space (by the seismic survey), while maintaining a sufficient resolution. Otherwise, the interpretation by DC models cannot be applied[23] to alternating and variable low frequency currents (Patra and Mallick 1980).

Of course, except in special cases (using specific injection configurations) the quasistatic approximation can only be considered for prospecting methods using amplitude or attenuation measurements, etc. It cannot be considered for methods using as a source of information the phase shift (frequency domain) and the time (time domain), measures that remain fundamental to differentiation and separation of the primary and secondary effects due to the phenomenon of induction and also for taking into account the reservoir as a wave guide.

3.1.3 Third Reminder: Propagation in Dielectric Media

In dielectric media (insulating) such as oil contained in the reservoir, the displacement currents can be considered as superior to conduction currents. We have then:

$$\vec{J} = \vec{J}_D \tag{5.16}$$

In this case there are physical phenomena of relaxation affected in the time domain that no longer allow us to apply Ohm's law. In these circumstances, the use of Maxwell's equations[24] for modeling is then needed.

3.1.4 Fourth Reminder: Equation of Diffusion

At low frequencies, when the ratio of the product of the angular frequency ω and the dielectric permittivity ε to the electric conductivity σ is very small,[25] i.e., when we have:

$$\omega\varepsilon/\sigma \ll 1 \tag{5.17}$$

[22]Under these conditions, we introduce the time factor, which must be strictly considered.

[23]In these cases, the devices then undergo phenomena of induction (a temporal change of flux, creating secondary currents).

[24]It was in 1932 that Maxwell equations were introduced (for the first time) in the interpretation of geophysical data (Peters and Bardeen 1932).

[25]The theory is recalled in the Appendix to Chap. 2 (cf. Eq. A2.1.4).

we use then the so-called diffusion equation:

$$\nabla^2 \vec{e} = \mu_0 \sigma \frac{\partial \vec{e}}{\partial t} \tag{5.18}$$

such as was defined in the Chap. 2 (see Sect. 2.5.8).

3.1.5 Complexity and Choice of Models

The diversity and complexity of the propagation media (imperfect dielectrics and conductors) do not allow us a priori to develop a single model. The already very complex analytical or numerical modeling in DC prospecting cannot be directly transposed into alternating current exploration. However, for lack of anything better, some models developed from theories for continuous currents, such as those described in the paragraphs that follow and in the Appendices A5, can be adapted with relative precision, with some factual tricks.

This is the case, for example, in the assessment of the thickness of sedimentary conductive layers placed above resistant *bedrock*[26] (which can be compared to a resistive thick layer) or even for the detection and location of a conductive or insulating anomaly (disk, sphere, cylinder, etc.) placed in a conductive medium or an electrolyte. In these particular cases, the simulations led from these models are in agreement with the measures whether they are field or experimental measurements (rheological tank). However, these simple models do not allow comprehensive lighting of the anomaly of conductivity at depth. It is then necessary to perform other approaches.

The interpretation of a resistivity profile is obtained mostly by data from a nondirectional device (e.g., an acquisition line, called a flute, where usually measures and injection points, defined by some electrodes, are aligned and dragged on the bottom of the sea; with few exceptions, the device has a fixed geometry → quasistatic approximation).

The interpretation then corresponds to an estimation of the thickness of superimposed geological horizons, delivered according to the vertical distribution of the resistivities associated with these underlying layers.

In fact the geophysical apparatus can only provide an apparent resistivity at each measuring point—a kind of average resistivity—integrating in depth the parameters of the different geological layers. In these situations, the mathematical calculation then establishes—depending on various values of apparent resistivity, estimated (based on the values of potential difference ΔV_{MN} and of current intensity I_{AB}) and

[26]Infinitely resistant layer, with also infinite thickness, being under the sedimentary substratum. This case answers the question of the presence or of the absence of hydrocarbons. It doesn't give any information about the layer thickness.

calculated (such as $K \times \Delta V_{MN}/I_{AB}$)—a geometric model in order to follow, according to a predetermined profile (the measures), associated variations in resistivity and thickness.

This model, to have a physical meaning, also has to define itself by a specific depth of investigation, which depends then on the geometry of the evaluation device (spacing between the various electrodes forming the measurement quadrupole and expressed by the K factor). Changes in K and especially in the current injection dipole then allow us to reach more or less deep layers, developing what is called an electrical sounding.[27] This type of investigation (acquisition and interpretation of geophysical criteria), initially associated with the assumption that a geological layer has laterally a substantially equal thickness (geological criterion), thereby allows us to get information only on the vertical component of the profile (1D model) and therefore is not suitable for the detection of discrete geological objects (deposits, massive sulfides, veins, etc.) defined in particular by their side dimensions (2D model).[28]

3.2 *Theory of Electrical Images*

One of the first ideas that made it possible to construct a synthetic image of the underlying seabed was that of electrical images[29] formerly developed to interpret onshore ground electrical vertical soundings or EVs.[30] This technique is still used (Grellier 2005).

It is particularly effective:

- When the model (tabular) has a conductive cover, such as the sea, which tends then to hide the presence of conductivity heterogeneities in depth

---

[27]Provided that the depth of water does not exceed a priori, according to some authors, 15 m (Lagabrielle et al. 2001).

[28]It is possible to acquire this information by performing crossword profiles.

[29]Theory introduced by Lord Kelvin in 1845 for electrostatics (Thomson 1884; Ramsey 1937) and more generally in electrodynamics by Maxwell in 1848, for which a definition of the electrical image is given on page 246 of his treatise on electricity and magnetism (Maxwell 1892).

[30]This was the first interpretation method employed in the 1930s by the German Hummel (series expansion), showing that under the reciprocity theorem (see Chap. 2, Appendix A2.2) the question to be addressed was similar to an optical reflection problem. This technique was proposed for the first time by the physicist Thomson (1845–1884). It preceded the methods of the Schlumberger school and particularly those developed by Sabba Stefanescu using, in particular, integration methods for a half-space:

$V(r) = RI/2\pi r\ [1 + 2r \int A(t) J_o(rt) dt]$.

Although less elegant than the latter, the Hummel series is more suitable for numerical calculations. We do not detail here the Stefanescu/Schlumberger method, which the reader will find in all books on applied geophysics more particularly about electrical methods (Kunetz 1966; Telford et al. 1978) or more generally about wave propagation in a stratified medium (Bannister 1968).

– When it is required, for example, to accelerate the process of numerical calculations in the interpretation of any heterogeneities present in stratified subsoil, of which the distribution is more or less known (Poirmeur 1986).

Details of the method for the specific interpretation of submarine surveys (acquisition devices immersed in seawater) can be found in the Appendix 5.1.

3.3 Method of Abacuses (1D Formulation)

For the interpretation of tabular models[31], different methods have been suggested in the past by many geophysicists (Stefanescu and Schlumberger 1930; Flath 1955; Kunetz 1966; Koefoed 1968). Other researchers (Terekhin 1962; Patella 1974) have made calculations more specifically for marine environments. These resolution methods partly stopped being practiced in the early 1980s with the advent of numerical methods using powerful computers (HP 1000 Digital series, VAX™, CRAY™, etc.) or later workstations (Apollo™, Sun™, etc.) working under operating systems like Unix™, for example.

Generally, the integration of the Laplace equation is performed in two steps:

– Firstly, by calculating a primary field, i.e., the one that would be in a homogeneous medium (including the source)
– Secondly, by calculating a secondary field that would produce a conductivity anomaly, considering at this time only the continuity conditions at the separation surfaces (source not included)

Through this discriminating process, singularities that specifically characterize the source are then eliminated.

These models are based on mathematical formulations, and more specifically on appropriate differential equations, whose solutions after resolution allow us to build abacuses on which the acquisition diagrams[32] are superimposed for different

---

[31]The problem of calculating the response of a laminate subsoil to electromagnetic solicitation was resolved early in the twentieth century (Sommerfeld 1909; Weyl 1919).

[32]*Reminder:* we can distinguish by electrical sounding only sets powerful enough to affect diagrams in a measurable way. The most favorable structures in this regard are series formed by alternating resistant and conductive sedimentary layers whose thickness increases with depth. The diagram is then formed by a sequence of hollows and bumps that allow us to read:

– The horizontal conductances (h/ρ) of the conductor sets
– The transverse resistances ($h\rho$) of the resistant sets

Each survey thus provides:

– At one point, valuable information on the conductance of the subsoil at this point
– From one point to another, the changes (deformation of the diagram) in conductors and resistants in space, i.e., in the thicknesses if the facies do not vary

superposed layers. These are then used either manually as in the past (superimposing of data) or by a computer[33] by successive and progressive semi-automatic adjustments (the least squares method, for example). These models can also be used in inversion data programs and in numerical packs when the researched solutions match simple models (superposition of horizontal layers among other).

3.3.1 Horizontal Devices with Controlled Sources

The potential distribution in a stratified medium composed of a stack of conductive and more or less thick layers, well delimited according to a good resistivity contrast, can be analytically solved. This relates to the *in-line* configuration, i.e., where the transmitter and the receiver are in the same vertical plane. This type of model can only applied for a thick resistive layer.

The potential in each layer then meets the Laplace equation $\nabla^2 \mathrm{V} = 0$, such that in cylindrical coordinates (r, z) we have:

$$\frac{\partial^2 \mathrm{V}}{\partial r^2} + \frac{1}{r}\frac{\partial \mathrm{V}}{\partial r} + \frac{\partial^2 \mathrm{V}}{\partial z^2} = 0 \tag{5.19}$$

Applying the classical method of separation of the variables r and z with:

$$\mathrm{V}(r, z) = \mathrm{R}(r)\,\mathrm{Z}(z) \tag{5.20}$$

we in general obtain solutions of the form:

$$\mathrm{e}^{-\lambda z} J_0(\lambda r) \quad \text{and} \quad \mathrm{e}^{+\lambda z} J_0(\lambda r) \tag{5.21}$$

where λ is an arbitrary constant and J_0 is the Bessel function of the first kind and zero order.

The linear combination of these solutions, multiplied by the undetermined functions of λ, gives the expression of the potential in any point of the stratification such that:

$$\mathrm{V_i} = \int_0^\infty \left[\mathrm{A_i}(\lambda)\,\mathrm{e}^{-\lambda z} + \mathrm{B_i}(\lambda)\,\mathrm{e}^{+\lambda z}\right] J_0(\lambda r)\,\mathrm{d}\lambda \tag{5.22}$$

The functions $\mathrm{A_i}$ (λ) and $\mathrm{B_i}$ (λ) are obtained thanks to the criteria that define the boundary conditions in the periphery of the electrodes and on the interfaces between the different natural media.

[33]First (in the 1960s) implemented in big systems (IBM 1620), mathematical models were gradually installed (in the 1970s) on portable microcomputers of the HP85 type (64KB of memory, expandable to 128KB using a specific programming language: HP basic) and then the compatible PC type (in the 1980s).

Thus, in a marine model, consisting, for example, of two layers of conductivity σ_1 (seawater) and σ_2 (marine sediments), the potential at any point of the stratification, distant by r from the source supplying a current intensity I, is given by the following integral:

$$V = \frac{I}{4\pi} \frac{2}{(\sigma_1 + \sigma_2)} \left\{ \frac{1}{r} + \int_0^\infty [A_1(\lambda) + B_1(\lambda)] J_0(\lambda r) \, d\lambda \right\} \tag{5.23}$$

Under these conditions, after calculations (cf. Appendix A5.2), the apparent resistivity has the value:

$$\rho_a = \frac{4\pi r^2}{I} \frac{\partial V}{\partial r} \tag{5.24}$$

This type of approach is also suitable for models in three layers (seawater, marine sediments and deep substratum), the last one of which is often considered as very resistant and of infinite depth (bedrock or a thick reservoir). On the same principle, more sophisticated analytical models consisting of a greater number of geological horizons were devised in the years 1960 to 1970 (Terekhin 1962; Patella and Schiavone 1974).

3.3.2 Vertical Devices with Controlled Sources

The same approach can be conducted for vertical geometry investigation devices.

In this case the potential in seawater (medium 1) is given by:

$$V_1 = \frac{I}{4\pi\sigma_1} \left\{ \int_0^\infty e^{-\lambda|z - z_0|} J_0(\lambda p) \, d\lambda + \int_0^\infty (A_1 e^{-\lambda z} + B_1 e^{\lambda z}) \, J_0(\lambda p) \, d\lambda \right\} \tag{5.25}$$

If we now pay more attention to the concrete case of marine exploration, i.e., when measuring a potential difference ΔV or a vertical field along the direction of z, we have then in this specific situation according to the different depths (see Appendix A5.3):

$$\Delta V = \frac{I}{4k\pi\sigma_1} \int_0^\infty e^{-\lambda(H - z_1 - Z)} \times \frac{1 - e^{-2\lambda Z} - k\left[e^{-\lambda(z_1 - z_2)} - e^{-\lambda(z_1 - z_2 + 2Z)} - e^{-\lambda(z_1 + z_2)} - e^{-\lambda(z_1 + z_2 + 2Z)}\right] - e^{-2\lambda z_1} + e^{-2\lambda(z_1 + Z)}}{1 + k^{-1} e^{-\lambda(H - h)} - e^{-2\lambda H} - e^{-2\lambda(H - h)}} d\lambda \tag{5.26}$$

The method requires the use of a measuring line (see Appendix A5.3, Fig. A5.1) whose symmetry must be perfect. Indeed, the potential difference due to an asymmetry of the device in seawater (AM: injection point A and measurement point M) is equal to:

$$\Delta V_d = \frac{I}{4\pi\sigma_1}\frac{\Delta(AM)}{AM^2} \tag{5.27}$$

an amount that can then no longer be considered negligible in this specific configuration.

For example (in a case study), this leads, for a misalignment of 1 cm on an AM length of 1 m, and for an injection of 1 A, to a ΔV_d of 0.2 mV then equivalent to a vertical movement of the line of approximately 50 cm.

3.3.3 Magnetotelluric Device

Magnetotelluric investigation requires us to measure one of the components of the electric field E_x and magnetic field E_y horizontal vectors in two perpendicular directions x and y to obtain the apparent conductivity (see Chap. 2, Eq. 2.65).

$$\sigma_a = \omega\mu_0 \left|\frac{H_y}{E_x}\right|^2 \tag{5.28}$$

From Maxwell's equations, a system of equations (see Appendix A5.5) is then formed, which allows us to go back to the specific resistivities of the layers according to depths z, such that we obtain for both components of these fields:

- For the layer n-1:

$$\begin{cases} B_x = \left(A_{n-1}e^{-k_{n-1}z_{n-1}} + B_{n-1}e^{k_{n-1}z_{n-1}}\right) \\ E_y = \dfrac{k_{n-1}}{\mu_0\sigma_{n-1}}\left(-A_{n-1}e^{-k_{n-1}z_{n-1}} + B_{n-1}e^{k_{n-1}z_{n-1}}\right) \end{cases} \tag{5.29}$$

- For the layer (n−1)−n:

$$\begin{cases} B_x = A_n e^{-k_n z_{n-1}} \\ E_y = -\dfrac{k_n}{\mu_0\sigma_n}A_n e^{-k_n z_{n-1}} \end{cases} \tag{5.30}$$

The interpretation of the magnetotelluric underwater soundings for tabular structures has been treated so far in the same manner as that used for measurements performed onshore. However, we can now find different approaches to analyze more difficult cases especially in 2D and 3D configurations (see Appendix A5.6).

3.4 Analytical Model of the Sphere (3D Formulation)

Modelings such as an electrical image or other, developed for exploration of the seabed, concern tabular models (stacking layers of different resistivity), generally corresponding to a resistivity distribution only in the vertical plane (1D model). Now, if you want to have lateral information, i.e., in the horizontal plane, then it is necessary to design the model in 2D or even in 3D (2D + depth). In this case, the analytical modeling can only be envisaged with elementary geometric patterns, isometrics such as spheres, disks, cylinders, and plates in particular being more similar to a resistive infinite layer (using in addition specific investigative devices).

To simplify the resolution problem, when the geological object does not present in its form singular morphological characteristics (the ratio of the sizes between them, for example), we commonly use structures whose axes of symmetry can match the model. This is the case, for example, with cylindrical symmetries simulating anticline folds. To go even further, we can imagine that the target has finite dimensions, characterized by an envelope on which all points, for example, are equidistant from its geometric center. This is the unique case of the sphere, which considerably simplifies the calculations and makes it possible to obtain excellent similar lateral resolutions in 3D.

In what follows, contrary to what is generally found in the literature, namely responses due to effects concerning uniform field excitations, i.e., corresponding to sources placed to infinity (Zaborovski 1936; Hansen et al. 1967; Telford et al. 1978), a solution is given in continuous mode (fixed geometry device → quasistatic approximation) in reaction to "punctual" stress materialized in this case by a discrete emissive dipole (finite dimension antenna).[34] This type of modeling can have interest, for example, for evaluation of the galvanic contribution compared with that due to the vortex effect or even the modeling of abnormal fields due to massive conductive anomalies in oil fields (wellhead, pipeline manifold, etc.).

As a general rule, and briefly, the resolution of such a problem is made in three stages:

- Firstly, by calculating the (primary) field off the anomaly corresponding to the response in a homogeneous and isotropic infinite medium for each contribution (injection point)
- Secondly, by considering this time the anomaly and including in addition some particular singularities (potential in $1/r$ in the near the source, for example)
- Thirdly, by summing the potentials

The existence of this heterogeneity finally produces an additional (secondary) field, which will not need to replicate in the medium the singularities relating to the

[34]The theoretical response of a spherical anomaly in an alternative regime is extremely complex to carry out. Therefore, as a first approximation (low frequency approximation), we prefer calculation in a continuous regime (Telford et al. 1978). Under these conditions, the equation of diffusion is reduced to the Laplace equation.

source but which is, however, subject to the boundary conditions (continuity conditions at the surfaces of separation between the two media).

The interpretation of the variations of abnormal fields can then be conducted in different ways either by comparing the measurements with a pre-established theoretical model (called a forward problem) or, when possible, by using an inversion model (dated back to the original parameters).

One of them (the forward problem) consists of solving (analytical method) a partial differential equation in a coordinate system adapted to the problem posed by introducing in addition conditions on the source[35] and on the limits on both sides of the different interfaces (resistivity contrast).

The simplest approach to move into 3D electromagnetic modeling is calculation of the anomaly caused by the presence of an isotropic sphere placed in a uniform electric field and immersed in a homogeneous medium of different resistivity.

We provide here an example of a solution after solving the *Laplace equation* ($\nabla^2 V = 0$) in a spherical coordinate system, for a sphere of radius R, resistivity ρ_s, immersed in a medium of resistivity ρ_m, with distances of r_o between its center and the injection point(s) and of r between its center and the measuring points (Sainson 1984). In these specific conditions (see Appendix A5.4), in low frequency approximation, the potential V_a due to this anomaly is equal to the difference between the potentials with and without the anomaly such that we have:

$$V_a = \frac{\rho_m I}{4\pi r_o} \sum_{n=1}^{\infty} \frac{n(\rho_s - \rho_m)\, P_n(\cos\theta)\, R^{2n+1}}{r_o^n [n(\rho_m + \rho_s) + \rho_s]\, r^{n+1}} \tag{5.31}$$

where $P_n(\cos\theta)$ then represents the *Legendre polynomials* of order n and θ as the angle between r and r_o.

For field variations following small angles θ, the expression of the value of the transverse component (perpendicular to the injection line AB and compared to the current density J), in the direction of the anomalous body (z), is, by far, in the detection, the most significant:

$$E_{T(z)} = -\frac{\partial V}{\partial y} J \tag{5.32}$$

Considering two points of injection of current A and B relatively close, such that $[A, B] \perp E_{T(y)}$, the electric transverse component of the anomalous field is then equivalent to the sum of the transverse components of the fields induced by each point A and B such that:

$$E_{T(z)} = E_{T(z)}^{A} + E_{T(z)}^{B} \tag{5.33}$$

or even:

[35]It may be assumed that the injection of an alternating current whose injection poles (EM source) are opposed in phase, for example, is similar to the injection of a polarized DC (+/−).

$$E_{T(z)} = \overbrace{\frac{-\rho_m I}{4\pi r_{OA}} \sum_{n=1}^{\infty} \frac{n(\rho_s - \rho_m) R^{2n+1}}{r_{0A}^n [n(\rho_m - \rho_s) + \rho_s]} \left[\frac{\partial P_n(\cos\theta)_A}{\partial(\cos\theta)_A} \frac{\partial(\cos\theta)_A}{\partial y} \frac{1}{r^{n+1}} + P_n(\cos\theta)_A \frac{\partial}{\partial y} \frac{1}{r^{n+1}} \right]}^{E^A_{T(z)}} + E^B_{T(z)} \tag{5.34}$$

This expression gives, after calculation of the derivatives, the values of the variations in the anomalous transverse field, directly usable for comparative interpretation of the data after correction and calibration.

Identical modelings in Cartesian or cylindrical coordinate systems can be similarly conducted for tabular or semicylindrical geometries corresponding to different geological architectures (vein, stratum, lens, anticline, dome, etc.).

This type of modeling is all the more effective when the resistivity contrast is important. The resolution is maximal when this is considered infinite—that is to say, more precisely, when the resistivity anomaly is zero or infinite (perfect conductors or insulators).

With an equivalent resistivity contrast, the conductive sphere has an anomalous response greater than the resistant sphere. For the latter, the detectability threshold lies at a depth approximately two times lower than that of the conductive sphere (Sainson 1984).

This type of modeling would be particularly well adapted, for example, to research into a conductive anomaly on the seabed (Andreis 2008).

3.5 Conclusion on Analytical Models

Analytical interpretation models, in the absolute, may seem outdated or obsolete. However, some of them are still used today (Bailey 2008) because they are the only ones to provide accurate solutions to the problems, provided that they remain confined to relatively simple structures and that the phase information is essential (see Sect 3.1.2).[36]

Furthermore, analytical models can provide, in some cases, important contributions as, for example, in:

- Evaluation of a method, technique or experimental material
- Equipment calibration, measure correction
- Comparison and certification of results from different types of modeling (numerical, inversion, etc.)

[36]This can be modeled by arranging some subtleties (see Appendix A5.4).

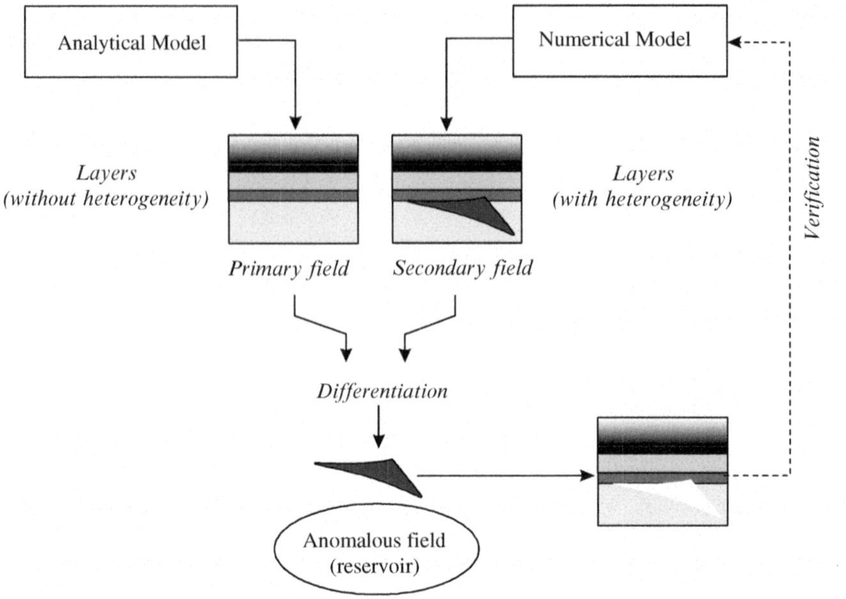

Fig. 5.11 Example of mixed models, juxtaposing an analytical model (tabular with *n* layers) and a numerical model (+ heterogeneity) to greatly shorten the time calculations

- Interpretation of isolated resistivity data
- Interpretation of simple geological structures (tabular configurations, for example)

Of course, these models provided for interpretation in continuous, inverted or pulsed current (skin effect neglected) may also be involved in the interpretation of investigations carried out at low frequencies (quasistatic approximation).

Finally, as we shall see later, they can be included in chains of numerical software as supplementary calculations (subroutines), making the execution times shorter, thus limiting iterations (stratified medium) (Fig. 5.11).

4 Numerical Models

Despite the strong integration of the method, the geometry of hydrocarbon reservoirs cannot be reduced to spheres, disks, cylinders or plates. For geological interpretation of acquisition data, it is then necessary to have access to models with geometries in accordance with gitology, especially if we use then inversion processes stalled on seismic data, for example.

However, since the 1980s, we have still used in the frequency domain 1D models for rapid data interpretation (Chave and Cox 1982) and in the time domain some models elaborated by groups such as the Toronto team (Edwards and Chave 1986; Cheesman et al. 1987; Flosadottir and Constable 1996). This type of simplified approach is particularly useful if we are looking for an acquisition located close to a tabular anomaly (a relatively representative case within the limits of current prospecting).

Beyond the second dimension and in some cases the first dimension, even for simple models, analytical analysis becomes too hard to provide a solution. We use then *numerical analysis*, which has much greater flexibility when complex structures in several dimensions are approached, but presents on the other hand approximate solutions that require, to be useful, a rigorous approach (well-defined model) and the important calculation means that only powerful computers and, more particularly, supercomputers offer. By these means, it is possible then to consider obtaining frequency information (phases). Concerning the 2D electric dipole problem, numerical models were developed at the same time (1993) for the time domain by Everett and Edwards and for the frequency domain by Unsworth et al. The first 2D inversion of real data was proposed in 2000 by MacGregor et al. (2001). A finite element forward code has been distributed by Professors Li and Key (Li and Key 2007). A 2D finite difference forward program with inverse code was written by Abubakar (Abubakar et al. 2008). Since 2008, a lot of codes have been written by different contractors and tested with a canonical oil field model but have not been completely validated in the field. These codes are propitious to detect resistive thin layers.

The reader well trained in mathematical analysis and, in particular, in numerical analysis may consult the numerous papers listed in the complementary bibliography and see the evolution of the technique in the list of geophysical journals located at the end of volume.

4.1 Generalities on Numerical Resolution Methods

In general, numerical models are conceived in several stages. In the interpretation of electromagnetic surveys, finite difference methods (FDm), finite element methods (FEm) and, more recently, integral techniques (MIt) have been mainly used.

For FDm, the model is chronologically built from:

- A step of dividing the propagation medium in a grid and nodes taking into account the different heterogeneities (and particularly their size and geometry), a mesh on which the finite difference method is then applied
- A step of approximating the differential equations using the finite difference, by linking the value of a variable (solution) at a point of the mesh to the neighboring points
- A step of solving equations by considering the initial conditions and boundary conditions on the considered domain

For FEm, the model is chronologically built from:

- A step of meshing (geometric division of space) and of discretization of the medium in similar elements

- A step of deriving for each point
- A step of assembling all the elements (solutions) in the concerned domain
- A step of solving the equations.

For MIt, the model is chronologically built from:

- A step of choosing the initial medium and the heterogeneities
- A step of surface meshing (geometric division of the space) and of discretization of the medium into similar elements,
- A step of resolution of the linear system (calculation, matrix filling, etc.)
- A step of calculating the surface potential anomaly at the measuring points

These models for solving partial differential equations provide approximate solutions, which must meet certain criteria such as uniqueness, stability, etc. These criteria can be put into default by the presence of previous errors that can occur in the modeling, the discretization, etc., then causing some problems of convergence of the system.

To not overload the presentation and excessively complicate reading, we give in what follows an example of resolution of the problem by the method of integral equations, which seems to be more and more used today. The interested reader, with good mathematical literacy, may consult the extensive literature devoted to all these methods (FDm, FEm and MIt in particular), thanks to the bibliographic references at the end of chapter (Key 2007). An excellent introduction to these methods is also given in the work of Professors Patra and Mallick (Patra and Mallick 1980) (Fig. 5.12).

4.2 *Integral Equation Method: An Example of Numerical Resolution*

Several numerical methods have been developed and a number of them can also be applied in the context of seabed logging, such as those developed in the USA

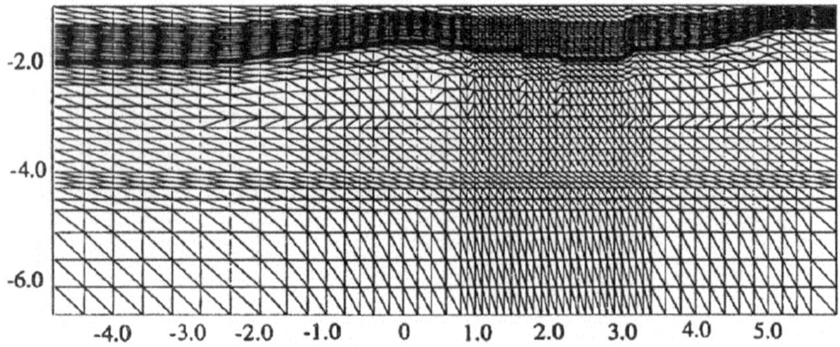

Fig. 5.12 Example of meshing of the geological structure. According to MacGregor et al. (1998)

(Hohmann 1989). Many of them are based on the calculation of two resulting fields, one a primary field, corresponding to the response of the subsoil without any anomalous heterogeneity (homogeneous or not), the other a secondary field, only containing the latter (Eskola 1992). The differentiation of the two responses then provides a response that is anomalous in phase and amplitude. An example of numerical modeling by integral equations is given in Appendix A5.7.

More recently, the Norwegian School (University of Trondheim) proposed a numerical method for the resolution of *Maxwell's equations*, specifically adapted to investigations on the seabed, with a formalism of integral equations (forward problem) followed by a nonlinear data inversion. It finally delivers (at the receptor level) some values of the electromagnetic field diffracted by the conductivity anomaly (reservoir). The integral formulation implies in this case an equation of the *Lippmann–Schwinger* type, which associates, via a *dyadic Green's function*[37] of the anomalous structure, the incident electromagnetic field (at the transmitter level) existing in the anomaly (Morse and Feshbach 1953; Wiik et al. 2010).

The medium is thus defined by three electromagnetic properties, which are in this case its electrical conductivity, its dielectric permittivity and its magnetic permeability. Calculations are performed first in the time domain.

- The electrical conductivity must reflect the spatial disparities (anisotropy). It is defined in the horizontal plane ($\sigma_{0,h}$) and in the vertical plane ($\sigma_{0,v}$) by a matrix $\boldsymbol{\sigma}_0$ such as:

$$\boldsymbol{\sigma}_0 = \begin{pmatrix} \sigma_{0,h} & 0 & 0 \\ 0 & \sigma_{0,h} & 0 \\ 0 & 0 & \sigma_{0,v} \end{pmatrix} \quad \text{or} \quad \begin{pmatrix} \Upsilon & 0 & 0 \\ 0 & \Upsilon & 0 \\ 0 & 0 & 1 \end{pmatrix} \tag{5.35}$$

- The permeability μ and the permittivity ε are given compared with those of a vacuum (cf. Chap. 3, Sects. 2.2 and 2.3) such that:

$$\varepsilon = \varepsilon_0 \mathbf{I} \tag{5.36}$$

and:

$$\mu = \mu_0 \mathbf{I} \tag{5.37}$$

where **I** is the identity matrix.

The spatial dependence of these parameters is provided by Ohm's law such that:

$$\vec{J} = \boldsymbol{\sigma} \vec{E} \tag{5.38}$$

[37]See Professor Tai's work (Tai 1971).

The electric field $\vec{E}$ and magnetic field $\vec{H}$ also satisfy Maxwell's equations such that:

$$\vec{\nabla} \wedge \vec{E} = -i\omega\mu_0 \vec{H} \tag{5.39}$$

and:

$$\vec{\nabla} \wedge \vec{H} = \tilde{\sigma}_0 \vec{E} + \vec{J}^S \tag{5.40}$$

where the complex conductivity is $\tilde{\sigma}_0 = \sigma_0 - i\omega\varepsilon_0\mathbf{I}$, ω is the pulsation, $i = \sqrt{-1}$ and $\vec{J}^S$ corresponds to the source of the electric current.

The vector nature of the electric and magnetic fields allows us to write Green's function in a tensor form[38] (Green 1850). Under these conditions, $\underline{G}^E$, $\underline{G}^H$ Green's tensors, which represent the electric and magnetic fields at a point (positioned in x), from a source point (placed in x_0), are then the solutions of the equations:

$$\nabla \wedge \underline{G}^E(x, x_0) = -i\omega\mu_0 \underline{G}^H(x, x_0) \tag{5.41}$$

and:

$$\nabla \wedge \underline{G}^H(x, x_0) = \tilde{\sigma}_0 \underline{G}^H(x, x_0) + \mathbf{I}\boldsymbol{\delta}(x, x_0) \tag{5.42}$$

where $\nabla\wedge$ corresponds to the dyadic operator and $\boldsymbol{\delta}$ is the Dirac function.

The solutions to Eqs. (5.39) and (5.40), for the fields $\vec{E}$ and $\vec{H}$ according to the Green's tensors (Zhdanov 2002b) are then equal to:

$$\vec{E}_i(x) = \int_{\mathfrak{R}^3} \underline{G}^E_{ij}(x, x')\, \vec{J}^S_j(x')\, dx' \tag{5.43}$$

and:

$$\vec{H}_i(x) = \int_{\mathfrak{R}^3} \underline{G}^H_{ij}(x, x')\, \vec{J}^S_j(x')\, dx' \tag{5.44}$$

with $i, j \in \{x, y, z\}$, according to the Einstein convention of notation on vectors and tensors.

Green tensors depend on the complex wave number k (where $k = \sqrt{i\omega\mu\tilde{\sigma}}$), whose sign must be chosen to satisfy the Sommerfeld radiation condition[39] (Sommerfeld 1912; Colton and Kress 1992; Zhdanov 2002a, b, c).

[38]Also called a *dyadic Green's function*, this function then connects the scalar equations with vector equations. The Green functions, sensu stricto, are used among others in the methods of numerical resolution by integrals (Roach 1970; Eskola 1992). Their definition is given in Footnote 28 of Chap. 3: Metrology.

[39]The energy of the source must disperse to infinity (no return).

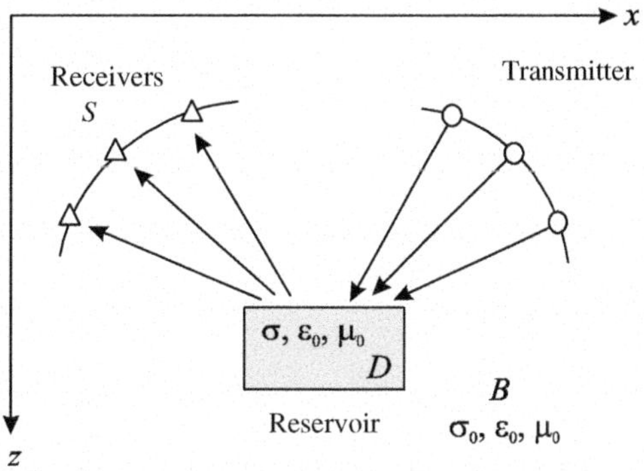

Fig. 5.13 Schematic diagram where the incident fields undergo a dispersion in contact with a conductivity anomaly ($\sigma \neq \sigma_0$)

If we now consider a medium (σ_0, ε_0, μ_0) in which there is a heterogeneity (reservoir) with a different conductivity (σ), we can calculate the EM incident fields $\vec{E}^{inc}$ and $\vec{H}^{inc}$ with the above Eqs. (5.43) and (5.44) (Fig. 5.13).

In the domain D (anomaly), the electric field satisfies the following equation (Abubakar et al. 2004):

$$\vec{E}_i(x) = \vec{E}_i^{inc}(x) + \int_D \underline{G}_{ij}^E(x, x')\, \sigma_{0,v}(x')\, \chi_{jj}(x')\, \vec{E}_j(x')\, dx' \quad \text{with } x \in D \tag{5.45}$$

This equation is equivalent to that of *Lippmann–Schwinger* for the *Helmholtz* scalar equation (Colton and Kress 1992). It is the sum of the incident electric field (first term) and the diffracted field (second term: integral). The contrast is given by the following matrix:

$$\boldsymbol{\chi} = \begin{pmatrix} \chi_h & 0 & 0 \\ 0 & \chi_h & 0 \\ 0 & 0 & \chi_v \end{pmatrix} \tag{5.46}$$

where $\chi_h = (\sigma_h/\sigma_{0,v}) - \Upsilon$ and $\chi_v = (\sigma_v/\sigma_{0,v}) - 1$ (referencing Eq. 5.35).

When the field is established in the domain D, the diffracted EM fields can be calculated at the receptor level in the domain S such that:

$$\vec{f}_i^E(x) = \int_D \underline{G}_{ij}^E(x, x')\, \sigma_{0,v}(x')\, \boldsymbol{\chi}_{jj}(x')\, \vec{E}_j(x')\, dx' \quad \text{with } x \in S \tag{5.47}$$

and:

$$\vec{f}_i^H(x) = \int_D \underline{G}_{ij}^H(x, x')\, \sigma_{0,v}(x')\, \boldsymbol{\chi}_{jj}(x')\, \vec{E}_j(x')\, dx' \quad \text{with } x \in S. \tag{5.48}$$

For simplicity, we can introduce the vectors $\vec{E} = \left(\vec{E}_x, \vec{E}_y, \vec{E}_z\right)^T$, $\vec{f}^E = \left(\vec{f}_x^E, \vec{f}_y^E \vec{f}_z^E\right)^T$, $\vec{f}^H = \left(\vec{f}_x^H, \vec{f}_y^H \vec{f}_z^H\right)^T$ and then rewrite the above equations, ultimately giving according to the matrix $\boldsymbol{\chi}$:

$$\vec{E} = \vec{E}^{inc} + G^{E,D}\chi\vec{E} \tag{5.49}$$

and:

$$\vec{f}^{E} = G^{E,S}\chi\vec{E} \tag{5.50}$$

and:

$$\vec{f}^{H} = G^{H,S}\chi\vec{E} \tag{5.51}$$

where $G^{E,D}, G^{E,S}, G^{H,D}$ are the integration operators.

4.3 Extrapolation Methods

To simplify the processing of data and the interpretation that follows, some authors as Dr MacGregor (2007) propose models of extrapolation reducing the initial 3D problem to the simpler resolution of a 2D or 1D problem.

For example, in the frequency domain, the extrapolation of the wave field can be reduced to solving the wave equation in 1D (a classical problem) such that:

$$\frac{\partial^2 E}{\partial z^2} = [ik(z)]^2 E \tag{5.52}$$

whose solutions, depending on the depth z for a plane wave, are of the well-known form:

$$E(z) = Ae^{ik(z)z} + Be^{-ik(z)z} \tag{5.53}$$

where A and B are the constants of scale and $k(z)$ is the complex wave number as a function of z equal to:

$$k(z) = \sqrt{\omega^2\mu_0\varepsilon - i\omega\mu_0\sigma} \tag{5.54}$$

In these circumstances, the extrapolated field can then be described by a linear combination of exponential terms as ultimately:

$$\begin{cases} E^d(z) = E^d(z=0)\ e^{i\Sigma_j k_j dz_j} \\ E^u(z) = E^u(z=0)\ e^{i\Sigma_j k_j dz_j} \end{cases} \tag{5.55}$$

where the E^d and E^u functions of z are respectively the primary field (extrapolated from the source) and the secondary field (extrapolated from the receiver) for a series of layers (from 1 to j) of thickness d_{zj} and wave number k_j.

While 1D extrapolation gives good results when we just study the conductivity variations in depth only (as in a sounding), 2D extrapolation is essential for data lateral interpretation. Several authors have then provided solutions extrapolated from the 2D field, calculated using finite difference computer codes (Claerbout 1976; Lee et al. 1987; Zhdanov et al. 1996).

4.4 Problems and Limitations of Numerical Methods

Although numerical methods are able to model anomalous heterogeneities of many and varied forms and often in 3D in a simple enough geological environment (generally homogeneous, even stratified), the same is not true when this environment is itself complex (heterogeneous, stratifications with dip or intersecting, folds, anticlines, lateral facies changes, etc.).

In these extreme mathematical demands, which actually correspond to geological reality, the use of more sophisticated methods is required. Logic dictates that they now rely on data exogenous to the electromagnetic models, which allow us to identify and describe this complicated but real environment with some precision to mathematically extract with more certainty and precision the anomalous values, i.e., those interesting the prospector. We then apply a so-called inversion method, which can come under a big diversity of techniques.

5 Inversion Techniques

The poor conditions related, strictly speaking, to the acquisition of data do not solve, except in rare exceptions, the forward problem for interpretation.

These include:

- Constraints and operational uncertainties, such as variations in bathymetry or in the position of the source (lateral displacements, elevation, rotation, etc.) or that of the sensors (orientation, inclination, etc.)
- Instrumental errors in fields measurements that are of very low amplitude
- Limitations in the number of the data and the quality of the sampling, etc.

Indeed, these techniques may not:

- Provide a single model
- Prevent the propagation of hazards and errors in the parameters of the model
- Deliver after all a solution to the problem unequivocally

It would be the same in absolute terms if we considered the inverse operation (inverse problem in the strict sense) without the provision of additional data. To compensate for this lack of information, inversion techniques therefore use external data, imported and acquired by other methods of exploration and investigation such as seismic reflection especially (MacGregor and Tomlinson 2014a).

As a more general plan, the reader can get acquainted with these specific techniques by consulting the relevant authors' books (Twomey 1977; Menke 1984, 1989; Zhdanov 2002a, b, c; Tarantola 1981, 2005; Glasko 1988; Meju 1994; Oldenburg 1992; Parker 1994; Troyan and Kiselev 2009) and, for mMt, the book of Professors Alan D. Chave and Alan G. Jones (Chave and Jones 2012).

The main limitations of these methods have recently been highlighted by several researchers (Pellerin and Wannameker 2005) (Fig. 5.14).

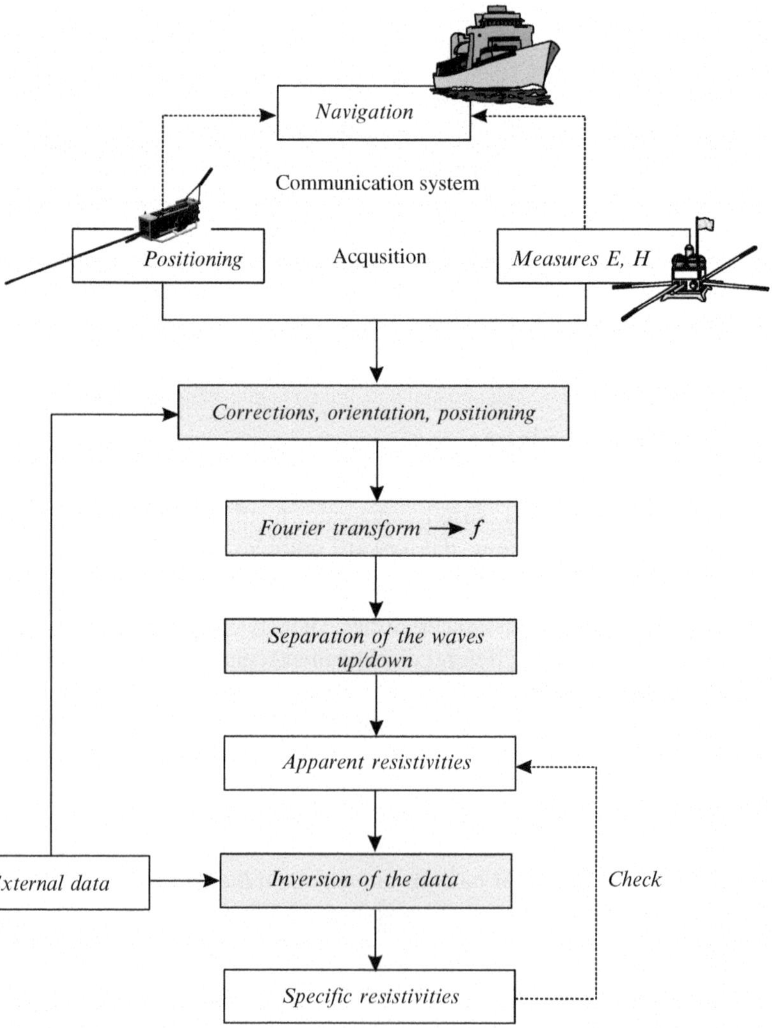

Fig. 5.14 General overview diagram: place of acquisition, processing and inversion data (interpretation) in the exploration process

5.1 Generalities on the Inversion of Data

With regard to the direct, older and less diversified methods, the processing of the inverse problem especially has been developed and intensified over the past two decades. It uses relatively complex mathematics and a very eclectic algorithm often representing some original operating points of view, frequently combining deterministic models (using the results of the forward problem resolution) and stochastic models, moreover asserting the degree of uncertainty of the resolutions. This approach is understandable, because of the versatile acquisition of the instrumentation on one hand and secondly because of the use of varied sources of data (seismic, well logging, geological surveys, etc.), forming an eclectic set that is difficult to assimilate by a monolithic system (MacGregor 2012; MacGregor et al. 2012).

But data inversion has a number of major difficulties, of which the most important are:

- The nonlinearity of the problem, related to changes in conductivity in the subsoil and especially laterally
- The existence or not of a solution
- The nonuniqueness of the solution of the problem if it exists

To resolve these troubles, theories of regulation were proposed in the 1940s to restore stability to the solution, allowing us to partly resolve these drawbacks and particularly in 3D interpretation (Tikhonov 1943; Tikhonov and Arsenin 1977).

5.2 Inversion Principle

Introduced in the nineteenth century for the study of the distribution of electrostatic fields (Maxwell 1884), and generalized in the late 1960s (Backus and Gilbert 1967), the inversion principle consists of, once the data have been measured and migrated[40] (i.e., those particularly concerning apparent resistivities based on electrical field measurements), finding local geological parameters, and more exactly those for inverted resistivities, then close to specific resistivities (Zhdanov 2015). Clearly it means, knowing the apparent conductivity, to calculate the distribution of conductivities supposed to be those of the grounds (cf. Eq. 5.2).

Inversion models are even more efficient when a priori the knowledge of the subsoil structure is accurate. The latter can be determined, for example, by multidimensional seismic survey reports (detailed seismics) that signal and then precisely describe the presence and geometry of traps (structure) and potential hydrocarbon reservoirs, and/or where appropriate by electric logs (well logging) directly offering the log of "true" resistivities.

[40]Before the inversion operation, the data undergo, if necessary, a migration (see Sect. 4.5.5.1), the results of which can then be compared to the inverted data (Zhdanov et al. 1996; Zhdanov and Portniaguine 1997; Mittet et al. 2005).

Theoretically, the purpose of the inversion is to define the parameters of an optimal model such that the sum of the residuals between the measured data d_i and calculated data $g_i(\vec{m})$ is minimal. This residual function, also referred to as an objective function, is expressed by the norm of the error vector $\vec{e}$ such that:

$$\| \vec{e} \| = \left[\sum_{i=1}^{N} | e_i |^p\right]^{p^{-1}} \quad \text{with} \quad e_i = d_i - g_i(\vec{m}) \tag{5.56}$$

When $p = 2$, the obtained norm, corresponding to the use of the least squares method (quadratic norm) absolutely assumes a Gaussian distribution of the error in the data (least squares criterion).[41]

In practical terms, the resolution of the inverse problem first requires a phase of modeling using fundamental laws of physics (deterministic part) to provide, from predetermined causes, observable effects.

Firstly, this approach corresponds to the analytical or numerical resolution of the forward problem that theoretically describes how the parameters of the (theoretical) model[42] evolve in space (or time).

Then begins a phase of comparison and adjustment (analytical or numerical) between the synthetic data (calculated from the resolution of the forward problem) and the field data (measures). This may be, for example, a minimization of an objective function (depending on differences between the measured and calculated data), squared, for example, in the case of minimizing by the least squares method.[43] The operation is carried out iteratively, as the relations linking the model parameters and the calculated data are not linear, and until the process converges (→ error criterion no longer fluctuating).[44]

To be carried out, it requires in its original definition (inversion algorithm) a deep understanding of the involved physical phenomena, the measurement equipment and the technical conditions of the operation. Indeed, even if the inverse problem has an optimal global minimum, there is no guarantee that it does not correspond then to a local minimum (Fig. 5.15).

Furthermore, the resolution of this problem is often made difficult because the acquisition data are generally not sufficient to determine all the parameters of the model (the case of the "badly put problem"). To fill these information gaps, it is then necessary to add additional constraints and some a priori from other sources of

---

[41]It is therefore important to remove the data that do not check this distribution, thereby improving the variance of the solution. For a series of noisy data ($p = 1$ or less), other laws are applied.

[42]We get an idea of the theoretical model, thanks to external elements supplied by seismic or geological data.

[43]The methods of Levenberg–Macquardt (see below) or those using Gauss/Newton-type optimization techniques, for example, can also be used (Chave and Jones 2012; Amaya 2015). These algorithms (Marquardt 1963) are often complemented by regulation and smoothing techniques to stabilize the inversion process. The latter are based on calculation of the sensitivity matrix at the measurement point. This matrix is calculated numerically by methods such as those for the moments, the finite differences or finite elements.

[44]The credibility of the model with regard to the measurements is determined by the mean square error or RMS (root mean square).

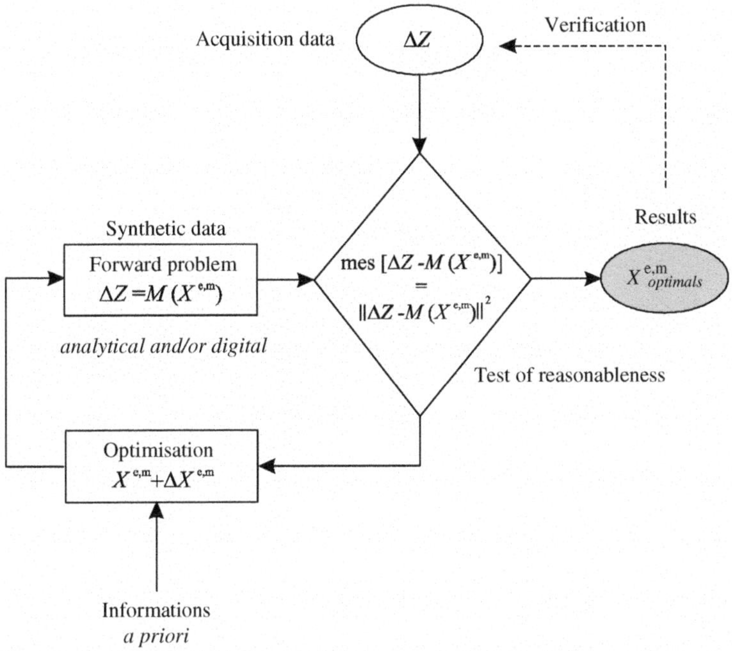

Fig. 5.15 Simplified organizational chart of the principle of data inversion (inverse modeling) built around a plausibility test, including at least a forward modeling of the studied physical phenomena, optimized by providing external information such as seismic data, for example

quantifiable data (geophysical, petrophysical, etc.), or even from data and information coming from experiments and possibly case histories (geological databases). In these particular conditions, the solution can then move toward uniqueness without however reaching it. For this reason, in recent years, various stochastic methods have been developed, thus allocating to the studied models a greater or lesser degree of certainty (probabilistic law) allowing us to accept or reject the proposed model (maximum-likelihood test).

Now we can finally admit that the model is a plausible representation of the geological reality because it can generate calculated apparent resistivities very close to the measured resistivities (Fig. 5.16).

5.3 *Inversion Methods*

There are thus many methods of resolution of the inverse problem, or inversion techniques.[45] The most common use:

[45]We can say, as Professor Claerbout pointed out in his book, that there are as many inversion methods as there are article writers (Claerbout 1992).

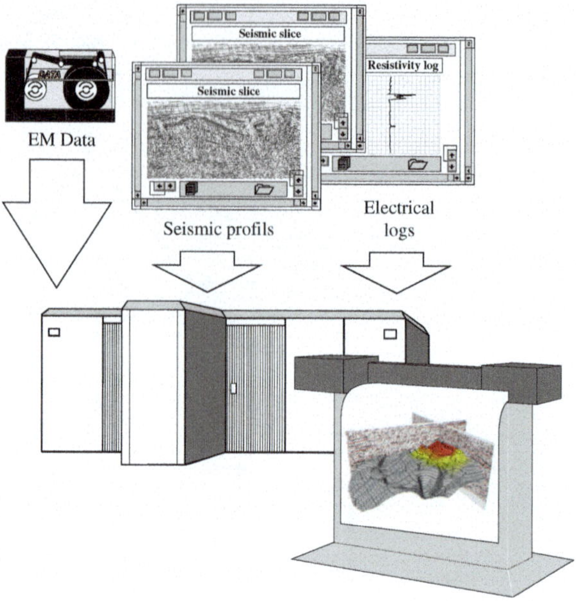

Fig. 5.16 Fabrication of a 2D inversion model (geological section, *right*) from seismic data (profile) and well logging (log resistivity). The setting of the section in depth can be done by sonic logging (3D Inversion Plug EM Bridge for Petrel™ software)

- Deterministic approaches such as derived methods, which have the merit of presenting a rapid convergence close to a minimum, but still remain very sensitive to the initial model and local minima
- Stochastic or even heuristic approaches to the random processes with the advantage of not computing derivatives while avoiding certain local minima such as those, for example, applying Monte Carlo techniques, the simplest but also the most limited (Anderson et al. 1972; Press 2007), or even those involving probabilistic aspects such as the Bayesian concept, for example (Jouanne 1991; Rakoto et al. 1997; Buland and More 2003; Hou et al. 2006; Zhangshuan et al. 2006; Idier 2008; Gunning et al. 2010)

It would take too long in this chapter to literally develop these specific methods, which only insiders can ordinarily understand.[46] However, users can obtain (1D) commercial inversion codes like Occam developed from previous works (Constable et al. 1987) by the Scripps Institution of Oceanography (Key 2009). As for some examples, we therefore limit ourselves here to mentioning a few of them (the simplest).

[46]We refer the reader to the specialized literature (Hohmann and Raiche 1987; Glasko 1988; Parker 1994; Colton and Kress 1992; Buland et al. 1996; Zhdanov 2002a, b, c; Hoversten et al. 2006; Stefano and Colombo 2006; Abubakar et al. 2007, 2008; Gribenko and Zhdanov 2007; Bornatici et al. 2007; Roth and Zach 2007; Plessix and Van der Sman 2007; Zeng et al. 2007; Mittet et al. 2007a, 2008; Carrazone et al. 2008; Jing et al. 2008; Plessix and Van der Sman 2008; Price et al. 2008; Zach et al. 2008a, b, c, d; Troyan and Kiselev 2009; Nguyen and Roth 2010; Kumar 2010).

5.3.1 Inversion by the Bayesian Approach

At the moment, the Bayesian method seems a priori the most appropriate for inversion of geoelectric data (Rakoto et al. 1997; Mitsuhata 2004; Idier 2008; Gunning and Glinsky 2009; Glinsky and Gunning 2011).

Theoretically, the technique is based on *Bayes' theorem* (Bayes 1763; Laurent 1873; Jeffreys 1939; Korchounov 1975). This shows that the estimated probability distribution $f(m)$ (given a posteriori) is proportional to the product of the function a priori by the likelihood function, such that, from the data m and the provided information I, we have:

$$f(m) \propto f\left(d^{*}|m, I\right) f\left(m|I\right) \tag{5.57}$$

Practically, this approach consists of integrating all available information concerning the subsoil, from any available source, and expressing it in the form of a probability law. These come mainly:

- From the m electromagnetic acquisition data, and more particularly the apparent resistivities ρ_a defined by:

$$Y = [\rho_a(k), k = 1, m] \quad \text{with} \quad Y \in D \subset R^d \tag{5.58}$$

- From the a priori data on the n layers proposed model (geological, seismic data, etc.) such as:

$$A = \left[\mathrm{a}^i = \rho(i), \mathrm{d}(i), i = 1, n\right] \quad \text{with} \quad A \in E \subset R^p \quad \text{and} \quad p = 2n - 1 \tag{5.59}$$

- From the physical law characterizing the forward problem with:

$$Y_{th} = G(A) = [\rho_a, t, h(k), k = 1, n] \tag{5.60}$$

where G is the functional for calculating the apparent resistivity (Y_{th}) as a function of the parameters X-A of the model.

Information on the geological structure (resistivity and thickness of the layer) is then translated using a uniform law or law known a priori as $P(X\text{-}A)$, and is updated according to the acquisition of the apparent resistivities. The probability distribution a posteriori (*Bayes' theorem*) is then defined by:

$$P(X = A/Y) = \frac{P(Y/X = A)\, P(X = A)}{\sum P(Y/X = A)\, P(X = A)} \tag{5.61}$$

This formula represents the updating of a priori knowledge about the model parameters, given the new observations X (measurements).

The conditional probability of the data, when the true model is $X = A$, is then of the form:

$$P(A/Y = A) = K\exp\left\{-[Y - G(A)]^T \mathbf{C}_Y^{-1}[Y - G(A)]\right\} \quad (5.62)$$

where K is the normalization constant and $\mathbf{C}_Y$ is the covariance matrix of Y (Bayes 1763).

The resolution of the Bayesian inverse problem is then based on the calculation of the probability a posteriori. Its estimation uses the principle of iterative methods like *Monte Carlo*. In this case, each iteration is associated with a drawing. The evolution of the system after each drawing is a random process then similar to a *Markov* chain.

5.3.2 Inversion by the Least Squares Methods

Another also used method is the most famous weighted least squares method.[47] Without going into details (Marquardt 1963; Marescot 2006), the method uses, unlike that of generalized least squares,[48] a linear unbiased estimator as detailed in Fig. 5.17.

5.3.3 Inversion by the Levenberg–Marquardt Method

Another algorithm also successfully used is that of Levenberg–Marquardt. This is an iterative optimization algorithm of the second order, which provides a numerical solution to the problem of minimizing a criterion or a nonlinear function dependent on several variables. The update formula for small step values is similar to that of *Newton/Gauss* while for larger values the formula tends to that of the simple gradient (cf. Fig. 5.18).

5.4 Actual Interpretation Process

The interpretation procedure itself, or simulation, involves several phases, which, according to the computer programs used, are more or less automated. We give here a typical example (1D modeling) of the greatly simplified procedure to follow to interpret data acquisition.

[47]The first inversion method to be used for electrical soundings.

[48]Developed by Legendre (1805) and Gauss (1809).

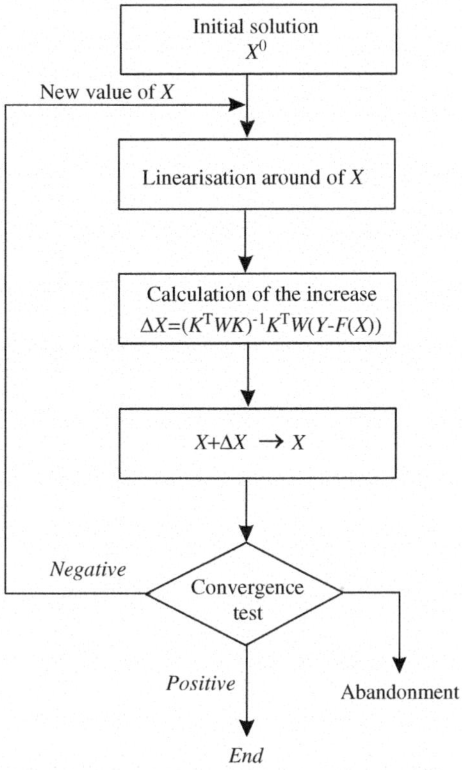

Fig. 5.17 Organizational chart of the inversion algorithm by the weighted least squares method (unbiased linear estimator)

First, from the apparent resistivity curves (measurements according to a given profile), a 1D resistivity model is established (forward problem) as a function of depth, which can correspond here, for example, to a three-layer model (cf. Fig. 5.19).

Once the model is entered and saved in the machine, the inversion phase itself begins. According to the available algorithms (weighted least squares, Levenberg–Marquardt, etc.), the procedure may be somewhat different. Generally, only the thickness and the resistivity of the geological horizons can be adjusted. These exogenous parameters come from external data (seismics and well logging, or even more rarely geological sections). They provide the basis for iterations required to develop models up to the final model that will be validated according to the confidence criteria that were initially established (convergence, uncertainties, margin of error, etc.).

In another way, for example, for multifrequency investigations and for resistive thin targets, authors such as Professor K. Key (2009) have demonstrated that inverting the mCSEM amplitudes and phases at two well-spaced frequencies produces much better sensitivity than inverting them at single frequencies. Key has also shown that inverting radial electric field data from a horizontal EM transmitter is better than inverting any other combinations of sources and receivers.

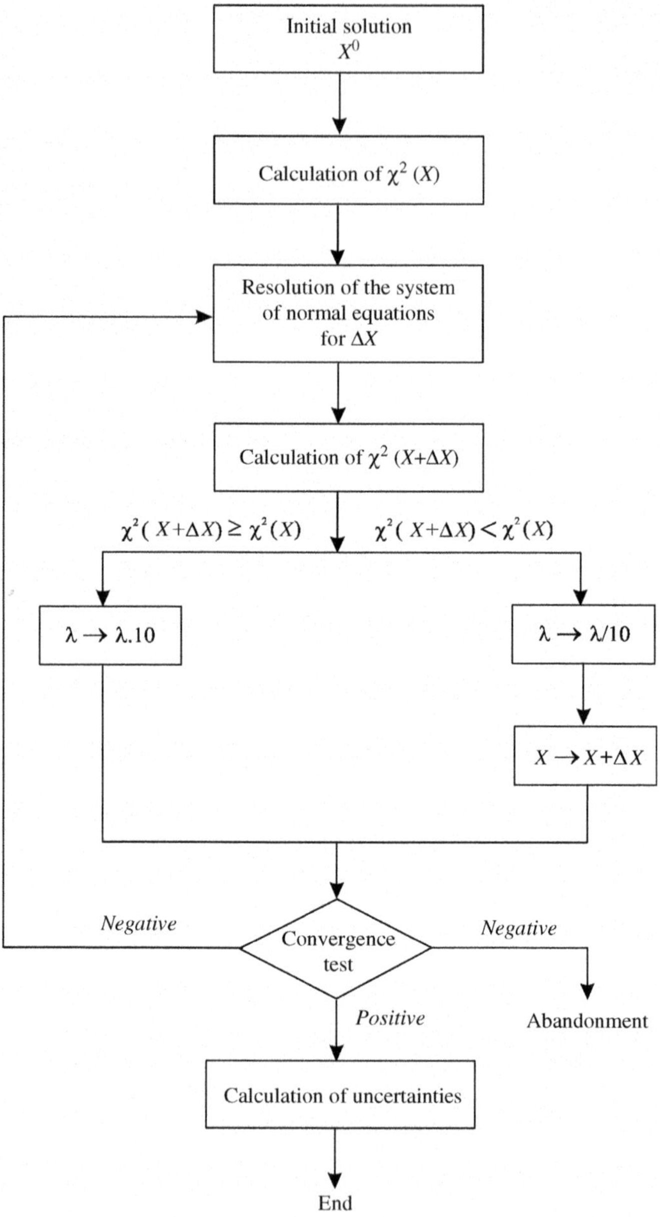

Fig. 5.18 Chart of *the Levenberg–Marquardt* inversion algorithm (example of a nonlinear adjustment corresponding to the improvement of the *Newton/Gauss* method)

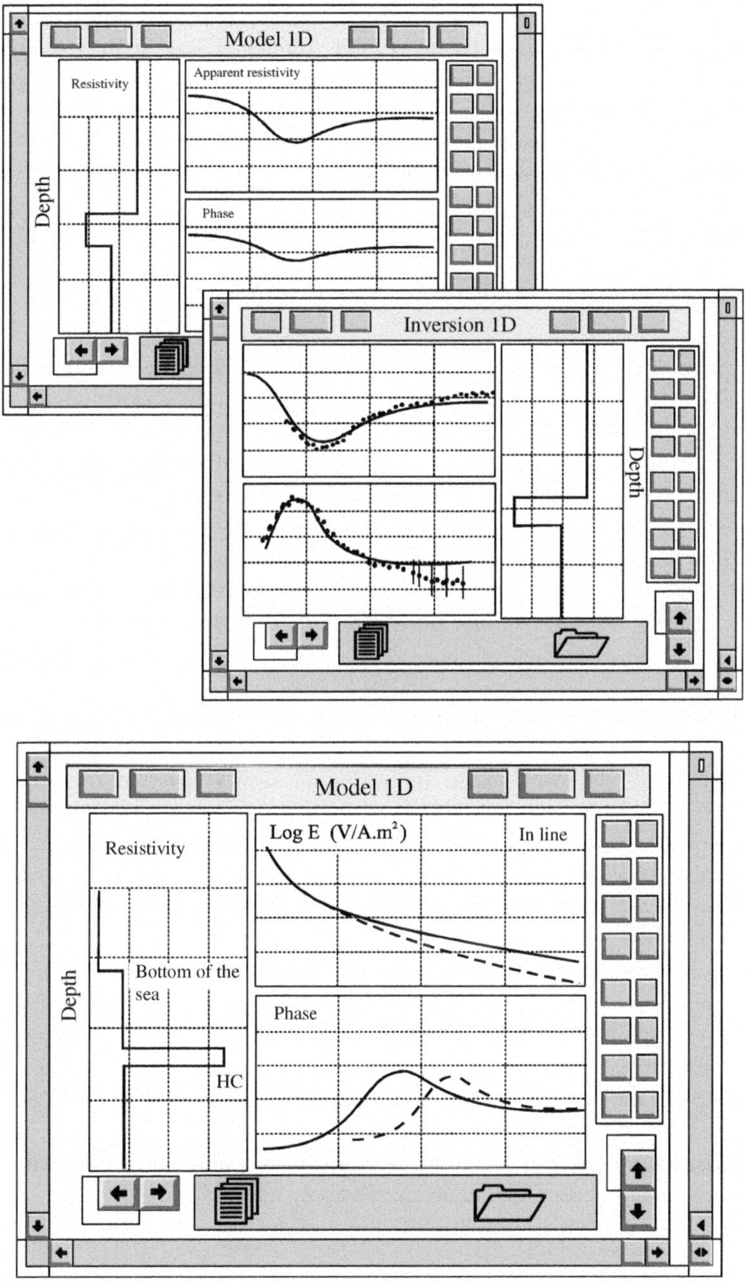

Fig. 5.19 Interpretation procedures (*top*): 1D model then 1D inversion for conventional sounding (ΔV). From *right* to *left* are a panel of control buttons, the resistivity curves and the geological model. SBL interpretation of field and phase measurements with and without HC anomaly (*bottom*)

Today other authors (Dieter Werthmüller et al. 2014) propose for multitransient electromagnetic technics to jointly invert high frequency seismic data and mCSEM data to obtain better rock physics relationships, especially to predict EM responses from seismic velocities.

It is sometimes necessary to perform field data with inversion to optimize results such as sensitivity and resolution.

5.5 Imagery: Electromagnetic Holography

Unlike mapping (2D), imaging is typically used for 3D representations and more specifically those that can be explored and viewed from different angles, points of view and perspectives.

For 1D or even 2D, data inversion is done by iterative approximations of the model (successive resolutions of the forward problem) until finally the model is acceptable. However in 3D, this practice is no longer possible (too complicated and too long), so it is necessary to use other means. These can be based, for example, on data migration then on the search for an optimization (i.e., stabilization) function, or on other approaches such as those practiced in mMT (Hautot and Tarits 2010).[49]

5.5.1 Electromagnetic Migration

The principle of electromagnetic migration is similar to that used in optical holography, which consists of reconstructing an image of an object in volume from data or measures of the amplitude and phase of the electric field at a specific time. In this sense, the migration is an operation similar to inversion. It is the information on the phase and, more precisely, that on the phase difference with a reference signal (generally a source) that can then reconstruct the image a posteriori, thereby accessing the third dimension.

In seabed logging the reconstructed virtual image is obtained by simultaneously recording, at each instant t, the information on amplitude A and phase φ for each receiver and for each position (x, y, z) of the EM energy source (cf. Fig. 5.20). Thereby we obtain a collection of signals localized in space and calibrated in time (A, φ, x, y, z, t).[50]

Thus, from these various elements of acquisition, a number of physical changes and mathematical operations is then successively applied.

For example (Zhdanov et al. 2010):

[49]Software sold by the French company IMAGIR (Brest).

[50]A similar "holographic" process and details of the calculations are described, for example, in a US patent (Zhdanov 2001).

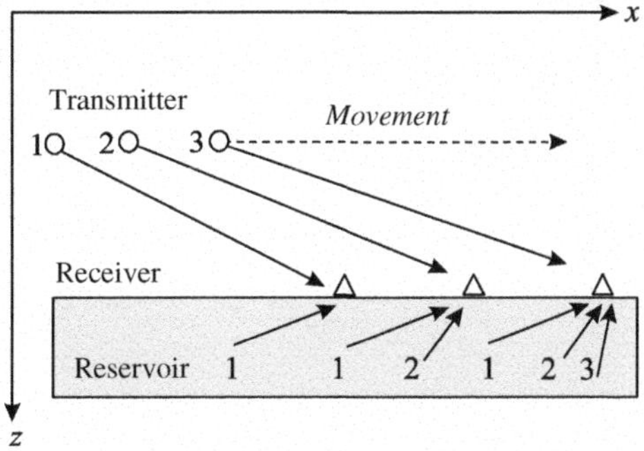

Fig. 5.20 Schematic diagram of an acquisition involving, a posteriori, a measured data migration process

- We calculate the *predicted field*, at each iteration of the 3D model for each sensor.
- We then make, thanks to the positioning data, a minimization of differences by applying the *reciprocity theorem*, i.e., in reversing respectively the functions of the transmitter and the receiver (see Chap. 2, Appendix A2.3).
- We finally form the *residual field* from the difference between the measured field and the predicted field.

This residual field is then migrated[51] and stabilized. In field this technique require accurate velocity data and massive computer processing.

5.5.2 Choice of the Stabilizing Function

The choice of a stabilizing function is very important because without it the model may introduce artifacts or even oscillations. To be effective, this choice has first to pass through thorough knowledge of the geological parameters and particularly the resistivities, whose distribution is generally discontinuous.

Such a parametric function, $P^{\alpha}(m)$, which should tend to a minimum, then can be defined as the sum of two functions of the model m such that:

$$P^{\alpha}(m) = \phi(m) + \alpha S(m) \rightarrow \min \tag{5.63}$$

where $\phi(m)$ is the function linking the models measured (a priori) and predicted (a posteriori), $S(m)$ being the stabilization function, and α being the regularization parameter.

The stabilization function must implicitly incorporate information used in the inversion process and is developed then from recognized and verified geological data.

Several types of function can be used and selected according to different criteria. We can adopt in this case a minimum standard criterion (*N*m), which minimizes the

[51]For more information, the general principles of migration are treated in various articles and books to which the reader can refer (Zhdanov et al. 1996; Zhdanov 2009). Firstly, as in seismic prospecting, the migration technique was used to correct seabed relief effect and bathemetry raw data errors.

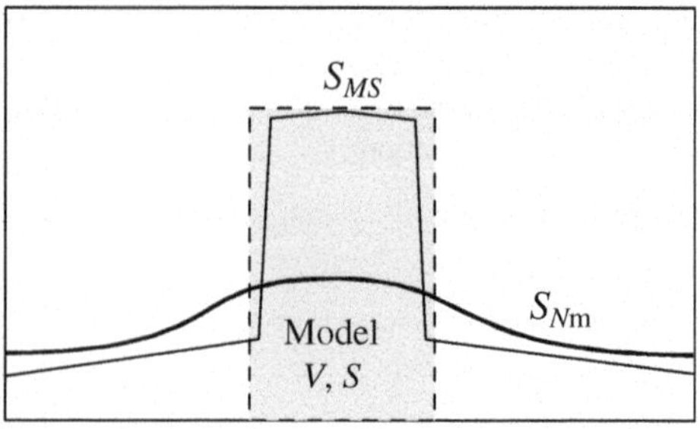

Fig. 5.21 Example of two S curves representing two stabilization functions around a simple prismatic model (According to Zhdanov et al. 2010)

difference between the model m of volume V and the a priori model m_{apr}, so that we have formally:

$$S_{N\mathrm{m}}(m) = \int_V \left(m - m_{\mathrm{apr}}\right)^2 \mathrm{d}v \tag{5.64}$$

We can also use a function based on the principle of *Occam's razor*[52] (OC_r), which then minimizes the norm of the first derivative (gradient: $\vec{\nabla}$) as:

$$S_{OC_r}(m) = \int_V \left(\vec{\nabla} m - \vec{\nabla} m_{\mathrm{apr}}\right)^2 \mathrm{d}v \tag{5.65}$$

These functions represent resistivity curves that are more or less flattened (cf. Fig. 5.21). To achieve a more acceptable result, we can similarly minimize the function on the volume v and introduce to avoid the singularity $m = m_{\mathrm{apr}}$, a focusing parameter e, such that:

$$S_{MS}(m) = \int_V \frac{\left(m - m_{\mathrm{apr}}\right)^2}{\left(m - m_{\mathrm{apr}}\right)^2 + e^2} \mathrm{d}v \tag{5.66}$$

Similarly, if it is desired to minimize the function on the volume thickness with, at the starting point of the a priori model, a nonzero value, we realize on the vertical component the integration on the surface such that:

$$S_{MVS}(m) = \int_V \frac{\left(m - m_{\mathrm{apr}}\right)^2}{\int_S \left(m - m_{\mathrm{apr}}\right)^2 \mathrm{d}s + e^2} \mathrm{d}v \tag{5.67}$$

where S is the horizontal section of the volume.

---

[52]The principle of Occam's razor states that if two hypotheses have the same degree of probability, then we favor the simplest hypothesis (the *parsimony principle*). The procedure is a direct application of *Bayes' theorem*, where the simplest hypothesis then presents the higher probability (see Sect. 5.3.1). The predictive techniques that use the principle of Occam do not guarantee the accuracy of the model.

Finally, we can also use the minimum gradient stabilization function:

$$S_{MGS}(m) = \int_V \frac{(\vec{\nabla} m - \vec{\nabla} m_{\mathrm{apr}}) \cdot (\vec{\nabla} m - \vec{\nabla} m_{\mathrm{apr}})}{(\vec{\nabla} m - \vec{\nabla} m_{\mathrm{apr}}) \cdot (\vec{\nabla} m - \overline{\nabla} m_{\mathrm{apr}}) + e^2} \mathrm{d}v \qquad (5.68)$$

which this time minimizes the volume of the model in three dimensions with a nonzero gradient.

This type of approach is discussed, for example, in a real case of controlled source electromagnetic prospecting (in the Shtokman gas field in the Barents Sea), where the convergence of the various functions of stabilization has been particularly studied (Cuma et al. 2010).

Migration/stabilization technics and their "holographic" results (Fig. 5.22) are heavy to implement. For instant, they require relatively long calculation times on the computer, which can exceed several tens of hours. In contrast, when the models are established once and for all, their representations using modern tools of visualization (3D theaters) are quite remarkable and very useful to geologists, allowing them to perceive the entire deposit (several reservoirs) in its structural context, clearly superimposing the different sources of geological information (cf. Fig. 5.23). Later, these holographic views are particularly interesting in viewing drillings when they have been profiled and mapped in 3D (Sainson 2010).

Today these virtual representations are also available on graphic workstations with single or multiple monitors (small size) or a widescreen (large size). Their

Fig. 5.22 3D "holographic" image (mMT) of a salt dome in the Gemini Prospect (Gulf of Mexico) (According to Key 2003)

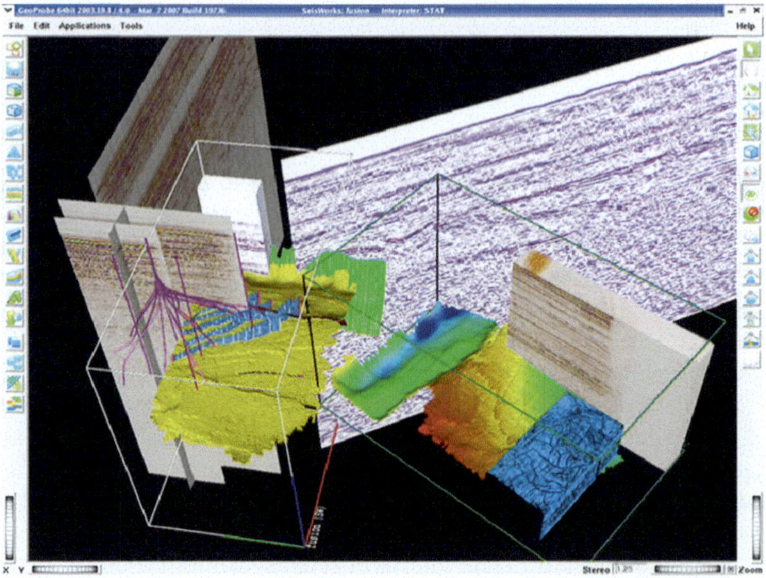

Fig. 5.23 Software packages allow us today to view a set of very diverse data from well acquisition (well logging), surface (seismic) and seabed logging (SBL)

Fig. 5.24 Graphic processor (GPU) Tesla C1060, with the power of TFLOPS (an acronym for a trillion floating-point operations per second) (4 GB RAM) (According to Nvidia Corporation)

performances have been greatly improved in recent years by the emergence of new graphic cards and processors provided by the manufacturer Nvidia Corporation, for example, and by the increased capacity of RAM memory reaching 6 GB (Neri 2011). These technological advances have been created through the miniaturization of computers (see Moore's law)[53] and with the advent of new associated programming languages such as OpenCL[54] run from CUDA (*Compute Unified Device Architecture*) technology (cf. Fig. 4.24).[55]

---

[53]Characterizes the evolution of computer power. Probably changed in the future by the appearance of quantum microprocessors.

[54]Free software originally developed by Apple.

[55]Technique using the graphic processor unit (GPU) to perform all the calculations instead of the central processor unit (CPU), which is slower.

6 Analog Models

For specific issues, when analytical and numerical models meet their limits or simply do not exist, it is possible to use analog models. Some of them are heavy to implement, and they cannot virtually replace automatic and mathematical interpretation, which remains the only alternative usable in the context of industrial activities (ease of use).

6.1 *Vocation of Analog Models*

Analog models are essential intermediaries to validate the principles and exploratory methods. Specifically, they are the direct link between measurements and theoretical interpretation models, and allow us to objectively assess the chances of success without going through a field phase, which is generally expensive in means and often inconclusive (uncontrolled parameters).

More specifically, the analog models intervene, under the cover of some simplifying assumptions (Utzmann 1954; Cagniard and Neale 1957; Wait 1971) to:

- Reinforce some new principles and techniques of detection
- Test some instrumentation (detection limit, resolution, etc.)
- Design correction systems
- Identify the topology of anomalous responses (forms and orders of magnitude)
- Simulate specific conditions of detection and localization
- Compare the results with analytical or numerical computer simulations
- Connect field studies to laboratory experiments for any other special applications, requiring to obtain orders of magnitude with a specific material in a known and controlled environment.

6.2 *Equipment*

Analog models are in the form of small scale models, which must allow easy access and manipulation. In their simplest form, they consist of a rheostatic tank,[56] also called a rheographic basin, (using graduated scales) of a few square meters in area (3 m × 2 m, for example), a few decimeters deep, and filled with a liquid conductor of electricity (salt water, for example).

[56]Formerly called an electrolytic tank or, more commonly, a bathtub in memory of the first experiments performed in 1911 by Conrad Schlumberger at the Ecole des Mines in Paris (Allaud and Martin 1976; Gruner Schlumberger 1982).

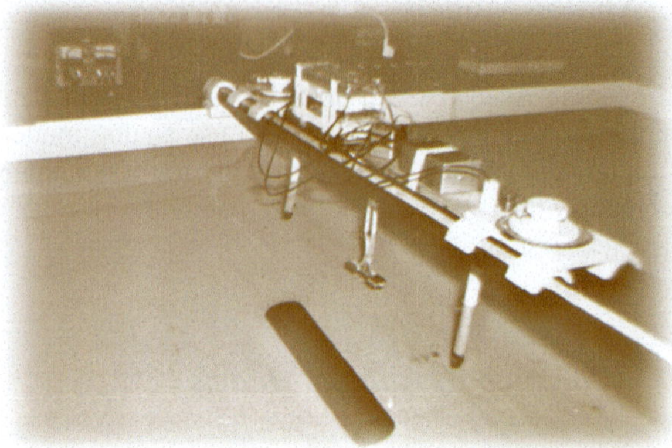

Fig. 5.25 Photographic detail of a rheographic tank in PVC used by the author and a measurement acquisition system (prototype) for transverse fields. Tank of the Ecole Normale Supérieure of Paris (According to Sainson 1982)

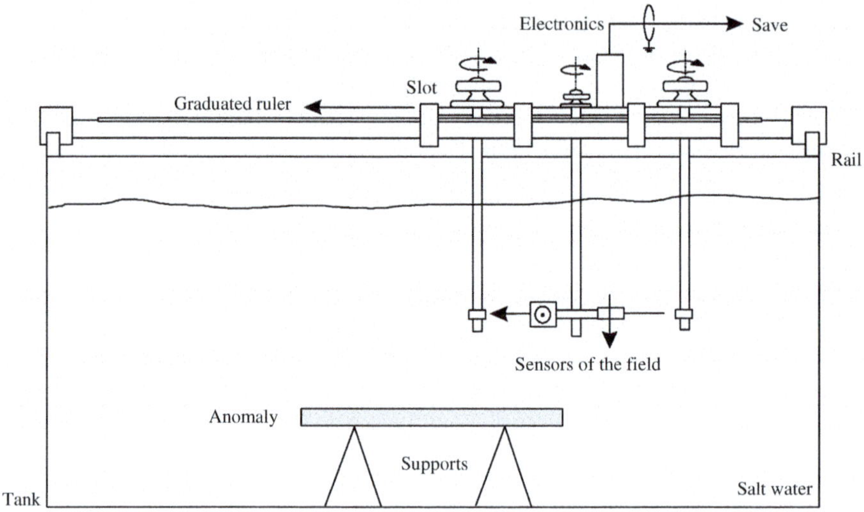

Fig. 5.26 Section view of a so-called rheographic tank and device for measuring the components of transverse fields. The conductivity anomaly is placed on the *bottom*. (According to Sainson 1984)

This basin is generally equipped with emerged and submerged handling means, with systems of moving and identifying the measuring instruments, thus ensuring the development as accurately as possible of measures or mapping profiles. Heavier equipment (a stabilized power supply, oscilloscope, spectrum analyzer, recorders, computer, etc.) are then deported to a nearby dry laboratory bench (Figs. 5.25, 5.26, and 5.27).

The analog models designed for the rheographic tank are the most successful analog models still being used (Loseth 2007). Other experimental arrangements built with discrete electronic components, forming a network, such as resistances and electrical capacities, were used for a long time until the appearance in the 1990s of wiring software and calculations using analog electronic circuits.

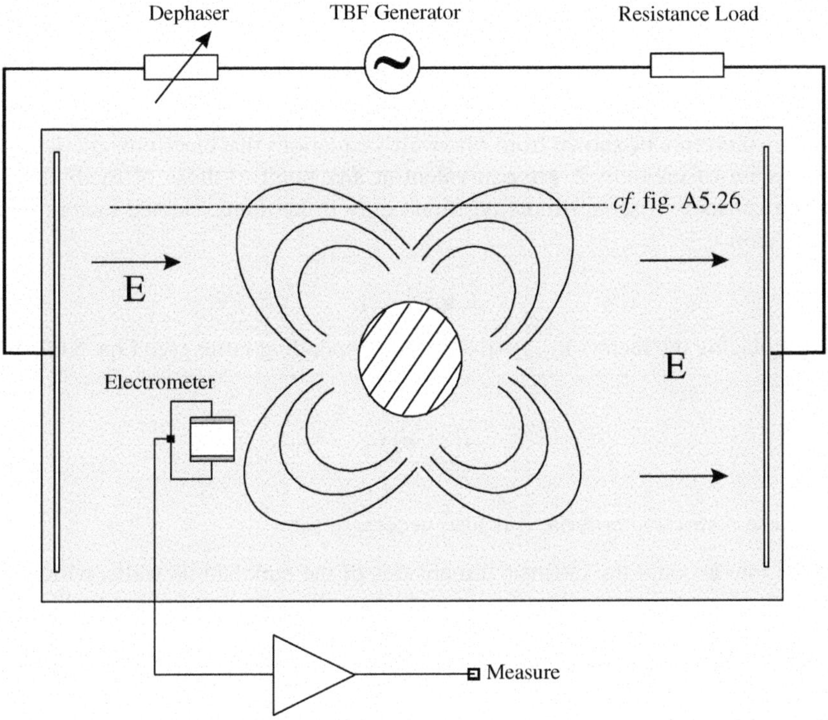

Fig. 5.27 Rheographic basin: top view with symbolic representation of equi-field lines around a spherical anomaly, issued by an acquisition system in TE mode in an in-line configuration (injection/measurement) (According to Sainson 1984)

6.3 *Precautions to Be Taken into Account During the Manipulations*

The use of an electromagnetic device in a confined space (finite dimensions) needs to comply with a number of criteria and requirements. For example, the behavior of the fields in the tank or basin should be similar to the one we might find in terms of size, i.e., on the ground. In this case a similarity rule must be ensured and verified. The correctness of this law is directly related to Maxwell's equations for variable currents or equations governing DC (the Laplace equation).

It must also respect (between the survey reality and the model):

- The homothety of the geometric quantities $l_{1,2}$ for all active systems (measuring devices), such that, to nearly a factor k_l, we have:

$$l_1 = k_l l_2 \tag{5.69}$$

- the ratio k_σ of the electrical conductivities $\sigma_{1,2}$ of the two present media such as:

$$\sigma_1 = k_\sigma \sigma_2 \tag{5.70}$$

– the ratio k_f of the frequencies $f_{1,2}$, such that:

$$f_1 = k_f f_2 \tag{5.71}$$

It may therefore be shown from Maxwell's equations that electromagnetic fields, at any point of medium 2, are equivalent at any point to those of medium 1, to nearly a constant of proportionality, if and only if the three selected factors verify the equation:

$$k_f k_\sigma k_l^2 = 1 \tag{5.72}$$

or, by replacing the factors $k_{l,\sigma,f}$ with their corresponding ratios (see Eqs. 5.69, 5.70, and 5.71):

$$\frac{l_1^2}{l_2^2} = \frac{\sigma_2 f_2}{\sigma_1 f_1} \tag{5.73}$$

With these restrictive criteria, it is also necessary to:

- Take into account the intrinsic dimensions of the tank and its walls, which may introduce stationary wave effects[57] and edge effects that alter the distribution of the field lines[58] (Goudswaard 1957)
- Respect the ratio of the measurement system dimensions (injection/reception) to those of the anomalies

7 Tutorial and Practice

This past decade has allowed many experiments to bring out the rules of good practice. Initially it was not easy as the models were numerous and varied. It depends on EM methods, technics and engaged companies in the offshore market. However, we can identify some guidelines.

1D forward models or inversions locally derived at receiver locations can be stitched together to derive resistivity pseudosections (Chave and Cox 1982). This limited approach aims to establish lateral variations in resistivity and anisotropy, then to start modeling for higher dimensions.

A 2D inversion assumes that the subsoil is invariant along one horizontal direction, usually taken to be perpendicular to the survey line (Unsworth et al. 1993; MacGregor et al. 1998; Key and Ovall 2011). With good structural information such as seismic data, the 2D approach can be sufficient without recourse to full 3D inversion.

---

[57] At this level, it important to take into account the wavelength in order to choose the transmission frequency (tank), while meeting the above criteria.

[58] In this case, it is recommended to choose not too small a tank and to perform the measurement at its center. One can also choose walls in a conductive material so that they are at the same potential: this was the case with C. Schlumberger's bath.

3D forward modeling and inversion approaches (Maao 2007; Commer and Newman 2008b) provide the best approximation of the true resistivity structure because the subsoil and the source fields are assumed to be 3D.

This paragraph is very short but important. For the moment, the processes are continuously improving and may quickly become obsolete. For more specific information, the interesting paper by Drs. Lucy MacGregor and James Tomlinson, which is an overview on mCSEM methods, is a good introduction to a tutorial and practice in data interpretation (MacGregor and Tomlinson 2014b).

8 Conclusion

In applied geophysics, compared to other exploratory investigations, the methods of interpretation of electromagnetic data are by far the most difficult to learn and lead. Some industrialists operating in the exploration sector have developed interpretation software in collaboration with universities or high level research labs. This is the case, for example, with the software CSEMoMatic v1.6 proposed by the Scripps Institute in San Diego and that developed by the NGI (Norwegian Geotechnical Institute).

Today many efforts are being undertaken to obtain quantitative answers (reservoir size, depth, thickness, lateral extension, form, etc.). using theoretical models and mathematical technics. However the integrative nature of the method itself makes it a relatively imprecise method (qualitative).

In the future, it is probably on the side of mathematical modeling that the most significant progressions in SBL are expected, with the development of more efficient 3D models (Loseth 2007; Shantsev and Maaø 2015; Piyoosh et al. 2015) for resolution of the forward problem, and also with more systematic use of stochastic analysis for the inverse problem. The ideal thing, of course, would be to eventually have information not only on the anomaly but also on the entire prospect, then incorporating the structuralist appearance as happens in seismics. This currently remains inaccessible compared to the contrasts in conductivity badly defined for the whole geological structure.

References

Several references in this following list correspond to appendices A5.1 to A5.7

Abubakar A, van den Berg PM, Mallorqui JJ (2004) Imaging a biomedical data using a multiplicative regularized contrast source inversion method. IEEE Trans Microw Theory Techn 50:1761–1771

Aboubakar A et al (2007) A two and half dimensional model based inversion algorithm for the controlled source electromagnetic data. In: SPE annual technical conference and exhibition California, 11–14 November, SPE 110054

Abubakar A, Habashy TM, Druskin VL, Knizhnerman L, Alumbaugh D (2008) 2.5D forward and inverse modeling for interpreting low frequency electromagnetic measurements. Geophysics 73(4):F165–F177
Allaud LA, Martin MH (1976) Schlumberger. The story of a technique. Ed. Wiley, pp 15–17
Amaya M (2015) High-order optimization methods for large-scale 3D CSEM data inversion. PhD thesis, NTNU, Trondheim, October, pp 53–83
Anderson TW et al (1972) An introduction to multivariate statistical analysis. Series in probability and statistics. Ed. Wiley, New York
Andreis D (2008) Electromagnetic surveying for resistive or conductive bodies. US patent no. 7362102
Angot A (1982) Compléments de mathématiques à l'usage des ingénieurs de l'électronique et des télécommunications. Ed. Masson, Paris
Arfken B (1968) Mathematical methods for physicists. Academic, New York
Backus G, Gilbert F (1967) Numerical applications of a formation for geophysical inverse problems. GJ 13, pp 247–276
Bailey RC (1970) Inversion of the geomagnetic induction problem. Proc R Soc A 315:185–184
Bailey RC (2008) The electrical response of an insulating circular disk to uniform field. Progress Electromag Res 88:241–254
Bannister PR (1968) The image theory quasi-static fields of antennas above the earth surface. Naval Underwater Systems Center report
Bannister PR et al (1965) Quasi-static electromagnetic fields. Scientific engineering studies. Naval Underwater Systems Center report
Bannister PR et al (1987a) Quasi-static electromagnetic fields. Scientific and engineering studies, Naval Underwater Systems Center, New-London
Bannister PR (1987b) Simplified expressions for the electromagnetic fields of elevated, surface or buried dipole antennas. Compiled 87 vol. 1, Scientific and Engineering Studies, Naval Underwater Systems Center, New-London, Connecticut
Baranov W (1951) Interprétation quantitative des mesures en prospection par courants telluriques. 3ème congrès mondial du pétrole, The Hague
Bayes (1763) An essay towards solving a problem in the doctrine of chances. By the Late Rev. Mr. Bayes, F. R. S communicated by Mr. Price, in a Letter to John Canton, A. M. F. R. S. Philos Trans, pp 370–418
Beaufort L (1944) Etude des champs téllurique en baignoire. Note technique no. 44 (tellurique). Ed. CGG
Boissonnas E (1946) Note sur les mesures en baignoire. Note technique no. 74 (tellurique). Ed. CGG
Bornatici L, Mackie R, Watts MD (2007) 3D inversion of marine CSEM data and its application to survey design. In: EGM conference proceedings, Capri, Italy
Boursian WR (1933) La théorie des champs électromagnétiques employés dans la prospection électrique. 1ère partie, GTTI (in Russian)
Bracewell RN (1986) The Fourier transform and its applications. McGraw-Hill, New York, 474 p
Brillouin M (1933) Fonctions sphériques. Formules générales de récurrence. Développement des fonctions non antipodes en séries de polynômes de Legendre et de Laplace. Ed. Gauthier-Villars, Paris, 4 p
Brillouin M (1938) Qu'apprend-on de l'intérieur du globe par les mesures faites à sa surface? Pesanteur, magnétisme. Annales de l'I.H.P, tome 8, no. 4.Ed. Gauthier-Villars, pp 151–192
Buland A, More H (2003) Bayesian linearized AVO inversion. Geophysics 68: 185–198
Buland A et al (1996) AVO inversion of Troll field data. Geophysics 61: 1589–1602
Butkov E (1968) Mathematical physics. Ed. Addison-Wesley, Reading
Cagniard L (1950) La prospection géophysique. Ed. Presses universitaires de France, Paris, p 32
Cagniard L, Neale RN (1957) Technique nouvelle de modèles réduits pour la prospection électrique. Geophys Prospect 5:259–271
Carrazone JJ et al (2008) Inversion study of a large marine CSEM survey. SEG 2008 Expanded Abstracts, Las Vegas

Chave AD (1983) Numerical evaluation of related Hankel transforms by quadrature and continued fraction expansion. Geophysics 48(12):1671–1686

Chave AD (2009) On the electromagnetic fields produced by marine frequency domain controlled sources. Geophys J Int 179, 29 p

Chave AD, Cox CS (1982) Controlled electromagnetic sources for measuring electrical conductivity beneath the ocean 1. Forward problem and model study. J Geophys Res 87 (B7):5327–5338

Chave AD, Jones AG (2012) The magnetotelluric method: theory and practice, pp 382–383 and 337–420

Cheesman SJ, Edwards RN, Chave AD (1987) On the theory of sea floor conductivity mapping using transient electromagnetic systems. Geophysics 52:204–217

Chisholm JSR, Morris RM (1965) Mathematical methods in physics. W B Saunders Company, Philadelphia, 719 p

Claerbout JF (1976) Fundamentals of geophysical data processing with applications to petroleum prospecting. MacGraw-Hill, New York

Claerbout JF (1992) Earth sounding analysis. Processing versus inversion. Ed. Blackwell Scientific Publications, Boston, p xiv

Colton D, Kress R (1992) Inverse acoustic and electromagnetic inverse scattering theory. Ed. Springer, Berlin

Commer M, Newman GA (2008a) Optimal conductivity reconstruction using three-dimensional joint and model-based inversion for controlled-source and magnetotelluric data. SEG 2008 Expanded Abstracts, Las Vegas

Commer M, Newman G (2008b) New advances in three dimensional controlled source electromagnetic inversion. Geophys J Int 172:513–535

Commer M et al (2008) Massively parallel electrical conductivity imaging of hydrocarbons using the Blue Gene/L supercomputer. University of California. Paper LBNL 63009. 34 p

Constable SC, Weiss CJ (2006) Mapping thin resistors and hydrocarbons with marine EM methods: insights from 1D modelling. Geophysics 71(2)

Constable SC, Parker RL, Constable CG (1987) Occam's inversion: a practical algorithm for generating smooth models from electromagnetic sounding data. Geophysics 52:289–300

Cuma M et al (2010) Multiple inversion scenarios for enhanced interpretation of marine CSEM data using iterative migration: a case study for the Shtokman gas field, Barents Sea. SEG Denver, annual meeting, pp 583–587

de Stefano M, Colombo D (2006) Geophysical modeling through simultaneous joint inversion of seismic, gravity and magneto-telluric data. WesternGeco EM, Geosystem, Milan

Durand E (1966) Electrostatique. 3 tomes. 1: Les distributions. 2: Problèmes généraux conducteurs. 3: Méthodes de calcul diélectriques. Ed. Masson, Paris

Eaton PA, Kaufman AA (2001) The theory of inductive prospecting. Methods in geochemistry and geophysics. Ed. Elsevier, Amsterdam

Edwards RN, Chave AD (1986) A transient electric dipole–dipole method for mapping the conductivity of the sea floor. Geophysics 51:984–987

Ehrenbourg DO, Watson RJ (1931) Mathematical theory of electrical flow in the stratified strata. AIME Geophys Prospect 1932:423–442

Eskola L (1992) Geophysical interpretation using integral equations. Chapman & Hall, London, 191 p

Everett ME, Edwards RN (1993) Transient marine electromagnetics—the 2.5-D forward problem. Geophys J Int 113(3):545–561

Flath H (1955) A pratical method of calculating geoelectrical model graphs for horizontally stratified media. Presented at the eighth meeting of the European Association of Exploration Geophysicists, held in Paris, 18/20 May

Flosadottir AH, Constable S (1996) Marine controlled source electromagnetic sounding: modelling and experimental design. J Geophys Res 101(4):5507–5517

Gallardo LA, Meju MA (2003) Characterization of heterogeneous near-surface materials by joint 2D inversion of DC resistivity and seismic data. Geophys Res Lett 30:1658

Glasko VB (1988) Inverse problems of mathematical physics. American Institute of Physics, New York, 97 p
Glinsky ME, Gunning JS (2011) Resolution and uncertainty of inversion CSEM remote sensing data can now be estimated using Bayesian statistic. World Oil, January, pp 87–62
Goudswaard W (1957) On the effect of the tank wall material in geo-electrical model experiments. Geophys Prospect 5(3):272–281
Grand FS, West GF (1965) Interpretation theory in applied geophysics. McGraw-Hill, New York
Gray A, Mathews GB (1922) Treatise on Bessel function and their application to physics. Macmillan, London, 291 p
Green G (1850) An essay on the application of mathematical analysis to the theory of electricity & magnetism. J Reine Angew Math 39:73–79
Grellier S (2005) Suivi géologique des centres de stockage de déchet bio-réacteurs par mesures géophysiques. Doctoral thesis, Université Pierre et Marie Curie
Gribenko A, Zhdanov M (2007) Rigorous 3D inversion of marine CSEM data based on the integral equation method. Geophysics 72:WA73–WA84
Gruner Schlumberger A (1982) The Schlumberger adventure. Ed. Arco Publishing, New York, p 1
Gunning JS, Glinsky ME (2009) Bayesian approaches to resolution in CSEM. In: 71st EAGE conference and exhibition, Amsterdam, 8–11 June
Gunning JS, Glinsky ME, Heddith J (2010) Resolution and uncertainty in 1D CSEM inversion: a Bayesian approach and open-source implementation. Geophysics 75:F151–F171
Habashy TM (1985) Quasi-static electromagnetic fields due to dipole antennas in bounded conducting media. EEE Trans Geosci Remote Sens GE-23(3):325–333
Hansen DA et al (1967) Mining geophysics, vol B: theory. Ed. Society of Exploration geophysicists, Tulsa, pp 63–83
Hautot S, Tarits P (2010) 3D magnetotelluric inversion of 2D profiles: application to land and marine data for shallow crustal investigation. ASEG, Sydney
Heaviside O (1899) Electromagnetic theory, 3 vols. Ed. Electrician Printing and Publishing, London
Heiland CA (1931) The department of geophysics. Vol xxvi no. 1, Suppl. A. Ed. Colorado School of Mines, Golden, p 21
Hohmann GW (1975) Three dimensional induced polarization and electromagnetic modeling. Geophysics 71:309–324
Hohmann GW (1989) Numerical modeling for electromagnetic methods of geophysics. Electromagnetic methods in applied geophysics, vol 1 Theory. Ed. Society of Exploration Geophysicists, Tulsa, pp 313–318
Hohmann GW, Raiche AP (1987) Inversion of controlled source electromagnetic data in electromagnetic methods in applied geophysics. Geophysics 1:469–503
Hou Z et al (2006) Reservoir parameter identification using minimum relative entropy based Bayesien inversion of seismic AVA and marine CSEM. Geophysics 71(6):077–088
Hoversten GM et al (2006) Direct reservoir parameter estimation using joint inversion of marine seismic AVA and CSEM data. Geophysics 71(3):1–13
Hummel JK (1928) Über die Tiefenwirkung bei geoelektrischer Potentiallaienmethoden., Zeitschrift für Geophysik
Hummel JK (1929a) Der scheinbare spezifische Widerstand. Zeitschrift für Geophysik. T.5, no. 3–5, p 89
Hummel JK (1929b) Der scheinbare spezifische Widerstand bei vier plan parallelen Schichten. Zeitschrift für Geophysik. T.5, no. 5–6, p 228
Idier J (2008) Bayesian approach to inverse problems. ISTE—Wiley, London
Inman JR, Ryu J, Ward SH (1973) Resistivity inversion. Geophysics 38:1088–1108
Jackson JD (1965) Classical electrodynamics. Ed. Wiley, New York, 641 p
Janovskij BM, Molocnov G (1959) Modellversuche mit niederfrequenten elektrischen Feld. Freib Forsch, pp 17–29

Jaysaval P, Shantsev VD, de la Kethulle de Ryhove S (2015) Efficient 3-D controlled-source electromagnetic modelling using an exponential finite-difference method. Geophys J Int 203:1541–1574

Jeans JH (1933) The mathematical theory of electricity and magnetism. Cambridge University Press, Cambridge

Jeffreys H (1939) Theory of probability. Clarendon, Oxford, 380 p

Jeffreys H, Jeffreys B (1956) Methods of mathematical physics. Cambridge University Press, Cambridge, 714 p

Jegen MD, Edwards RN (2000) On the physics of marine magnetotelluric sounding. Geophys Int 142

Jing C, Green KE, Willen DE (2008) CSEM inversion: impact of anisotropy, data coverage, and initial models. SEG 2008 Expanded Abstracts, Las Vegas

Johansen S et al (2008) How EM analysis validates current technology, processing and interpretation methodology. First Break 26:83–88

Jouanne V (1991) Application des techniques statistiques bayésiennes à l'inversion de données électromagnétiques. Doctoral thesis, Université de Paris

Kaufman AA (1994a) Geophysical field theory and method, part B, volume 48: electromagnetic fields I, international geophysics. Ed. Elsevier, Amsterdam

Kaufman AA (1994b) Geophysical field theory and method, part C, volume 49: electromagnetic fields II, international geophysics. Ed. Elsevier, Amsterdam, 335 p

Kaufman AA, Eaton PA (2001) The theory of inductive prospecting. Elsevier, Amsterdam, 681 p

Kaufman AA, Hoekstra P (2001) Electromagnetic sounding. Elsevier, Amsterdam

Kaufman AA, Keller GV (1983) Frequency and transient soundings. Ed. Elsevier, Amsterdam, pp 411–450

Keller GV, Frischknecht FC (1966) Electrical methods in geophysical prospecting. Ed. Pergamon Press, Oxford, p 1104

Key KW (2003) Application of broadband marine magneto-telluric exploration to a 3-D salt structure and a fast spreading ridge. PhD dissertation, University of California, San Diego

Key KW (2007) Adaptive finite element modeling for 2D electromagnetic problems using unstructured grids. Scripps Institution of Oceanography, La Jolla

Key KW (2009) 1D inversion of multicomponent, multifrequency marine CSEM data: methodology and synthetic studies for resolving thin resistive layers. Geophysics 74(2):F9–F20

Key K, Ovall J (2011) A parallel goal-oriented adaptive finite element method for 2.5-D electromagnetic modelling. Geophys J Int 186(1):137–154

Koefoed O (1968) The application of the kernel function in interpreting geoelectrical resistivity measurements. Geoexploration monographs. Series 1, no. 2. Ed. Gebruder Borntraeger, Berlin

Koefoed O (1980) Geosounding principles, 1. Resistivity sounding measurements. Methods in geochemistry and geophysics 14 A. Ed. Elsevier, Amsterdam

Korchounov Y (1975) Fondements mathématiques de la cybernétique. Ed. Mir, Moscou, pp 282–290

Kraichman MB (1970) Handbook of electromagnetic propagation in conducting media. US Government Printing Office, Washington, DC

Kumar A (2010) Inversion of magnetotelluric data in offshore hydrocarbon exploration. VDM Verlag Dr. Mueller, Saarbrücke, 72 p

Kunetz G (1966) Principles of direct current resistivity prospecting. Geopublication associates. Geoexploration monographs. Series 1, no. 1. Ed. Gebruder Borntraeger, Berlin

Lagabrielle R et al (2001) Reconnaissance par prospection électrique du sous-sol marin pour le nouveau port du Havre. Geofcan, Géophysique des sols et des formations superficielles. Actes du 3ème colloque, Orléans, 25–26 September. Ed. INRA, pp 149–153

Lamb H (1924) Evolution of mathematical physics. Cambridge University Press, Cambridge

Langer RE (1933) An inverse problem in differential equations. Am Soc Math J 39:14–28

Langer RE (1936) On determination of earth conductivity from observed surface potentials. Am Soc Math J vol 42

Lasfargues P (1957) Prospection électrique par courants continus. Manuel de prospection géophysique. Ed. Masson, Paris, pp 76–81
Laurent H (1873) Traité de calcul des probabilités. Ed. Gauthier-Villars, Paris, pp 56–61
Lee S et al (1987) Phase-field imaging: the electromagnetic equivalent of seismic migration. Geophysics 52(5):678–693
Li Y, Key K (2007) 2D marine controlled-source electromagnetic modeling: part 1—an adaptive finite-element algorithm. Geophysics 72(2): WA51–WA62
Litvinov SY (1941) Electrical sea prospecting.. Gostoptekhizdat
Lorentz HA (1927) Lectures on theoretical physics, vol 3. Ed. Macmillan, London
Loseth LO (2007) Modeling of controlled source electromagnetic data. PhD thesis, Université norvégienne, 186 p
Maao F (2007) Fast finite-difference time-domain modelling for marine-subsurface electromagnetic problems. Geophysics 72(2):A19–A23
MacGregor LM (2012) Integrating seismic, CSEM and well log data for reservoir characterisation. The Leading Edge, pp 268–277
MacGregor LM, Tomlinson J (2014a) Marine controlled-source electromagnetic methods in the hydrocarbon industry: a tutorial on method and practice. SEG Interpretation 2(3): SH24–SH25
MacGregor LM, Tomlinson J (2014b) Marine controlled-source electromagnetic methods in the hydrocarbon industry: a tutorial on method and practice. Interpretation 2(3):SH13–SH32
MacGregor LM, Constable SC, Sinha MC (1998) The RAMESSES experiment-III. Controlled-source electromagnetic sounding of the Reykjanes Ridge at 57° 45'N. Geophy J Int 135(3):773–789
MacGregor L, Sinha M, Constable S (2001) Electrical resistivity structure of the Valu Fa Ridge, Lau Basin, from marine controlled-source electromagnetic sounding. Geophys J Int 146:217–236
MacGregor LM, Bouchrara S, Tomlinson JT, Strecker U, Fan J, Ran X, Yu G (2012) Integrated analysis of CSEM, seismic and well log date for prospect appraisal: a case study from West Africa. First Break 30: 77–82
Maillet R (1947) The fundamental equations of electrical prospecting. Geophysics 12(4): 529–556
Marescot L (2006) Un algorithme d'inversion par moindres carrés pondérés: application aux données géophysiques par méthodes électromagnétiques en domaine fréquence. Bulletin de géologie de l'Université de Lausanne, no. 357, pp 277–300
Marquardt DW (1963) An algorithm for least-squares estimation of non linear parameters. J Soc Ind Appl Math 11:431–441
Maxwell JC (1873, 1892) A treatise on electricity and magnetism, vol 1. Ed. Clarendon, Oxford, pp 245–280
Maxwell JC (1884) Traité élémentaire d'électricité. Ed. Gauthier-Villars, Paris, pp 91–92
Medkour, M. (1984). Prospection électrique par la technique dipôle-dipôle: interprétation quantitative et application à la géophysique structurale et minière. Doctoral thesis, Université de Nancy
Meju MA (1994) Geophysical data analysis. Understanding inverse problem theory and practice. Society of Exploration Geophysicists course notes series, vol 6. Ed. SEG Publishers, 296 p
Menke W (1984) Geophysical data analysis: discrete inverse theory. Academic, San Diego
Menke W (1989) Geophysical data analysis: discrete inverse theory, Edition réviséeth edn. Ed. Elsevier, Amsterdam, 289 p
Mitsuhata Y (2004) Adjustment of regularization in ill-posed linear inverse problems by the empirical Bayes approach. Geophys Prospect 52(3):213–239
Mittet R, Schaug-Pettersen T (2007) Shaping optimal transmitter waveforms for marine CSEM surveys. In: San SEG 2007 conference proceedings, San Antonio
Mittet R et al (2005) A two-step approach to depth migration of low frequency electromagnetic data. Paper presented at the 75th SEG conference & exhibition, Houston, USA, 6–11 November
Mittet R et al (2007a) On the orientation and absolute phase of marine CSEM receivers. Geophysics 72:145–155

Mittet R et al (2007b) CMP inversion of marine CSEM data. In: EMG international workshop, innovation in EM, Grav. and Mag. methods, a new perspective for exploration, Capri, Italy, 15–18 April

Mittet R et al (2008) CMP inversion and post inversion modelling for marine CSEM data. First Break 26:59–57

Morse PM, Feshbach H (1953) Methods of theoretical physics. Ed. McGraw-Hill, New York, Tome 2, p 1769

Nabighian MN (1987) Electromagnetic methods in applied geophysics. Theory, vol 1. Ed. Society of Exploration Geophysicists, Tulsa

Neri P (2011) GPU rendering for volume visualization. New capabilities improve accessibility quality of renderings and accelerate workflow. Offshore. Ed. Pennwell, pp 60–63

Newman GA, Alumbaugh DL (1997) Three-dimensional massively parallel electromagnetic inversion part I. Theory. Geophys J Int 128:345–354

Newman GA, Høversten, GM (2000) Solution strategies for toward three-dimensional electromagnetic inverse problems. Inverse Prob 16: 1357–1375

Nguyen AK, Roth F (2010) Application of the frequency differenced field to 3D inversion of shallow water CSEM data. In: EGM international workshop. Adding new value to electromagnetic, gravity and magnetic methods for exploration, Capri, 11–14 April

Oldenburg (1992) Inversion of magneto-telluric data for a one-dimensional conductivity. Geophysical monograph. Series 1, no. 5. Ed. Gebruder Borntraeger, Berlin

Oritaglio M, Spies B (2000) Three-dimensional electromagnetics. Geophys Dev Ser V V:7

Panofsky WKH, Phillips M (1955) Classical electricity and magnetism. Ed. Addison-Wesley, Reading

Papoulis A (1962) The Fourier integral and its application. Ed. McGraw-Hill, New York, 318 p

Parker RL (1970) The inverse problem of electric conductivity in the mantle. Geophys J 22:121–138

Parker RL (1994) Geophysical inverse theory. Ed. Princeton University Press, Princeton, 400 p

Patella D, Schiavone D (1974) A theoretical study of the kernel function for resistivity prospection with a Schlumberger apparatus in water. Annali di geofisica 27(1–2):295–313

Patra HP, Mallick (1980) Geosounding principles, 2. Time varying geo electric soundings. Methods in geochemistry and geophysics 14 B. Ed. Elsevier, Amsterdam, pp 17–45

Pellerin L, Wannameker PE (2005) Multi-dimensional electromagnetic modelling and inversion with application to near surface earth investigation. Computers and electronics in agriculture. 46. Ed. Elsevier, New York, pp 71–102

Peters LJ, Bardeen L (1932) Some aspects of electrical prospecting applied in locating oil structures. Physics 2:1–20

Piyoosh J, Daniil VS, Sébastien de la Kethulle de Ryhove (2015) Efficient 3-D controlled-source electromagnetic modelling using an exponential finite-difference method. Geophys J Int 203(3): 1541–1574

Plessix RE, Van der Sman P (2007) 3D CSEM modeling and inversion in complex geological settings. Expanded Abstracts, SEG 2007 annual meeting, San Antonio

Plessix RE, Van der Sman P (2008) Regularized and blocky controlled source electromagnetic inversion. PIERS 2008, Cambridge, MA

Plonsey R, Collin RE (1961) Principles and applications of electromagnetic fields. Ed. McGraw-Hill, New York, 555 p

Poirmeur C (1986). Modélisation tridimensionnelle en courant continu. Méthode et applications à la prospection géophysique. Doctoral thesis, Université de Montpellier, 232 p

Porstendorfer G (1960) Tellurik. Grundlagen, Messtechnik und neue Einsatzmoglichkeiten. Ed. Akademie Verlag, Berlin

Porstendorfer G (1975) Principles of magneto-telluric prospecting. Ed. Gebruder Borntraeger, Berlin, pp 47–51

Press SJ (2007) Subjective and objective Bayesian statistics: principles, models and applications. Ed. Willey, Hoboken, 558 p

Price A et al (2008) 1D, 2D and 3D modeling and inversion of 3D CSEM data offshore West Africa. SEG 2008 Expanded Abstracts, Las Vegas, NV, USA
Rakoto H et al (1997) Réinterprétation par inversion bayésienne des sondages électriques sur le lac Tritrivakely. Colloque Geofcan, 11–12 September, Bondy, France,
Ramsey AS (1937) Electricity and magnetism. An introduction to the mathematical theory. Ed. Cambridge University Press, Cambridge, p 115
Roach GF (1970) Green's functions. Introductory theory with applications. Ed. Cambridge University Press, Cambridge, 279 p
Roman I (1934) How to compute tables for determining electrical resistivity of underlying beds and their application to geophysical problems. US Bur. Of Mines. Tech. Pub. 502. AIME Geophysical Prospecting, pp 183–200
Roth F, Zach JJ (2007) Inversion of marine CSEM data using up-down wavefield separation and simulated annealing. In: SEG 2007 conference proceedings, San Antonio
Sainson S (1984) Etude d'une méthode de détection électrique d'anomalies conductrices voisines d'un forage. Thèse de doctorat en géophysique, Université d'Orléans, 177 p
Sainson S (2010) Les diagraphies de corrosion. Acquisition et interprétation des données. Ed. Tec et Doc Lavoisier, Paris, pp 77–81
Sarkisov GA, Andreyev LI (1964) Results and potential of marine electrical prospecting in the Caspian Sea. Int Geol Rev 6(9):1573–1584
Scales et al (2001) Introductory geophysical inverse theory. Ed. Samizdat Press, Golden, 208 p
Schiavone D, Patella D (1974) A direct interpretation method for Schlumberger resistivity soundings with the apparatus immersed in water. In: 19th annual international geophysical symposium, Torun, Poland
Shantsev D, Maaø F (2015) Rigorous interpolation near tilted interfaces in 3-D finite-difference EM modelling. Geophys J Int 200:745–757
Shen LC, Kong JA (1983) Applied electromagnetism. Ed. Brooks, Monterey, pp 326–339
Sinha M, Macgregor LM (2003) Electromagnetic surveying for hydrocarbon reservoirs. University of Southampton. GB patent no. 2 382 875
Slichter LB (1933) The interpretation of the resistivity prospecting method for horizontal structures. Physics 4:307–322
Smith WH, Wessel P (1990) Gridding with continuous curvature splines in tension. Geophysics 55 (3):293–305
Smythe WR (1939) Static and dynamic electricity. McGraw-Hill, New York
Sommerfeld A (1909) Über die Ausbreitung der Wellen in der drahtlosen Telegraphie. Annalen des Physik 386:1135–1153
Sommerfeld A (1912) Die Greensche Funktionen der Schwingungsgleichung. Jahresber Deutsch Math Verein 21:309–353
Sommerfeld A (1952) Electrodynamics. Lectures of theoretical physics, vol 3. Ed. Academic, New York
Stefanescu S, Schlumberger C (1930) Sur la distribution électrique potentielle autour d'une prise de ponctuelle dans un terrain à couches horizontal, homogènes et isotropes. J Physique et le Radium I 4:132–140
Stokes GG (1880) Mathematical and physical papers. Cambridge University Press, Cambridge
Strack KM (1999) Exploration with deep transient electromagnetics. Methods in geochemistry and geophysics. Ed. Elsevier, Amsterdam
Stratton JA (1941) Electromagnetic theory. McGraw-Hill, New York, 615 p
Tai CT (1971) Dyadic Green's function in electromagnetic theory. Ed. Intext Educational Publishers, Scranton, 246 p
Tarantola A (1981) Essai d'une approche générale du problème inverse. Ed. Université Pierre et Marie Curie, Institut de Physique du Globe de Paris
Tarantola A (2005) Inverse problem theory and methods for model parameter estimation. Society for Industrial and Applied Mathematics, Amsterdam, 358 p
Telford WM et al (1978) Applied geophysics. Ed. Cambridge University, London, pp 647–649

Terekhin EI (1962) Theoretical bases of electrical probing with an apparatus immersed in water. Applied geophysics USSR. Ed. Pergamon Press, Oxford, pp 169–195
Thomson W (1884) Reprint of papers on electrostatics and magnetism. Ed. Macmillan, London, p 144
Tikhonov AN (1943) On the stability of inverse problems. Doklady 39(5): 195–198 (in Russian)
Tikhonov AN, Arsenin VY (1977) Solution of ill-posed problems. V H Winston and Sons, Washington
Tompkins M (2005) Electromagnetic surveying for hydrocarbon reservoir. US patent no. 7,191,063
Troyan V, Kiselev Y (2009) Statistical methods of geophysical data. World Scientific Publishing, London
Twomey S (1977) An introduction to the mathematics of inversion in remote sensing and indirect measurements. Elsevier, Amsterdam
Unsworth MJ, Travis BJ, Chave AD (1993) Electromagnetic induction by a finite electric dipole source over a 2-D earth. Geophysics 58:198–214
Utzmann R (1954) Prospection électrique et tellurique: études sur modèles réduits. Bulletin de l'AFTP no. 107. 60 p
Vannamaker et al (1984) Electromagnetic modeling of three dimensional bodies in layers media earths using integral equations. Geophysics 49: 60–74
Verma Rajni K (1986) Offshore seismic exploration: data acquisition, processing, interpretation. Gulf Publishing, Houston, 591 p
Wait JR (1971) Electromagnetic probing in geophysics. Ed. The Golem Press, Boulder, pp 291–293
Ward SH et al (1967) Mining geophysics, vol 2. Theory. Ed. Society of Exploration Geophysicists Mining Geophysics, Tulsa
Weaver JT (1967) The quasi-static field of an electric dipole embedded in a two-layer conducting half-space. Can J Phys 45:1981–2002
Weaver JT (1994) Mathematical methods for geo-electromagnetic induction. Research Studies Press, Somerset
Wenner F (1916) A method of measuring earth resistivity. US Dept. Com. Cur. of Stand. Sc. paper no. 258
Werthmüller D, Ziolkowski A, Wright D (2014) Predicting controlled-source electromagnetic responses from seismic velocities. Interpretation 2(3):SH115–SH131
Weyl H (1919) Ausbreitung elektromagnetischer Wellen über einen ebenen Leiter. Annalen des Physik 60:481–500
Wiik et al (2010) TIV contrast source inversion of nCSEM data. Norwegian University of Science and Technology, pp 6–9
Wijewardena K (2007) Introduction to Fourier transforms in physics. Cambridge University, Cambridge
Wilson W (1933) Theoretical physics. Electromagnetism and optics. Maxwell, Lorentz, vol 2. Ed. Methuen, London, 315 p
Wu X, Sandberg S, Roper T (2008) Three-dimensional marine magneto-telluric resolution for sub-salt imaging and case study in the Gulf of Mexico. SEG 2008 Expanded Abstracts, Las Vegas
Zaborovski A (1936) Problème de sphère dans la théorie de la prospection électrique. Bulletin géophysique pétrolière 2, 8 p
Zach JJ et al (2008a) 3D inversion of marine CSEM data using a fast finite-difference time-domain forward code and approximate Hessian-based optimisation, Las Vegas
Zach JJ, Roth F, Yuan H (2008b) Pre-processing of marine CSEM data and model preparation for frequency-domain 3D inversion. PIERS In: Proceedings, Cambridge, USA, 2–6 July, pp 144–148
Zach JJ, Bjørke AK, Støren T, Maaø F (2008c) 3D inversion of marine CSEM data using a fast finite-difference time-domain forward code and approximate Hessian-based optimization. SEG 2008 Expanded Abstracts, Las Vegas, NV, USA

Zach JJ, Roth F, Yuan H (2008d) Data pre-processing and starting model preparation for 3D inversion of marine CSEM surveys. EAGE 2008 Expanded Abstracts, Rome, Italy
Zach JJ et al (2009) Salt mapping using 3D marine CSEM surveys. CSPG CSEG CWLS convention. Frontiers + innovation, Calgary, Canada, pp 428–431
Zeng ZF et al (2007) The electromagnetic responses of under seabed layer and inversion method study. PIERS Online 3(2):197–201
Zhangshuan H et al (2006) Reservoir-parameter identification using minimum relative entropy-based Bayesian inversion of seismic AVA and marine CSEM data. Geophysics 71(6):77–88
Zhdanov MS (2001) Method of broad band electromagnetic holographic imaging. US patent no. 6253.100, June 26
Zhdanov MS (2002a) Geophysical inverse theory and regularization problems. Methods in geochemistry and geophysics. Ed. Elsevier, Amsterdam
Zhdanov MS (2002b) Three dimensional electromagnetics. Methods in geochemistry and geophysics. Ed. Elsevier, Amsterdam
Zhdanov MS (2002c) Geophysical inverse theory and regularization problems. Volume 36. Methods in geochemistry and geophysics. Ed. Elsevier, Amsterdam
Zhdanov MS (2009) Geophysical electromagnetic theory and methods. Methods in geochemistry and geophysics, vol 43. Ed. Elsevier, Amsterdam, 848 p
Zhdanov MS (2015) Inverse theory and applications in geophysics. Ed. Elsevier, Amsterdam, pp 1–15
Zhdanov MS, Keller GV (1994) The geoelectrical methods in geophysical exploration. Elsevier, Amsterdam
Zhdanov MS, Portniaguine O (1997) Time-domain electromagnetic migration in the solution of inverse problems. Geophys J Int 131:293–309
Zhdanov MS, Traynin P, Booker JR (1996) Underground imaging by frequency-domain electromagnetic migration. Geophysics 61(3):666–682
Zhdanov MS et al (2010) Exploring multiple 3D inversion scenarios for enhanced interpretation of marine CSEM data: an iterative migration analysis of the Shtokman gas field. First Break 28:95–101

Chapter 6
Geological Applications

The character of a science is recognized in its own way to question nature.

(L. Elie de Beaumont)

Abstract This last chapter is devoted to geological applications. It corresponds to a tour of geologically favorable cases. After a succinct recollection of the data-processing process, seven prospecting campaigns borrowed from the technical literature are exemplified. They reflect very different (historically, geographically and geologically) types of prospects. To simplify reading and comprehension, these studies are presented following the same scheme—with their geographical, geological and technical context—and then the interpretation results are presented as profiles, sections or maps. The chapter concludes with an outline of the ongoing developments and an epilogue dealing with commercial success since 2000.

Keywords Case histories • Geological interpretation • Regional study

Preamble

In applied geophysics, the most interesting prospecting results are usually gathered in the form of case histories. These concern the most significant geological interpretations (first achievements) or those with a special interest (regional study) or singularities, or even difficulties in the analysis, where the relevance of operational procedures is then discussed. These case histories are often presented as papers in international conferences and seminars and are published in the form of proceedings in a thematic monograph.

In what follows, we try to simply give the reader a very limited overview of these geological interpretations. The choice to treat only a few examples is deliberate, as advances in modeling and in its corollary, interpretation, are constantly changing. Moreover, we do not want to replace the geologist in digressions that would have no meaning in the absence of specific elements and, especially, sufficient geological knowledge.

S. Sainson, *Electromagnetic Seabed Logging*, DOI 10.1007/978-3-319-45355-2_6

The examples presented here were chosen for their educational and historic qualities. They were selected in very different geological areas (global tectonics, oil and gas exploration, detection of diapirs, surveillance or monitoring of deposits, water research, etc.). They are provided in addition to the examples briefly and partially covered in previous chapters to present the range of possibilities offered by the different methods and techniques currently available.

In addition to this brief presentation, a nonexhaustive set of titles of publications on this subject is given in the bibliography, allowing readers to deepen their knowledge of specific exploration cases. Readers may also regularly consult specialized journals, a list of which is given at the end of volume. A good introduction to these methods is given, for example, in an article written by Professor Kerry Key (2012).

1 Introduction

The interpretative models based on physical and mathematical concepts being finalized (see Chap. 5), the phase of geological interpretation[1] itself can be conducted. This is done on computers or, more precisely, on powerful computing systems and builds after the preprocessing phase on:

- A computer program or code (*software*) specifically developed for the application, or a software package (see Chap. 5)
- A set of electromagnetic acquisition data (*hardware*) directly resulting from the survey
- A set of seismic and geological data from sources outside the survey, adapted to the format

All this information is gathered by the geologist responsible for the project. It is entered in the machine (physically or by download) by a geophysicist operator according to a very precise procedure.

In general, except for the electromagnetic data, much information has already been collected during the seismic survey that has preceded the SBL operation and is stored on data disks or *read-only memory* (ROM) storage such as hard disks[2] (Sheriff and Geldart 1984). Taken together, they concern, in particular, information relating to:

- The geology at large, local and regional (historical geology, stratigraphy, sedimentology, tectonics, petrography, etc.) of the studied prospect
- The traces, profiles and *seismic sections* and their interpretation in 2D or 3D

[1]Not to be confused with the interpretation/modeling that matches the calculations (see Chap. 5).
[2]For a few years the data were stored on magnetic tapes.

- The *sonic logs* (geological formation velocity) for calibration in depth of the seismic horizons

It goes without saying that the more numerous and reliable this information is, the more faithful the interpretive model will be. If, however, the interpreter has *porosity* or *density logs* (electric, nuclear logs, NMR,[3] etc.) the work will be considerably improved.[4]

Currently, most of the models used in geological interpretation correspond to tabular plate-like shape reservoirs where the reservoir rock has a large expansion in surface (lateral dimensions significant but limited however), and whose thickness, usually high, is well demarcated from other strata by the roof and the wall of the layer (see Chap. 1, Fig. 1.1a). This typology, in oil vocabulary, precisely corresponds to the deposit[5] in clusters where oil/water or gas/oil contacts are at the same hypsometric quotation over all the deposit (Abrikossov and Goutman 1986).

Often because of a lack of specific lithological knowledge, the heterogeneity of the tabular reservoir is very difficult to understand and therefore difficult to model. Concretely to a certain extent, only its degree of horizontal anisotropy (direction of the strata when it is identified) can be taken into account in 3D simulations (see Chap. 2, Sect. 4.3).

Numerical models allow us, by a well-defined mesh, to take into account these more unusual morphologies such as in bevel or periclinal endings specified by the seismics. Disk- or sphere-type topologies will be used more rarely, reserved for specific gitological configurations (small deposits or bigger at the end of the exploitation where oil is then concentrated at the apex of the geological structure).

Since 2000, the date of the first industrial operation, hundreds of surveys have taken place across the seas in water depths ranging from 100 to 3200 m.

Here will be treated successively seven examples of different prospects from various geological regions and application areas such as:

- The North Atlantic: RAMESSES III experiment, earth physics
- The Gulf of Guinea: Angolan Basin (oil), first commercial operation
- The North Sea: Troll fields (gas) and Frigg fields (with gas pipelines)
- The Barents Sea: salt diapir
- Malaysia: Sabah field, field monitoring
- Norway: littoral zone, resolution of an environmental problem (freshwater)

Excellent examples of mMT for petroleum prospecting are given in the book of AD Chave and AG Jones (2012).

---

[3]Nuclear magnetic resonance.

[4]All methods of appreciation in situ of the petrophysical parameters such as those from the drill: instantaneous and well logs (*well logging*), downhole and lateral core drillings (drill coring, side wall coring) mud analysis (*mud logging*) and drill cuttings are to be considered especially in this step.

[5]Not to be confused with that defining the mining clusters (*sulfide ores*, for example) forming spheroid-shaped deposits typically interlinked in a metamorphic or volcanic complex geological environment.

2 North Atlantic: RAMESSES III (Oceanic Ridge)

One of the first very convincing experiments was undoubtedly the RAMESSES III operation. It was effected off Iceland on the Reykjanes Ridge (57° 45′ N) in 1997 (MacGregor et al. 1998).

The experiment took place in a context of basic research in marine tectonophysics (submarine volcanism), following several geophysical surveys previously conducted on the site (1993–1996), and was also included in a more general program studying the Mid-Atlantic Ridge.

2.1 *Geological Context*

The Reykjanes Ridge is representative of an area of lithospheric divergence (horizontal plate spacing of 20 mm/year). Besides its geodynamic importance (global tectonics), this ridge attracts particular interest in the sense that one of its parts is emerged (Iceland) giving direct access to valuable information about its structure and composition. Such a rare geological event can thus be regarded as convenient in the comparison of terrestrial and marine results.

2.2 *Technical Context*

About 15 km of profiles, all in all, were carried out with a DASI-type power source (a deep-towed active source instrument) delivering an intensity of about 300 A peak to peak, with an antenna with a moment of 10^4 A.m (100 m length) transmitting in the band of 0.35–11 Hz. The profiling was performed at an elevation of 60 m above the seabed.

Three types of receivers were used:

- ELFR-type (electric field recorder) electric fields receivers and electromagnetic field (LEMUR type)[6] receivers with a *short base* (13 m antenna), measuring the two horizontal components, whose offset was limited to 15 km
- A receiver (LEM type)[7] with a *long base* (300 m antenna), with a single horizontal component to increase the sensitivity (greater field amplitude), with an offset of 90 km

The measuring devices were placed along and perpendicular to the axis of the ridge. The acquisition was carried out at a rate of 64 Hz. The background noise in

---

[6] An acronym for *low frequency electromagnetic underwater recorder*, able to record fields below 10^{-11}V/m.

[7] For a *long antenna EM recorder* (Webb et al. 1985).

the band of 0.1–10 Hz was evaluated over a period of 96 h (time of acquisition) and was estimated in variations depending on the location, in the range of 10^{-12} to 10^{-16} V/A.m^2. In this specific area, the highest values of electromagnetic noise could be currently attributed to the microseismic activity of the ridge.[8]

2.3 *Results: Interpretation*

The technique of geological interpretation consisted of:

- Creating a realistic model of the ridge (see Chap. 5, Fig. 5.12), which can be morphologically inspired from the emerged part represented by the volcanoes of Iceland
- Solving the forward problem by a 2D finite element code adapted to the predefined model
- Making an inversion of the data thanks to the important geological and geophysical information previously acquired (Occam algorithm)

The results are presented in the form of an mEM survey of the seafloor at this structural level, i.e., of resistivity profiles given as a function of depth.

On the interpretative level, they show a gradual increase in resistivity in the first 100 m to reach 10 Ω.m at a 500 m depth (cf. Fig. 6.1). The relatively low resistivity values correspond to a fractured geomorphological complex where conductive infiltration waters flow.

This type of survey also highlighted levels of even higher electrical conductivity with actually, on the scene, the expression of a dominant *galvanic effect*. This could correspond to privileged circulation channels of electric current of large scale, interpreted as major geodynamic structures with a complex network of fractures and faults of significant dimensions, interconnected, which makes the method more attractive to study the oceanic active macrostructures.[9]

In this kind of large scale investigation (the study of ridges more generally), the interpretation can be distorted by two types of problem, directly related to the hydrothermal activity, that substantially alter the apparent conductivity. The first one concerns the very high temperature at this point (350 °C). The second one is the presence of large amounts of *metallic sulfide deposits* escaping through hydrothermal vents (fumaroles). It is therefore necessary in the interpretation models to consider and correct these effects by using, for example, the Becker formula or changing the interpretation algorithms such as those proposed, among others, by Fofonoff and Millard (cf. Chap. 3, Sect. 2.1.1).

[8] Phenomena still poorly understood today, but which may be correlated.

[9] A similar approach was led on the ground in the 1980s (EM measurements from continent to continent: Europe/Africa) by Professor J. Mosnier's team from the ENS in Paris, and thus can be considered on lithospheric compression zones (see Chap. 1, Sect. 6.3), on overlap, collision and, above all, subduction zones (Wheelock et al. 2010).

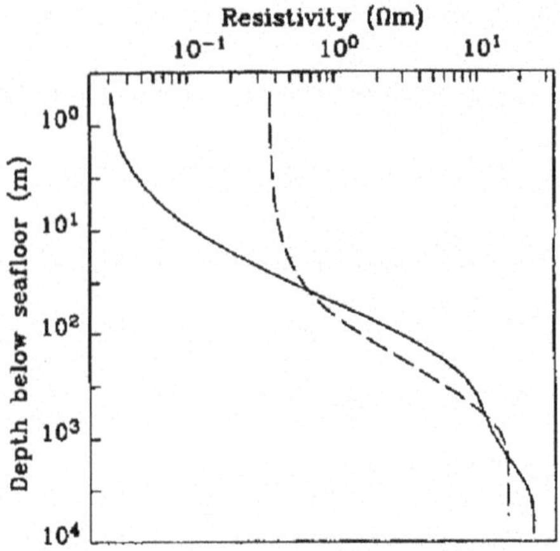

Fig. 6.1 Representative result of the modeling: sounding or resistivity profile depending on the depth (According to MacGregor et al. 1998)

3 West Africa: Angolan Basin (Oil)

The first commercial operation, or more precisely the one conducted on an industrial scale, took place in November 2000 in West Africa. It was initiated and led by the Norwegian oil company Statoil in collaboration with the Norwegian Geotechnical Institute (NGI), the Scripps Institution of San Diego (California) and the Oceanographic Centre of Southampton (UK), and took place in the Gulf of Guinea (Ellingsrud et al. 2002).

3.1 Geological Context

The prospected Angola Basin (block 17 located in deep water, 135 km of the Angolan coast), more than 10 km in thickness and identified as potentially promising in hydrocarbons, is overall present in the form of a thick sedimentary secondary series composed of alternating sand and shale layers of high thicknesses (cf. Fig. 6.2).

The survey focused on an area with identified (seismic) traps, slightly buried ($\approx$100 m), due to the presence of sandy and salty allochthonous channels present in the Aptian layers (lower Cretaceous: -120 My).

3.2 Technical Context

The survey, just over 300 km of unwound profile, took place under approximately 1200 m of water and under a cover of marine sediments around 1000 m thick. All in all, 26 bottom receivers were deployed on 17 lines (cf. Fig. 6.3a).

The acquisition was made at a rate of 25 Hz for the mMT measures and at a frequency of 32 Hz for mCSEM measures totaling more than 110 h of continuous

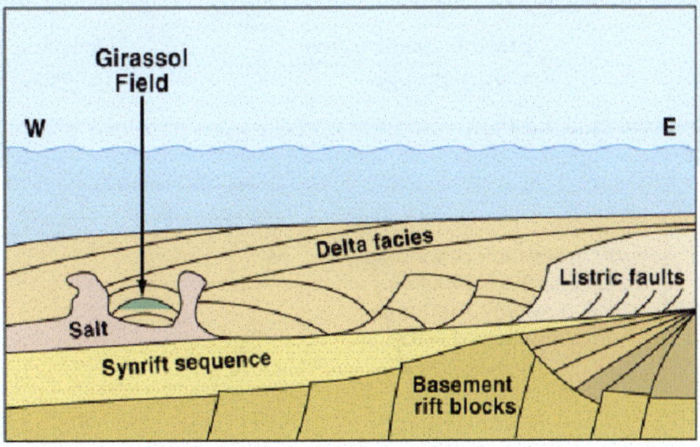

Fig. 6.2 Girassol deep water oil field operated by Total and BP (West Africa), discovered in 1996 with a production of 200,000 barrels a day with a total proven reserve estimated around 700 million barrels

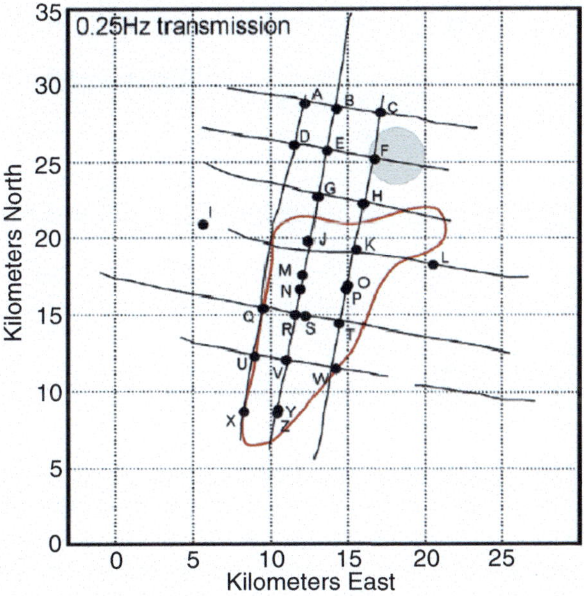

Fig. 6.3a Detail of the prospect and the conductivity anomaly (According to Ellingsrud et al. 2002)

recording. The EM source used (0.25–0.75 Hz, 300 A) was provided with a 55 m antenna (moment of 16,500 A.m).

To prepare the model and perform the necessary corrections for the interpretation of electromagnetic acquisition data, additional measurements were previously made during the EM acquisition and a posteriori with:

- Immersion of a current meter buoy in the center of the operating device floating in the water a few meters away from the seafloor

Table 6.1 Values of oceanographic parameter seabed data by complementary measures

Temperature	Electric conductivity	Oceanic current velocity
4.00–4.15°C	3.36–3.42 S/m	10–25 cm/s

According to Ellingsrud et al. (2002)

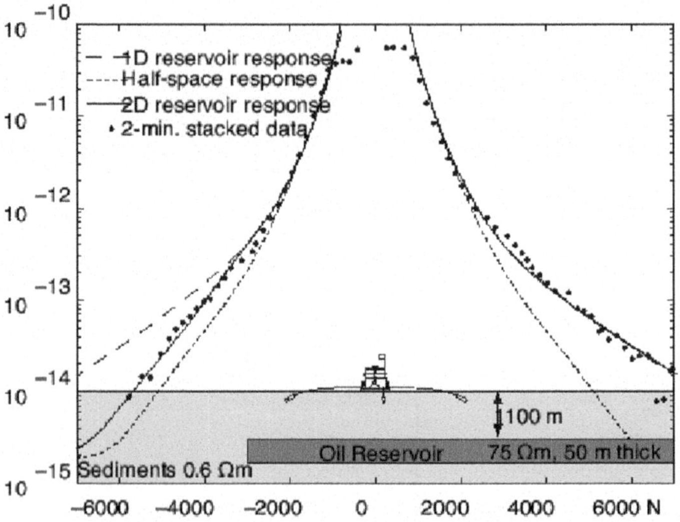

Fig. 6.3b Radial electric field in V/(A.m^2): 1D and 2D responses: Girassol reservoir offshore from Angola (According to Constable and Srnca 2007 and Ellingsrud et al. 2002)

– Electrical logging (logs) of a few places in the area to precisely estimate the electrical resistivity of the marine sediments (Table 6.1)

3.3 Results: Interpretation

The results are presented as a 1D profile and as a 2D map of the reports of standard fields for different depths. Conductivity anomalies are characterized by the sensitivity of the measurements in *in-line* and *broadside* configurations on their radial and azimuthal horizontal components. These maps show then the emergence of a salt dome more specifically marked on the radial component, thus specifying the position of the reservoir (cf. Figs. 6.3b and 6.3c).

4 North Sea: Troll Field (Gas)

Discovered in the 1980s in the North Sea, the strategic Troll field, located 61 miles off the coast of Norway, presented so far a very demonstrative operation theater for the development of the SBL method.

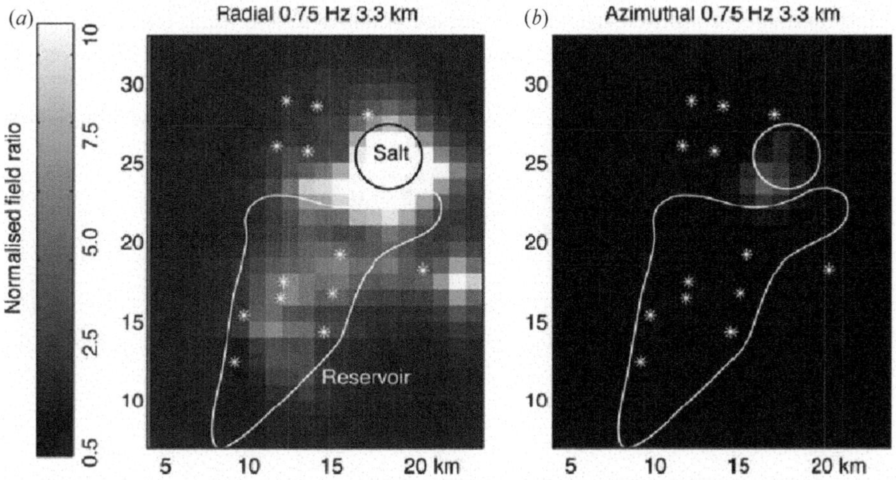

Fig. 6.3c Results: maps of the normalized electric fields allowing us to individualize in the plan the reservoir of hydrocarbons of a saliferous structure in the Angola Basin. They show the predominance of the radial field in the in-line configuration (*a*) and, to a lesser extent, that of the azimuthal field in the broadside configuration (*b*) (According to Ellingsrud et al. 2002)

This field was operated in December 2003 by a consortium composed of the Norwegian oil company Norsk Hydro and the Norwegian services company EMGS. The survey highlighted the presence of hydrocarbons in a potential trap previously spotted by 3D seismics (Johnstad et al. 2005).

4.1 Geological Context

The Troll field has one of the largest gas fields in the Norwegian North Sea concession. Located in the most northern part of the zone under a cover of water of 340 m on average, the deposit is in a Jurassic sandstone formation 100 m thick. The reservoir at 1000 m depth has a resistivity of 250 Ω.m, contrasting sharply with the surrounding sedimentary layers (1–2.5 Ω.m).

4.2 Technical Context

A total of 41 field sensors (24 above the structure) were deployed across the prospect, along a profile SW–NE about 20 km long, overflowing widely on both sides of the structure. The electromagnetic stimulation is made from a 1000 A controlled source emitted from an immersed antenna 230 m long, positioned by a short base ultrasonic communication device (Amundsen et al. 2004; Farrelly et al. 2004a).

Various transmission frequencies were tested to optimize the EM responses and thus obtain a multidepth investigation (survey) with:

- A monochromatic continuous square wave of 0.25 Hz
- A pulse consisting of three sines (0.12 Hz, 0.48 Hz and 0.96 Hz)
- A pulse of 0.1 Hz

The recording of electric fields (horizontal components) and magnetic fields for certain sensors ($\times 12$) was performed in the frequency domain with an offset of about 6.5 km.

4.3 Results: Interpretation

The following figures show a series of views allowing us to see the whole prospect in section and relief. Advances in imaging now make it possible to visualize the entire field by virtual 3D or holographic observation methods. This type of representation then allows precise positioning and implement of the drillings (cf. Fig. 6.4).

We would like to remark that the seismic survey generally offers little convenience for the distinction between oil and gas. In contrast, the use of SBL in a well-established geological context and in some favorable conditions can help remove the uncertainty. Today this aspect of the question has major interest in the context of an increasingly evident decoupling of oil and gas prices.

5 North Sea: Frigg Field (Gas and Pipelines)

The Frigg field, operated since the early 1970s by, among others, the French company Elf Aquitaine (now Total) is the oldest field discovered in the North Sea (1971). This gas field is one of the largest offshore deposits in the world and is interesting in the sense that it already has a subsea infrastructure including a network of pipelines (collection and transport) dense enough to create, by its presence, significant interferences in electromagnetic measurements.

5.1 Geological Context

Situated between Norway and the Shetland Islands, located west of the Viking Ocean Trench in a water depth of 100 m, the Frigg reservoir is in the Eocene

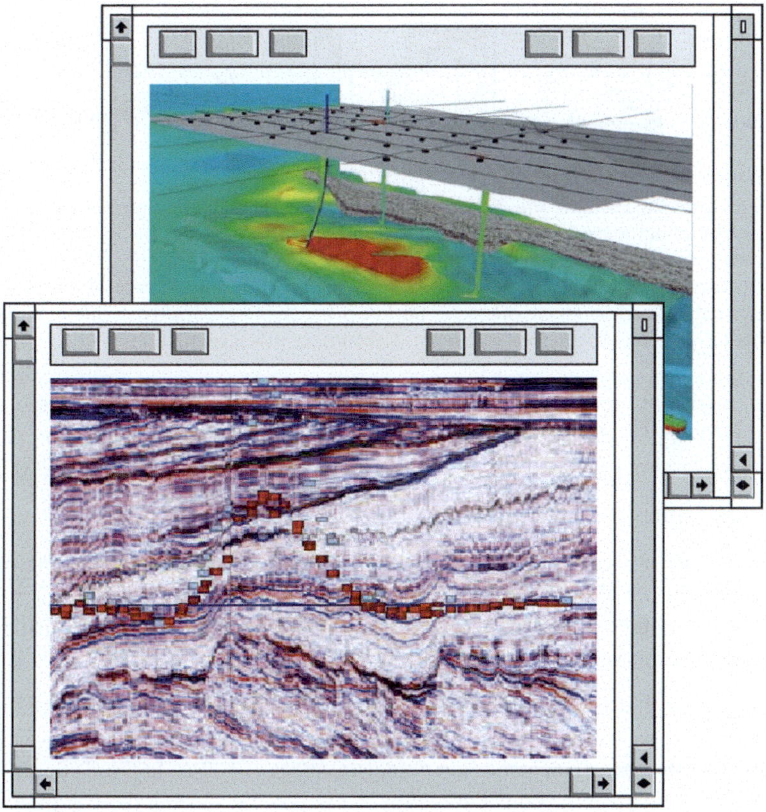

Fig. 6.4 Modeling result of electromagnetic prospecting (plan and profile) of the Troll gas field in the North Sea (TWGP oil deposit) and geological interpretation wedged on the seismic section (structural model). The normalized amplitude values ■ (x axis) correspond to an offset of 6.5 km. The presence of oil in the reservoir has been confirmed by drilling and resistivity logs (According to EMGS and Farrely et al. 2004a, b)

sandstones (−45 My) nearly 1800 m deep. The structure has a lobed shape known as the "Chinese Butterfly" (Héritier et al. 1979; Perrodon 1989).

5.2 *Technical Context*

The presence of a large metal underwater complex for operation (pipelines, risers, manifolds, flow lines, wellheads, etc.) required de facto an extensive and tight coverage of receivers around the production platform, especially taking into account physically the subsea infrastructures in electromagnetic recordings.

In this type of marginal technical context, it is often essential to make beforehand a cartographic or even geographical statement of the precise location of the network of pipelines.

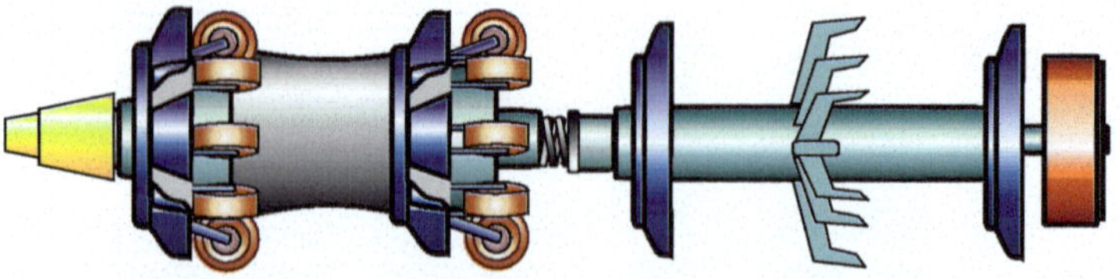

Fig. 6.5 Instrumented scraper (*intelligent pig*) for pipeline 3D mapping (internal in-line inspection). The tool is introduced upstream and received downstream of the line by two pressurized stations (sas) placed in bypass on the pipe. This type of scraper is equipped with an *inertial platform* (generally consisting of three accelerometers stabilized by mechanical gyroscopes), a mechanical caliper and an electromagnetic thickness imager to eventually evaluate mechanical constraints through deformation measurements (According to Sainson 2007)

It is made by passing inside the production lines an instrumented pig[10] (cf. Fig. 6.5), by far the preferable solution, or, for lack of anything better, outside by underwater sonar (AUV type)[11] and/or surface sonar (Sainson 1990, 2007).

The acquisition is carried out using:

- A source of power of 1250 A emitting a pulse composed of a fundamental frequency of 0.2 Hz and even harmonics (0.4 Hz, 0.6 Hz, 1 Hz and 1.2 Hz)
- 106 electrometers spread on ten lines among which approximately a third are for measures of the vertical component (cf. Fig. 6.6)

We can note that in the North Sea, for protection against corrosion, pipelines in addition to their protective coating[12] are provided with soluble zinc anodes,[13] which can deliver, in the case of an insulation defect, currents of 1–2 A. In these conditions of a galvanic coupling, for the considered pipe diameters, the current density in this case is approximately equal to 10 mA/cm^2, which then corresponds at most to electric fields of a few μV/m (Sainson 2007). Cathodic protection by a sacrificial anode is established under a continous regime (DC) or pulsed regime (alternative rectifier), which means that theoretically, the latter, when it produces (in the presence of an insulation defect), should not influence the mEM measurements. However, it is important to note that continuous fields located in mobile environments (a moving

[10]*Internal and automatic technics of oil pipeline auscultation*: Control equipment (NDT to measure the thicknesses of the pipeline or inertial equipment for their mapping) involves a specific scraper (*pig*), which moves with the fluid. This elegant means of line inspection (*intelligent pigging*) is done when the pipeline is in charge (dynamic), which gives this method the advantage of not stopping production.

[11]*Autonomous underwater vehicle*: This type of investigation now comes at a higher cost than that offered by *instrumented pigs*, because the number of these vehicles is restricted and the operation setting is more complex (immersion of acoustic localization databases). However, it is very useful for the localization of wellheads, manifolds and other subsea equipment, which can only be done by this means. In this case, we can use the US echo localization device of the ongoing EM investigation (see Chap. 4, Sect. 6.2).

[12]Materials conforming to the international standards ISO 21809–2: 2007 and ISO 21809–2: 2014.

[13]Zinc is a more electronegative metal than steel, which allows it to give its electrons more easily in the medium.

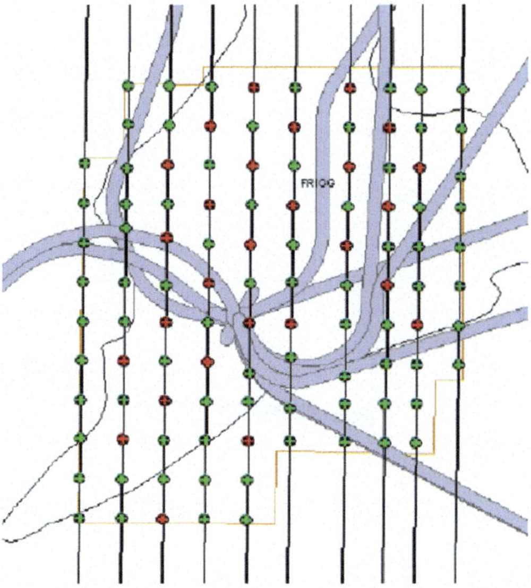

Fig. 6.6 Map of the location of seafloor EM sensors above a network of subsea pipelines. The lines are separated by 990 m and the sensors between them by about 1000 m (According to Price et al. 2010)

water mass, for example) can cause extra variable fields in the range of low and very low frequencies (see Chap. 3, Sect. 6.9.2). The few surveys conducted so far have not revealed this type of parasite event. In the case of a problem, if the existence of these electrodynamic phenomena is proven, then it is possible to estimate their value in situ by potential measurements carried out, for example, by underwater vehicles (ROVs).

5.3 *Results: Interpretation*

The electromagnetic signature of submerged metal structures can be theoretically evaluated by specific modeling (synthetic responses) taking into account the physical and topological characteristics of the pipeline on the site (cf. Fig. 6.7).

This anomalous mapping is then subtracted from multifrequency acquisition data previously processed by a 3D model set from precise knowledge of the characteristics of the deposit, i.e., the available seismic, well-logging and petrographic information (see Chap. 3, Fig. 3.46).

6 Barents Sea: Nordkapp Basin (Diapir)

Salt diapirs are geological structures (salt tectonics) that are convenient in hydrocarbon trapping.[14] On the electrical plan, these can be considered as rather resistant (1000 Ω.m), even very resistant, but at their periphery some

[14]Favorable geological structures discovered in the 1930s in the southern USA (Gulf Coast oil fields, 1936).

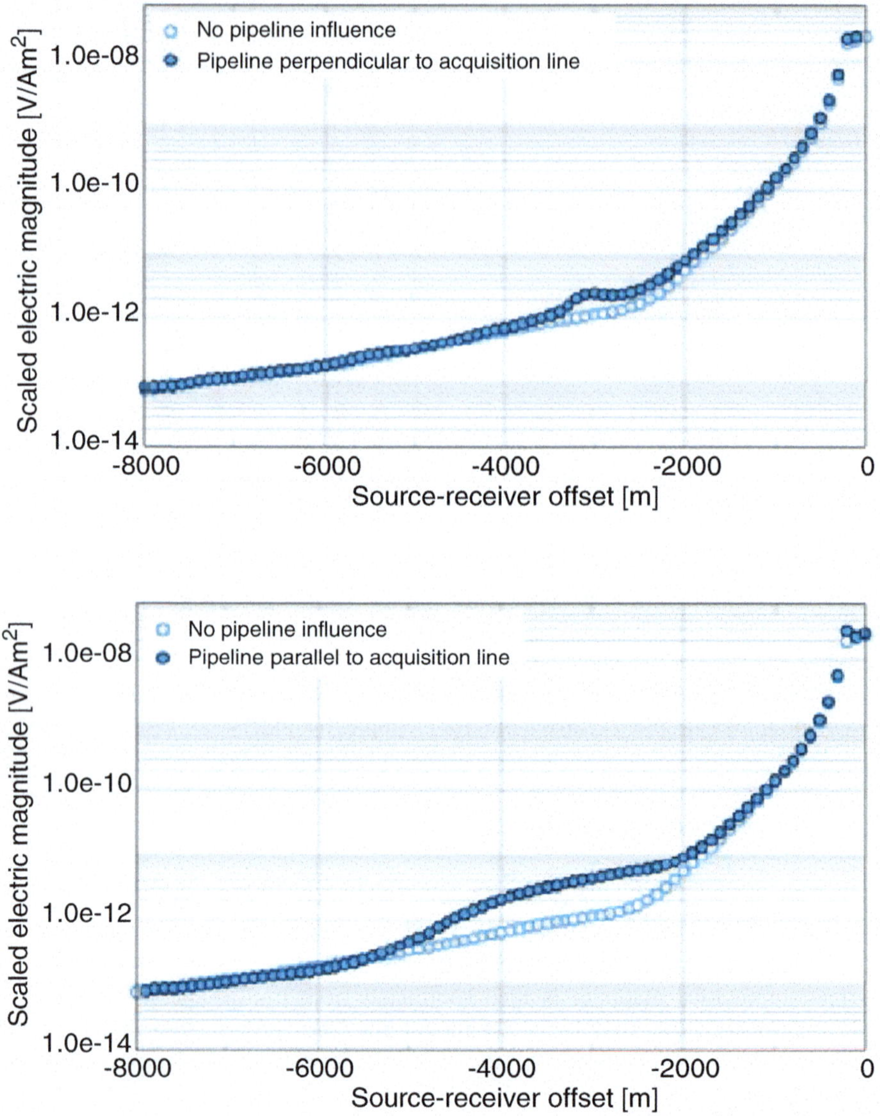

Fig. 6.7 Effect of a pipeline on electric field measurements for two receivers (e.g., 1.2 Hz). In this case the measure profiles are perpendicular to the pipeline (3 km) (According to Hamilton et al. 2010)

areas (?) are generally more conductive[15] (cf. Fig. 6.8). In general it is difficult from the surface to "illuminate" the lower parts of these domes. Only appropriate well seismics (a *walkaway vertical seismic profile*) allow us, in some favorable

[15]Once wrongly attributed to salt dissolution in the periphery (leaching), the highest conductivity would rather come from a superficial phenomenon mainly due to the continuing rise of the dome during recent geological times (Schlumberger 1930).

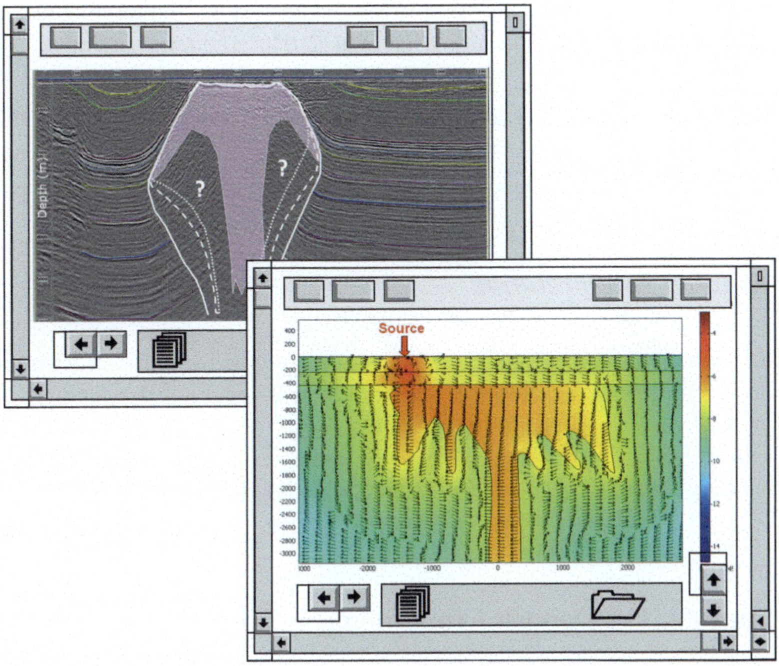

Fig. 6.8 Seismic section and visualization after modeling of the strength field at the level of the Uranus salt diapir (Barents Sea) (According to Vöge et al. 2010)

cases, to obtain satisfactory results after, of course, an extensive gravimetric study.[16]

6.1 Geological Context

The Uranus diapir is in the Barents Sea and, more specifically, in the Nordkapp Basin located north of Norway. According to experts, it should contain about thirty similar salt structures. These domes would have formed from Permian salt, then mobilized during a Triassic sedimentation phase, with concomitant formation of hydrocarbons. Then it would have suffered a subsequent erosion phase followed by Cretaceous sedimentary cover in its terminal phase (Nielsen et al. 1995).

6.2 Technical Context

The survey was carried out at a water depth of 230 m. The acquisition chain consisted of three measurement stations placed on the seabed above the diapir,

[16]In most cases, the gravimetric anomaly cannot be individualized—in other words, many causes may give the same answers.

each comprising a seismic unit (four seismometers and a remotely located air gun[17]) and an electromagnetic unit (Vöge et al. 2010). The latter included:

- Two electric lines (dipoles of 20 m), separated by 180 m
- Three triaxial fluxgate magnetometers
- A 150 m dipolar electromagnetic source, emitting a sweep (0.01–250 Hz) at 100 A and placed at a distance of 300 m from the downhole sensors

6.3 Results: Interpretation

The data (field values) were analyzed using an automatic inversion program using data coming from seismics for the determination of geological structures and a modeling EM code (2.5D) of finite elements.

Resistivity values were provided by data directly coming from the Uranus well (resistivity log), which partly crosses the salt dome (Hokstad et al. 2009; Vöge et al. 2010).

7 Malaysia: Oil Deposit Monitoring

An interesting application and one that is still little spread, which does not enter within the framework of the prospecting itself, is that of deposit time tracking or *monitoring* (see Fig. 6.9). This type of approach already exists in active and passive

Fig. 6.9 Oil field monitoring (According to CGGVeritas 2008)

[17] An offshore seismic source.

(listening) seismics and has had huge success in the evaluation monitoring oil field (production).

7.1 Principles of the Method

The principle is to perform, on a field already in operation, some measures over time (*time-lapse technics*) and to compare them retrospectively, hence the common name mCSEM 4D (3D + time).

7.2 Objectives Reached or to Be Achieved

Under favorable conditions (the same locations of sensors, same material used for exploration, same calibration, suitable interpretation software, etc.) we assume that the values of electrical conductivity and their variations can highlight the evolution in space and time of the contact limit water/oil[18] and gas/oil. These fundamental data about reservoir production, only given by the mEM method, then allow us more or less to:

- Follow (in nonreal time) precisely the exhaustion of the oil deposit by comparing the mEM data between them and with those of the wells (bottom and surface)
- Check and control (in real time) the phases of *fluid injection* (gas or water)
- Determine the groundwater charge in a *water-drive* system[19]
- Analyze the effectiveness of the gas lift in a *gas-cap* system
- Identify changes in water/oil/gas limits (relative data) and, if appropriate, changes in their contacts (absolute data)
- Report changes in hypsometric ratings
- Increase the accuracy of material balances (inputs, outputs)
- Appreciate the lithological cohesion factors characterizing the links from bench to bench and the continuity factors characterizing the degree of extension of these benches
- Establish the hydraulic degree of dependence of the oil deposits in the same field
- Observe the effectiveness of *well stimulations*, *reservoir acidification*, etc.
- Guide *hydraulic fracturing* operations (performance analysis)
- Detect *leaks* and losses of the product
- Assess changes in the extent of *polluted areas*
- Build and especially complete *piezometric* and *bathymetric maps*

[18]Difficult to establish with the usual methods (seismic and wireline logging especially).

[19]For these techniques, see M. Muskat's work (Muskat 1949).

- Characterize *formation waters*
- Map the effective thickness of the zone soaked with oil and its internal and external outlines (isopachous lines)
- Identify variations in the walls and roof of the productive layer (in charge)
- Assist in the calculation of reserve evaluation
- Predict volumes and their migration (Vieugue and Tinder 2015)
- Optimize well production and the oil recovery ratio
- Establish forward operating systems
- Propose laws of distribution for the various types of oil deposit to organize and clarify future campaigns of detail prospecting

Surely, with such a tool at their disposal, improved if necessary by microseismic monitoring (common acquisition chain as proposed by CGGVeritas cf. Fig. 6.9) or gravimetric and clinometric monitoring (for study of *subsidence*, for example), reservoir engineers will not fail to find for this nonexhaustive list other applications equally useful to their job with regard to well data. These may be already checked by simulation of synthetic data (Andreis and MacGregor 2011).

7.3 Technical Context

The technique presently consists of performing several surveys at different times in the life of the reservoir with the existing prospecting equipment. However, we can consider in the future permanent equipment adapted to the problem, which is more compact with greater durability and real-time telemetry (cf. Sect. 9).

7.4 Results: Interpretation

Figure 6.10 shows an example of monitoring on a deposit in Malaysia operated by the Shell company in the Sabah region, where we can observe the progress of the water in the reservoir (Black and Zhdanov 2009).

Permanent mEM or semipermanent (time-lapse mEM) monitoring should take a more and more important part in the future management of the exploitation of offshore hydrocarbon deposits (Kasahara et al. 2010) and particularly in that using enhanced recovery means[20] (Ziolkowski et al. 2010; Salako et al. 2015).

[20]Dynamic means to pressurize the reservoir by injecting water (*water flooding*) or gas (*gas flooding*) into secondary wells. This secondary energy intake (excluding the deposit) must be decided at the appropriate time to maintain acceptable operating conditions.

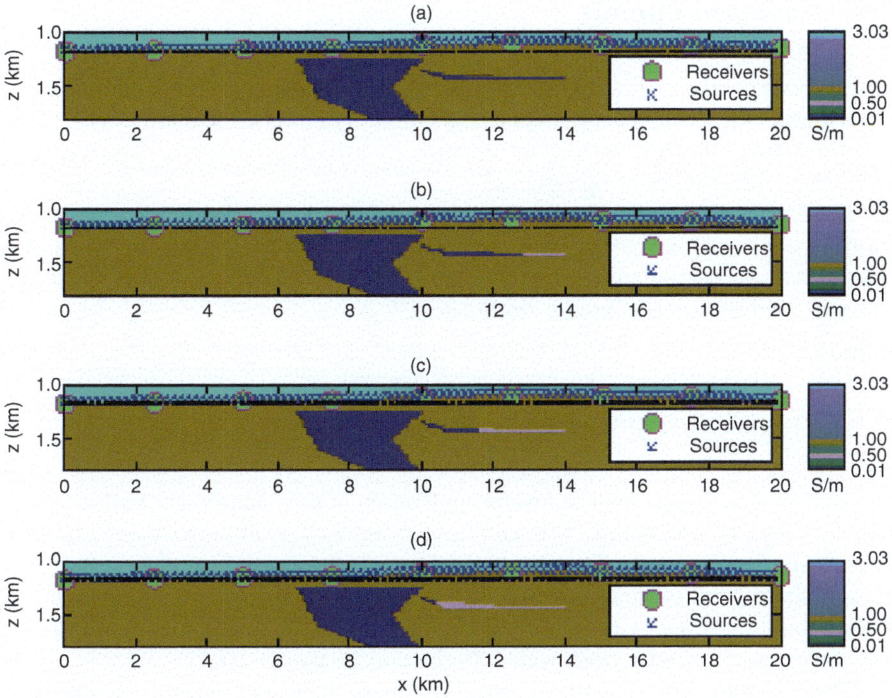

Fig. 6.10 Evolution of a vertical section of electric conductivity over time (technical time-lapse) on a deposit associated with a saliferous structure (Sabah, Malaysia) (According to Black and Zhdanov 2009)

8 Example of Prospecting by Direct Current

Prospecting by direct (DC) or pulsed (inverted) current (to avoid polarization effects at the measurement electrodes) is reserved only for operations under a low water cover (see Chap. 2, Sect. 2.7.1.2). In the past this type of survey was used for recognition of the extension of onshore oil fields.

8.1 Geological Context

These methods are the oldest to have been developed (see Chap. 1, Sect. 6.2). They have been so far little used and often reserved for very specific auscultation cases. This is the case, for example, with geological, hydrogeological, and geotechnical recognition of lakesides, harbors or even coastal areas.

8.2 Technical Context

This technique then makes it possible, usually due to previous or concomitant land acquisition, easy to implement, to extend the investigations to the marine area while benefiting from all the necessary information (resistivity, geological sections, etc.) in the use of data inversion programs.

We then obtain maps and resistivity sections, which can make appear, in the form of isoanomalous curves, some specific geological areas with, for example, risks or dangers of contamination to the environment (seawater infiltration, silting areas, ground subsidence, polder study, etc.).[21]

8.3 Results: Interpretation

Figure 6.11 is an example of maritime application of a Schlumberger half-inverted survey materially composed of seven lines carried by a 60 m long submerged *flute* (see Chap. 2, Fig. 2.19). The data inversion is performed in real time at each measurement point geographically seated by GPS and provides, thanks to a three-layer geometric model (seawater/intermediate layer/clay layer), the depth of the sand layer saturated with fresh water (Baron and Chapelier 2003).

These electrical investigations are all the more interesting when it is relatively difficult to highlight these sedimentary characteristics with conventional acoustic surveys (costal sediment sounder) and when this is virtually impossible with mechanical soundings (destructive drilling), technologies also with a much higher cost.[22]

8.4 Other Mining Applications

Marine exploration with DC, as its equivalent onshore, may also have mining applications such as finding *hydrothermal ore* deposits in accretion zones (Francis 1985), *seabed massive sulfide* deposits (SMS) in subduction areas, evaluating the thickness of the permafrost in arctic regions (Corwin 1983) or detecting massive methane hydrates (about several tens of Ω.m) in the deep seafloor (Goto et al. 2008). Indeed, few attempts have been made so far, but with the decrease of rare metal resources, in sulfides, research in this direction is likely to intensify in the future.

[21]The first study of this type was conducted on Lake Ijssel in the Netherlands (Volker and Dijkstra 1955).

[22]The penetrometer can sometimes be used with profit.

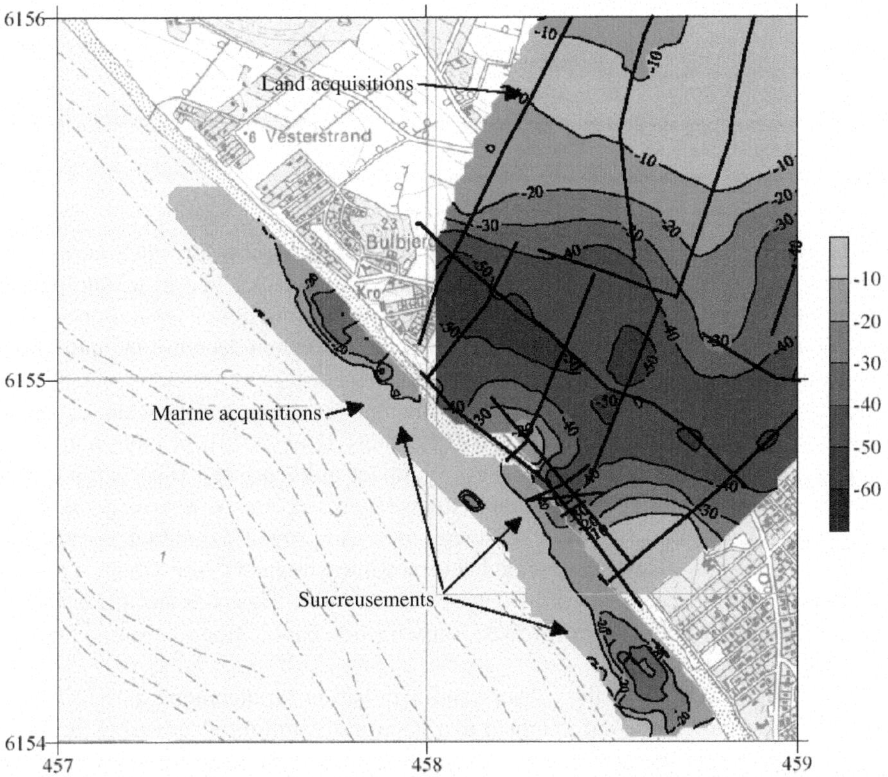

Fig. 6.11 Map of resistivity values extended to the marine domain to define the geometry of a freshwater aquifer, assess its lower rating and highlight its level of digging (−30 m) (According to Baron and Chapelier 2003)

9 Ongoing Developments

The examples that have been developed above show that mEM precision is relative and is obtained in advance thanks to seismic data. This is especially due to the specific defect in the EM marine method, which will always remain integrative, but also to the important dimensions and the limited amount of acquisition equipment.

It goes without saying that the separated and simultaneous measurement of the six components in phase and amplitude of the mEM fields (associated electric and magnetic fields) will be crucial in the future in improving the knowledge of the subsoil (Constable and Key 2005; Nichols and Cuevas 2009; Kjerstad 2011). These measurements will then offer much broader exploration possibilities going from retail prospecting to deposit monitoring through very low frequency investigations for deep layers. Much smaller targets can also be achieved with greater precision, as is the case in the context of mineral exploration (Boda et al. 2014).

Fig. 6.12 Example of application: OBC (ocean bottom cable) modified with a seismic sensor on the *left* (Sercel-type) and an mEM current density sensor on the *right* (Sainson-type). Low frequency acquisition allows telemetry common to both systems

Based on these findings, progress will inevitably require the integration and miniaturization of electric and magnetic field sensors using hybrid technologies. These allow us then to obtain more reliable transfer functions (separation of the electric and magnetic fields, for example) and a consequent decrease in equipment congestion (ASIC).[23]

Current density vector sensors with magnetic sensors today seem to be a privileged research path (Boda et al. 2014; Baicry et al. 2015). In large numbers, arranged in a *phased array,* these very compact elements can then equip more complex linear or surface acquisition systems (such as *synthetic aperture antennas* and multipole panels) but more efficient ones (improved azimuthal coverage). These sensors can also be integrated into seismic streamers (see Chap. 4, Sect. 5.3 and following), on seabed OBC-type cables[24] (cf. Figs. 6.9 and 6.12) or on downhole optical fiber cables (probes), on wire line cables (logging tools) and on coiled tubing, for example.

The miniaturization of the sensors and acquisition arrangements now offers a field of investigation and enforcement that is virtually unlimited, covering the field of 3D imaging at large, such that one can envisage it as with seismics in holography and tomography.

10 Epilogue

This very brief overview, which is of course not exhaustive, of geological applications with electromagnetic methods at sea should not cause us to forget, apart from DC techniques, the need to use first seismic information (Fanavoll et al. 2014) and eventually electrical information when it is available (logs), essential in the processes of data inversion and therefore interpretation (Tseng et al. 2015).

[23]*Application Specific Integrated Circuits*: IC assembling a set of discrete and active components specifically adapted to a precise electronic function (detection, amplification, etc.) enabling the miniaturization of the sensors.

[24]Cables on which are mounted velocimeters (geophones), accelerometers or hydrophones (pressure). The recording of the vibratory characteristics of the waves (displacement fields, speeds, pressures) from the reservoir and then the calculation of the connected acoustic impedances allow us, through petromechanical models, to establish variations in time in the elastic properties of the reservoir, the latter being directly related to the dynamic state of the oil deposit.

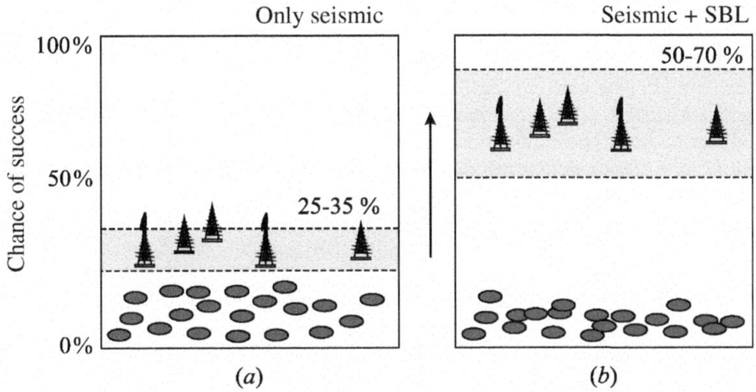

Fig. 6.13 The combination of seismic reflection and SBL statistically increases the chances of success of discovering a rise from 30 % (*a*) to 70 % (*b*) (According to Ridyard and Heshammer 2011)

Combining the concepts of feasibility indices (see Chap. 3, Sect. 6.8.1), of operational risks and probabilities of discovery, some authors (Ridyard and Heshammer 2011) have recently been able to identify in a set of oil operations (86 in total) some statistical results of quite promising success, showing then an increase in the chances of success from 30 to 70 % (cf. Fig. 6.13).

Since the early 2000s, all operations conducted so far have shown that SBL methods indeed provide a relatively reliable indication of the presence of hydrocarbons in proven traps, thus confirming and reinforcing the purpose of this method (Kong et al. 2002; Lindon et al. 2007; Mohr et al. 2008; Buland et al. 2010; Hesthammer et al. 2010; Furuholt 2012).

References

Abrikossov I, Goutman I (1986) Géologie du pétrole. Généralités, prospection, exploitation. Ed. Mir. pp 204–205

Amundsen HE, Johansen S, Rosten T (2004) Seabed logging calibration survey over the Troll gas field. EAGE 66th Conference & Exhibition, Paris, June

Andreis D, MacGregor LM (2011) Using CSEM to monitor production from a complex 3D reservoir. A synthetic case study. SEG paper, pp 1070–1079

Baicry M, Le Prado M, Le Frou C, Rouve LL (2015) Calibration of a device for measuring and electric field in a conducting medium. PCT Patent no. WO2015067884

Baron L, Chapelier D (2003) Application des méthodes électriques lacustres et terrestres à l'étude d'un aquifère côtier. 4ème colloque GEOFCAN. Orléans

Black N, Zhdanov MS (2009) Monitoring of hydrocarbon reservoirs using marine CSEM method. In: SEG Houston International exposition and annual meeting

Boda M, Coillot C, Sainson S (2014) Electromagnetic sensor for SMS geophysical survey. Concours mondial de l'innovation proposal

Buland et al (2010) The value of CSEM data in exploration. Paper C006. In: 72nd EAGE conference & exhibition. Barcelona. 14–17 June

Constable S, Key K (2005) Three-axis marine electric field sensor for seafloor electrical resistivity measurement US patent no. 7482813

Constable S, Srnca LJ (2007) An introduction to marine controlled-source electromagnetic methods for hydrocarbon exploration. Geophysics 72(2):WA3–WA12

Corwin RF (1983) Marine permafrost detection using galvanic electrical resistivity methods. In: Proceedings of offshore technology, conference Paper OTC 4480

Ellingsrud S, Sinha MC, Constable S, MacGregor LM, Eidesmo T, Johansen S (2002) Remote sensing of hydrocarbon layers by sea bed logging (SBL): results from a cruise offshore Angola. Lead Edge 21:972–982

Farrelly B et al. (2004a) Remote charaterization of hydrocarbon filled reservoir at the Troll field by sea bed logging. EAGE fall research workshop on advances in seismic acquisition technology. Rhode, 23 September

Farrelly B, Ringstad C, Johnstad, SE, Ellingsrud S (2004b) Remote characterization of hydrocarbon filled reservoirs at the troll field by sea bed logging. EAGE Fall Research Workshop on Advances in Seismic Acquisition Technology, Greece, 19–23 September

Francis TJG (1985) Electrical methods in the exploration of seafloor mineral deposits. In: Teleki PG (ed) Marine minerals. D. Reidel, Dordrecht, pp 413–420

Furuholt V (2012) How to use CSEM to rank prospects. Offshore 72(4):50–61

Goto T, Kasaya T, Machiyama H, Takagi R, Matsumoto R, Okuda Y, Satoh M, Watanabe T, Seama N, Mikada H, Sanada Y, Kinoshita M (2008) A marine deep-towed DC resistivity survey in a methane hydrate area, Japan Sea. Explor Geophys 39:52–59

Hamiltion MP, Mikkelsen G, Pourjadieu R, Price A (2010) CESEM survey over the Frigg gas field, North Sea. In: 72nd EAGE conference and exhibition incorporating SPE EUROPEC. Barcelona, 14–17 June

Héritier FE, Lossell P, Wathne E (1979) Frigg field. Large submarine fan trap in lower Eocene rock of North Sea. AAPG Bull 63(11):1999–2020

Hesthammer J et al (2010) CSEM performance in light of well results. Lead Edge 29(34):258–264

Hokstad K, Myrlund EA, Fotland B (2009) Salt imaging in the Nordkapp Basin with electromagnetic data. In: AAPG 3- P Arctic Conference and Exhibition, Moscou

Johnstad SE et al (2005) Seabed logging on the North Sea Troll field. In: Offshore technology conference. OTC 17661. Houston. 2–5 May

Kasahara J, Korneev V, Zhdanov MS (2010) Active geophysical monitoring. In: Handbook of geophysical exploration, vol 10. Ed. Elsevier, Amsterdam, pp 135–159

Kjerstad JK (2011) Subsea vertical electromagnetic signal receiver for a vertical field component and also a method of placing the signal receiver in uncompacted material. US patent no. 20110210742

Kong FN, Westerdahl H, Ellingsrud S, Eidesmo T, Johansen S (2002) Seabed logging: a possible direct hydrocarbon indicator for deep sea prospects using EM energy. Oil Gas J :30–38, May 13

Lindon et al (2007) Electromagnetic prospect scanning. Seabed logging moves risk reduction to value creation. SPE paper 108631. In: International oil conference and exhibition, Vera Cruz

MacGregor LM, Constable SC, Sinha MC (1998) The RAMESSES experiment—III: controlled source electromagnetic sounding of the Reykjanes Ridge at 57 to 45′N. Geophys J Int 135:773–789

Mohr J et al (2008) Seabed logging acquisition as a tool in exploration decision making. Lead Edge 27(4):532–536

Muskat M (1949) Physical principles of oil production. Ed. McGraw-Hill, New York, pp 528–644

Nichols E, Cuevas N (2009) Deriving an electromagnetic field in one direction based on measurement data of one or more sensing elements for measuring an electromagnetic field in another direction. US patent no. 8289025

Nielsen KT, Vendeville BC, Johansen JT (1995) Influence of regional tectonics on halokenesis in the Norkapp Basin, Barents Sea. AAPG Mem 65:413–436

Perrodon A (1989) Le pétrole à travers les âges. Ed. Boubée, p 207
Price A, Mickkelsen G, Hamilton M (2010) 3D CSEM over Frigg dealing with cultural noise. SEG annual meeting, Denver
Ridyard D, Heshammer J (2011) Value creation using electromagnetic imaging. World Oil 234 (3):51–54
Sainson S (1990) Gyrovoyager: an instrumented pig for pipeline positioning. Ed. European Commission/THERMIE. European oil and gas technology projects
Sainson S (2007) Inspection en ligne des pipelines. Principes et méthodes. Ed. Lavoisier. pp 231–246, pp 284–285, pp 159–160
Salako O, MacBeth C, MacGregor L (2015) Potential applications of time-lapse CSEM to reservoir monitoring. First Break 33:35–46
Schlumberger C, Schlumberger M (1930). La méthode de la carte des résistivités et ses applications pratiques. Annales des mines. 12éme série, tome 18. Ed. Dunod, p 107
Sheriff RE, Geldart LP (1984) Traité de prospection sismique. Tome 2. Traitement Interprétation. Ed. ERG, La Barbannerie. pp 140–234
Vöge M, Pfaffhuber AA, Hokstad K, Fotland B (2010) A broadband marine CSEM demonstration survey to map the Uranus salt structure. In: EGM 2010 international workshop. Adding new value to electromagnetic, gravity and magnetic methods for exploration. Capri, Italy, April 11–14
Volker A, Dijkstra J (1955) Détermination des salinités des eaux dans le sous-sol du Zuiderzee par prospection géophysique. Geophys Prospect 3:111–124
Webb et al (1985) A seafloor electric filed instrument. J Geomag Geoelectric 37:115–1129
Wheelock BD, Constable S, Key KW (2010) Electromagnetic imaging the of the Pacific–North American Plate boundary in central California, USA. American Geophysical Union, Fall Meeting

Additional References (Not Considered in the Writing of This Chapter)

Cairns G, Evans RL, Edwards RN (1996) A time domain electromagnetic survey of the TAG hydrothermal mound. Geophys Res Lett 23:3455–3458
Constable S (2002) Remote sensing of hydrocarbon layers by seabed logging: results from a cruise offshore Angola. Lead Edge 21:972–982
Constable SC, Key K (2009) Marine EM studies over Scarborough gas field. Scripps Institution of Oceanography, San Diego
Constable S, Key K, Myer D (2009) Marine electromagnetic survey of Scarborough gas field, preliminary cruise report
Cudrak CF, Clowes RM (1993) Crustal structure of Endeavour Ridge segment, Juan de Fuca Ridge, from a detailed seismic refraction survey. J Geophys Res 98:6329–6349
Darnet M, Choo MC, Plessix RE, Rosenquist ML, Cheong KY, Sims E, Voon J (2007) Detecting hydrocarbon reservoirs from CSEM data in complex settings: application to Deepwater Sabah, Malaysia. Geophysics 72(2):WA97–WA103
Ellis M, Evans R, Hutchinson D, Hart P, Gardner J, Hagen R (2008) Electromagnetic surveying of seafloor mounds in the northern Gulf of Mexico. Mar Pet Geol 25(9):960–968
Evans RL, Sinha MC, Constable SC, Unsworth MJ (1994) On the electrical nature of the axial melt zone at 13° N on the East Pacific Rise. J Geophys Res 99:577–588
Evans RL, Law LK, Louis B, Cheesman SJ (2000) Buried Paleo-channels on the New Jersey continental margin: channel porosity structures from electromagnetic surveying. Mar Geol 170:381–394
Fanavoll S, Gabrielsen PT, Ellingsrud S (2014) The impact of CSEM on exploration decisions and seismic: two case studies from the Barents Sea. First Break 32:105–110

Fischer PA (2005) New EM technology offerings are growing quickly. After decades of languishing in scientific use, a variety of geo-electromagnetic techniques are coming into the commercial arena. World Oil 226(6)
Hoversten GH, Constable S, Morrison HF (2000) Marine magnetellurics for base salt mapping: Gulf of Mexico field-test at the Gemini structure. Geophysics 65:1476–1488
Key K (2012) Marine electromagnetic studies of seafloor resources and tectonics. Surv Geophys 33:135–167. doi:10.1007/s10712-011-9139-X
Key KW, Constable SC, Weiss CJ (2006) Mapping 3D salt using the 2D marine magnetelluric method: case study from Gemini Prospect, Gulf of Mexico. Geophysics 71(1):B17–B27
Lien M, Mannseth T (2008) Sensitivity study of marine CSEM data for reservoir production monitoring. Geophysics 73(4):F151–F163
Lovatini AM, Watts KE, Umbach A, Ferster A, Patmore S, Stilling J (2009) Application of 3D anisotropic CSEM inversion offshore west of Greenland. In: 79th annual international meeting SEG, expanded abstracts, pp 830–834
MacGregor LM, Sinha MC (2000) Use of marine controlled source electromagnetic sounding for sub-basalt exploration. Geophys Prospect 48:1091–1106
MacGregor L, Sinha M, Constable S (2001) Electrical resistivity structure of the Valu Fa Ridge, Lau Basin, from marine controlled-source electromagnetic sounding. Geophys J Int 146:217–236
MacGregor LM, Andreis D, Tomlinson J, Barker N (2006) Controlled-source electromagnetic imaging on the Huggets-1 reservoir. Lead Edge 84–992
Orange A, Key K, Constable S (2009) The feasibility of reservoir monitoring using time-lapse marine CSEM. Geophysics 74(2):F21–F29
Røsten T, Johnstad SE, Ellingsrud S, Amundsen HE, Johansen S, Brevik I (2003) Seabed logging calibration survey over the Ormen Lange gas field. In: EAGE 65th Conference and exhibition
Schwalenberg K, Willoughby E, Mir R, Edwards RN (2005) Marine gas hydrate electromagnetic signatures in Cascadia and their correlation with seismic blank zones. First Break 23:57–63
Sinha MC, Navin DA, MacGregor LM, Constable SC, Peirce C, White A, Heinson G, Inglis MA (1997) Evidence for accumulated belt beneath the slow spreading Mid-Atlantic Ridge. Phil Trans R Soc Lond A 355:233–253
Tseng HW, Stalnaker J, MacGregor LM, Ackermann RV (2015) Multi-dimensional analyses of the SEAM controlled source electromagnetic data the story of a blind test of interpretation workflows. Geophys Prospect 63:1383–1402
Vieugue V, Tinder J (2015) 3D CSEM can play vital role in offshore exploration. Offshore, April, pp 52–54
Wannamaker PE, Booker JR, Jones AG, Chave AD, Filloux JH, Waff HS, Law LK (1989) Magnetotelluric observation across the Juan de Fuca subduction system in the EMSLAB project. J Geophys Res 94:14111–14125
Wicklund TA, Fanavoll S (2004) Norwegian sea. SBL case study. In: EAGE 66th Conference & Exhibition. Paris, France, 7–10 June
Young PD, Cox CS (1981) Electromagnetic active source sounding near the East Pacific Rise. Geophys Res Lett 8:1043–1046
Ziolkowski A, Parr R, Wright D, Nockles V, Limond C, Morris E, Linfoot J (2010) Multi-transient electromagnetic repeatability experiment over the North Sea Harding field. Geophys Prospect 58:1159–117

General Conclusion and Perspectives

Though seabed logging techniques have their origin in the early days of applied geophysics, relayed then in earth physics, they have especially developed over a decade with a priori undeniable success in the search for hydrocarbons. This therefore justifies an introductory book on these new methods and prospecting technologies, then answering questions that any innovative technology awakens as soon as its appears, and especially when it seems to question established industrial processes.

Petroleum geophysics, at sea and on land, has relied very readily on geological knowledge and more particularly on that concerning the structure of the subsoil, qualifying these indirect geophysics as structural or even stratigraphic today, intimately connected with the tectonics of sedimentary basins, which gives rise to the trap and then eventually to the reservoir of hydrocarbons.

On the other hand, petroleum geophysics fitted in early and almost exclusively with seismic methods, refraction and reflection especially, which have continued to evolve with advances in digital electronics, computer science and especially signal processing.

In contrast, the probability of finding a productive trap, despite continuing efforts in research and development, has so far remained relatively low (25 % chance of success on average), with no hope of immediate improvement, leading to very important investments in terms of well logging and especially exploration drilling which, let us remember, are the heaviest budget items in a marine exploration campaign and more particularly in the deep sea.

Although indirect geophysics has thus so far dominated oil exploration for nearly 70 years, direct prospecting has always been of very understandable interest, which has often led over time and events to stormy and passionate debates that have not always been objective. Ironically, the common point of these techniques was the systematic use, more or less wisely, of radio or high frequency electromagnetic waves, whose interpretation for the detection and localization of oil fields was very often outside the scope of the demonstrated physics.

Currently announced with great fanfare as a method for direct detection of hydrocarbons, seabed logging in its commercial versions (mCSEM and mMT)

S. Sainson, *Electromagnetic Seabed Logging*, DOI 10.1007/978-3-319-45355-2

can be understood as such, if we only look at first the rate of discovery, which currently approximates about 90 %.

However, looking at it closer, we firstly see that electromagnetic investigation can only be considered and carried out on geologically recognized traps and secondly that these techniques are all the more effective when the external data used to run the interpretation models (inversion data) are numerous and varied.

In these particular circumstances, which are virtually unique in the field of applied geophysics, EM seabed logging, without the contribution of this additional and complementary information, cannot be considered in the current state of knowledge, and objectively, as a full direct method. However, coupled with a structural geophysics technique, such as reflection seismics, which has in addition much superior resolution, seabed logging can be seen and then accepted as one of the key elements of a direct method provided that:

- The electrical conductivity of the hydrocarbon reservoir contrasts very strongly with that of the surrounding sedimentary environment.
- The dimensions of the target are large enough so that at a distance the disturbing action is meaningful or more accurately measurable.

In these favorable conditions, where:

- A priori reflection seismics gives information precisely on the form and structure of the terrains, in this case on the presence of potential traps
- A posteriori seabed EM logging provides information on the horizontal and lateral evolution of the properties of the various geological strata, i.e., facies

we can then assume that the coupling of the two techniques forms a full direct prospecting method, or at least that EM seabed logging for its part provides a good indicator of the presence of hydrocarbons (direct hydrocarbon indicator or DHI) when the existence of traps has been previously proven.

On the other hand, one can imagine, with the progress of seismic analysis with offset (AVO), that electromagnetic and seismic techniques jointly applied (joint acquisition) will provide in the future a better DHI, thus reducing economic risks.

Similarly, we can also hope that research on the seismo-electric effect (creation of an electric field concomitant to important mechanical stimulation) may be included this time with offshore equipment. In a marine environment (incompressible), saturated with water (sediments), the method should a priori lead to much better results than that initiated in the 1930s–1960s on land in an undoubtedly more complex geological context (presence of overburden).

Field monitoring will also be greatly improved by permanently placed instrumentation, allowing us to monitor virtually and in real time the evolution of the reserve, in addition to well data (pressure, temperature, flow, etc.), production logs and 4D seismics for example. This time-lapse control will be all the more accurate when by measures are propped up once and for all in time and space.

In addition, mMT may also play a key role in cases of the presence of seismic masks that prevent sound waves to penetrate deeper. In this case we would then turn to the EM methods only, or then in support of the gravimetric method.

In an increasingly tense energy context, where one barrel of oil is discovered as three are consumed, where production techniques reach their limits (considerable water depths), it seems well established that seabed logging is at the point of becoming a truly disruptive technology (breakthrough) now impacting the entire exploration sector and can be regarded as a promising technology just as much as 4D coverage and multiple-azimuth seismic surveys a few years ago (Sanière et al. 2010).

Does this represent a revolution in the world of exploration?

Probably yes, because if in the future the success is confirmed, this highly innovative advance will likely have immediate repercussions for drilling and logging activities and a significant impact on offshore exploration in general, undoubtedly completely changing the situation in the oil service companies of the sector (Anonymous 2011; Ridward and Hesthammer 2011).

The future of seabed logging will inevitably be in the miniaturization of the sensors, and in the increase of their performance (sensitivity, accuracy, etc.), allowing us in particular to increase the depth of investigation and resolution. The immediate future is already in sensors with very low dimensions (EM streamers being commercialized), which may for example be integrated later on seismic acquisition streamers or OBC cables, giving then, thanks to this combined technology (seismics/EM), direct and immediate access (in real time) to geological, and why not economical, information.

The flexibility of use of this unique technology can also be a sign of many more geological applications with the search for various types of mineral wealth such as:

- Reserves of *gas hydrates* as economically viable methane resources (Edwards 1997; Hyndman et al. 1999; Yuan and Edwards 2000; Schwalenberg et al. 2005, 2009; Thakur and Rajput 2011)
- *Ore deposits* (Mero 1965; Gibbons 1987; Fouquet and Lacroix 2012), sulfur cluster type (Wolfgram 1986), polymetallic or hydrothermal nodules, for example related to submarine volcanism (the Kulolasi volcano off the Wallis and Futuna Islands)[1]
- *Rare-earths deposits*,[2] recently found in the Pacific (Wallis and Futuna), the need for which in the electronic industry has become urgent (strategic minerals)
- *Freshwater aquifers* present in coastal areas allowing some countries to have resources that they previously lacked

[1]These rare-earth reserves are among the largest in the world (Planchais, 2011). With its marine areas, the French territory is larger than Europe (11.1 million km^2 against 9 million km^2) and is second behind the US (11.3 million km^2). China, which now has a virtual monopoly on lanthanides (La, Ce, Pr, Nd, Pm, Sm, Eu, Gd, Tb, Dy, Ho, Er, Tm, Yb, Lu), with 97% of the world production, regularly files license applications for prospecting and exploitation of these deposits with the ISA (International Seabed Authority).

[2]Rare earths are not found in nature, but are combined with other elements (minerals and ores such as monazite and bastnaesite). On land, their geophysical exploration is done by magnetic and gravimetric methods, or even in some cases by radiometric methods. At sea no method has yet been proposed.

or why not, in the longer term, in addition to other techniques such as:

- "Forecasting" of seismic hazards (earthquakes and especially tsunamis and tidal waves, underwater landslides due to hydrates and shallow gas, etc.), taking into account submarine warning signals (Rikitake 1976; Kornprobst and Laverne 2011; Surkov and Hayakawa 2014)
- Or monitoring of CO_2 storage sites

We can hope that with this new method, the number of production wells, for better resource management, will decrease, relatively speaking, thus reversing the exponential trend that began in the year 1990 (Hesthammer et al. 2010). This will also allow us to reduce the ecological footprint in sensitive areas that already suffer enough constraints due to noise pollution (Lurton and Antoine 2007), risks of oil spills, etc., and to support, if necessary, moratoria on certain areas such as, for example, the Gulf of Mexico or the Arctic territories, and the prospect of new frontiers.

Finally, we hope that this introduction to EM seabed logging techniques, where many pages remain to be written, will have aroused interest. We also hope not to have failed *Paul Valery's* maxim, "*what is simple is false, what is complicated is unusable*" by more or less skillfully transcribing through our words the ideas and concepts of the promoters of this original technique for marine exploration.

References

Anonymous (2011) The future of marine CSEM. First Break, vol. 29, April

Edwards RN (1997) On the resource evaluation of marine gas hydrate deposits using a seafloor transient electric dipole–dipole method. Geophysics, 62:63–74

Fouquet Y, Lacroix D (2012) Les ressources minérales marines profondes: Étude prospective à l'horizon 2030

Gibbons et al (1987) Marine minerals: exploring our new ocean frontier. Ed. US Congress, Washington, DC, 349 p

Hesthammer et al (2010) CSEM performance in light of well results. Lead Edge 29(34):258–264

Hyndman RD, Yuan T, Moran K (1999) The concentration of deep sea gas hydrates from downhole electrical resistivity measurements. Earth Planet Sci Lett 172:167–177

Kornprobst J, Laverne C (2011) A la conquête des grands fonds. Ed. Quae. p 172

Lurton X, Antoine L (2007) Analyse des risques pour les mammifères marins liés à l'emploi des méthodes acoustiques en océanographie. Rapport Ifremer. DOP/CB/NSE/AS/07-07. 88 p

Mero JL (1965) The mineral ressources of the sea. Ed. Elsevier, Amsterdam, 312 p

Planchais B (2011) Les ruptures stratégiques dans l'espace maritime. Centre d'étude de prospection stratégique

Ridward DR, Hesthammer J (2011) Value creation using electromagnetic imaging. World Oil 234 (3):51–54.

Rikitake T (1976) Earthquake predictions. Ed. Elsevier, Amsterdam, 337 p

Sanière A et al (2010) Les investissements en exploration-production et en raffinage. IFP Energie nouvelle, p 15

Schwalenberg K, Willoughby E, Mir R, Edwards N (2005) Marine gas hydrate electromagnetic signatures in Cascadia and their correlation with seismic blank zones. First Break 23:57–64

Schwalenberg K, Haeckel M, Poort J, Jegen M (2009) Evaluation of gas hydrate deposits in an active seep area using marine controlled source electromagnetics: results from Opouawe Bank, Hikurangi Margin, New Zealand. Mar Geol doi:10.1016/j.margeo/2009.07.006

Surkov V, Hayakawa M (2014) Ultra and extremely low frequency electromagnetic fields. Ed. Springer, Tokyo, 300 p

Thakur NK, Rajput S (2011) Exploration of gas hydrates. Ed. Springer, Berlin, 281 p

Wolfgram PA (1986) Stanford exploration project report polymetallic sulfide exploration on the deep seafloor. The feasibility of the MINI-MOSES experiment. Geophysics 51(9):1808–1818

Yuan J, Edwards RN (2000) The assessment of marine hydrates through electrical remote sounding: hydrate without a BSR? Geophys Res Lett 27(16):2397–2400

Postface

Our friends, the readers, may have noticed the fierce and bitter struggle, barely concealed, between the oil majors and, in particular, oil services companies regarding innovation (technological watch, war of patents, trials concerning forgery, etc.). The stakes are huge. We can then wonder what means the authors of university works, small businesses, independent engineers and many other inventors can have to assert their rights over intellectual property. Of course, the filing of a patent is a guarantee if indeed one is then able to carry out an action in terms of justice and if a circumvention of claims (now a rule) has not yet been made. All that remains to us unfortunate innovators is the publication of an article or, if we are brave, the writing of a book.

A priori, this fact is not new. Here is what was written in the 1930s by *Robert Esnault-Pelterie*, a French industrialist and metrologist, but especially a pioneer in aviation and astronautics, as a preamble to a note on his scientific work, particularly in a paragraph on the spoliation of his ideas, titled:

Why and how others have used my inventions more than myself.

> The introduction of this chapter in the history of my work caused me some perplexity. Having written it a first time, I thought to suppress it, not to risk giving me the air of a righter of wrongs or a martyr, states for which I really have no taste. Then I thought that in our time of reversal of the values, where scholars and professors are treated as pariahs while entertainers of crowds earn millions, too many people are inclined to judge on what they call the results, without worry about the means leading to it, not even their real value… (E. P. November 21, 1931)

The author then mentions some of the plagiarism of which he was the victim.

The reader may well find that, despite the enormous scientific progress, and the more and more important involvement of people of science in our modern world, our society, however, has not fundamentally changed… (Stéphane Sainson, September 8, 2011)

S. Sainson, *Electromagnetic Seabed Logging*, DOI 10.1007/978-3-319-45355-2

Complementary Bibliography

This nonexhaustive list, not taken into account in the writing of this book, contains references (articles, theses, dissertations, reports, patents, etc.) that complement those that helped to partly develop each chapter.

The publications prior to 2000 mainly concern works in earth physics and many of them are results from US teams who have worked on international oceanic geophysics programs. Since then, the number of items, especially those relating to offshore exploration, has significantly increased each year (more than 50 in 2011).

Alcocer JA, Garcia MV, Soto HS, Baltar D, Paramo VR, Gabrielson P, Roth F (2013) Reducing uncertainty by integrating 3D CSEM in the Mexican deepwater workflow. First Break 31:75–79

Alekseev DA, Palshin NA, Varentsov IM (2009) Magnetotelluric dispersion relations in a two-dimensional model of the coastal effect. Izv Phys Solid Earth 45(2):167–170

Andreis D, MacGregor L (2011) Using CSEM to monitor production from a complex 3D gas reservoir: a synthetic case study. Lead Edge 30:1070–1079

Ashour AA (1965) The coast line effect on rapid geomagnetic variations. Geophys J RAS 10 (2):147–161

Baba K, Chave AD, Evans RL, Hirth G, Mackie RL (2003) Melt generating processes at the southern East Pacific Rise revealed by the electrical conductivity structure. In: Proceedings 23nd general assembly of the international union of Geodesy and Geophysics, Sapporo, Japan, 30 June–11 July, paper SS03/07A/A02-011, B.460

Baba K, Tarits P, Chave AD, Evans RL, Hirth G, Mackie RL (2006) Electrical structure beneath a magma-poor segment of the East Pacific Rise at 15°45′S. Geophys Res Lett 33, L22301, doi:10.1029/2006GL027538

Bachrach R (2011) Elastic and resistivity anisotropy of shale during compaction and diagenesis: joint effective medium modeling and field observations. Geophysics 76(6):E175–E186

Bakr SA, Mannseth T (2009) Feasibility of simplied integral equation modelling of low frequency marine CSEM with a resistive target. Geophysics 74(5):F107–F117

Bannister PR (1984) ELF propagation update. IEEE J Ocean Eng OE-9:179–188

Barker ND, Morten JP, Shantsev DV (2012) Optimizing EM data acquisition for continental shelf exploration. Lead Edge 31:1276–1284

Belash VA (1981) Characteristic features of undersea electromagnetic sounding. Geophys J 3 (6):860–875

Berre I, Lien M, Mannseth T (2011) Identification of three-dimensional electric conductivity changes from time-lapse electromagnetic observations. J Comput Phys 230(10):3915–3928

Bhuiyan AH (2009) Three-dimensional modelling and interpretation of CSEM data from offshore Angola. Petrol Geosci 15(2):175–189

Bhuyian AH, Thrane BP, Landro M, Johansen SE (2010) Controlled source electromagnetic three-dimensional grid-modelling based on a complex resistivity structure of the seafloor effects of acquisition parameters and geometry of multi-layered resistors. Geophys Prospect 58 (3):505–533

Bindoff NL, Filloux JH, Mulhearn PJ, Lilley FE, Ferguson IJ (1986) Vertical electric field fluctuations at the floor of the Tasman abyssal plain. Deep Sea Res Part A Oceanogr Res Pap 33(5):587–600

Bindoff NL, Lilley F, Filloux JH (1988) A separation of ionospheric and oceanic tidal components in magnetic fluctuation data. J Geomag Geoelec 40:1445–1467

Bostick FX, Cox CS, Field EC (1978) Land-to-seafloor electromagnetic transmissions in 0.1 to 15 Hz band. Radio Sci 13(4):701–708

Brown V, Hoversten M, Key K, Chen J (2012) Resolution of reservoir scale electrical anisotropy from marine CSEM data. Geophysics 77(2):E147–E158

Brock-Nannestad L (1965a) EM phenomena in the ELF range. EM fields of submerged dipoles and finite antennas. Technical report

Brock-Nannestad L (1965b) Determination of the electric conductivity of the seabed in shallow waters with varying conductivity profile. Electron Lett 1:274–276

Bhuiyan A, Sakariassen R, Hallanger $\tilde{A}$, McKay A (2013) Modelling and interpretation of CSEM data from Bressay, Bentley and Kraken area East Shetland platform, North Sea: 83rd Annual International Meeting, SEG, Expanded Abstracts.

Cagniard L (1953) Basic theory of the magneto-telluric method of geophysical prospecting. Geophysics 37:605–635

Cairns G (1997) Development of a short-baseline transient EM marine system and its application in the study of the TAG hydrothermal mound. PhD thesis, University of Toronto

Chan GH, Dosso HW, Law LK (1981) Electromagnetic induction in the San-Juan Bay region of Vancouver-Island. Phys Earth Planet Inter 27(2):114–121

Chan E, Dosso HW, Law LK, Auld DR, Nienaber W (1983) Electromagnetic induction in the Queen Charlotte Islands Region—analog model and field station results. J Geomag Geoelec 35 (11–1):501–516

Chang-Sheng L, Jun L (2006) Transient electromagnetic response modelling of magnetic source on seafloor and the analysis of seawater effect. Chin J Geophys 49(6):1891–1898

Chang-Sheng L, Everett, ME, Jun L, Feng-Dao Z (2010) Modelling of seafloor exploration using electric-source frequency-domain CSEM and the analysis of water depth effect. Chin J Geophy 53(8):1940–1952

Chave AD, Tarits P, Evans RL, Booker JR (2000a) Asymmetric electrical structure beneath the southern East Pacific Rise, EOS, 81 (48), Fall Meet. Suppl., Abstract V12B-03

Chave AD, Evans R L, Tarits P (2000b) Asymmetric mantle electrical structure beneath the East Pacific Rise at 17S. In: Proceedings of 15th Workshop on Electromagnetic Induction in the Earth, Cabo Frio, Brazil, 19–26 August

Chave AD (1983a) On the theory of electromagnetic induction in the earth by ocean currents. J Geophys Res 88(B4):3531–3542

Chave AD (1983b) Numeric evaluation of related Hankel transforms by quadrature and continued fraction expension. Geophysics 48(12):1671–1686

Chave AD (1984) On the electromagnetic fields induced by oceanic internal waves. J Geophys Res 89(C6):10.519–10.528

Chave AD, Cox CS (1983) Electromagnetic induction by ocean currents and the conductivity of the oceanic lithosphere. J Geomag Geoelectr 35:491–499

Chave AD, Filloux JH (1984) Electromagnetic induction field in the deep ocean of California: oceanic and ionospheric sources. Geophys J R Astr Soc 77:143–171

Chave AD, Filloux JH (1985) Observation and interpretation of the sea floor vertical electric field in the eastern north Pacific. Geophys Res Lett 12(12):793–796

Chave AD, Constable SC, Edwards RN (1986) Electrical exploration methods for the sea floor. In: Nabighian MN (ed) Electromagnetic methods vol 2: applications. Ed. Society of Exploration Geophysicists, Tulsa

Chave AD et al (1989a) Observations of motional electromagnetic fields during EMSLAB. J Geophys Res 94(B10):14.153–14.166

Chave AD, Thomson DJ (1989b) Some comments on magnetotelluric response function estimation. J Geophys Res 94(B10):14.215–14.225

Chave AD, Luther DS (1990) Low-frequency, notionally induced electromagnetic field in the ocean. J Geophys Res 95(C5):7185–7200

Chave AD, Tarits P (1992) Passive electromagnetic methods in the ocean. Ridge Events 3:5–6

Chave AD et al (1990a) Report of a workshop on the geoelectric and geomagnetic environment of continental margins; NTIS AD-A223 743, Scripps Institution of Oceanography report 90–20, University of Califomia, San Diego

Chave AD et al (1990b) Some comments on seabed propagation of ULF/BLF electromagnetic fields. Radio Sci 25(5):825–836

Chave AD, Evans RL, Hirth JG, Tarits P, Mackie RL, Booker JR (2001) Anisotropic electrical structure beneath the East Pacific Rise at 17S. In: Proceedings of Ocean Hemisphere Program/ International Ocean Network Joint Symposium, Mt. Fuji, Japan, 21–27, 4 pp

Cheesman SJ, Edwards RN, Law LK (1990) A short baseline transient electromagnetic method for use on the sea floor. Geophy J Int 103:431–437

Cheesman SJ, Law LK, Edwards RN (1991) Porosity determinations of sediments in Knight Inlet using a transient electromagnetic system. Geomar Lett 11:84–89

Chen J, Hoversten GM (2012) Joint inversion of marine seismic AVA and CSEM data using statistical rock physics and Markov random fields. Geophysics 77(1):R65–R80

Chen JP, Alumbaugh DL (2001) Three methods for mitigating air waves in shallow water marine controlled-source electromagnetic data. Geophysics 76(2):F89–F99

Chen JP, Dickens TA (2009) Effects of uncertainty in rock-physics models on reservoir parameter estimation using seismic amplitude variation with angle and controlled-source electromagnetics data. Geophys Prospect 57(1):61–74

Chen JP, Oldenburg DW (2006) A new formula to compute apparent resistivities from marine magnetometric resistivity data. Geophysics 71(3):G73–G81

Chen J, Haber E, Oldenburg DW (2002) Three-dimensional numerical modelling and inversion of magnetometric resistivity data. Geophys J Int 149:679–697

Chen JP, Oldenburg DW, Haber E (2005). Reciprocity in electromagnetics: application to modelling marine magnetometric resistivity data. Phys Earth Planet Inter 150(1–3):45–61

Chen JP, Hoversten GM, Vasco D, Rubin Y, Hou, Z (2007) A Bayesian model for gas saturation estimation using marine seismic AVA and CSEM data. Geophysics 72(2):WA85–WA95

Christensen NB, Dodds K (2007) 1D inversion and resolution analysis of marine CSEM data. Geophysics 72(2):WA27–WA38

Coggon JH, Morrison HF (1970) Electromagnetic investigation of sea floor. Geophysics 35 (3):476–489

Colombo D, MacNiece G, Curiel ES, Fox A (2013) Full tensor CSEM and MT for subsalt structural imaging in the Red Sea: implications for seismic and electromagnetic integration. Lead Edge 32:436–449

Commer M, Newman GA (2008) New advances in three-dimensional controlled-source electromagnetic inversion. Geophys J Int 172(2):513–535

Commer M, Newman GA (2009) Three-dimensional controlled-source electromagnetic and magnetotelluric joint inversion. Geophys J Int 178(3):1305–1316

Commer M, Newman GA, Caraone JJ, Dickens TA, Green KE, Wahrmund LA, Willen DE, Shiu J (2008) Massively parallel electrical conductivity imaging of hydrocarbons using the IBM Blue Gene/L supercomputer. IBM J Res Dev 52:93–103

Connell D (2011) A comparison of marine time-domain and frequency-domain controlled source electromagnetic methods. Master's thesis, University of California, San Diego

Constable S (2006) Marine EM methods—a new tool for offshore exploration. Lead Edge, April, pp 438–444

Constable S, Cox C (1996) Marine controlled source electromagnetic sounding—II: the PEGASUS experiment. J Geophys Res 97:5519–5530

Constable S, Weiss CJ (2006) Mapping thin resistors and hydrocarbons with marine EM methods: insights from 1D modelling. Geophysics 71:G43–G51

Constable SC, Parker RL, Constable CG (1987) Ocean's inversion: a practical algorithm for generating smooth models from EM sounding data. Geophysics 52:289–300

Constable S, Orange AS, Hoverston GM, Morrison M (1998) Marine magnetotellurics for petroleum exploration part I: a sea-floor equipment system. Geophysics 63:816–825

Constable S, Kannberg P, Callaway K, Ramirez Mejia D (2012) Mapping shallow geological structure with towed marine CSEM: 82nd Annual International Meeting, SEG, Expanded Abstracts

Cox CS (1980) Electromagnetic induction in the ocean and interference on the constitution of the Earth. Geophys Surv 4:137–156

Cox CS, Filloux JH (1974) Two dimensional numerical models of the califomia electromagnetic coastal anomaly. J Geomag Geoelectr 26:257–267

Cox CS, Filloux JH, Larsen JC (1971) Electromagnetic studies of ocean currents and electrical conductivity below the ocean floor. The Sea, vol. 4, part 1. Ed. Willey Interscience, New York, pp 637–693

Darnet M, Choo MC, Plessix RE, Rosenquist ML, Yip-Cheong K, Sims E, Voon JW (2007) Detecting hydrocarbon reservoirs from CSEM data in complex settings. Application to Deep-water Sabah, Malaysia. Geophysics 72(2):WA97–WA103

Daud H, Yahya N, Asirvadam V (2011) Development of EM simulator for sea bed logging applications using MATLAB. Indian J Geo-Mar Sci 40(2):267–274

Davydycheva S, Rykhlinski N (2011) Focused-source electromagnetic survey versus standard CSEM: 3D modelling in complex geometries. Geophysics 76(1):F27–F41

de Groot-Hedlin C, Constable S (2004) Inversion of magnetotelluric data for 2D structure with sharp resistivity contrasts. Geophysics 69(1):78–86

DeLaurier JM (1985) First results of the MOSES experiment: sea sediment conductivity and thickness determination, Bute Inlet, British Columbia, by magnetometric offshore electrical sounding. Geophysics 50:153–160

DeLaurier JM, Auld DR, Law LK (1983) The geomagnetic response across the continental-margin off Vancouver Island—comparison of results from numerical modelling and field data. J Geomag Geoelec 35(11–1):517–528

Dell'Aversana P, Vivier M (2009) Expanding the frequency spectrum in marine CSEM applications. Geophys Prospect 57:573–590

Dmitriev VI, Mershchikova NA (2010) Mathematical modelling of the influence of the coastal effect on marine magnetotelluric soundings. Izv Phys Solid Earth 46(8):717–721

Du Frane WL, Stern LA, Weitemeyer KA, Constable S, Pinkston JC, Roberts JJ (2011) Electrical properties of polycrystalline methane hydrate. Geophys Res Lett 38

Duran PB (1987) The use of marine electromagnetic conductivity as a tool in hydrogeologic investigations. Ground Water 25(2):160–166

Edwards RN (1988) Two-dimensional modelling of a towed electric dipole–dipole EM system: the optimum time delay for target resolution. Geophysics 53:846–853

Edwards RN (1997) On the resource evaluation of marine gas hydrate deposits using a sea floor transient electric dipole–dipole method. Geophysics 62:63–74

Edwards RN, Law LK, DeLaurier JM (1981) On measuring the electrical conductivity of the oceanic crust by a modified magnetometric resistivity Method. J Geophys Res 86:11609–11615

Edwards RN, Wolfgram PA, Judge AS (1988) The ICE-MOSES experiment: mapping permafrost zones electrically beneath the Beaufort Sea. Mar Geophys Res 9:265–290

Eidesmo T, Ellingsrud S, Kong FN, Westerdahl H, Johansen S (2000) Method and apparatus for determining the nature of subterranean reservoirs, Patent application number WO 00/13046

Eidesmo T et al (2002) Seabed logging (SBL) a new method for remote and direct identification of hydrocarbon filled layers in deepwater areas, First Break

Eidesmo T, Ellinsrud S, Kong FN, Westerdahl H, Johansen S (2003) Method and apparatus for determining the content of subterranean reservoirs. US Patent no. 6628119B1. Den Norske Stats Oljeselskap

Eidsvik J, Bhattacharjya D, Mukerji T (2008) Value of information of seismic amplitude and CSEM resistivity. Geophysics 73(4):R59–R69

Ellingsrud S, Eidesmo T, Johansen S, Sinha MC, MacGregor LM, Constable SC (2002a) Remote sensing of hydrocarbon layers by seabed imaging (SBL): results from a cruise offshore Angola. Lead Edge 21:972–982

Ellingsrud S, Eidesmo T, Johansen S, Sinha MC, MacGregor LM, Constable S (2002b) Remote sensing of hydrocarbon layers by seabed logging (SBL): results from a cruise offshore Angola. Lead Edge 21:972

Ellingsrud S, Eidesmo T, Westerdahl, H, Kong FN (2003) Method and apparatus for determining the nature of subterranean reservoirs. U. S. Patent no. 2003/0052685

Ellis M, Evans RL, Hutchinson D, Hart P, Gardner J, Hagen R (2008) Electromagnetic surveying of seafloor mounds in the northern Gulf of Mexico. Mar Pet Geol 25(9):960–968

Ellis MH, Sinha MC, Minshull TA, Sothcott J, Best AI (2010) An anisotropic model for the electrical resistivity of two-phase geologic materials. Geophysics 75(6):E161–E170

Endo M, Cuma M, Zhdanov MS (2009) Large-scale electromagnetic modelling for multiple inhomogeneous domains. Commun Comput Phys 6(2):269–289

Ershov S, Mikhaylovskaya I, Novik O (2006) Theory of EM monitoring of sea bottom geothermal areas. J Appl Geophys 58(4):330–350

Evans RL, Webb SC, Jegen M, Sananikone K (1998) Hydrothermal circulation at the Cleft-Vance overlapping spreading center: results of a magnetometric resistivity survey. J Geophys Res 103:12321–12338

Evans RL, Tarits P, Chave AD, White A, Heinson G, Filloux JH, Toh H, Seama N, Utada H, Booker JR, Unsworth M (1999) Asymmetric mantle electrical structure beneath the East Pacific Rise at 17S. Sci, 286:752–756

Everett M (1991) Active electromagnetics at the mid-ocean ridge. PhD thesis, University of Toronto

Everett ME, Constable S (1999) Electric dipole fields over an anisotropic seafloor: theory and application to the structure of 40 Ma Pacific ocean lithosphere. Geophys J Int 136(1):41–56

Everett ME, Edwards RN (1989) Electromagnetic expression of axial magma chambers. Geophys Res Lett 16(9):1003–1006

Everett ME, Edwards RN (1991) Theoretical controlled-source electromagnetic responses of fast-spreading mid-ocean ridge models. Geophys J Int 105(2):313–323

Fan Y, Snieder R, Slob E, Hunziker J, Singer J, Sheiman J, Rosenquist M (2010) Synthetic aperture controlled source electromagnetics. Geophys Res Lett 37

Fanavoll S, Ellingsrud S, Gabrielsen PT, Tharimela R, Ridyard D (2012) Exploration with the use of EM data in the Barents Sea: the potential and the challenges. First Break 30:89–96

Farrelly BC, Ringstad SE, Johnstad, Ellingsrud S (2004) Remote characterization of hydrocarbon filled reservoirs at the Troll field by sea bed logging, EAGE Fall Research Workshop Rhodes, Greece, 19th–23rd September

Ferguson IJ (1988) The Tasman project of seafloor magnetotelluric exploration. PhD thesis, Australian National University

Ferguson IJ, Edwards RN (1994) Electromagnetic mode conversion by surface-conductivity anomalies—applications for conductivity soundings. Geophys J Int 117(1):48–68

Ferguson I, Filloux JH, Lilley F (1985) A seafloor magnetotelluric sounding in the Tasman Sea. Geophys Res Lett 12:545–548

Ferguson I, Lilley F, Filloux JH (1990) Geomagnetic induction in the Tasman Sea and electrical conductivity structure beneath the Tasman seafloor. Geophys J Int 102:299–312

Filloux JH (1967) Oceanic electric currents, geomagnetic variations and the deep electrical conductive structure of the ocean-continent transition of central California. PhD thesis, University of California, San Diego

Filloux JH (1978) Observation of VLF electromagnetic signals in the ocean in U. Schmuker. Adv Earth Planet Sci 9:1–12

Filloux JH (1980) Observation of very low frequency electromagnetic signals in the ocean. J Geomag Geoelec 32:SIl–SI12

Filloux JH, Tarits P, Chave AD (1988) EM sounding of oceanic upper mantle in BEMPEX area (42N, 162W). In: Proceedings of Ninth Workshop on EM Induction in the Earth and Moon, Sochi, USSR, 24–31 October

Filloux JH, Chave AD, Tarits P, Petitt RA, Bailey J, Moeller HH, Debreule A, Petiau G, Banteaux L (1994) Southeast Appalachians experiment: offshore component. In: Proceedings of 12th Workshop on Electromagnetic Induction, Brest, France, 7–14 August 1

Flosadottir AH, Constable S (1996) Marine controlled-source electromagnetic sounding, 1. Modelling and experimental design. J Geophys Res Solid Earth 101:5507–5517

Fonarev G (1982) Electromagnetic research in the ocean. Geophys Surv 4(4):501–508

Franke A, Borner RU, Spitzer K (2007) Adaptive unstructured grid finite element simulation of two-dimensional magnetotelluric fields for arbitrary surface and seafloor topography. Geophys J Int 171(1):71–86

Fujii I, Chave AD (1999) Motional induction effect on the planetary scale geoelectric potential in the eastern North Pacific. J Geophys Res Oceans 104(C1):1343–1359

Gaillard F, Malki M, Iacono-Marziano G, Pichavant M, Scaillet B (2008) Carbonatite melts and electrical conductivity in the asthenosphere. Sci 322(5906):1363–1365

Giannini JA, Thayer DL (1982) Extremely low-frequency quasi-static propagation measurements from a calibrated electric-field source in the ocean. IEEE Trans Antennas Propag 30 (5):825–831

Goldman M (1987) Forward modelling for frequency domain marine electromagnetic systems. Geophys Prospect 35:1042–1064

Goldman M, Levi E, Tezkan B, Yogeshwar P (2011) The 2D coastal effect on marine time domain electromagnetic measurements using broadside dBz/dt of an electrical transmitter dipole. Geophysics 76(2):F101–F109

Greenhouse JP (1972) Geomagnetic time variations on the sea floor of Southern California. PhD thesis, University of California, San Diego

Greenhouse JP, Parker R, White A (1973) Modelling geomagnetic variations in or near an ocean using a generalized image technique. Geophys J Royal Astron Soc 32:325–338

Gribenko A, Zhdanov M (2007) Rigorous 3D inversion of marine CSEM data based on the integral equation method. Geophysics 72(2):WA73–WA84

Groom RW, Bailey RC (1988) A decomposition of the magnetotelluric impedance tensor which is useful in the presence of channelling, J Geophys Res 94b:1913–1925

Gui-Bo C, Hong-Nian W, Jing-Jin Y, Zi-Ye H (2009) Three-dimensional numerical modelling of marine controlled-source electromagnetic responses in a layered anisotropic seabed using integral equation method. Acta Physica Sinica 58(6):3848–3857

Gunning J, Glinsky ME, Hedditch, J (2010) Resolution and uncertainty in 1D CSEM inversion: a Bayesian approach and open-source implementation. Geophysics 75(6):151–F171

Harvey RR (1972) Oceanic water motion derived from the measurement of the vertical electric field report. Hawaii Institute of Geophysics, HIG-72-7, Honolulu, Hawaii

Harvey RR (1974) Derivation of oceanic water motions from measurement of the vertical electric field. J Geophys Res 79(30):4512–4516

Harvey RR, Larsen JC, Montaner R (1977) Electric field recording of tidal currents in the strait of Magellan. J Geophys Res 82(24):3472–3476

Heinson G (1999) Electromagnetic studies of the lithosphere and asthenosphere. Surv Geophy 20:229–255

Heinson G, Constable S (1992) The electrical conductivity of the oceanic upper mantle. Geophys J Int 110:159–179

Heinson G, Lilley F (1993) An application of thin-sheet electromagnetic modelling to the Tasman Sea. Phys Earth Planet Inter 81:231–251

Heinson G, Segawa J (1997) Electrokinetic signature of the Nankai Trough accretionary complex: preliminary modelling for the Kaiko-Tokai program. Phys Earth Planet Inter 99:33–54

Heinson G, White A (2005) Electrical resistivity of the Northern Australian lithosphere. Crustal anisotropy or mantle heterogeneity. Earth and Planet Sci Lett 232(1–2):157–170

Heinson G, Constable S, White A (1993a) The electrical conductivity of the lithosphere and asthenosphere beneath the coastline of Southern California. Explor Geophys 24:195–200

Heinson G, Constable S, White A (1993b) EMRIDGE: the electromagnetic investigation of the Juan de Fuca Ridge. Mar Geophys Res 15:77–100

Heinson G, Constable S, White A (1996a) Seafloor magnetotelluric sounding above axial sea mount. Geophys Res Lett 23:2275–2278

Heinson G, Constable S, White A (1996b) Seafloor magnetotelluric sounding above axial seamount. Geophys Res Lett 23(17):2275–2278

Heinson G, White A, Constable S, Key K (1999) Marine self potential exploration. Expl Geophys 30(1/2):1–4

Heinson G, Constable S, White A (2000) Episodic melt transport at mid-ocean ridges inferred from magnetotelluric sounding. Geophys Res Lett 27(15):2317–2320

Heinson G, White A, Lilley FEM (2005a) Rifting of a passive margin and development of a lower-crustal detachment zone: evidence from marine magnetotellurics. Geophys Res Lett 32(12): L12305

Heinson G, White A, Lilley FEM (2005b) Marine self-potential gradient exploration of the continental margin. Geophysics 70:G109–G118

Hesthammer J, Stefatos A, Boulaenko M, Vereshagin A, Gelting P, Wedberg T, Maxwell G (2010) CSEM technology as a value driver for hydrocarbon exploration. Mar Pet Geol 27 (9):1872–1884

Hoefel FG, Evans RL (2001) Impact of low salinity porewater on seafloor electromagnetic data: a means of detecting submarine groundwater discharge. Estuar Coast Shelf Sci 52(2):179–189

Hoehn GL, Warner BN (1983) Magnetotelluric measurements in the Gulf of Mexico at 20 m ocean depths. In: Geryer RA, Moore JR (eds) CRC handbook of geophysical exploration at sea. CRC Press, Boca Raton, pp 397–416

Hou Z, Rubin Y, Hoversten GM, Vasco D, Chen J (2006) Reservoir parameter identification using minimum relative entropy-based Bayesian inversion of seismic AVA and marine CSEM data. Geophysics 71(6):O77–O88

Hoversten GM, Cassassuce F, Gasperikova E, Newman GA, Chen JS, Rubin Y, Hou ZS, Vasco D (2006a) Direct reservoir parameter estimation using joint inversion of marine seismic AVA and CSEM data. Geophysics 71(3):C1–C13

Hoversten GM et al (2006b) 3D modelling of a deepwater EM exploration survey. Geophysics 71 (5):G239–G248

Hoversten GM, Morrison HF, Constable SC (1998) Marine magnetotellurics for petroleum exploration, part II: numerical analysis of subsalt resolution. Geophysics 63(03):826–840

Hoversten GM, Constable SC, Morrison HF (2000) Marine magnetotellurics for base-of-salt mapping: Gulf of Mexico field test at the Gemini structure. Geophysics 65(5):1476–1488

Hu W, Abubakar A, Habashy TM (2009) Joint electromagnetic and seismic inversion using structural constraints. Geophysics 74(6):R99–R109

Hunziker J, Slob E, Mulder W (2011) Effects of the airwave in time-domain marine controlled-source electromagnetics. Geophysics 76(4):F251–F261

Hyndman RD, Yuan T, Moran K (1999) The concentration of deep sea gas hydrates from downhole electrical resistivity logs and laboratory data. Earth Planet Sci Lett 172:167–177

Ichiki M, Baba K, Obayashi M, Utada H (2006) Water content and geotherm in the upper mantle above the stagnant slab: interpretation of electrical conductivity and seismic P-wave velocity models. Phys Earth Planet Inter 155(1–2):1–15

Ichiki M, Baba K, Toh H, Fuji-ta K (2009) An overview of electrical conductivity structures of the crust and upper mantle beneath the northwestern Pacific, the Japanese Islands, and continental East Asia. Gondwana Res 16(3–4):545–562

Inan AC, Fraser OC, Villard J (1986) ULF/ELF electromagnetic fields generated along the seafloor interface by a straight current source on infinite length. Radio Sci 21(3):409–420

Jegen M, Edwards RN (1998) The electrical properties of a two-dimensional conductive zone under the Juan de Fuca Ridge. Geophys Res Lett 19:3647–3651

Johansen SE, Amundsen HE, Rosten T, Ellingsrud S, Eidesmo T, Bhuyian AH (2005) Subsurface hydrocarbons detected by electromagnetic sounding. First Break 23:31–36

Joseph EJ, Toh H, Fujimoto H, Iyengar RV, Singh BP, Utada H, Segawa J (2000) Seafloor electromagnetic induction studies in the Bay of Bengal. Mar Geophys Res 21(1–2):1–21

Kasaya T, Goto TN (2009) A small ocean bottom electromagnetometer and ocean bottom electrometer system with an arm-folding mechanism. Explor Geophys 40(1):41–48

Kasaya T, Goto TN, Mikada H, Baba K, Suyehiro K, Utada H (2005) Resistivity image of the Philippine Sea Plate around the 1944 Tonankai earthquake zone deduced by marine and land MT surveys. Earth Planet Space 57(3):209–213

Kasaya T, Mitsuzawa K, Goto TN, Iwase R, Sayanagi K, Araki E, Asakawa K, Mikada H, Watanabe T, Takahashi I, Nagao T (2009) Trial of multidisciplinary observation at an expandable sub-marine cabled station “Off-Hatsushima Island Observatory” in Sagami Bay, Japan, Sens 9(11):9241–9254

Kaufman et al (1981) Ocean floor electrical surveys. 51st annual international meeting of the Society of exploration geophysicists

Kellett R, Lilley F, White A (1991) A 2-Dimensional interpretation of the geomagnetic coast effect of southeast Australia, observed on land and sea floor. Tectonophysics 192(3–4):367–382

Key K, Weiss C (2006) Adaptive finite element modelling using unstructured grids: the 2D magnetotelluric example. Geophysics 71(6):G291–G299

Key K, Constable S, Behrens J, Heinson G, Weiss C (2005) Mapping the northern EPR magmatic system using marine EM. Ridge 2000 Events 3:35–37

Key KW, Constable SC, Weiss CJ (2006) Mapping 3D salt using the 2D marine magnetotelluric method: case study from Gemini Prospect, Gulf of Mexico. Geophysics 71(1):B17–B27

King J (2004) Using a 3D finite element forward modelling code to analyze resistive structures with controlled-source electromagnetics in a marine environment. Master’s thesis, Texas A&M University

Kong FN (2007) Hankel transform filters for dipole antenna radiation in a conductive medium: Geophys Prospect 55(1):83–89

Kong FN, Johnstad SE, Rosten T, Westerdahl H (2008) A 2.5D finite element modelling difference method for marine CSEM modelling in stratified anisotropic media. Geophysics 73(1):F9–F19

Kong FN, Roth F, Olsen PA, Stalheim SO (2009) Casing effects in the sea-to-borehole electromagnetic method. Geophysics 74(5):F77–F87

Kong FN, Johnstad SE, Park J (2010) Wavenumber of the guided wave supported by a thin resistive layer in marine controlled-source electromagnetics. Geophys Prospect 58(4):711–723

Korotaev SM, Trofimov IL, Shneyer VS (1981) Integral conductivity determination of sea sediments in some world ocean areas by the sea currents electric field. Ann Géophys t. 37, fasc. 2:321–325

Korneva LA (1951) The anomalous geomagnetic field and the equivalent system of currents in the worlds ocean. Dok Akad Nauk 76(1):49–52

Kwon MJ, Snieder R (2011) Uncertainty analysis for the integration of seismic and controlled source electro-magnetic data. Geophys Prospect 59(4):609–626

Lanzerotti LJ, Chave AD, Sayres CH, Medford LV, Maclennan CG (1993) Large-scale electric-field measurements on the Earths surface—a review. J Geophys Res Planet 98 (E12):23525–23534

Larsen JC (1968) Electric and magnetic field induced by deep sea tides. Geophys J R Astr Soc 16:47–70

Larsen JC (1971) The electromagnetic field of long and intermediate water waves. J Mar Res 69 (20):28–45

Larsen JC (1975) Low frequency (0.1–6.0 cpd) electromagnetic study of deep mantle electrical conductivity beneath the Hawaiian Islands. Geophys J R Astr Soc 43:17–46

Larsen JC (1991) Transport measurements from in-service undersea telephone cables. IEEE J Ocean Eng 16:313–318

Larsen J, Cox C (1966) Lunar and solar daily variation in magnetotelluric field beneath ocean. J Geophys Res Solid Earth 71(18):4441–4445

Larsen JC, Sanford TB (1985) Florida curent volume transports from voltage measurements. Science 227:302–304

Launay L (1974) Conductivity under the oceans: interpretation of a magnetotelluric sounding 630 km off the Californian coast. Phys Earth Planet Inter 8:83–86

Law L (1983) Marine electromagnetic research. Surv Geophys 6:123–135

Law LK, Greenhouse JP (1981) Geomagnetic variation sounding of the asthenosphere beneath the Juan de Fuca Ridge. J Geophys Res 86:967–978

Law LK, Edwards RN (1986) The determination of resistivity and porosity of the sediment and fractured Basalt layers near the Juan de Fuca Ridge. Geophys J R Astr Soc 86:289–318

Lei H, Nobes DC (1994) Resistivity structure of the under consolidated sediments of the Cascadia Basin. Geophys J Int 118(3):717–729

Lewis L (2005) A marine magnetotelluric study of the San Diego Trough, Pacific Ocean, USA. Master's thesis, San Diego State University, San Diego State University

Lezaeta PF, Chave AD, Evans RL (2005) Correction of shallow-water electromagnetic data for noise induced by instrument motion. Geophysics 70(5):G127–G133

Li Y, Constable S (2007) 2D marine controlled-source electromagnetic modelling. Part 2: the effect of bathymetry. Geophysics 72(2):WA63–WA71

Li Y, Constable S (2010) Transient electromagnetic in shallow water. Insights from 1d modelling. Chin J Geophys 53(3):737–742

Li Y, Dai S (2011) Finite element modelling of marine controlled-source electromagnetic responses in two-dimensional dipping anisotropic conductivity structures. Geophys J Int 185:622–636

Li Y, Key K (2007) 2D marine controlled-source electromagnetic modelling: part 1—an adaptive nite element algorithm. Geophysics 72(2):WA51–WA62

Li Y, Pek J (2008) Adaptive finite element modelling of two-dimensional magnetotelluric fields in general anisotropic media. Geophys J Int 175(3):942–954

Li S, Booker JR, Aprea C (2008) Inversion of magnetotelluric data in the presence of strong bathymetry/topography. Geophys Prospect 56(2):259–268

Li M, Abubakar A, Habashy TM, Zhang Y (2010) Inversion of controlled-source electromagnetic data using a model-based approach. Geophysi Prospect 58(3):455–467

Lilley FE, Hitchman A (2004) Sea-surface observations of the magnetic signals of ocean swells. Geophys J Int 159:565–572

Lilley FE, Filloux JH, Bindoff N, Ferguson IJ (1986) Barotropic flow of a warm-core ring from seaoor electric measurements. J Geophys Res Solid Earth 91:979–984

Lilley FE, Filloux JH, Ferguson IJ, Bindoff NL, Mulhearn PJ (1989) The Tasman project of seaoor magnetotelluric exploration—experiment and observations. Phys Earth Planet Inter 53 (3–4):405–421

Lilley FE, Filloux JH, Mulhearn PJ, Ferguson IJ (1993) Magnetic signals from an ocean eddy. J Geomag Geoelec 45(5):403–422

Lilley F, White A, Heinson G (2000) The total field geomagnetic coast effect. The CICADA97 line from deep Tasman Sea to inland New South Wales. Explor Geophys 31:52–57

Lilley F, White A, Heinson G (2001) Earth's magnetic field: ocean current contributions to vertical profiles in deep oceans. Geophys J Int 147:163–175

Lilley F, White A, Heinson G (2004) Seeking a seafloor magnetic signal from the Antarctic Circumpolar Current. Geophys J Int 157:175–186.

Lizarralde D, Chave A, Hirth G, Schultz A (1995) Northeastern Pacific mantle conductivity profile from long-period magnetotelluric sounding using Hawaii-to-California submarine cable data. J Geophys Res Solid Earth 100(B9):17837–17854

Loseth LO (2011) Insight into the marine controlled-source electromagnetic signal propagation. Geophys Prospect 59(1):145–160

Loseth LO, Ursin B (2007) Electromagnetic fields in plenary layered anisotropic media. Geophys J Int 170(1):44–80

Loseth L, Pedersen HM, Schaug-Pettersen T, Ellingsrud S, Eidesmo T (2008) A scaled experiment for the verification of the sea bed logging method. J Appl Geophys 64(3–4):47–55

Loseth LO, Amundsen L, Jenseen AJ (2010) A solution to the airwave-removal problem in shallow-water marine EM. Geophysics 75(5):A37–A42

Loseth LO, Wiik T, Olsen PA, Becht A, Hansen JO (2013) CSEM exploration in the Barents Sea, part 1: detecting Skrugard from CSEM: 75th Annual Conference and Exhibition, EAGE, Extended Abstracts

Luanxiao Z, Jianhua G, Shengye Z, Dikun Y (2008) 1-D controlled sourceelectromagnetic forward modelling for marine gas hydrates studies. Appl Geophys 5(2):121–126

Luther DS, Filloux JH, Chave AD (1998) Low-frequency, motionally induced electromagnetic fields in the ocean. 2. Electric field and eulerian current comparison. J Geophys Res Solid Earth 96(12):79712

Maao FA and Nguyen AK (2010) Enhanced subsurface response for marine CSEM surveying. Geophysics 75(3):A7–A10

MacBarnet A (2004) All at sea with EM. Offshore Eng 29:20–22

MacGregor LM (2012) Integrating seismic, CSEM and well log data for reservoir characterisation. Lead Edge 31:268–277

Matsuno T, Seama N, Baba K, Goto T, Chave A, Evans R, White A, Boren G, Yio Yoneda A, Iwamoto H, Tsujino R, Baba Y, Utada H, Suyehiro K (2007a) Preliminary results of magnetotelluric analysis across the central Mariana transect. In: Proceedings of Japanese Geoscience Union, Chiba, Japan, May 19–24

Matsuno T, Seama N, Baba K (2007b) A study on correction equations for the effect of seafloor topography on ocean bottom magnetotelluric data. Earth Planet Space 59(8):981–986

Matsuno T, Seama N, Evans R, Chave A, Baba K, White A, Goto TN, Heinson G, Boren G, Yoneda A, Utada H (2010) Stanford exploration project report upper mantle electrical resistivity structure beneath the central Mariana subduction system. Geochem Geophys Geosyst 11 (9):Q09003

Mattsson J, Lund L, Lima J, Engelmark F, McKay A (2010) Case study: a towed EM test at the Peon discovery in the North Sea. EAGE Meeting, pp 1–5

Mattsson J, Lindqvist P, Juhasz R, Bjornemo E (2012) Noise reduction and error analysis for a towed EM system: 72nd Annual International Meeting, SEG, Expanded Abstracts, pp 1–5

Mattsson J, Englemark F, Anderson C (2013) Towed streamer EM: the challenges of sensitivity and anisotropy. First Break 31:155–159

Menvielle M, Tarits P (1994) Adjustment of the electromagnetic-field distorted by 3-D heterogeneities. Geophys J Int 116(3):562–570

Meunier J (1965) Enregistrements telluriques en mer. Bull Inst Océano de Monaco 65, no. 1339, 28 p

Mittet R (2010) High-order finite-difference simulations of marine CSEM surveys using a correspondence principle for wave and diffusion fields. Geophysics 75(1):F33–F50

Mittet R, Schaug-Pettersen T (2008) Shaping optimal transmitter waveforms for marine CSEM surveys. Geophysics 73(3):F97–F104

Mittet R, Aakervik OM, Jensen HR, Ellingsrud S, Stovas A (2007) On the orientation and absolute phase of marine CSEM receivers. Geophysics 72(4):F145–F155

Morten JP, Roth F, Karlsen SA, Tampko D, Pacurar C, Olsen PA, Nguyen AK, Gjengedal J (2012) Field appraisal and accurate resource estimation from 3D quantitative interpretation of seismic and CSEM data. Lead Edge 31:448–456.

Mosnier J (1982) Induction in the Earth's crust. Observational methods on land and sea. Geophys Surv 4(4):353–371

Mulder WA (2006) A multigrid solver for 3D electromagnetic diffusion: Geophys Prospect 54 (5):633–649

Mulder WA, Wirianto M, Slob EC (2008) Time-domain modelling of electromagnetic diffusion with a frequency-domain code. Geophysics 73(1):F1–F8

Muller H, Dobeneck T, Nehmiz W, Hamer K (2010) Near-surface electromagnetic, rock magnetic, and geochemical finger printing of submarine freshwater seepage at Eckernforde Bay (SW Baltic Sea). Geo-Mar Lett 31:1–18

Myer D, Constable S, Key K (2010) A marine EM survey of the Scarborough gas field, Northwest Shelf of Australia. First Break 28:77–82

Myer D, Constable S, Key K (2011) Broad-band waveforms and robust processing for marine CSEM surveys. Geophys J Int 184:689–698

Myer D, Constable S, Key K (2012) Marine CSEM survey of the Scarborough gas field, part 1: experimental design and data uncertainty. Geophysics 77:E281–E299

Newman GA, Commer M, Carazzone JJ (2010) Imaging CSEM data in the presence of electrical anisotropy. Geophysics 75(2):F51–F61

Nobes DC, Law LK, Edwards RN (1986a) The determination of resistivity and porosity of the sediment and fractured basalt layers near the Juan-De-Fuca Ridge. Geophys J R Astron Soc 86 (2):289–317

Nobes DC, Law LK, Edwards RN (1986b) Estimation of marine sediment bulk physical-properties at depth from sea-floor geophysical measurements. J Geophys Res Solid Earth Planet 91 (B14):14033–14043

Nobes DC, Law LK, Edwards RN (1992) Results of a sea-floor electromagnetic survey over a sedimented hydrothermal area on the Juan-De-Fuca Ridge. Geophys J Int 110(2):333–346

Nolasco R, Tarits P, Filloux, JH, Chave A (1998) Magnetotelluric. Imaging of the Society Island hotspot, J Geophys Res 103(B12):30287–30309

Nordskag JI, Amundsen L (2007) Asymptotic airwave modelling for marine control led-source electromagnetic surveying. Geophysics 72(6):F249–F255

Nordskag JI, Amundsen L, Loseth L, Holvik E (2009) Elimination of the water-layer response from multi-component source and receiver marine electromagnetic data. Geophys Prospect 57 (5):897–918

Palshin NA (1996) Oceanic electromagnetic studies. A review. Surv Geophys 17(4):455–491

Pardo D, Nam MJ, Torres-Verdin C, Hoversten MG, Garay I (2011) Simulation of marine controlled source electromagnetic measurements using a parallel fourier hp-finite element method. Comput Geosci 15:53–67

Peter W (2007) Guided waves in marine CSEM. Geophys J Int 171(1):153–176

Petitt RA, Chave AD, Filloux JH, Moeller HH (1994) Electromagnetic field instrument for the continental-shelf. Sea Technol 35(9):10–13

Pirjola R, Viljanen A (2000) Electric field at the seafloor due to a two dimensional ionospheric current. Geophys J Int 140:286–294

Pistek P (1977) Conductivity of the ocean crust. PhD thesis, University of California, San Diego

Plessix RE, Mulder WA (2008) Resistivity imaging with controlled-source electromagnetic data depth and data weighting. Inverse Prob 24(3):1–21

Plessix RE, Darnet M, Mulder WA (2007) An approach for 3D multisource, multifrequency CSEM modelling. Geophysics 72(5):SM177–SM184

Popkov I, White A, Heinson G, Constable S, Milligan P, Lilley FEM (2000) Electromagnetic investigation of the Eyre Peninsula conductivity anomaly. Expl Geophys 31:187–191

Price AT (1962) The theory of magnetotelluric methods when the source field is considered. Geophys Res Lett 67:4309–4329

Ramananjaona C, MacGregor L, Andreis D (2011) Sensitivity and inversion of marine electromagnetic data in a vertically anisotropic stratified earth. Geophys Prospect 59(2):341–360

Ramaswamy V, Nienaber W, Dosso HW, Jones FW, Law LK (1975) Numerical and analog model results for electromagnetic induction for an island situated near a coastline. Phys Earth Planet Inter 11(2):81–90

Rossignol JC (1966) Sur les variations transitoires du champ magnétiques terrestre en bordure de la manche. Dess université de Paris. 101 p

Sampaio EE (2009) A moving magnetic dipole in a conductive medium. Geophys Prospect 57 (4):591–600

Sanford TB (1986) Recent improvements in ocean currents measurement from motional electric fields and currents. In: Proceedings of IEEE third conference on current measurement. pp 1–12

Sasaki Y, Meju MA (2009) Useful characteristics of shallow and deep marine CSEM responses inferred from 3D finite difference modelling. Geophysics 74(5):F67–F76

Schmucker U (1970). Anomalies of geomagnetic variations in the south western United States: Bull. Scripps Inst. Oceanogr., 13, pp.1–170.

Scholl C, Edwards RN (2007) Marine down hole to seafloor dipole–dipole electromagnetic methods and the resolution of resistive targets. Geophysics 72(2):WA39–WA49

Schuleikin VV (1958) Telluric currents in an ocean and magnetic declination. Dokl AN URSS 119 (2):257–260

Schwalenberg K, Edwards RN (2004) The effect of seafloor topography on magnetotelluric fields. An analytic formulation confirmed with numerical results. Geophys J Int 159:607–621

Schwalenberg K, Willoughby EC, Mir R, Edwards RN (2005) Marine gas hydrate electromagnetic signatures in Cascadia and their correlation with seismic blank zones. First Break 23:57–63

Schwalenberg K, Haeckel M, Poort J, Jegen M (2010a) Evaluation of gas hydrate deposits in an active seep area using marine controlled source electromagnetics. Results from Opouawe Bank, Hikurangi Margin, New Zealand. Mar Geol 272:79–88

Schwalenberg K, Haeckel M, Poort J, Jegen M (2010b) Preliminary interpretation of electromagnetic, heat flow, seismic, and geochemical data for gas hydrate distribution across the Porangahau Ridge, New Zealand. Mar Geol 272:89–98

Seama N, Yamazaki T, Evans R, Goto T, Utada H, Chave A, Suyehiro K (2002) Tectonics of the Mariana Trough and magnetotelluric transects across the central Mariana subduction system. In: Proceeding of NSF/IFREE MARGINS workshop on the Izu-Bonin-Mariana-Subduction system, Honolulu, HI, September 8–12

Seama N, Baba K, Utada H, Toh H, Tada N, Ichiki M, Matsuno T (2007) 1-D electrical conductivity structure beneath the Philippine Sea. Results from an ocean bottom magnetotelluric survey. Phys Earth Planet Inter 162:2–12

Segawa J, Koizumi K (1994) An attempt to detect a long-term change of geomagnetic total force at fixed observation stations on the sea-floor. J Geomag Geoelec 46(5):381–391

Segawa J, Toh H (1990) Electromagnetometry at the seafloor. Proc Indian Acad Sci Earth Planet Sci 99(4):639–656

Segawa J, Toh H (1992) Detecting fluid circulation by electric-field variations at the Nankai Trough. Earth Planet Sci Lett 109(3–4):469–476

Sheard SN, Ritchie TJ, Christopherson KR, Brand E (2005) Mining, environmental, petroleum, and engineering industry applications of electromagnetic techniques in geophysics. Surv Geophys 26(5):653–669

Shimizu H, Koyama T, Baba K, Utada H (2010) Revised 1-D mantle electrical conductivity structure beneath the North Pacific. Geophys J Int 180(3):1030–1048

Shneyer VS, Trofimov I, Abramov YM, Zhdanov MS, Machinin VA, Shabelyansky SV (1991) Some results of gradient electromagnetic sounding in Doldrums Mid-Atlantic Ridge fracture. Phys Earth Planet Inter 66:259–264

Singer BS (2008) Electromagnetic integral equation approach based on contraction operator and solution optimisation in Krylov subspace. Geophys J Int 175(3):857–884

Sinha MC, Navin DA, MacGregor LM, Constable S, Peirce C, White A, Heinson G, Inglis MA (1997) Evidence for accumulated melt beneath the slow spreading Mid-Atlantic Ridge. Philos Trans R Soc Lond Series Math Phys Eng Sci 355(1723):233–253

Sinha MC, Constable SC, Peirce C, White A, Heinson G, MacGregor LM, Navin DA (1998) Magmatic processes at slow spreading ridges: implications of the RAMESSES experiment at 5745′N on the Mid-Atlantic Ridge. Geophys J Int 135(3):731–745

Slob E, Hunziker J, Mulder WA (2010) Green's tensors for the diffusive electric field in a VTI half space. Prog Electromagnetics Res-Pier 107:1–20

Spence GD, Hyndman RD, Chapman NR, Walia R, Gettrust J, Edwards RN (2000) North Cascadia deep sea gas hydrates. Gas hydrates. Challenges future 912:65–75

Streich R, Becken M (2011) Electromagnetic fields generated by finite length wire sources: comparison with point dipole solutions. Geophys Prospect 59(2):361–374

Snrka LJ (1986) Method and apparatus for offshore electromagnetic sounding utilizing wavelength effects to determine optimum source and detector position. US patent no. 4617518. Oct 14

Srnka L (2003). Remote reservoir resistivity mapping. US. Patent no. 6603313

Srnka LJ, Carazzone JJ, Ephron MS, Eriksen EA (2006) Remote reservoir resistivity mapping. Lead Edge August, pp 972–975

Streich R (2009) 3D finite difference frequency-domain modelling of controlled source electromagnetic data: direct solution and optimisation for high accuracy. Geophysics 74(5):F95–F105

Swidinsky A, Edwards RN (2009) The transient electromagnetic response of a resistive sheet: straightforward but not trivial. Geophys J Int 179(3):1488–1498

Swidinsky A, Edwards RN (2010) The transient electromagnetic response of a resistive sheet: an extension to three dimensions. Geophys J Int 182(2):663–674

Swidinsky A, Edwards RN (2011) Joint inversion of navigation and resistivity using a fixed transmitter and a towed receiver array: a transient marine CSEM model study. Geophys J Int 186:987–996

Tada N, Seama N, Goto TN, Kido M (2005) 1-D resistivity structures of the oceanic crust around the hydrothermal circulation system in the central Mariana Trough using magnetometric resistivity method. Earth Planet Space 57(7):673–677

Tarits P (1984) Algorithme de modélisation magnéto-tellurique en deux dimension: rapport interne, Beicip

Tarits P, Chave AD, Evans R (2000) Along ridge changes in the mantle electrical conductivity structure beneath the EPR between 17 and 15S. In: Proceeding of 15th workshop on electromagnetic induction in the earth, Cabo Frio, Brazil, 19–26 August

Tarits P, Mandea M, Chave AD, Garcia X, Calzas M, Drezen C, Dubreule A, Bailey J, Filloux JH (2003) The French-US program of seafloor geomagnetic observatories. In: Proceeding of 23nd general assembly of the international union of Geodesy and Geophysics, Sapporo, Japan, 30 June-11 July, paper JSS03/03P/A13-005, A-169

Tarits P, Filloux JH, Chave AD (1986a) Constraints on conductivity structure of Juan de Fuca plate deduced from EM fields tidally induced over offshore EMSLAB area. In: Proceedings of eigth workshop on electromagnetic induction in the earth and moon, Neuchatel, Switzerland, 24–31 August

Tarits P, Filloux JH, Chave AD (1986b) EM induction by tides in EMSLAB experiment: tidal modes separation and electrical conductivity of the seafloor, EOS, 67, 919, AGU meeting

Tarits P, Filloux JH, Chave AD, Menvielle M, Sichler B (1990a) Seafloor electromagnetic sounding of the Tahiti hot spot. In: Proceeding of tenth workshop on electromagnetic induction in the earth, Ensenada, B.C., Mexico, 22–29 August

Tarits P, Filloux JH, Chave AD, Menvielle M, Sichler B (1990b) Sondages électromagnétiques sous-marins sur le point chaud de Teahitia. In: Proceeding of Symposium on Intraplate Volcanism of the Reunion Hot Spot, Reunion Island, 12–17 November

Tarits P, Filloux JH, Chave AD, Menvielle M (1991) Structure of the Tahiti hot spot inferred from sea floor electromagnetic soundings In: Proceeding of 20th general assembly of the international union of Geodesy and Geophysics, Vienna, Austria, 11–24 August

Tarits P, Chave AD, Schultz A (1993) Comment on "the electrical conductivity of the oceanic upper mantle" by G. Heinson and S. Constable. Geophys J Int 114:711–716

Tehrani AM, Slob E (2010) Fast and accurate three-dimensional controlled source electromagnetic modelling. Geophys Prospect 58(6):1133–1146

Thayer DL, Scheer L, Tossman BE (1982) A triaxial coil receiver system for the study of subsurface electromagnetic propagation. IEEE J Ocean Eng 7(2):75–82

Toh H, Baba K, Ichiki M, Motobayashi T, Ogawa Y, Mishina M, Takahashi I (2006) Two dimensional electrical section beneath the eastern margin of Japan Sea. Geophys Res Lett 33 (22):L22309

Toh H, Hamano Y, Goto T, Utada H (2010) Long term seafloor electromagnetic observation in the Northwest Pacific may detect the vector geomagnetic secular variation. Data Sci J 9 IS -: IGY100–IGY109

Tompkins MJ, Srnka LJ (2007) Marine controlled-source electromagnetic methods—introduction. Geophysics 72(2):WA1–WA2

Ueda T, Zhdanov MS (2008) Fast numerical methods for marine controlled source electromagnetic (EM) survey data based on multigrid quasi-linear approximation and iterative EM migration. Explor Geophys 39(1):60–67

Um ES (2005) On the physics of galvanic source electromagnetic geophysical methods for terrestrial and marine exploration. Master's thesis, University of Wisconsin-Madison

Um ES, Alumbaugh DL (2007) On the physics of the marine controlled source electromagnetic method. Geophysics 72(2):WA13–WA26

Unsworth M, Oldenburg D (1995) Subspace inversion of electromagnetic data: application to mid-ocean-ridge exploration. Geophys J Int 123(1):161–168

Unsworth MJ (1991) Electromagnetic exploration of the oceanic crust with controlled sources, PhD thesis, University of Cambridge

Unsworth MJ (1994) Exploration of mid-ocean ridges with a frequency domain electromagnetic system. Geophys J Int 116:447–467

Utada H, Booker, JR, Unsworth MJ (1999) Asymmetric electrical structure in the mantle beneath the East Pacific Rise at 17 deg. S. Sci 286:752–756

Vanyan LL, Cox CS (1983) Comparison of deep conductivities beneath continents and oceans. J Geomag Geoelec 35(11–1):805–809

Vanyan L, Kharin EP, Osipova IL, Spivak VA (1976) Interpretation of a deep electromagnetic sounding of the ocean floor near the Californian coast. Phys Earth Planet Inter 13(2):119–122

Vitale S, de Santis A, Di Mauro D, Cafarella L, Palangio P, Beranzoli L, Favali P (2009) GEOSTAR deep seafloor missions: magnetic data analysis and 1D geoelectric structure underneath the Southern Tyrrhenian Sea. Ann Geophys 52(1):57–63

Vozoff K (1988) The magnetotelluric method, in Electromagnetic methods in applied geophysics—Application, Part B, Ed. M.N. Nabighian, SEG, Tulsa, pp 641–712

Wang Z, Gelius LJ, Kong FN (2009) Simultaneous core sample measurements of elastic properties and resistivity at reservoir conditions employing a modified tri-axial cell: a feasibility study. Geophys Prospect 57(6):1009–1026

Wang Z, Deng M, Chen K, Wang M (2013) An ultralow-noise Ag/AgCl electric field sensor with good stability for marine EM applications. In: Seventh International Conference on Sensing Technology (ICST), pp 747–750

Wannamaker PE, Booker JR, Filloux JH, Jones AG, Jiracek GR, Chave AD, Tarits P, Waff HS, Egbert GD, Young CT, Stodt JA, Martinez M, Law LK, Yukutake T, Segawa JS, White A, Green AW (1989a) Magnetotelluric observations across the Juan de Fuca subduction system in the EMSLAB project. J Geophys Res Solid Earth 94(B10):14111–14125

Wannamaker PE, Booker JR, Filloux JH, Jones AG, Jiracek GR, Chave AD, Tarits P, Waff HS, Egbert GD, Young CT, Stodt JA, Martinez M, Law LK, Yukutake T, Segawa JS, White A, Green AW (1989b) Resistivity cross section through the Juan de Fuca subduction system and its tectonic implications. J Geophys Res Solid Earth 94(B10):14127–14144

Wapenaar K, Slob E, Snieder R (2008) Seismic and electromagnetic controlled-source interferometry in dissipative media. Geophys Prospect 56:419–434

Webb SC, Cox C (1982) Electromagnetic fields induced at the seafloor by Rayleigh Stoneley waves. J Geophys Res Solid Earth 87:4093–4102

Webb SC, Edwards RN (1995) On the correlation of electrical conductivity and heat flow in Middle Valley, Juan-De-Fuca Ridge. J Geophys Res Solid Earth 100(B11):22523–22532

Webb SC, Constable SC, Cox CS, Deaton T (1985) A seafloor electric field instrument. J Geomag Geoelec 37:1115–1130

Webb S, Edwards RN, Yu LM (1993) 1st measurements from a deep-tow transicnt electromagnetic sounding system. Mar Geophys Res 15(1):13–26

Weiss C (2007) The fallacy of the shallow-water problem in marine CSEM exploration. Geophysics 72(6):A93–A97

Weiss CJ, Constable S (2006) Mapping thin resistors in the marine environment, part II: modelling and analysis in 3D. Geophysics 71(6):G321–G332

Weitemeyer K (2008) Marine electromagnetic methods for gas hydrate characterization. PhD thesis, University of California, San Diego

Weitemeyer KA, Constable SC, Key KW, Behrens JP (2006a) First results from a marine controlled-source electromagnetic survey to detect gas hydrates offshore Oregon. Geophys Res Lett 33(3):L03304

Weitemeyer KA, Constable SC, Key KW, Behrens JP (2006b) Marine EM techniques for gas-hydrate detection and hazard mitigation. The Leading Edge, 25, no. 5, pp. 629–632.

Weitemeyer K, Gao G, Constable S, Alumbaugh D (2010) The practical application of 2D inversion to marine controlled-source electromagnetic data. Geophysics 75(6):F199–F211

Wen-Bo W, Ming D, Zhen-He W, Zi-Li Z, Gao-Feng Y, Sheng J, Qi-Sheng Z, Jian-En J, Meng W, Kai C (2009) Experimental study of marine magnetotellurics in southern Huanghai. Chin J Geophys 52(3):740–749

Wertheim G (1953) Studies of the electric potential between Key West, Florida, Havana, Cuba. Technical report

White A (1979) A sea floor magnetometer for the continental shelf. Mar Geophys Res 4:105–114

White SN, Chave AD, Filloux JH (1997) A look at galvanic distortion in the Tasman Sea and the Juan de Fuca plate. J Geomag Geoelec 49(11–12):1373–1386

Wiik T, Loseth LO, Ursin B, Hokstad K (2011) TIV contrast source inversion of mCSEM data. Geophysics 76(1):F65–F76

Wirianto M, Mulder WA, Slob EC (2011) Applying essentially nonoscillatory interpolation to controlled-source electromagnetic modelling. Geophys Prospect 59(1):161–175

Wolfgram PA (1985) Development and application of a short-baseline electromagnetic technique for the ocean floor. PhD thesis, University of Toronto, Toronto

Worzewski T, Jegen M, Kopp H, Brasse H, Castillo WT (2010) Magnetotelluric image of the fluid cycle in the Costa Rican subduction zone. Nat Geosci 3(12):1–4

Wu X, Sandberg S, Roper T (2008) Three-dimensional marine magnetotelluric resolution for subsalt imaging and case study in the Gulf of Mexico. SEG Technical Program Expanded Abstracts, 27, no. 1, pp 574–578

Yang JW, Edwards RN (2000) Controlled source time-domain electromagnetic methods for seafloor electric conductivity mapping. Transactions of nonferrous metals. Society of China, 10, no. 2, pp 270–274

Yin C (2006) MMT forward modelling for a layered earth with arbitrary anisotropy. Geophysics 71(3):G115–G128

Young PD, Cox CS (1981) Electromagnetic active source sounding near the East Pacific Rise. Geophys Res Lett 8:1043–1046

Yu LM, Edwards RN (1992a) Algorithms for the computation of the electromagnetic response of a multi layered, laterally anisotropic sea floor to arbitrary finite sources. Geophys J Int 111 (1):185–189

Yu LM, Edwards RN (1992b) The detection of lateral anisotropy of the ocean floor by electromagnetic methods. Geophys J Int 108(2):433–441

Yu LM, Edwards RN (1996) Imaging axi-symmetric TAG-like structures by transient electric dipole sea floor electromagnetics. Geophys Res Lett 23(23):3459–3462

Yuan J, Edwards RN (2000) The assessment of marine gas hydrate through electrical remote sounding: hydrate without a BSR. Geophys Res Lett 27:2397–2400

Yukutake T, Filloux JH, Segawa J, Hamano Y, Utada H (1983) Preliminary report on a magnetotelluric array study in the Northwest Pacific. J Geomag Geoelec 35(11–1):575–587

Zach JJ, Brauti K (2009) Methane hydrates in controlled-source electromagnetic surveys—analysis of a recent data example. Geophys Prospect 57(4):601–614

Zaslavsky M, Druskin V, Davydycheva S, Knizhnerman L, Abubakar A, Habashy T (2011) Hybrid finite difference integral equation solver for 3D frequency domain anisotropic electromagnetic problems. Geophysics 76(2):F123–F137

Zhanxiang H, Strack K, Gang Y, Zhigang W (2008) On reservoir boundary detection with marine CSEM. Appl Geophys 5(3):181–188

Zhan-Xiang H, Zhi-Gang W, Cui-Xian M, Xi-Ming S, Xiao-Ying H, Jian-Hua X (2009) Data processing of marine CSEM based on 3D modelling. Chin J Geophys 52(8):2165–2173

Zhao G, Yukutake T, Filloux JH, Law LK, Segawa J, Hamano Y, Utada H, White T, Chave AD, Tarits P (1989) Two-dimensional modelling of the electrical resistivity structure of the Juan de Fuca plate. In: Proceedings of Sixth Scientific Assembly of the International Association of Geomagnetism and Aeronomy, Exeter, UK, 24 July – 4 August

Zhao GZ, Yukutake T, Hamano Y, Utada H, Segawa J, Filloux JH, Law LK, White T, Chave AD, Tarits P (1990) Investigation on magneto-variational data of the Defuca Juan Plate in Eastern Pacific ocean. Acta Geophys Sin 33(5):521–529

Zhdanov MS, Lee SK, Yoshioka K (2006) Integral equation method for 3D modelling of electromagnetic fields in complex structures with inhomogeneous background conductivity. Geophysics 71(6):G333–G345

Zhdanov MS, Wan L, Gribenko A, Cuma M, Key K, Constable S (2009) Rigorous 3D inversion of marine magnetotelluric data in the area with complex bathymetry. SEG Technical Program Expanded Abstracts, 28, no. 1, pp 729–733

Zhdanov MS, Wan L, Gribenko A, Cuma M, Key K, Constable S (2011) Large-scale 3D inversion of marine magnetotelluric data: case study from the Gemini Prospect, Gulf of Mexico. Geophysics 76(1):F77–F87

Zhdanov MS, Endo M, Cuma M, Linfoot J, Cox L, Wilson G (2012). The first practical 3D inversion of towed streamer EM data from the Troll field trial: 82nd Annual International Meeting, SEG, Expanded Abstracts

Ziolkowski A, Hobbs BA (2003) Detection of subsurface resistivity contrasts with application to location of fluids. International patent WO 03/023452 A1. Edinburgh University

Ziolkowski A, Parr R, Wright D, Nockles V, Limond C, Morris E, Linfoot J (2010) Multi-transient electromagnetic repeatability experiment over the North Sea Harding field. Geophys Prospect 58(6):1159–1176

Ziolkowski A, Wright D, Mattsson J (2011) Comparison of pseudo random binary sequence and square wave transient controlled source electromagnetic data over the Peon gas discovery, Norway. Geophys Prospect 59:1114–1131

To follow technical information about this technology, particularly that relating to prospecting, the reader may consult the following monthly and bimonthly journals:

- *Geophysical Prospecting*. Ed. Wiley
- *Journal of Applied Geophysics*. Ed. Elsevier
- *Marine and Petroleum Geology*. Ed. Elsevier
- *Applied Geophysics*. Ed. Springer
- *Surveys in Geophysics*. Ed. Springer
- *First Break*. Ed. EAGE
- *Petroleum Geoscience*. Ed. EAGE
- *Geophysics*. Ed. SEG
- *The Leading Edge*. Ed. SEG
- *Journal of Exploration Geophysics*. Ed. CSEG
- *Hydrographic and Seismic*. Ed. Engineer Live
- *Oil & Gas Engineer*. Ed. Engineer Live
- *Offshore*. Ed. PennWell
- *World Oil*. Ed. Gulf Publishing
- *Pipeline and Gas Journal*. Ed. Oildom Publishing

available in print (magazine) or digital (Internet) versions.

For the aspects concerning earth physics, the reader can also read more fundamental articles in the leading journals:

- *Geophysical Research Letters*
- *Annals of Geophysics*
- *Izvestiya (in Russian)*
- *Journal of Geomagnetism and Geoelectricity*
- *Geophysical Journal International*
- *Chinese Journal of Geophysics*
- *Physics of the Earth and Planetary Interiors*
- *Journal of Geophysical Research Solid Earth*
- *Geophysical Journal of the Royal Astronomical Society*
- *Oceanographic Research*
- *Earth Planets and Space*
- *Nature Geoscience*
- *Marine Geology*
- *Marine Geophysical Research*
- *Journal of Geophysical Research Planets*
- *Earth and Planetary Science Letters*
- *Journal of Oceanic Engineering*
- *Oceanographic Research. Papers*

Appendices

S. Sainson, *Electromagnetic Seabed Logging*, DOI 10.1007/978-3-319-45355-2

Appendix A1.1

Book References (in Order of Publication)

To deepen this very informative aspect of the industrial *history of applied geophysics* in general, as well as that of the techniques and practices related to general electrical and electromagnetic prospecting, the reader may refer to the chronological list of a few books:

- *Etude sur la prospection électrique du sous-sol* (Schlumberger 1920)
- *Electrical prospecting in Sweden* (Sundberg et al. 1925)
- *Methoden der angewandten Geophysik* (Ambronn 1926)
- *Geologische Einführung in die Geophysik* (Sieberg 1927)
- *Conférences sur la prospection géophysique* (Charrin 1927)
- *Los metodos geofisicos de prospection* (Sineriz 1928)
- *Elektrische Bodenforschung* (Heime 1928)
- *Les méthodes géophysiques de prospection appliqués à la recherche du pétrole* (Boutry 1929)
- *Geophysical prospecting* (AIME 1929)
- *Applied geophysics* (Eve and Key 1929)
- *Geophysical methods of prospecting* (Heiland 1929)
- *Les méthodes de prospection du sous-sol* (Rothé 1930)
- *Angewandte Geophysik* (Angenheister et al. 1930)
- *Principle and practice of geophysical prospecting* (Broughton Edge and Laby 1931)
- *Applied geophysics* (Shaw et al. 1931)
- *Geophysical prospecting* (AIME 1932)
- *Traité pratique de prospection géophysique* (Alexanian 1932)
- *Geophysical prospecting* (AIME 1934)
- *Lehrbuch der angewandten Geophysik* (Haalck 1934)
- *Angewandte Geophysik für Bergleute und Geologen* (Reich 1934)
- *Geophysics* (AIME 1940)
- *Geophysical exploration* (Heiland 1940)
- *Exploration geophysics* (Jakovky 1940)
- *Taschenbuch der angewandten Geophysik* (Reich and Zwenger 1943)
- *Praktische Geophysik* (Messer 1943)
- *Geophysics* (AIME 1945)
- *La prospection électrique du sous-sol* (Poldini 1947)
- *Grundzuge der angewandte Geoelektrik* (Fritsch 1949)
- *Introduction to geophysical prospecting* (Dobrin 1952)
- *Die physikalisch technischen Fortschritte der Geoelektrik* (Muller 1952)
- *Prospection géophysique* (Rothé 1952)
- *Tellurik, Grundlagen und Anwendungen* (Porstendorfer 1954)
- *Grundlagen der Geoelektrik* (Krajew 1957)
- *Lehrbuch der allgemeinen Geophysik* (Toperczer 1960)

- *Principles of applied geophysics* (Parasnis 1962)
- *Applied geophysics USSR* (Rast 1962)
- *Essai d'un historique des connaissances magnéto-telluriques* (Fournier 1966)
- *Interpretation of resistivity data* (Van Nostrand and Cook 1966)
- *The history of geophysical prospecting* (Sweet 1969)
- *Zur Geschichte der Geophysik* (Birett et al. 1974)
- *Schlumberger : histoire d'une technique* (Allaud and Martin 1976)
- *Schlumberger: the history of a technique* (Allaud and Martin 1977)
- *La boite magique* (Gruner 1977)
- *The Schlumberger adventure* (Gruner 1982)
- *Geophysics in the affairs of man* (Bates et al. 1982)
- *A short history of electrical techniques in petroleum exploration* (Hughes 1983)
- *60 ans de géophysique en URSS* (Itenberg 1994)
- *Science of the run* (Bowker 1994)
- *Les aventuriers de la terre CGG: 1931–1990...* (Castel et al. 1995)
- *Geschichte der Geophysik* (Kerz and Glassmeier 1999)
- *CGG 1931–2006, 75 ans de passion* (Chambovet et al. 2006)
- *Le sens du courant, la vie d'Henri Georges Doll* (Dorozynski and Oristaglio 2007)
- *A sixth sense, the life and science of Henri Georges Doll* (Dorozynski and Oristaglio 2009)

The reader will also find short histories in the many monographs devoted to specific electrical and electromagnetic prospecting.

Appendix A2.1

References to authors in the following texts can be found in Chapter 2 references

Theoretical electromagnetism recollections[3]

Continuous Currents

For continuous currents (DC), in an heterogeneous but isotropic medium, considering the conservation of charges in the medium (charge density q) expressed by the conservation equation:

[3]The reader will find more detailed presentations in the literature on theoretical electromagnetism (Stratton 1961; Gardiol 1979) and more specifically on the Maxwell equations (Hulin et al. 1993; De Becherrawy 2012). The latter are also contained in more or less specialized books about applied geophysics (Keller and Fischknecht 1966; Nabigian 1987).

$$\vec{\nabla}.\vec{J}+\frac{\partial q}{\partial t}=0 \quad \text{(A2.1.1)}$$

the electric field, the current density and the electric conductivity then obey the following three laws.

The electric field $\vec{E}$ drifts from a scalar potential V such that:

$$\vec{E}=-\vec{\nabla}V \quad \text{(A2.1.2a)}$$

The current is said to be continuous or stationary when there is no accumulation of charges, that is, when the flow of $\vec{J}$ (current density) through a closed surface is zero:

$$\vec{\nabla}.\vec{J}=0 \quad \text{(A2.1.2b)}$$

In the local conditions, the current density $\vec{J}$ is proportionately related to the electric field $\vec{E}$ by the constant of the medium, i.e., in this case its conductivity σ such that:

$$\vec{J}=\sigma\vec{E} \quad \text{(A2.1.2c)}$$

a generic term that formalizes then *Ohm's law*.

From these three Eqs. (A2.1.2a, A2.1.2b, A2.1.2c) is thus easily deduced $\nabla^2\vec{E}=0$, which then means the *Laplace equation* whose solution will give, after taking into account the limit conditions on the electrical discontinuities, the field values (solutions).

Alternative and Variable Currents

In what follows, for alternative or periodic currents (AC), we consider a plane wave (far field criterion) with a sinusoidal time variation ($e^{-i\omega t}$) moving in the direction of propagation z (see Fig. A2.1). This signal S is characterized by its amplitude A, its frequency or its pulsation ω and phase φ such that $S=A\ e^{-i\omega t+\varphi}$.

This description can be generalized to any waveform, thanks to Fourier analysis, which allows us to decompose signals of any form in a sum of elementary sinusoids (cf. Chap. 4, Sect. 4.3).

We would like to recall that solving the problem of wave propagation can often only be done in the time domain. Indeed, in many cases, it is necessary to introduce the concept of time and especially to concretely define its direction ($\pm\ t$). We then set additional or initial conditions to get the uniqueness of the solution. Then the transition from the time domain to the frequency domain is carried out by a Fourier transform (the variable t then disappears).

We can recall that the electromagnetic wave is defined by its pulsation:

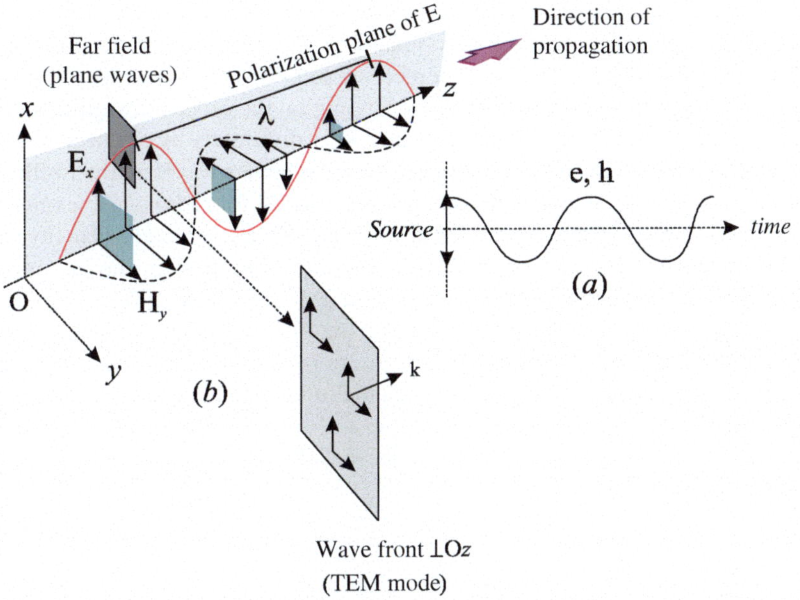

Fig. A2.1 In a continuous isotropic medium, for a sinusoidal plane wave (*a*) the electric field and magnetic field vectors are orthogonal to each other and oscillate in phase everywhere. They find themselves in a plan perpendicular to the direction of propagation (*b*). The wave has its electric field vector invariably headed in the direction of O*x*. This direction remains constant throughout the propagation (O*z* axis). The wave is said to be plane polarized and its plan of polarization is *x*O*z*. In this case the components E_z and B_z are zero and the wave propagates in a TEM mode

$$\omega_{[\mathrm{rad/s}]} = 2\pi f \qquad \text{(A.2.1.3a)}$$

It is related to the propagation velocity *c* ($m.s^{-1}$) in the medium by the wave number[4]:

$$k = \omega/c = 2\pi/\lambda \qquad \text{(A.2.1.3b)}$$

The intervals T (defining the period) or the frequency *f* (number of beats or cycles per second equivalent to 1/T) are characterized by the wavelength λ (m) and the propagation velocity as:

$$T_{[s]} = \lambda/c \quad \text{or} \quad f_{[\mathrm{Hz}]} = c/\lambda \qquad \text{(A.2.1.3c)}$$

The wavelength reflects the spatial interval between two points of the medium animated by the same vibratory state (with a phase shift of 2π) or the distance

[4]Physically the wave number counts the number of "peaks" over a given distance and is calculated by dividing the latter by the length of the wave.

traveled by the wave during one period of the signal (or one complete oscillation). It is just an intermediate quantity only related to the speed of propagation in the medium, external to the source and the receiver and stays the same regardless of the distance it is from the emission point. Depending on the order of magnitude of the frequency, the vibratory movement manifests very differently by electric, magnetic, chemical, calorific, light effects characterizing the higher wavelengths (millimeter to kilometers). In electromagnetic prospecting, except for radiometric exploration (γ radiation), the spectrum covers the band from 10^9 Hz (ground radar investigation: GPR) to 10^{-3} Hz (magnetotelluric sounding: MT) through the intermediate frequency methods with a controlled source (TEM and CSEM) (Fig. A2.2).

For variable currents, the distribution of the electric, magnetic fields and the induced currents in the conductors of electricity is obtained by solving the general equation of wave propagation coming from the fundamental equations of Maxwell, which themselves express, at a macroscopic scale, except for the limit conditions, the passage relations (media at rest) of these fields in the different materials or media (Maxwell 1865).

A2.1 Homogenous Maxwell's Equations

The wave propagation is governed by Maxwell's unified theory, which brings the laws and theorem of Faraday, Ampère and Gauss together and amounts in the time domain to the four following equations respectively:

$$\vec{\nabla} \wedge \vec{e} = -\frac{\partial \vec{b}}{\partial t} \tag{A2.1.4}$$

$$\vec{\nabla} \wedge \vec{h} = \vec{j} + \frac{\partial \vec{d}}{\partial t} \tag{A2.1.5}$$

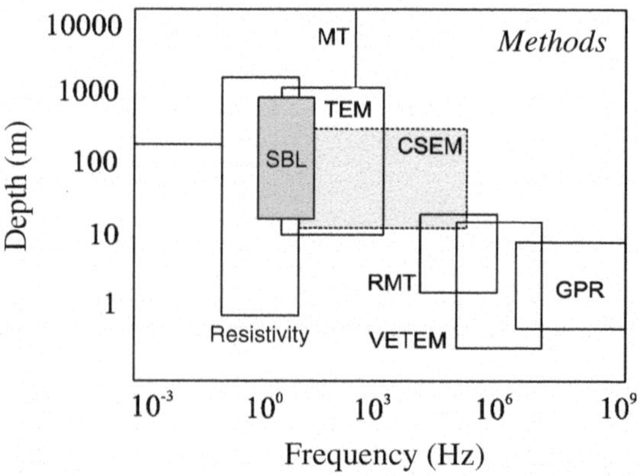

Fig. A2.2 Place of SBL methods depending on the frequency and depth of investigation in the wide range of terrestrial and marine electromagnetic survey techniques

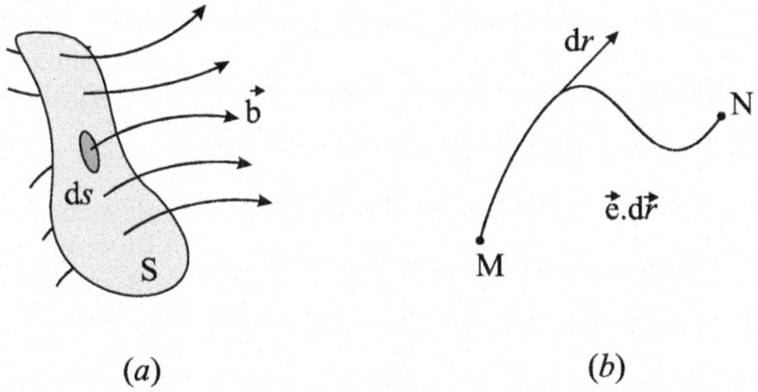

Fig. A2.3 Illustration of the nature of the magnetic and electric fields: vector $\vec{b}$ passing through a surface S (*a*) and vector $\vec{e}$ circulating on an MN curve (*b*)

$$\vec{\nabla} \cdot \vec{b} = 0 \tag{A2.1.6}$$

$$\vec{\nabla} \cdot \vec{d} = p \tag{A2.1.7}$$

where, in standardized SI units, $\vec{e}$ is the electric field (V/m), $\vec{b}$ the magnetic induction (Tesla), $\vec{h}$ the magnetic field (A/m), $\vec{d}$ the dielectric displacement (C/m^2), $\vec{j}$ the current density (A/m^2) and finally p the density of electric charge (C/m^3).

These equations state that any spatial variation of a field (electric or magnetic) at any point of the space leads to the existence of a time variation of another field at the same point and vice versa. These equations are presented here in their local form, i.e., a differential form (Fig. A2.3).

Maxwell's equations can also be made in the integral form, where they express then the relations between the electromagnetic fields in an area, rather than at a point (local form). Under these conditions the relations with rotationals are integrated on a surface using *Stokes' theorem* to obtain the flow of the vectors $\vec{e}$ and $\vec{h}$.

A2.2 Constitutive Relations

As Maxwell's equations are not coupled together, it is then necessary to connect the expressions of the fields, the charges and the currents by relations expressing behavioral laws depending this time on the frequency such that:

$$\vec{D} = \underline{\varepsilon}(\omega, \vec{e}, \vec{r}, t, T, P, \ldots) \cdot \vec{E} \tag{A2.1.8}$$

$$\vec{B} = \underline{\mu}(\omega, \vec{e}, \vec{r}, t, T, P, \ldots) \cdot \vec{H} \tag{A2.1.9}$$

$$\vec{J} = \underline{\sigma}(\omega, \vec{e}, \vec{r}, t, T, P, \ldots) \cdot \vec{E} \tag{A2.1.10}$$

where $\underline{\varepsilon}$, $\underline{\mu}$ and $\underline{\sigma}$ are respectively the dielectric permittivity, magnetic permeability and electrical conductivity tensors, and *t, T* and *P* the parameters of time, temperature and pressure. Theoretically these tensors are complex, involving phase shifts between the fields $\vec{D}$ and $\vec{E}$ and also between and $\vec{H}$, $\vec{J}$ and $\vec{E}$.

Practically, except in special cases, these tensors may be replaced by scalars, under the cover of simplifying assumptions:

- Linear propagation media, homogeneous and isotropic
- Electrical process no more dependent on time, temperature and pressure
- Nonmagnetic media where the permeability of the media is equivalent to the permeability of the vacuum ($\mu = \mu_{rock} = \mu_{water} = \mu_0 = 4\pi \times 10^{-7}$H.m^{-1}).

Under these conditions, the above constitutive equations reduce to:

$$\vec{D} = \left[\varepsilon'(\omega) - i\varepsilon''(\omega)\right] \vec{E} = \varepsilon\vec{E} \quad (A2.1.11)$$

$$\vec{J} = \left[\sigma'(\omega) + i\sigma''(\omega)\right] \vec{E} = \sigma\vec{E} \quad (A2.1.12)$$

and:

$$\vec{B} = \mu\vec{H} \quad (A2.1.13)$$

where the permittivity and conductivity are complex functions of frequency, when the permeability, which no longer depends on the frequency, is real (Fig. A2.4).

A2.3 Formation and Formulation of the Wave Equation

Taking the rotational $\left(\vec{\nabla} \wedge\right)$ of the first two equations of Maxwell, i.e.:

$$\vec{\nabla} \wedge \left(\vec{\nabla} \wedge \vec{e}\right) + \vec{\nabla} \wedge \frac{\partial \vec{b}}{\partial t} = 0 \quad (A2.1.14)$$

and:

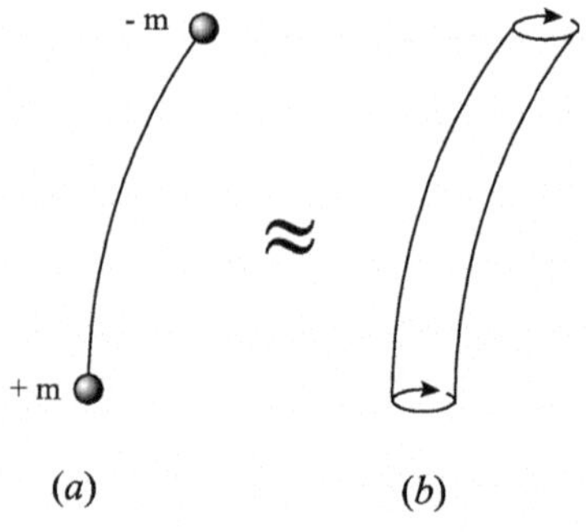

Fig. A2.4 In Maxwell's terminology, the lines of force (*a*) between two magnetic charges (+/−m) can be interpreted as tubes (*b*) formed from the current loops (see Fig. 2.3) (Maxwell 1861, 1862)

$$\vec{\nabla} \wedge (\vec{\nabla} \wedge \vec{h}) - \vec{\nabla} \wedge \frac{\partial \vec{d}}{\partial t} = \vec{\nabla} \wedge \vec{j} \tag{A2.1.15}$$

using by another way the constitutive relations in the time domain, where the constants ε, μ, et σ are then independent on time such that:

$$\vec{d} = \varepsilon \vec{e} \tag{A2.1.16}$$

and:

$$\vec{b} = \mu \vec{h} \tag{A2.1.17}$$

and:

$$\vec{j} = \sigma \vec{e} \tag{A2.1.18}$$

and replacing the latter in the former, we finally obtain:

$$\vec{\nabla} \wedge \vec{\nabla} \wedge \vec{e} + \mu \vec{\nabla} \wedge \frac{\partial \vec{h}}{\partial t} = 0 \tag{A2.1.19}$$

and:

$$\vec{\nabla} \wedge \vec{\nabla} \wedge \vec{h} - \varepsilon \vec{\nabla} \wedge \frac{\partial \vec{e}}{\partial t} = \sigma \vec{\nabla} \wedge \vec{e} \tag{A2.1.20}$$

By interchanging the derivative operators (as vector functions $\vec{h}$ and $\vec{e}$ and their first and second derivatives are continuous throughout the domain), we arrive at:

$$\vec{\nabla} \wedge \vec{\nabla} \wedge \vec{e} + \mu \frac{\partial}{\partial t} (\vec{\nabla} \wedge \vec{h}) = 0 \tag{A2.1.21}$$

and:

$$\vec{\nabla} \wedge \vec{\nabla} \wedge \vec{h} - \varepsilon \frac{\partial}{\partial t} (\vec{\nabla} \wedge \vec{e}) = \sigma \vec{\nabla} \wedge \vec{e} \tag{A2.1.22}$$

and then replacing $\vec{\nabla} \wedge \vec{h}$ and $\vec{\nabla} \wedge \vec{e}$ given by Maxwell's equations it remains that:

$$\vec{\nabla} \wedge \vec{\nabla} \wedge \vec{e} + \mu\varepsilon \frac{\partial^2 \vec{e}}{\partial t^2} + \mu\sigma \frac{\partial \vec{e}}{\partial t} = 0 \tag{A2.1.23}$$

and:

$$\vec{\nabla} \wedge \vec{\nabla} \wedge \vec{h} + \mu\varepsilon \frac{\partial^2 \vec{h}}{\partial t^2} + \mu\sigma \frac{\partial \vec{h}}{\partial t} = 0 \quad \text{(A2.1.24)}$$

Considering the remarkable vector identity $\vec{\nabla} \wedge \vec{\nabla} \wedge \vec{a} \equiv \vec{\nabla}\,\vec{\nabla} \cdot \vec{a} = \nabla^2 \vec{a}$ (given for any vector $\vec{a}$) with $\vec{\nabla}.\vec{h} = 0$ and $\vec{\nabla}.\vec{e} = 0$ for homogeneous media,[5] we can write then (Reitz and Milford 1962):

$$\nabla^2 \vec{e} - \mu\varepsilon \frac{\partial^2 \vec{e}}{\partial t^2} - \mu\sigma \frac{\partial \vec{e}}{\partial t} = 0 \quad \text{(A2.1.25)}$$

and:

$$\nabla^2 \vec{h} - \mu\varepsilon \frac{\partial^2 \vec{h}}{\partial t^2} - \mu\sigma \frac{\partial \vec{h}}{\partial t} = 0 \quad \text{(A2.1.26)}$$

Turning now to the Fourier domain, such that the field excitation varies over time in a sinusoidal manner, it eventually comes for the electric field[6] and for a monochromatic plane wave to the propagation/diffusion equation, which in the field frequency (ω) is written (Reitz and Milford 1962)[7]:

$$\nabla^2 \vec{E} + \left(\omega^2 \mu\varepsilon - i\omega\mu\sigma\right) \vec{E} = 0 \quad \text{(A2.1.27)}$$

more commonly called the Helmholtz equation, which also reflects an irreversible phenomenon.

A2.4 Helmholtz Equation: Discussion

By grouping the variational terms (μ, σ, ε) affecting the propagation medium and the frequency ω, the above equation in the frequency domain (cf. Eq. A2.1.27) can be reduced to:

$$\nabla^2 \vec{E} + k^2 \vec{E} = 0 \quad \text{(A2.1.28)}$$

with:

---

[5]In reality the rocks cannot be considered as homogeneous media and we have $\nabla \vec{e} \neq 0$. The resistivity contrasts then act as secondary sources.

[6]For convenience, we introduce here the complex notation that can express derivations such as $\partial/\partial t \to i\omega$ or $\nabla^2 \to -k^2$.

[7]To form the wave equation we can also rely on Maxwell's equations set out in the frequency domain. They can be found in all books of mathematical physics.

$$k^2 = \omega^2\mu\varepsilon - i\omega\mu\sigma \tag{A2.1.29}$$

where k represents the wave number or spreading factor.

This complex number consisting of a real part and an imaginary part, depending on the electromagnetic properties of the crossed media, will not have the same impact on the propagation/diffusion phenomena depending on the balance of its parts whether they are real or imaginary.

Discussion

The previous expression (cf. Eq. A2.1.29) therefore shows two scenarios:

- First case: $\omega^2\mu\varepsilon >> \omega\mu\sigma$ or $\omega\varepsilon/\sigma >> 1$. In this case we have:

$$k^2 \approx \omega\mu\varepsilon \tag{A2.1.30}$$

which is real as k is. By replacing k^2 with its value ($\mu\varepsilon$), the equation (cf. Eq. A2.1.25) in the time domain becomes then:

$$\nabla^2\vec{e} - \mu\varepsilon\frac{\partial^2\vec{e}}{\partial t^2} = 0 \tag{A2.1.31}$$

and describes, because of the presence of the second derivative, the wave propagation (*propagation equation*) moving at the speed $1/\sqrt{\mu\varepsilon}$ and dependent on the magnetic permeability μ and especially on the dielectric permittivity ε of the medium. In these circumstances, the displacement currents are dominant (→ insulating media).

- Second case: $\omega^2\mu\varepsilon << \omega\mu\sigma$ or $\omega\varepsilon/\sigma << 1$. In this case we have then:

$$k^2 \approx -i\omega\mu\sigma \tag{A2.1.32}$$

which is a pure imaginary and k a complex. By replacing k^2 with its value ($\mu\sigma$), the Eq. (A2.1.25) in the time domain becomes:

$$\nabla^2\vec{e} - \mu\sigma\frac{\partial\vec{e}}{\partial t} = 0 \tag{A2.1.33}$$

and describes the diffusion of a field (*diffusion equation*) whose amplitude, dependent on the electrical conductivity σ of the medium, decreases with the distance. In these circumstances the conduction currents dominate (→ conductive media). This equation, which neglects the second derivatives, is then applied to the phenomena slowly varying over time.

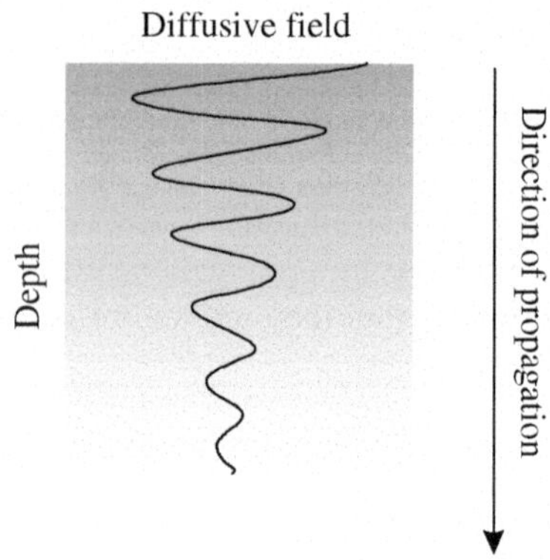

Fig. A2.5 Allure of a diffusive field marking an exponential diminution of energy (amplitude) with depth

Solution for the diffusion equation: considering a single dimension (z) in the direction of propagation, the equation becomes:

$$\frac{\partial^2 \vec{e}}{\partial z^2} - \mu\sigma \frac{\partial \vec{e}}{\partial t} = 0 \tag{A2.1.34}$$

The solution for the electric field is now of the form:

$$\vec{e} = \vec{e}_0^{\,+} e^{-i(kz-\omega t)} + \vec{e}_0^{\,-} e^{-i(-kz-\omega t)} \tag{A2.1.35}$$

where $\vec{e}_0^{\,+}$ and $\vec{e}_0^{\,-}$ respectively correspond to the upward and downward field.

If we now only consider the downward wave, there is in these circumstances:

$$\vec{e} = \vec{e}_0^{\,+} e^{-i\alpha z} e^{-\beta z} e^{i\omega t} \tag{A2.1.36}$$

which shows the weakening of an alternating field ($e^{-i\alpha z}$, $e^{i\omega t}$) with depth ($e^{-\beta z}$) (Fig. A2.5).

A2.5 Laplace Equations

Solving the Laplace equation allows us, whenever possible (quasistatic approximation), to overcome the effects of frequency as is the case at very low frequencies.[8] We can then solve complex problems of potential or field distribution as in DC stimulation with a minimum of calculations and good estimation.

The Laplace equation is obtained by taking the first Maxwell equation:

[8] At sufficiently low frequencies, the AC behaves in the subsoil and especially in conducting media as DC.

$$\vec{\nabla} \wedge \vec{e} + \frac{\partial \vec{b}}{\partial t} = 0 \qquad \text{(A2.1.37)}$$

and considering a quasistatic or steady regime with:

$$\frac{\partial \vec{b}}{\partial t} = 0 \qquad \text{(A2.1.38)}$$

where:

$$\vec{\nabla} \wedge \vec{e} = 0 \qquad \text{(A2.1.39)}$$

This expression shows that the electric field is conservative, a necessary and sufficient condition to demonstrate that it do derive of a scalar potential (gradient) such that:

$$\vec{E} = -\vec{\nabla} V \qquad \text{(A2.1.40)}$$

From the fourth law of Maxwell (ε does not vary with position) we then show that:

$$\vec{\nabla} . \vec{E} = q/\varepsilon \qquad \text{(A2.1.41)}$$

or that:

$$\vec{\nabla} . \vec{\nabla} V \equiv \nabla^2 V = q/\varepsilon \qquad \text{(A2.1.42)}$$

In each point free of charge ($q = 0$), we finally obtain the Laplace equation valid in an environment without a power source:

$$\nabla^2 V = 0 \qquad \text{(A2.1.43)}$$

which can be solved in different types of coordinates, Cartesian, cylindrical or spherical, according to the desired geophysical applications.

However, this partial derivative equation is not sufficient in itself to determine the function V to which it relates. This uncertainty reflects the fact that at a given electric field corresponds to not a potential but a group of potentials. To remove the indeterminacy, we then fix limit conditions that define the boundary elements on which restrictions may be imposed.

A2.6 Poisson Equation

The Poisson equation or equation of the potential vector is an equation of the same type as the Laplace equation (see Eq. A2.1.43) but whose second member this time is not zero:

$$\nabla^2 V = q/\varepsilon \tag{A2.1.44}$$

The general solution of this equation is obtained by adding to the solution of the equation without a second member (Laplace equation) a particular solution of the equation with a second member.

A2.7 Solution Unicity

If we consider, for distinct potentials V_1 and V_2, the equations:

$$\nabla^2 V_1 = q/\varepsilon \quad \text{and} \quad \nabla^2 V_2 = q/\varepsilon \tag{A2.1.45}$$

the unicity theorem proved for the Laplace equation is then transposable to the Poisson equation, such that by taking the difference we obtain again:

$$\nabla^2 (V_1 - V_2) = 0 \tag{A2.1.46}$$

A2.8 Passage Relations at the Interfaces

At the interfaces separating two different propagation media (1 and 2 for example), the components of the electric fields $\vec{E}$, $\vec{D}$ and magnetic fields $\vec{H}$, $\vec{B}$ must satisfy certain conditions of passage. These data are then given by integrating, on an elementary volume, the fundamental equations, such that we have:

$$\vec{n}_{12} \wedge \vec{E}_2 + \vec{n}_{21} \wedge \vec{E}_1 = \vec{n}_{12} \wedge \left(\vec{E}_2 - \vec{E}_1\right) = 0 \tag{A2.1.47}$$

$$\vec{n}_{12} \wedge \vec{D}_2 + \vec{n}_{21} \wedge \vec{D}_1 = \vec{n}_{12} \wedge \left(\vec{D}_2 - \vec{D}_1\right) = \vec{q}_s \tag{A2.1.48}$$

$$\vec{n}_{12} \wedge \vec{H}_2 + \vec{n}_{21} \wedge \overleftarrow{H}_1 = \vec{n}_{12} \wedge \left(\vec{H}_2 - \vec{H}_1\right) = \vec{J}_s \tag{A2.1.49}$$

$$\vec{n}_{12} \wedge \vec{B}_2 + \vec{n}_{21} \wedge \vec{B}_1 = \vec{n}_{12} \wedge \left(\vec{B}_2 - \vec{B}_1\right) = 0 \tag{A2.1.50}$$

where $\vec{n}_{12}$ and $\vec{n}_{21}$ are normal at the considered interfaces (respectively from 1 to 2 and from 2 to 1), $\vec{q}_s$ the surface charge density and $\vec{J}_s$ the actual density of surface current (Fig. A2.6).

More simply for a stationary current or equivalent, at a point P located at the interface of two media of different resistivities ρ_1 and ρ_2, the relations of passage through a plan (n, x) correspond, for some fields $\vec{E}$ such as $\vec{E}_{1,2} = -\vec{\nabla} V_{1,2}$ to:

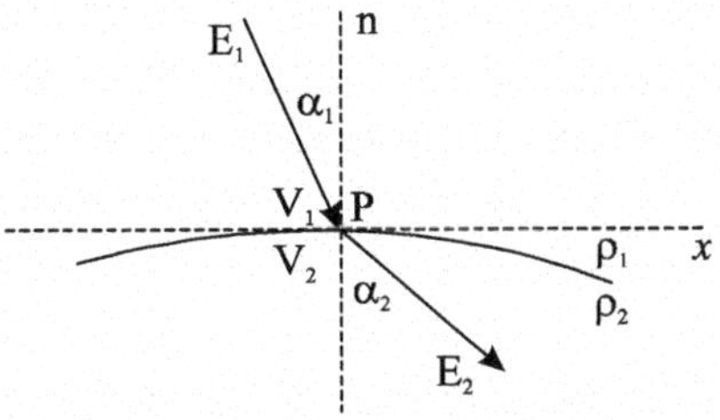

Fig. A2.6 Crossing relations at two interfaces of differents conductivity ρ_1 and ρ_2

– A continuity of the potential at the interface such that there is an equality of the potentials:

$$V_1 = V_2 \tag{A2.1.51}$$

and of the derivatives such that:

$$\frac{\partial V_1}{\partial x} = \frac{\partial V_2}{\partial x} \tag{A2.1.52}$$

– A continuity of the normal components in the plan of separation such that:

$$\frac{1}{\rho_1}\frac{\partial V_1}{\partial n} = \frac{1}{\rho_2}\frac{\partial V_2}{\partial n} \tag{A2.1.53}$$

– Equality of the angular relations on the fields $\vec{E}_{1,2}$ (cf. Fig. A2.6) such that:

$$\rho_1 \mathrm{tg}\alpha_1 = \rho_2 \mathrm{tg}\alpha_2 \tag{A2.1.54}$$

A2.9 Principle and Reciprocity Theorem

It has been shown (Landau and Lifshitz 1969) that, for two dipole sources (antenna) of separate currents $\vec{J}_{ext}^{(A)}$ and $\vec{J}_{ext}^{(M)}$ propagating in any medium, the fields $\vec{E}$ and the potentials V due to each source in the position of the other one (A or M), are then electrically equivalent and verify:

$$\int \vec{J}_{ext}^{(A)} \vec{E}_M dV_A = \int \vec{J}_{ext}^{(M)} \vec{E}_A dV_M \tag{A2.1.55}$$

This formulation, which corresponds to the reciprocity theorem,[9] where the Maxwell's equations satisfy these properties, is particularly important in electrical prospecting, especially in the interpretation algorithms using migration techniques (3D imaging) and the control of the quality of the acquisition.

A2.10 Static and Quasistatic Approximations

To simplify the calculations, it is possible in certain situations to establish some approximations. This is the case for instance:

- In DC prospecting where we practice static approximation, which consists of considering the potential differences or gradient as differences of electrostatic potentials[10] where electrical and magnetic phenomena are then independent.[11]
- In the investigations in low frequency alternating current when we practice the quasistatic approximation which consists of neglecting the induced effect[12] $(\vec{\nabla} \wedge \vec{e} = 0 \quad \Rightarrow \quad \vec{E} = -\vec{\nabla} V)$ in the limit of the skin depth (cf. Sect. 3.3.1 Chap. 3) where only the conduction currents are considered. In the case where the field change is sufficiently slow ($T >> \varsigma / c$ where ς is the size of the circuit) or in other words where this variability occurs on long time scales relative to the time characteristic of field adjustment, their distribution throughout space at any instant looks like that of a static field. The propagation velocity and time delay can be neglected. The field equations, also called pre-Maxwell equations (since they were discovered before Maxwell's equations were introduced), are invariant under the Galilean transformation. It follows that for quasistatic fields the differential equation is given by Eq. A2.1.33.

In those situations where we consider a uniform field, then we can use the results of the mathematical analysis (analytical or numerical) on the distributions established in the electrostatic field (Fig. A2.7).

---

[9]For the whole vector field see also (Kraichman, 1976) and for a detailed demonstration see the following Appendix.

[10]The electrical potential $V(r)$ is defined (Ellis and Singer, 2007) as the electrostatic potential $\upsilon(r)$ coming from the electrostatic field E (*Coulomb's law*), itself attributed to the electrostatic force field (q) such that: $\vec{E} = \frac{1}{4\pi\varepsilon_0}\frac{q}{r^2}\hat{r} \rightarrow \vartheta(r) = \frac{q}{4\pi\varepsilon_0}\frac{1}{r} = V(r)$

It is assumed in this case that the electrostatic laws still apply when electricity moves, i.e., when electrical currents appear as long as we are dealing with a steady state.

[11]Unlike electric masses at rest which do not engage any action on magnetic masses, electrical masses in motion engage one.

[12]See Chap. 5, Sect. 3.1.2

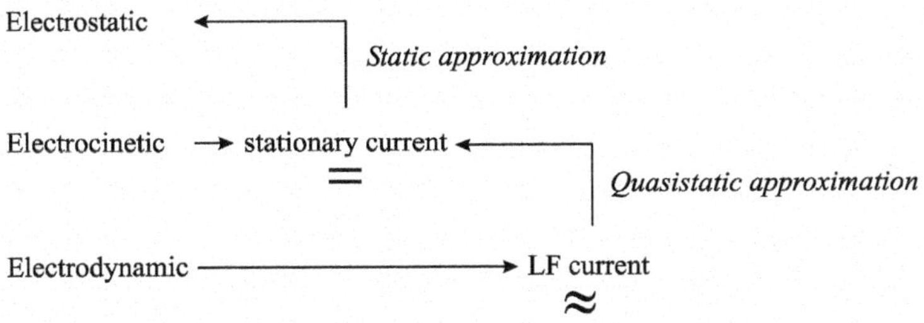

Fig. A2.7 Static and quasistatic approximations performed under exploration DC and under low frequency variable currents (LF)

Appendix A2.2

Demonstration of the reciprocity theorem

Preamble

In physics, the reciprocity theorem allocated to the principle of the same name takes on an important general character.[13] It can be applied both in the field of elastic wave propagation and in that of the diffusion and potential of electric currents.

Specifically, in acoustics for example (Landau and Lifchitz 1971), it governs the operation of piezoelectric sensors by using them both together or separately as a transmitter and a receiver (transducers or reciprocal sensors). It greatly also improves seismic data processing (Claerbout 1976).

In on-land or seabed electromagnetic prospecting, it allows us to diversify easily the geometric patterns as the arrays and instrumental arrangements according to the experimental stresses, and is of the greatest importance in the field of downhole well logging. The first one in 1915 to seize on the problem and use the reciprocity theorem (a priori without demonstrating this) was Frank Wenner from the US Bureau of Standards (Van Nostrand and Cook 1966).

In what follows, we take into account for the calculations the quasistatic approximation as DC computing.

A2.1. Principle

We prove for two fixed points A and M (electrodes) immersed in any homogeneous or heterogeneous medium, isotropic or anisotropic, that the potential V in M

[13]First formulated by Lord Rayleigh in his famous book *Theory of Sound*. It was H. A. Lorentz who enunciated in 1895 a reciprocity theorem for electromagnetic fields, which was completed in 1923 by J. R. Carson of Bell Laboratories for radio wave communication (Carson, 1923). See also P. Poincelot (Poincelot, 1961) and M. L. Burrows (Burrows 1978).

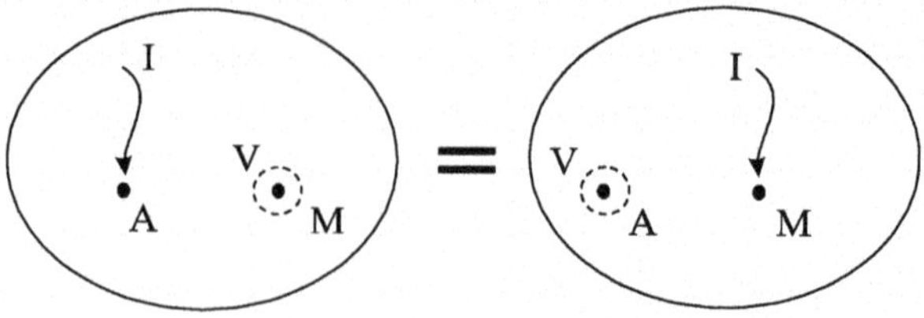

Fig. A2.8 Principle of reciprocity: an equivalence of potentials and currents

resulting from a given current I sent to A is equal to what would be the potential in A if the current was sent to M (cf. Fig. A2.8).

A2.2. Preliminary Formula Demonstration

In the proof of the theorem that follows, to avoid writing at great length, we use symbolic notations, which will be explained as and when they are introduced in the text.

A2.2.1. Partial Derivative Equation

The partial derivative equation, which satisfies the electric potential in a heterogeneous and anisotropic medium, and allows us to solve the problem, is built from three basic assumptions that:

– The electric field $\vec{E}$ is derived from a scalar potential V:

$$\vec{E} = -\vec{\nabla} V \tag{A2.2.1}$$

also given for its components in rectangular coordinates (x_1, x_2, x_3):

$$E_1 = -\frac{\partial V}{\partial x_1}, \quad E_2 = -\frac{\partial V}{\partial x_2}, \quad E_1 = -\frac{\partial V}{\partial x_3} \tag{A2.2.2}$$

– *Ohm's law*, which states that the current density vector $\vec{J}$ is deduced from the electric field vector $\vec{E}$ and the electrical conductivity σ by a symmetrical determinant linear transformation such that:

$$\vec{J} = \sigma \vec{E} \tag{A2.2.3}$$

or by considering the anisotropy of conductivity:

$$\begin{cases} J_1 = \sigma_{11}E_1 + \sigma_{12}E_2 + \sigma_{13}E_3 \\ J_2 = \sigma_{21}E_1 + \sigma_{22}E_2 + \sigma_{23}E_3 \\ J_3 = \sigma_{31}E_1 + \sigma_{32}E_2 + \sigma_{33}E_3 \end{cases} \quad \text{(A2.2.4)}$$

with $\sigma_{21} = \sigma_{12}$, $\sigma_{31} = \sigma_{13}$ and $\sigma_{23} = \sigma_{32}$
equations that can be symbolically written as:

$$J_k = \sigma_{kl}E_l \quad \text{with} \quad \sigma_{kl} = \sigma_{lk} \quad \text{(A2.2.5)}$$

– *Kirchhoff's law*, according to which the flow of the current density vector through a closed surface containing no power source is zero, results in:

$$\vec{\nabla}.\vec{J} = 0 \quad \text{(A2.2.6)}$$

or in rectangular coordinates:

$$\frac{\partial J_1}{\partial x_1} + \frac{\partial J_2}{\partial x_2} + \frac{\partial J_3}{\partial x_3} = 0 \quad \text{(A2.2.7)}$$

By transferring this time (A2.2.2) and (A2.2.4) into (A2.2.7) we obtain then:

$$\begin{aligned} &\sigma_{11}\frac{\partial^2 V}{\partial {x_1}^2} + \sigma_{12}\frac{\partial^2 V}{\partial x_1 \partial x_2} + \sigma_{13}\frac{\partial^2 V}{\partial x_1 \partial x_3} + \frac{\partial \sigma_{11}}{\partial x_1}\frac{\partial V}{\partial x_1} + \frac{\partial \sigma_{12}}{\partial x_1}\frac{\partial V}{\partial x_2} + \frac{\partial \sigma_{13}}{\partial x_1}\frac{\partial V}{\partial x_3} \\ &+\sigma_{21}\frac{\partial^2 V}{\partial x_1 \partial x_2} + \sigma_{22}\frac{\partial^2 V}{\partial {x_2}^2} + \sigma_{23}\frac{\partial^2 V}{\partial x_2 \partial x_3} + \frac{\partial \sigma_{21}}{\partial x_2}\frac{\partial V}{\partial x_1} + \frac{\partial \sigma_{21}}{\partial x_2}\frac{\partial V}{\partial x_2} + \frac{\partial \sigma_{23}}{\partial x_2}\frac{\partial V}{\partial x_3} \\ &+\sigma_{31}\frac{\partial^2 V}{\partial x_1 \partial x_3} + \sigma_{32}\frac{\partial^2 V}{\partial x_2 \partial x_3} + \sigma_{33}\frac{\partial^2 V}{\partial {x_3}^2} + \frac{\partial \sigma_{31}}{\partial x_3}\frac{\partial V}{\partial x_1} + \frac{\partial \sigma_{32}}{\partial x_3}\frac{\partial V}{\partial x_2} + \frac{\partial \sigma_{33}}{\partial x_3}\frac{\partial V}{\partial x_3} = 0 \end{aligned} \quad \text{(A2.2.8)}$$

an equation which is symbolically written:

$$\sigma_{kl}\frac{\partial^2 V}{\partial x_k \partial x_l} + \frac{\partial \sigma_{kl}}{\partial x_k}\frac{\partial V}{\partial x_{kl}} = 0 \quad \text{(A2.2.9)}$$

This is the basic equation which the electrical potential satisfies at any point where it is regular.

A2.2.2. Green Formula

From the preceding equation we can deduce another relation that satisfies the electrical potential.

If we call L(V) the first member of the Eq. (A2.2.9), U and V any two functions defined inside a closed domain D we find:

$$UL(V) = U\left[\sigma_{kl}\frac{\partial^2 V}{\partial x_k \partial x_l} + \frac{\partial \sigma_{kl}}{\partial x_k}\frac{\partial V}{\partial x_{kl}}\right] \quad (A2.2.10)$$

We have then:

$$\sigma_{kl}U\frac{\partial^2 V}{\partial x_k \partial x_l} = \frac{\partial}{\partial x_k}\left(\sigma_{kl}U\frac{\partial V}{\partial x_l}\right) - \frac{\partial}{\partial x_l}\left(\frac{\partial(\sigma_{kl}U)}{\partial x_k}V\right) + V\frac{\partial^2(\sigma_{kl}U)}{\partial x_k \partial x_l} \quad (A2.2.11)$$

and:

$$\frac{\partial \sigma_{kl}}{\partial x_k}U\frac{\partial V}{\partial x_l} = \frac{\partial}{\partial x_l}\left(\frac{\partial \sigma_{kl}}{\partial x_k}UV\right) - V\frac{\partial}{\partial x_l}\left(\frac{\partial \sigma_{kl}}{\partial x_k}U\right) \quad (A2.2.12)$$

Adding at once term by term we obtain:

$$M(U) = \frac{\partial^2(\sigma_{kl}U)}{\partial x_k \partial x_l} - \frac{\partial}{\partial x_l}\left(\frac{\partial \sigma_{kl}}{\partial x_k}U\right) \quad (A2.2.13)$$

and:

$$UL(V) - VM(U) = \frac{\partial}{\partial x_k}\left(\sigma_{kl}U\frac{\partial V}{\partial x_l}\right) - \frac{\partial}{\partial x_l}\left(\sigma_{kl}V\frac{\partial U}{\partial x_k}\right) \quad (A2.2.14)$$

Noting that we have M (U) = L (U) and interchanging the indices in the second term of the second member, we come finally to:

$$\begin{aligned} UL(V) - VL(U) &= \frac{\partial}{\partial x_k}\left(\sigma_{kl}U\frac{\partial V}{\partial x_l} - \sigma_{lk}V\frac{\partial U}{\partial x_l}\right) \\ &= \frac{\partial}{\partial x_k}\left(\sigma_{kl}U\frac{\partial V}{\partial x_l} - \sigma_{kl}V\frac{\partial U}{\partial x_l}\right) \end{aligned} \quad (A2.2.15)$$

as $\sigma_{kl} = \sigma_{lk}$.

If we now call P_{kl} the quantity in parentheses and if we integrate the two members of the equation (A2.2.14) in the domain D, then:

$$\iiint_D [UL(V) - VL(U)]\, dV = \iiint_D \frac{\partial P_{kl}}{\partial x_k} dx_1 dx_2 dx_3 = \iint_{S_D} P_{kl} n_k ds \quad (A2.2.16)$$

S_D being the surface that limits the domain D and n_k one of the director cosines of the normal to the surface facing outwardly.

If U and V are considered as electric potentials, the amount to be included in the second term taking into account the symbolic notations can be simplified as:

$$\sigma_{kl}\frac{\partial U}{\partial x_l} = J_k \text{ and } \sigma_{kl}\frac{\partial V}{\partial x_l}n_k = -J_k n_k = J_n \qquad \text{(A2.2.17)}$$

J_n then designating the normal component of the current density directed outward D.

If j_n is the analogous quantity for the potential U, we obtain:

$$\iiint_D [UL(V) - VL(U)]\, dV = \iint_S [UJ_n - Vj_n]\, ds \qquad \text{(A2.2.18)}$$

which thus represents the final *Green's formula.*

A2.3. Demonstration of the Reciprocity Theorem

Consider now two electrodes A and M respectively enclosed in two small areas S_A and S_M, all wrapped in a closed surface S. The domain D will be formed by the domain inside S and outside S_A and S_M (see Fig. A2.9).

First assume that a current is emitted by electrode A and call $V(x_1, x_2, x_3)$ the resulting potential at any point of D. Suppose then that electrode M also emits a current and that $U(x_1, x_2, x_3)$ is the potential that only results from the current at any point of D.

The domain D containing no power source since A and M are excluded, the potentials U and V then satisfy at any point of D the basic equation (cf. Eq. A2.2.9). We therefore have under these conditions:

$$L(U) = 0 \quad \text{and} \quad L(V) = 0 \qquad \text{(A2.2.19)}$$

Green's formula applied now to the domain D then gives:

$$\iint_S [UJ_n - Vj_n]\, ds = 0 \qquad \text{(A2.2.20)}$$

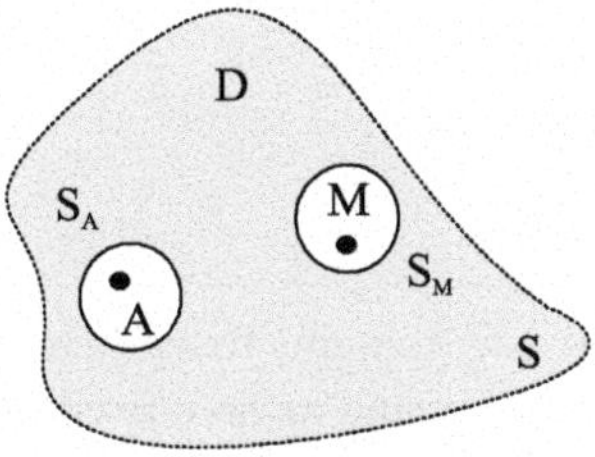

Fig. A2.9 Electrodes A and M enclosed in a domain D surrounded by the surface S of the envelope

where S denotes the set of the surfaces S_D, S_A, S_M and J_n, j_n are the normal components of the current crossing these surfaces to the interior of the domain D.

If we call i the first member of the above equation, i consists of three terms i_A, i_M and i_D related to the integrals extended respectively to the corresponding surfaces S_A, $S_{M,}$ S_D.

First considering the term:

$$i_A = \iint_{S_A} [UJ_n\text{-}Vj_n]\, ds \tag{A2.2.21}$$

the surface S_A then tends to the point A.

As the potential U (potential due to the electrode M) has no singularity at the A its value will tend to the value it has in A.

As on the other hand:

$$\iint_S J_n ds = J \tag{A2.2.22}$$

current emitted by A, we can easily recognize (see below) that the first part of i_A tends to JU(A) such that:

$$\lim_{S \to S_A} \iint UJ_n ds = JU(A) \tag{A2.2.23}$$

As for the second part of i_A:

$$\iint_{S_A} Vj_n ds \tag{A2.2.24}$$

it can be assumed that it tends to zero.

Indeed, the current from M remains naturally finite and as the area S_A also tends to zero as r^2 (r is the distance of any of its points to A) one need only assume that V tends to infinity as $1/r$ to see that the integral tends to zero as r.

For the demonstration to be complete it would still have to be proved that V is actually of the order of $1/r$. However, we can in a first step overcome this condition (cf. Sect. A2.4).

Ultimately we have for A:

$$\lim_{S_A \to 0} i_A = JU(A) \tag{A2.2.25}$$

and for M:

$$\lim_{S_M \to 0} i_M = -jV(M) \tag{A2.2.26}$$

It now remains to evaluate the integral on S. Assuming that S is the surface of the subsoil, and as no current flows through this area, we have in fact:

$$J_n = 0 \quad j_n = 0 \quad \text{so} \quad i_{S_D} = 0 \tag{A2.2.27}$$

The equation:

$$i = i_A + i_M + i_{S_D} = 0 \tag{A2.2.28}$$

therefore ultimately reduces to:

$$JU(A) - jV(M) = 0 \tag{A2.2.29}$$

If in addition the currents sent by A and M are equal, that is J = j, one will finally have:

$$U(A) = V(M) \tag{A2.2.30}$$

The potential in A which results from a certain current supplied by M and the potential in M which results from an equal current supplied by A are finally equal.

A2.4. Condition on the Potential

In the first part of this Appendix, to avoid overloading the scriptures, certain assumptions have been accepted and in particular those concerning the potential.

Thus, it can be shown:

- Firstly that:

$$\lim_{S_A \to 0} \iint_{S_A} UJ_n ds = JU\ (A) \tag{A2.2.31}$$

Let P be any point in S_A. The potential U from M being continuous next to A we have:

$$U(P) - U(A) \langle \xi \tag{A2.2.32}$$

ξ being very small.

If P – A is small enough, that is $(P - A) \langle \eta$, and on the other hand if we have whatever S_A:

$$\iint_{S_A} J_n ds = J \tag{A2.2.33}$$

we must show that:

$$i = \iint_{S_A} U(P)\, J_n ds - JU(A) \quad \rightarrow 0 \tag{A2.2.34}$$

Now we can write:

$$i = \iint_{S_A} [U(P) - U(A)]\, J_n ds \tag{A2.2.35}$$

and as $[U(P) - U(A)] \langle\, \xi$, $i \langle\, \xi J$, and as J remains finite, we then have:
$i \rightarrow 0$ when $(P - A) \rightarrow 0$

- Secondly, that:

$$\lim_{S_A \rightarrow 0} \iint_{S_A} V i_n ds = 0 \tag{A2.2.36}$$

Applying Green's formula to the domain D_A inside the surface S_A, L(V) is not zero, since V has a pole in A. We however have L (U) = 0. The formula becomes:

$$\iiint_{D_A} UL(V) = -\iint_{S_A} [UJ_n - Vj_n]\, ds \tag{A2.2.37}$$

The sign change comes from the fact that J_n and j_n now denote the normal components of the current leaving D_A.

We now show that:

$$\iint_{S_A} Vj_n\, ds = \iint_{S_A} UJ_n\, ds + \iiint_{D_A} UL(V)\, dV \rightarrow 0 \tag{A2.2.38}$$

but we have:

$$\iint_{S_A} UJ_n\, ds = \iint_{S_A} [U - U(A)]\, J_n\, ds + JU(A) \tag{A2.2.39}$$

and also:

$$L(V) = -\vec{\nabla}.\vec{J} \tag{A2.2.40}$$

Thus we obtain the equations:

$$\begin{aligned} \iiint_{D_A} UL(V)\, dV &= -\iiint_{D_A} U\vec{\nabla}.\vec{J}\, dV = -\iiint_{D_A} [U - U(A)]\, \vec{\nabla}.\vec{J} dV \\ &= -U(A) \iiint_{D_A} \vec{\nabla}.\vec{J}\, dV \end{aligned} \tag{A2.2.41}$$

As:

$$\iiint_{D_A} \vec{\nabla}.\vec{J}\, dV = \iint_{S_A} J_n ds = J \tag{A2.2.42}$$

then:

$$\iint_{S_A} Vj_n\, ds = \iint_{S_A} [U - U(A)]\, J_n\, ds - \iiint_{D_A} [U - U(A)]\; \vec{\nabla}.\vec{J}\, dV \tag{A2.2.43}$$

and therefore:

$$\left[\iint_{S_A} Vj_n\, ds\right] \leq 2\xi J \rightarrow 0 \tag{A2.2.44}$$

Appendix A2.3

Magnetic field produced by telluric currents

Preamble

To evaluate the magnetic field from the telluric currents flowing through the earth, we consider here a subsoil consisting of a stack of substantially horizontal geological strata.

A3.1. Magnetic Field Calculation

It is assumed here that the telluric field is uniform. We then consider a horizontal layer of thickness dz, where ρ is the electrical resistivity and the value of the current density flowing through the element is j (see Fig. A2.10).

In the x direction, a current tube lying in this layer with vertical (z, z + dz) and horizontal (x, x + dx) dimensions pierces the plan xOz at a point M. The horizontal magnetic field at a point P located at the ground surface, perpendicular to the current direction, is of the form:

$$dH = \frac{2j}{PM} dx\, dz\, \cos\, \alpha \tag{A2.3.1}$$

where α is the angle between the line joining the points P and M, and the z-axis.

Considering that:

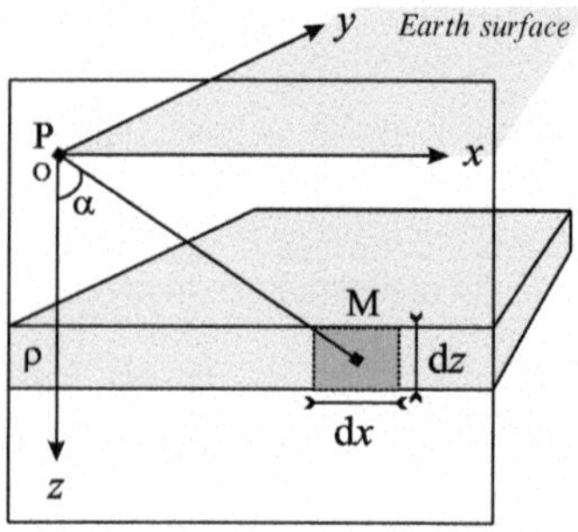

Fig. A2.10 Geometric model for the establishment of the calculation of the magnetic field associated with the flow of telluric currents

$$PM = \frac{z}{\cos\alpha} \quad \text{and} \quad x = z\,\mathrm{tg}\alpha$$
$$\rightarrow \quad dx = z\frac{d\alpha}{\cos^2\alpha} \tag{A2.3.2}$$

we draw:

$$dH = 2j\,dz\,d\alpha \tag{A2.3.3}$$

and then integrating from $-\pi/2$ to $+\pi/2$ we finally obtain:

$$dH = 2\pi j\,dz \tag{A2.3.4}$$

As:

$$j = \frac{1}{\rho}\frac{dV}{dy} \tag{A2.3.5}$$

where dV/dy is the potential gradient in the y direction,
thus we arrive at:

$$H = \int_0^z \frac{2\pi}{\rho}\frac{dV}{dy}dz \tag{A2.3.6}$$

As the field is uniform, the potential gradient dV/dy is then constant everywhere. We can therefore write:

$$H = 2\pi\frac{dV}{dy}\int_0^z \frac{dz}{\rho} \tag{A2.3.7}$$

which is representative of the total conductance of the field.

A3.2. Order of Magnitude of the Magnetic Field

If now V is expressed in volts, y, z and r in meters and ρ in Ω.m, we obtain the value of H (by dividing by 10^3) in Gauss:

$$H_{[G]} = \frac{2\pi}{10^3}\frac{dV}{dy}\int_0^z \frac{dz}{\rho} \qquad \text{(A2.3.8)}$$

To assess and establish an order of magnitude, the following is assumed in substance:

- A telluric field of 1 mV/km, i.e., a gradient dV/dy corresponding to 10^{-6} V/m
- A layer of conductive ground of 10 Ω.m and 1 km thick resting on insulating grounds ($z/\rho=10^2$)

In these local conditions we find after integration of (A2.3.8):

$$H_{[G]} = \frac{2\pi}{10^3}\left(\frac{dV}{dy}\right)\frac{z}{\rho} \qquad \text{(A2.3.9)}$$

or finally by substituting by the above values:

$$H = 2\pi 10^{-7}G \quad \text{or} \quad H = 2\pi 10^{-2}\gamma \qquad \text{(A2.3.10)}$$

a very low value[14] compared to the natural allochthonous variations present in the subsoil, but still measurable over a long period with sensitive variometers (see Chap. 4, Sect. 5.7.2).

However, if we admit that the telluric field varies with an angular velocity of about 1/5 rad for a period of 30 s, the maximum dH/dt will be then:

$$\frac{2\pi}{5}10^{-2}\gamma/s \qquad \text{(A2.3.11)}$$

This variation is actually of the same order of magnitude as that which corresponds to the diurnal variation of the earth's magnetic field and far greater than that which corresponds to the secular variation (see Chap. 4, Sect. 4.5.1).

However it can on one hand be measured with sufficient accuracy in the range of considered frequencies, and on the other hand be relatively easily separated from the natural variations as it precisely follows the variations of the telluric field and remains proportional to it.

---

[14]I hope that the younger generation of geophysicists will not be cross with our use of the old notation (CGS), i.e., the Gauss (G) and gamma (γ), which I think are more appropriate than the Tesla (T) at the magnitude orders of the measurements in geophysics (1 γ is equal to 10^{-9} T or to 10^{-5} G).

Appendix A2.4

Brief history of onshore electrotelluric or telluric prospecting

Preamble

The experience of Vitré[15] (Schlumberger 1930) demonstrated the efficiency of the electrical method for deep geological layer investigations, but also the inadequacy of the means implemented for a truly industrial application. The idea was then to use not an artificial current but the natural currents flowing through the earth's crust and well known by the earth scientists and telegraphists at this time.

This principle has the merit of significantly limiting the length of lines and consequently the emitted power. Considering the local variations in the density and azimuth of the telluric current sheets, all the emission devices disappear and the exploration equipment becomes much lighter. However, it no longer reaches vertical scaling as in conventional electric sounding, but there are structural variations in the horizontal direction with an ad hoc device (base and mobile station).

However, the major difficulty is that these fields are highly variable, depending on the time, direction and intensity. The comparisons of the fields at two distinct points on the surface earth's, to be valid, must be made between fields at the same time. We then observe, as the theory predicts, that the relation between the fields at two points is projective. This projection can be represented by two linear relations between the horizontal components of the two fields whose determinant is a number that only depends on the two points of measurement. This number, known as the area, corresponds to the ratio of the areas constructed from the two field vectors and is the basis of the interpretation of the telluric maps; the area map represents the map of the interferences caused by the geological structure.

By 1921, unidirectional observations in France were realized on each side of the Rhine fault (Léonardon 1928), followed in June 1922 by others executed in the French district of the Haute Marne with two identical measuring devices 2 km apart, each of them this time made with two perpendicular lines 100 long, leading to remarkable conclusions about possible correlations at a distance. At Val Richer in March 1934 (the Norman property of the Schlumberger family), it was found that

[15]The electrical sounding used at Vitré (the Ile et Vilaine French district) foreshadowed deep oil exploration. Before all experience of earth physics, this survey realized in 1928 in Normandy (France) aimed to establish the structure of the subsoil at depths allowing the researchers to reach the base of the Armorican block. The difficulty here lay in the fact that to achieve a sufficient depth of investigation it was necessary to have long lengths of line. For that, Conrad and Marcel Schlumberger and their collaborators used the telegraph line along the Rennes–Laval railway, available to them for the occasion from the local post and telecommunications. The sending power line (2 A at 200 V) was then a little over 200 km long. The DC was periodically reversed after a varying time of a few seconds. The expected depth of investigation was approximately 50 kilometers and the measured resistivities varied from 500 to 1800 $\Omega.m^2/m$. This unique experiment would be published a few months later in the renowned journal of the American Institute of Mining and Metallurgical Engineers (AIME).

with 500 m bases, correlations were possible between stations spaced by several tens of kilometers. At this time, it was suggested to apply the method to the exploration of the salt domes of the Oural–Emba region in the Soviet Union (now in Kazakhstan). To reinforce this idea some tests took place near the Colmar city in June 1934 and in July 1936 in the plain of Alsace near Hettenschlag on the very site where a diapir had been recognized. The map of tellurics as it was called later by the prospectors perfectly overlapped with that of the resistivities obtained by previous electrical soundings, marking then the first success of this method. On the eve of the Second World War a specific and compact device (the UR for Universal Recorder) was designed and manufactured clandestinely in France, collecting in the same "box", of some 20 kg weight, the entire measuring device: antivibration galvanometers of the Piccard type and two canals photographic recorder (argentic film).

Then experimented with in Morocco and Sumatra, the method mainly developed in occupied France. From 1941 to 1945, several teams from CGG covered the Aquitaine region totaling over 20,000 km^2, highlighting the major tectonic axes of the southern part of the sedimentary basin. After 1945, the technique spread in the French colonies and especially in the Sahara Desert, Equatorial Africa, Madagascar, Italy, and Sicily, but also in the USSR and Austria (Porstendorfer 1960), generally prior to detailed seismic operations or replacing them for some reasons of difficult topography or complex tectonics (Migaux 1948). For lack of market outlets, especially in the USA—a major consumer of geophysics, where seismic reflection was largely dominant in the market—the telluric method dwindled in the late 1950s (Allaud and Martin 1976).

Bibliographic References

Allaud L, Martin M (1976) Schlumberger: histoire d'une technique. Ed. Berger-Levrault. 348 p

Leonardon EG (1928) Some observations upon telluric currents and their application to electrical prospecting. Terr Magn Atmos Electr 33(2):91–94

Migaux L (1948) Une méthode nouvelle de géophysique appliquée: la prospection par courants telluriques. Tiré à part de la Compagnie générale de géophysique, 31 p

Porstendorfer G (1960) Tellurik. Grundlagen Messtechnik und neue Einsatzmöglichkeit. Ed. Akadenie Verlag, 186 p

Schlumberger C et M (1930) Electrical studies of the earth's crust at great depth. New York meeting. Geophysical prospecting AIME 1932, pp 134–140

Appendix A2.5

Definitions

As long as it is to make measurements, geophysics can be regarded as an exact science. It is no more true when it comes to interpreting the geological results of

these measurements, since it involves so many simplifying assumptions that are established a priori from the knowledge of the regional or local geology. This may seem paradoxical if we also think of the extremely sophisticated mathematical models that are used.

In this context of uncertainty, it is necessary to precisely define some concepts on natural environments. They allow us to fix once and for all the limitations of the geological interpretation that varies as we think from one prospect to another.

Homogeneous Medium. The geophysical definition of homogeneity has not the rigor of that used in physics or chemistry. A rock formation is considered electrically homogeneous if its resistivity does not substantially vary over the whole or a part of its extent (of the order of 1/20). It is enough that the layer is geologically monotonous and regular over the considered extent, i.e., its physical and chemical structure remains substantially constant in this space.

Isotropic Medium. A medium is said to be isotropic when its electrical properties are the same in all directions. In reality, however, the grounds are very often anisotropic. This anisotropy in the sedimentary layers of the same age, for example, is often linked to the stratification (due in particular to the conditions of sedimentation and tectonics). In this case, the electric current tends to flow along a preferred direction, in this case that of the strata whose conductivity is generally maximum.

Anomaly. We can affirm the existence of an anomaly when, after the use of a suitable investigation depth and a choice of suitable scale, the amplitude of the detected anomaly is on one hand consistent with what can be expected in a given geological environment (size, depth, electromagnetic characteristics, etc.), and on the other hand greater than the possible or probable error made in the acquisition measures.

Appendix A3.1

Abacus giving the resistivities (Ω.m) of different waters, depending on the concentration in G/G (grains per gallon) or in PPM of sodium chloride (NaCl) and temperature (°F)

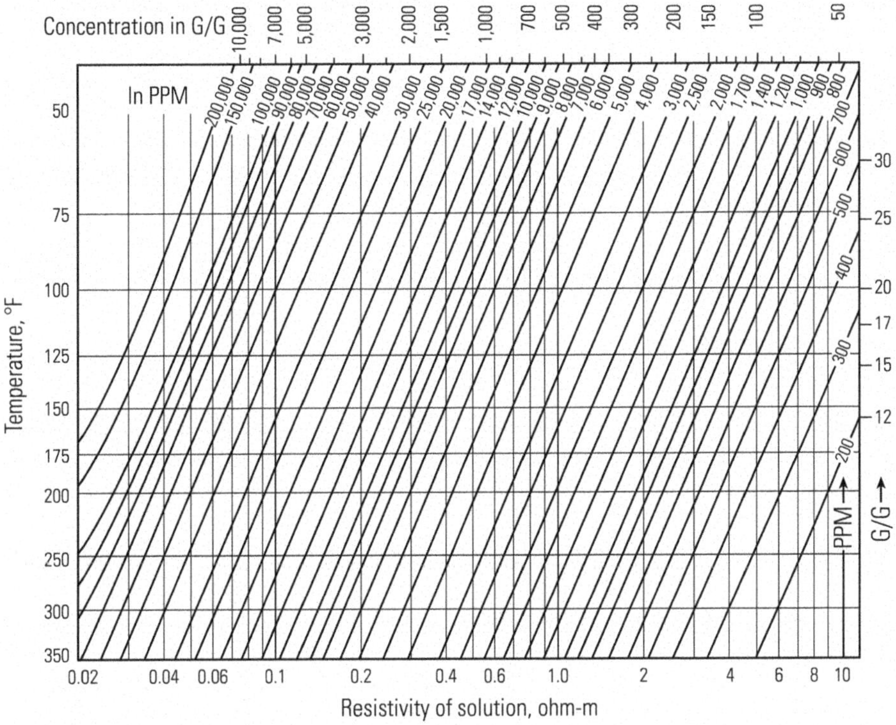

According to Ellis and Singer (2007).

Appendix A3.2

Calculation of the magnetic field of an electric current from one or two point electrodes and topological shape of this field in a electrically conductive environment

In low frequency approximation (quasistatic approximation), we usually consider the distribution of the electric current from a point electrode placed in O in an indefinite homogeneous medium as having a spherical symmetry, the current then regularly escaping in all directions of space.

If I is the intensity of the total current emitted by this electrode, the value of the current density $\vec{J}$ at any point M of space, located at a distance r of O, is radial (Sunde 1948) and equal in module to:

$$J = \frac{di}{ds} = \frac{I}{S} = \frac{I}{4\pi r^2} \qquad (A3.2.1)$$

This means, in other words, that as one moves away from the injection point O, the current density gradually decreases according to a geometric law of spherical divergence (in $1/4\pi r^2$) (Fig. A3.11).

Question Now, assuming the relative magnetic permeability of the medium is equal to one (not magnetic media), what is the magnetic field corresponding to such a distribution?

Response According to Maxwell's laws (Ampere's equation), the magnetic field $\vec{H}$ is linked to the current density $\vec{J}$ by the relation:

$$\vec{\nabla} \wedge \vec{H} = 4\pi\vec{J} \qquad (A3.2.2)$$

$\vec{H}$ is then a vector for which the rotational, at each point, is radial and equal to I/r^2. Moreover, we can also build $\vec{H}$ from the potential vector $\vec{A}$, which it is the rotational such that we have:

$$\vec{H} = \vec{\nabla} \wedge \vec{A} \qquad (A3.2.3)$$

The potential vector is then built itself from the current distribution such that:

$$\vec{A} = \int \frac{\vec{J}}{r_1} dv \qquad (A3.2.4)$$

where the integral is then extended to the entire space, where $\vec{J}$ is the current density at the point N surrounded by the volume element dv and at the distance r_1 from M. From this expression, it follows that $\vec{A}$ as $\vec{J}$ has a spherical symmetry centered on O and then its rotational $\vec{H}$ is zero at any point. But this result is inconsistent with equality:

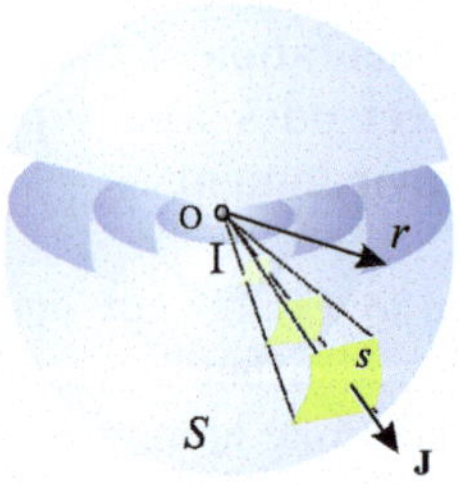

Fig. A3.11 Decrease in current density with the remoteness (spherical divergence) of the injection point

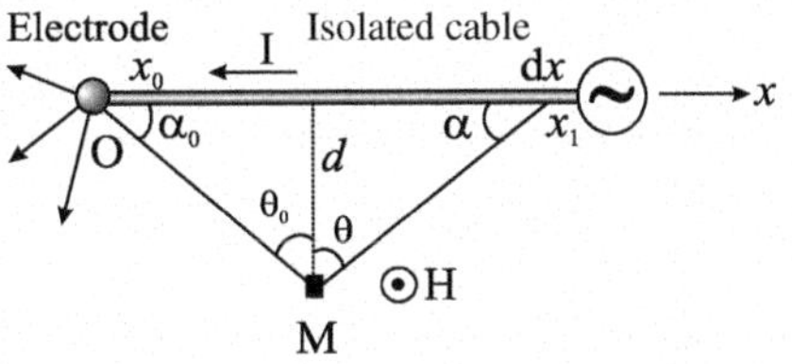

Fig. A3.12 Diagram of the antenna consisting of an electrically isolated conductor strand and a point O of the output current (injection site)

$$\vec{\nabla} \wedge \vec{\mathrm{H}} = \frac{\mathrm{I}}{r^2} \tag{A3.2.5}$$

as a zero vector everywhere cannot have a nonzero rotational. Such a contradiction can only come from a defect in the expression of the original problem. This is because the injected current I is ultimately brought by the wire (antenna) and the latter contributes among other things to the establishment of the other field. Under these conditions, the fundamental element is no longer the electrode O that disperses current but the rectilinear cable indefinite in a direction that brings current.

The magnetic field $\vec{\mathrm{H}}$ in M (see Fig. A3.12), located at a distance d from the cable ($\perp$) where a current I flows, follows Ampere's theorem such that a small part of current Ids (x_1 axis) creates a small field element dH perpendicular to the plane OxM:

$$\mathrm{dH} = \frac{\mathrm{I}ds \sin\alpha}{r_1^2} = \frac{\mathrm{I}d\mathrm{d}x}{r_1^3} \tag{A3.2.6}$$

where α is the angle between the directions of d and Mx_1.

By integrating the above equation (cf. Eq. A3.2.6) from x_o, we obtain the total field which is perpendicular to the plane MOx such that its intensity is equal to:

$$\begin{aligned} \mathrm{H} &= \mathrm{I}d\int_{z_0}^{\infty} \frac{\mathrm{d}x}{r_1^3} = \frac{\mathrm{I}}{d}\int_{\theta_0}^{\pi/2} \frac{\mathrm{d}(\tan\theta)}{(1+\tan^2\theta)^3} \\ &= \frac{\mathrm{I}}{d}\int_{\theta_0}^{\pi/2} \cos\theta\,\mathrm{d}\theta = \frac{\mathrm{I}}{d}(1-\sin\theta_0) = \frac{\mathrm{I}}{d}\left(1-\frac{x_0}{\sqrt{x_0^2+d^2}}\right) \end{aligned} \tag{A3.2.7}$$

If instead of taking the projection of M on Ox as the origin, we take O, and thus x is the abscissa of M, then we can write that:

$$\mathrm{H} = \frac{\mathrm{I}}{d}\left(1+\frac{x}{\sqrt{x^2+d^2}}\right) \tag{A3.2.8}$$

Under the relation (A3.2.2), we finally obtain a current density value equivalent to:

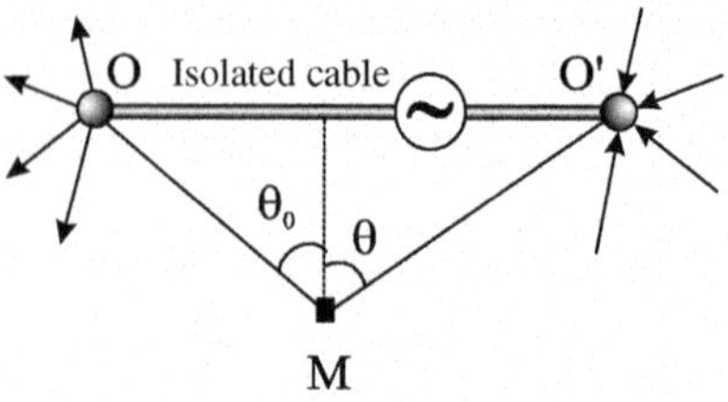

Fig. A3.13 Diagram of the antenna with two injection points OO′ (input and return current)

$$\mathrm{J}=\frac{\mathrm{I}}{4\pi r^{2}}=\frac{\mathrm{I}}{4\pi\left(x^{2}+d^{2}\right)} \tag{A3.2.9}$$

The magnetic field resulting from the current threads flowing in the medium is zero because of their spherical symmetry. The radial current can only exist because the total magnetic field has a nonzero rotational. This circumstance only occurs because the cable of the antenna Ox is interrupted in O. In other words, the current is spread in the environment only because the cable is interrupted. Conversely, the spherical distribution of the current from O cannot be conceivable without the Ox cable.

When the cable is interrupted at another point O′ serving for example as a return electrode, we then obtain in such circumstances the corresponding magnetic and electric fields, by superposing on the Ox cable (conducting I, interrupted in O), a cable O′x (conducting I, interrupted in O′) (Fig. A3.13).[16]

The value of the magnetic field, perpendicular in M to the plane MOO′, is then equal in these circumstances to:

$$\mathrm{H}=\frac{\mathrm{I}}{d}\left(\sin\theta_{0}^{\prime}-\sin\theta_{0}\right)=\frac{\mathrm{I}}{d}\left(\frac{x_{0}^{\prime}}{\sqrt{x_{0}^{\prime}+d_{2}}}-\frac{x_{0}}{\sqrt{x_{0}+d_{2}}}\right) \tag{A3.2.10}$$

In the case of a dipole, the field is reduced to:

$$\mathrm{H}=\frac{\mathrm{I}}{d}\cos\theta\,\mathrm{d}\theta=\mathrm{I}d\frac{\mathrm{d}x}{\left(x^{2}+d^{2}\right)^{3/2}} \tag{A3.2.11}$$

If instead of being rectilinear the cable OO′ is of any form, the electric field and the current density in the medium remain unchanged. In contrast, all magnetic fields corresponding to the various forms of the cable have the same rotational and thus correspond to the same current distribution in the medium.

[16]In AC, this condition can be made possible, for example, when the current flowing out of the first electrode is in phase opposition with the current entering the second electrode (→ polarized electric dipole ±).

The magnetic field of the rectilinear cable Ox interrupted in O has, as a rotational, a distribution of spherical current around O, with an own magnetic field equal to zero. We have just seen that the antenna consisting of the Ox cable with spherical emission through O is a solution of the equations of electricity. Is this the only one?

There may exist a current distribution with nonspherical symmetry around O that would not have a zero magnetic field, such that this magnetic field had then precisely as a rotational the difference between the real distribution and the spherical distribution of the current. We may think that such distributions exist, and that they even are the only real ones for the following reason.

A thread of current in the environment in reality consists of a discrete sequence of charged particles (electrons, ions, etc.) moving at a speed $\vec{v}$. Each particle with a charge q is subjected because of the electric field $\vec{E}$ to an electrostatic force $q\vec{E}$. But as this one moves in a magnetic field $\vec{H}$, it is also subjected to a Lorentz force as $(\vec{v} \wedge \vec{H})\,q$. The total electromotive force is then $\vec{E} + (\vec{v} \wedge \vec{H})$. Ohm's law in these circumstances is no longer written in the standard form $\vec{J} = \sigma\vec{E}$ but in the form:

$$\vec{J} = \sigma\left(\vec{E} + \vec{v} \wedge \vec{H}\right) \tag{A3.2.12}$$

where σ is the electrical conductivity of the medium.

The total force and the current density do have a cylindrical symmetry around the axis Ox. The transmitting device then has a priori no reason to have a spherical symmetry.

In fact, the Lorentz force is at each point perpendicular to the current thread and the magnetic field. It is contained in the meridian plane of the thread and gives it a curvature which deviates it from Ox. It performs no work and therefore does not alter the potential distribution. The potential, which E continues to be, except for the sign, the gradient ($\vec{E} = -\vec{\nabla}V$), results from the condition of conservation of electricity ($\vec{\nabla} \cdot \vec{J} = 0$) which is then written:

$$\vec{\nabla} \cdot \vec{E} + \vec{\nabla} \cdot \left(\vec{v} \wedge \vec{H}\right) = 0 \tag{A3.2.13}$$

It goes without saying that $\vec{v}$, in a homogeneous and isotropic medium, is proportional to $\vec{J}$: $\vec{v} = \xi\vec{J}$ where ξ is a factor dependent on the density of the working charges in the conductor (number of free electrons or ions per cm^3). The spherical symmetry current then corresponds to $\xi = 0$.

However, the experiments on ion mobility showed that $\vec{v}$ was still very small in electrolytes.[17] As a result, as long as $\vec{H}$ is low (a few Gauss), then the product $\vec{v} \wedge \vec{H}$ is very small in front of $\vec{E}$. The weakness of $\vec{v} \wedge \vec{H}$ compared to $\vec{E}$[18] thus enables us

[17]Depending on the nature of the ions, the speed varies from 4 to 33 $\mu m.s^{-1}$ in a total electric field of 1 V/cm (Lodge, 1892).

[18]Indeed if $E = 100\ V.m^{-1}$, v on average is about 10 $\mu m.s^{-1}$ and for $H = 1G$, (v.H) is at a maximum of 10^{-3}. Then E is 10^{11} times larger than (v.H). The total electric force only differs from E by 10^{-22}.

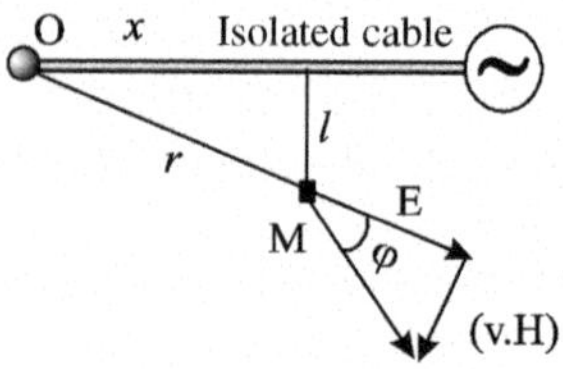

Fig. A3.14 Effect of the magnetic field H at a point M remote from the source

to treat the spherical symmetry field in a first approximation and then the effect of the magnetic field $\vec{H}$ as an additional disturbance (Fig. A3.14).

At any point M, $(\vec{v}.\ \vec{H})$ must be perpendicular to $\vec{v}$, then to $\vec{J}$, i.e., to the total electrical force. The latter therefore makes with $\vec{E}$ an angle a φ such that:

$$\sin \varphi = \frac{(\vec{v}.\ \vec{H})}{\vec{E}} \tag{A3.2.14}$$

But as a first approximation we have:

$$\vec{H} = \frac{I}{l}\left(1 + \frac{x}{r}\right) \qquad \text{with} \qquad \vec{v} = 10^{-11}\,\vec{E} \tag{A3.2.15}$$

so finally:

$$\sin \varphi = 10^{-11}\frac{I}{l}\left(1 + \frac{x}{r}\right) \tag{A3.2.16}$$

As the inside of the parentheses is between 0 and 2, we see that for low values of I (1 A for example), sin φ remains extremely small (of the order of 10^{-12}).

In conclusion we can neglect the difference between $\vec{E}$ and the total field. It could be said that the actual current distribution from a point electrode remains almost spherical in electrolytes (seawater) and grounds with a conductivity of electrolytic type (marine sediments).[19]

We can note that near the seafloor, above a conductor ground of thickness h overcoming a resistant horizon, the current threads from a source O (cf. Fig. A3.15) do not spread in a spherical manner as in an isotropic medium (cf. Eq. A3.2.1), but then follow a cylindrical symmetry such that the value of the current density is of the form:

$$J = \frac{I}{4\pi r h} \tag{A3.2.17}$$

In this particular condition, density will be r/h times larger than that present in an homogeneous undefined medium.

[19]Calculation of the electromagnetic field caused by an endless cable submerged in the sea is disclosed in the article by Von Aulock (Von Aulock, 1953).

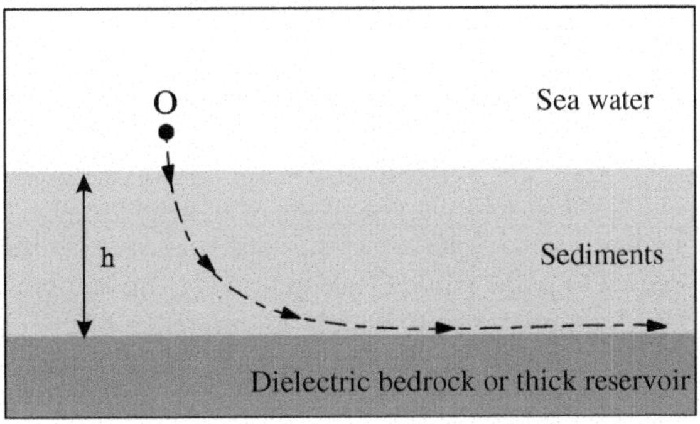

Fig. A3.15 Near an insulating horizon, the current density then assumes a cylindrical distribution

Appendix A4.1

Impedance and Resistance

This digression is only intended to show, as an exercise, at which frequency, in the marine environment, we must consider losses by capacity and if so, what form should then take the expression of the electrical conductivity.

When a pair of metallic electrodes is immersed in an electrically conductive environment and when we inject a DC, it is known that the resistance offered by the medium, that is its the reaction, only depends on the geometric shape of the electrode system and on the resistivity ρ of the immersion medium.

Nevertheless, when using an alternating current, the dielectric constant ε intervenes.[20] Resistance and capacitance then combine into one identity: the impedance. This corresponds to the apparent resistance to the passage of the alternating current but also takes into account the reactance, i.e., further opposition to the movements of the electric charges caused by changes in the electromagnetic fields.

Question

At what frequency f does the current leakage by capacity start to become significant compared to the current flowing directly by conduction between the two electrodes?

Answer

If it is considered that this system is equivalent to an electrical circuit (analog model) comprising in parallel a resistance R and a capacity C, then we know that for a sinusoidal current, with angular frequency ω ($\omega = 2\pi f$) passing through it, the impedance is equivalent to:

$$Z = \frac{1 + R}{1 + \omega^2 R^2 C^2} \qquad \text{(A4.1.1)}$$

So this impedance differs from the resistance in so far as the denominator differs from 1. If we are able to measure R within 5 %, the influence of C will be effective only when, in the denominator, $\omega^2 R^2 C^2$ is greater than 0.05, that is to say when:

[20]In a nonmagnetic medium, the magnetic permeability is not considered.

$$\omega \rangle \frac{\sqrt{0.05}}{\mathrm{RC}} \quad \rightarrow \quad f \rangle \frac{0.224}{2\pi\mathrm{RC}} \tag{A4.1.2}$$

As far as f does not reach this value, resistance and impedance are then equivalent.

For example, to simplify the discussion, if it is assumed that the electrode system is reduced to a single electrode, with a spherical shape, having a radius r, the other one being discharged to a very large distance (infinite), we know that in this case R is equal to $\rho/4\pi r$ while C is equal to εr. The amount 2πRC is then equal to $\rho\varepsilon/2$. The limit frequency below which the influence of the capacitance is negligible is then:

$$f_l = \frac{2\sqrt{0.05}}{\rho\varepsilon} \tag{A4.1.3}$$

Using the average electrical characteristics of seawater ($\rho_w = 0{,}2\ \Omega$.m and $\varepsilon_w = 80 \times 8{,}8.10^{-12}$ F/m) we then obtain, under these conditions, a limit frequency f_l of the order of 1 GHz which happens to be well beyond the frequencies used.

So it can be shown that regardless of the shape of the electrodes and their arrangement in the medium, there is always between C and R a relation of the form:

$$2\pi\mathrm{RC} = \frac{\rho\varepsilon}{2} \tag{A4.1.4}$$

For the Eq. (A4.1.4) to be a generic law, we must remember that the leakage by capacity is not anything other than the one that governs the *Maxwell displacement current*. If e is the value of the electric field that exists in the environment, the displacement current density has for general expression in the *time domain*:

$$\frac{\varepsilon}{4\pi}\frac{\partial \mathrm{e}}{\partial t} \tag{A4.1.5}$$

If the field is sinusoidal $\left(\mathrm{E} = \mathrm{E}_0 \mathrm{e}^{\mathrm{i}\omega t}\right)$, this is equivalent in the *frequency domain* to:

$$\frac{\partial \mathrm{e}}{\partial t} = \mathrm{i}\omega_0 \mathrm{E}_0 \mathrm{e}^{\mathrm{i}\omega t} = \mathrm{i}\omega\mathrm{E} \tag{A4.1.6}$$

Hence, if the values of the density of the displacement current $\mathrm{J_d}$ and of the conduction current $\mathrm{J_c}$ are respectively equal to:

$$\mathrm{J_d} = \frac{\mathrm{i}\omega\varepsilon}{4\pi}\mathrm{E} \quad \text{and} \quad \mathrm{J_c} = \frac{1}{\rho}\mathrm{E} \tag{A4.1.7}$$

and are phase shifted by 90°, the ratio of their absolute value is equivalent to:

$$\frac{\omega\varepsilon\rho}{4\pi} = \frac{\varepsilon\rho}{2} f \tag{A4.1.8}$$

a ratio that remains constant regardless of E for a frequency f.

Thus at any point in space, once f is determined, there is a unique and defined ratio between the displacement current and the conduction current. The ratio of the total current to the conduction current, which also represents the ratio of the resistance to the impedance, is given by:

$$1 + \left(\frac{\rho\varepsilon}{2}f\right)^2 \tag{A4.1.9}$$

This result can also be deduced from *Maxwell's equations*, which are the purely mathematical translation of the reasoning that has just been conducted.

In summary, when sinusoidal currents are used, the electromagnetic properties of the medium (ρ, σ, ε) are not involved independently in the phenomena but appear intimately related as:

$$\frac{1}{\rho} + \frac{i\omega\,\varepsilon}{4\pi} \quad \text{or either} \quad \sigma + i\frac{\varepsilon}{2}f \tag{A4.1.10}$$

These complex expressions replace the resistivity ρ or conductivity σ when you want to include capacitive phenomena that appear from a certain frequency f_l.

Appendices A5

References to authors in the following texts can be found in the references of Chapter 5

Preamble

This series of seven appendices refers to various sections of Chap. 5, whose writing without these annexes would have been overloaded and the subject less intelligible. More generally, the reader interested in the most basic aspects can refer, among others, to specialized books on electrostatics (Durand 1966), electricity/magnetism (Panofsky and Phillips 1955), electromagnetism (Wilson 1933; Stratton 1941), electrodynamics (Sommerfeld 1952; Plonsey and Collin 1961; Jackson 1965), mathematical physics (Stokes 1880; Lorentz 1927; Jeffreys and Jeffreys 1956; Morse and Feshbach 1953; Butkov 1968) or mathematics for physicists (Chisholm and Morris 1965; Arfken 1968; Angot 1982).[21] Regarding geophysics and EM applied geophysics we can more specifically consult the works of Professors Bannister (Bannister et al. 1965; Weaver 1994), Kraichman, the first to proposed a study based on asymptotic solutions (Kraichman 1970), and Nabighian (Nabighian et al. 1987).

All the appendices deal with the forward problem whose solutions can be used either alone for previous studies or evaluation studies for example, or more generally in association with the resolution of the inverse problem for the final geological interpretation.

[21] Most of these books, now considered classics, have been republished or have been the subject of many revisions. The reader may also refer to more recent works.

The first six appendices are devoted to analytical methods and techniques that have been chosen for their educational value, some relying on layered models (1D), others including heterogeneities or conductivity defects (3D isometrics) in a homogeneous medium (cf. Fig. A5.16). These models specifically adapted to the aquatic environment mostly have, to our knowledge, never been published. They give an overview of all the analytical techniques usable in this particular investigation area, which are defined as part of *quasistatic approximation* (equivalent to DC prospecting where the conductive effect is prepoderant). The last Appendix provides an outline of one of the many numerical methods applied to the treatment of the interpretation of more complex anomalous zones. Readers interested in these operating techniques can supplement their information with the articles and books whose references are indexed in the bibliography.

In detail, the appendices that follow are dedicated in the order of appearance to the interpretation of:

- Submarine electric soundings by the *theory of electrical images* (tabular model 1D)
- Submarine electric soundings for horizontal devices by the *integrals theory* (tabular model 1D)
- Submarine electric soundings for vertical devices by the *integrals theory* (tabular model 1D)
- Isometric analytical sphere-type anomalies for submarine vector electrical devices (transverse fields) by solving the *Laplace equation* (3D modeling)
- Submarine magnetotelluric surveys by solving *Maxwell's equations* (tabular model 1D)

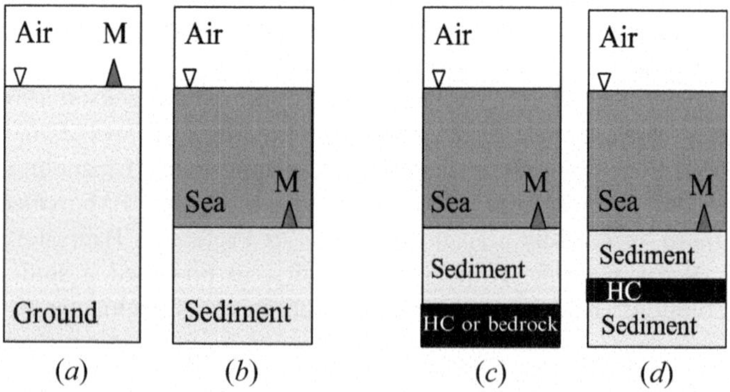

Fig. A5.16 With few exceptions (drilling measurements), almost all models have been established so far for surface exploration by considering air/ground models (*a*). The introduction of a liquid element conductor of electricity (sea) changes the conceptual approach (*b*) if the measures (M) are made in the medium, i.e., in this case in SBL at the interface of the two conductive layers (seawater/marine sediments). On the *right*, two geological canonic models are shown for analytical (*c*) and numerical (*d*) simulations corresponding to a thick (*c*) and a thin (*d*) dielectric substratum corresponding to a reservoir of hydrocarbon (HC) or a resistive bedrock

- Cylindrical anomalies for submarine vector magnetotelluric devices (transverse fields) by the *coefficients of reflection* method (2D modeling)
- Anomalies of any shape by the numerical method of *integral equations* (3D modeling)

The reader will also find after these modelings two computer programs corresponding to the calculations in the appendices at the end of the book (cf. program P5.1 and pg. P5.2).

Appendix A5.1

Interpretation of submarine electric soundings by the *theory of electrical images* (tabular model 1D)

Here we consider a model composed of horizontal layers of different thicknesses $h_{1,2}$ and resistivities $\rho_{1,2,}$ representing *per descensum* the air, the seawater, the marine sediments and the resistive thick layer (Fig. A5.17).

In electrical prospecting, it is customary to call A and B the points of current injection and M and N the measurement points. In what follows we assume accordingly the arrangement AN = 2AM. These points are materialized in reality by electrodes attached to a flute dragged horizontally on the seabed by the vessel or any other means of navigation.

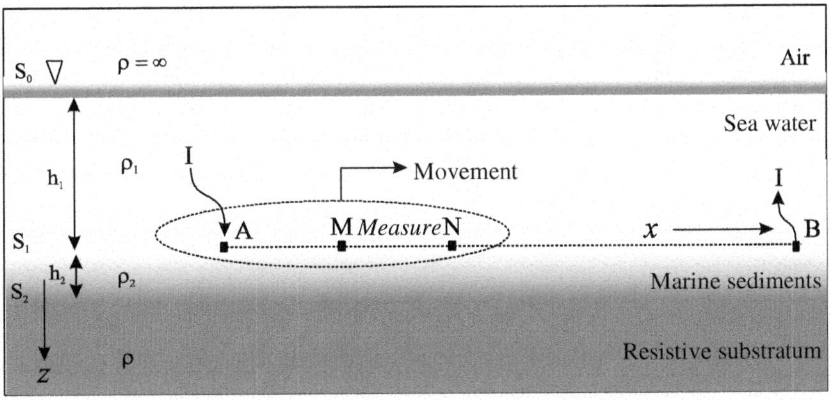

Fig. A5.17 Submarine geoelectric tabular model representing the movement of a submerged NBMA quadrupole called HES for horizontal electrical sounding (electrode B at "infinity"). The depth of investigation can vary in proportions depending on the electrodes spacing. In the case where the latter is fixed, the depth of investigation may be to some extent considered a constant. This gives a resistivity profile along x function of depth along z, corresponding to a 1D modeling

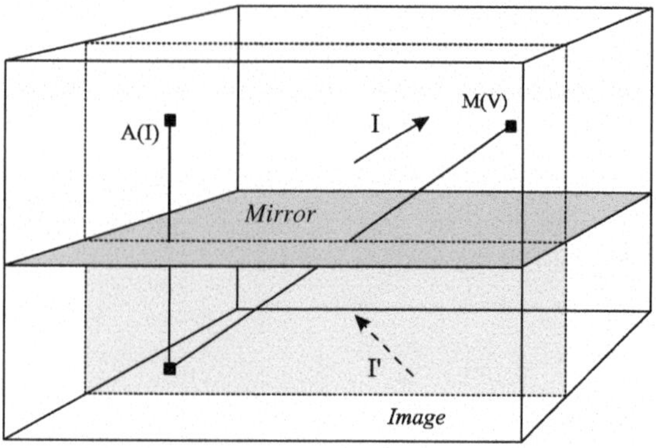

Fig. A5.18 Method of electrical images where the fictional mirror (*shaded*) corresponds to the plane of separation between two adjacent geological layers (analogy with geometric optics)

To establish an abacus considering this particular environment (a measuring device immersed in seawater),[22] it is necessary to find, based on the resistivities ρ_1 and ρ_2, on the depth ratio h_1/h_2 and spacing $AM = x$, the law of variation of the apparent resistivity ρ_a at depth (along z) measured at the bottom of the sea.

The value of the potential V_M at any point M of the space (see Fig. A5.18) created by injecting a DC I, output by an electrode A, can be calculated by replacing the effect of this current with that of an infinite number of fictive charges, concentrated in A and in points obtained by taking successive images of A compared to the fictive mirrors formed by the planes of separation of the different media (air, seawater, marine sediments and the resistive thick layer). The value of these successive potentials decreases according to the iterations. The summation is stopped when values below the measurement errors are reached.

In this case, these images are distributed according to a simple law when the ration $h_1/h_2 = p$ is an integer. In other words, the images are located on the perpendicular from A to the contact plane seawater/marine sediment, placed on successive points (depth value $2nh_1/p$ measured from A) where n is any of all successive integers. A being taken as the origin of coordinates, x being the distance AM, and $2x$ the distance AN, the images of A with respect to three fictional mirrors are placed at points of quotation:

$$z = 2n\frac{h_1}{p} = 2nh_2 \tag{A5.1.1}$$

where n is a positive or negative integer as the considered images are located above or below the plane going through A. Under these conditions the potential of M is given by the expression:

[22]The abacuses proposed for the interpretation of the so-called surface electrical surveys were calculated for quadrupole topologies arranged on the surface, where the upper medium is then considered infinitely resistant (air).

$$\frac{1}{\rho_1}V_M = \frac{S_0}{x} + \sum_{n=1}^{n=\infty} \frac{S_n}{\sqrt{x^2 + n^2(2h_2)^2}} \tag{A5.1.2}$$

where S_0 is the emissivity (or $I/4\pi$) of the electrode A and S_n the sum of the emissivities of the pair of images of quotations $+2nh_1$ and $-2nh_1$.

If now we pose:

$$u = 2\frac{h_2}{x} \tag{A5.1.3}$$

it comes at the point M:

$$\frac{1}{\rho_1}V_M = \frac{1}{x}\left(S_0 + \sum_{n=1}^{n=\infty} \frac{S_n}{\sqrt{1 + n^2h_2^2}}\right) \tag{A5.1.4}$$

as well as at the point N:

$$\frac{1}{\rho_1}V_N = \frac{1}{x}\left(\frac{S_0}{2} + \sum_{n=1}^{n=\infty} \frac{S_n}{\sqrt{4 + n^2h_2^2}}\right) \tag{A5.1.5}$$

from which we have for the potential difference ΔV_{MN}:

$$\frac{1}{\rho_1}\Delta V_{MN} = \frac{1}{x}\left(\frac{S_0}{2} + \sum_{n=1}^{n=\infty} S_n\left(\frac{1}{\sqrt{1 + n^2h_2^2}} - \frac{1}{\sqrt{4 + n^2h_2^2}}\right)\right) \tag{A5.1.6}$$

But to define the apparent resistivity, it is necessary to consider a fictitious homogeneous ground[23] giving the same ΔV_{MN} for the same values of x. In such a ground where $S_n = 0$, we can write to define ρ_a that:

$$\frac{1}{\rho_a}\Delta V_{MN} = \frac{1}{x}\frac{S_0}{2} \tag{A5.1.7}$$

when setting:

[23]Here we form (the method of Hummel) a fictional ground developed from the first layers, so that this one, electrically equivalent, forms with the underlying grounds a new interpretable set (new curve). This principle of sequential development of "auxiliary curves" called the *principle of reduction* can be repeated interactively until the desired number of layers (Hummel 1929) is obtained.

$$K_n = \frac{1}{\sqrt{1 + n^2 h_2^2}} - \frac{1}{\sqrt{4 + n^2 h_2^2}} \quad \text{(A5.1.8)}$$

we finally obtain the expression of the apparent resistivity:

$$\rho_a = \rho_1 \left(1 + 2 \sum_{n=1}^{n=\infty} \frac{S_n}{S_0} K_n \right) \quad \text{(A5.1.9)}$$

From this simple formulation, we can then construct abacuses. These are developed by successive approximations by summing a number of terms of the series $\sum(S_n/S_0)$ K_n after determining S_0, S_n and K_n.

The values of the K_n terms were calculated by the physicist Hummel (Hummel 1929).

The emissivity values S_0, S_n are then determined from the recurrence relations between the emissivities of the images n, n−1, n−2, etc. For this we express the values of the potential $V_{1,2}$ at any point of space (defined by its coordinates x and z) by considering the four media of resistivity ρ_0 for the air, ρ_1 for seawater, ρ_2 for the marine sediments and ρ_3 for the resistive thick layer. This potential is thus expressed according to the image emissivities.

Taking into account the conditions that must be met in the different environments, we obtain a number of equations from which can be derived a recurrence formula.

If we agree then to indicate by:

- a_n' and b_n' the emissivities of the images contributing to give the potential V_1
- a_0' and a_0'' the emissivities of the images located respectively below and above the contact between 1 and 2
- a_n'' and b_n'' the emissivities of the images used to calculate the potential V_2

thus we obtain:

$$\frac{1}{\rho_1} V_1 = \frac{a_0'}{\sqrt{x^2 + z^2}} + \sum_{n=1}^{n=\infty} \frac{a_n'}{\sqrt{x^2 + \left(z + \frac{2n}{p}\right)^2}} + \sum_{n=1}^{n=\infty} \frac{b_n'}{\sqrt{x^2 + \left(z + \frac{2n}{p}\right)^2}} \quad \text{(A5.1.10)}$$

and:

$$\frac{1}{\rho_2} V_2 = \frac{a_0''}{\sqrt{x^2 + z^2}} + \sum_{n=1}^{n=\infty} \frac{a_n''}{\sqrt{x^2 + \left(z + \frac{2n}{p}\right)^2}} + \sum_{n=1}^{n=\infty} \frac{b_n''}{\sqrt{x^2 + \left(z + \frac{2n}{p}\right)^2}} + \cdots \quad \text{(A5.1.11)}$$

Using the four classical boundary conditions adopted in the definition of the model with four layers, namely:

$$\begin{cases} \dfrac{\partial V_1}{\partial z} = 0 & \text{for} \quad z = h_1 + ph_2 = 1 \\ \dfrac{\partial V_2}{\partial z} = 0 & \text{for} \quad z = -h_2 = \dfrac{1}{p} \\ \dfrac{1}{\rho_1}\dfrac{\partial V_1}{\partial z} = \dfrac{1}{\rho_2}\dfrac{\partial V_2}{\partial z} & \text{for} \quad z = 0 \\ V_1 = V_2 & \text{for} \quad z = 0 \end{cases} \tag{A5.1.12}$$

thus we obtain in those situations:

$$\frac{a'_0}{\sqrt[3]{x^2+1}} + \sum_{n=1}^{n=\infty} \frac{\left(1+\frac{2n}{p}\right)a'_n}{\sqrt[3]{x^2+\left(1+\frac{2n}{p}\right)^2}} + \sum_{n=1}^{n=\infty} \frac{\left(1-\frac{2n}{p}\right)b'_n}{\sqrt[3]{x^2+\left(1-\frac{2n}{p}\right)^2}} = 0 \tag{A5.1.13}$$

and:

$$\frac{\frac{1}{p}a''_0}{\sqrt[3]{x^2+\frac{1}{p^2}}} + \sum_{n=1}^{n=\infty} \frac{\left(\frac{2n-1}{p}\right)a''_n}{\sqrt[3]{x^2+\left(\frac{2n-1}{p}\right)^2}} + \sum_{n=1}^{n=\infty} \frac{\left(\frac{2n-1}{p}\right)b''_n}{\sqrt[3]{x^2+\left(\frac{2n-1}{p}\right)^2}} = 0 \tag{A5.1.14}$$

and:

$$\rho_1\left[\frac{a'_0}{x} + \sum_{n=1}^{n=\infty} \frac{a'_n + b'_n}{\sqrt{x^2+\left(\frac{2n}{p}\right)^2}}\right] = \rho_2\left[\frac{a''_0}{x} + \sum_{n=1}^{n=\infty} \frac{a''_n + b''_n}{\sqrt{x^2+\left(\frac{2n}{p}\right)^2}}\right] \tag{A5.1.15}$$

and:

$$\frac{\left(a'_n - b'_n\right)\frac{2n}{p}}{\sqrt[3]{x^2+\left(\frac{2n}{p}\right)^2}} = \frac{\left(a''_n - b''_n\right)\frac{2n}{p}}{\sqrt[3]{x^2+\left(\frac{2n}{p}\right)^2}} \tag{A5.1.16}$$

On the other hand, the fact that the potential distribution near the injection electrode A is spherical (cf. Chap. 3 Appendix A3.2) now implies that:

$$\rho_1 a'_0 = \rho_2 a''_0 \tag{A5.1.17}$$

thereby allowing us to alleviate the above equations, by setting:

$$\rho_1 a_0' = \rho_2 a_0'' = 1 \qquad \text{and} \qquad A_n' = \rho_1 a_0' a_n' \ldots \tag{A5.1.18}$$

The basic equations then take the form:

$$\frac{1}{\sqrt[3]{x^2+1}} + \sum_{n=1}^{n=\infty} \frac{\left(1+\frac{2n}{p}\right) A_n'}{\sqrt[3]{x^2+\left(1+\frac{2n}{p}\right)^2}} + \sum_{n=1}^{n=\infty} \frac{\left(1-\frac{2n}{p}\right) B_n'}{\sqrt[3]{x^2+\left(1-\frac{2n}{p}\right)^2}} = 0 \tag{A5.1.19}$$

and:

$$\frac{-\frac{1}{p}}{\sqrt[3]{x^2-\frac{1}{p}}} + \sum_{n=1}^{n=\infty} \frac{\left(\frac{2n-1}{p}\right) A_n''}{\sqrt[3]{x^2+\left(\frac{1-2n}{p}\right)^2}} + \sum_{n=1}^{n=\infty} \frac{\left(\frac{2n+1}{p}\right) B_n''}{\sqrt[3]{x^2+\left(\frac{1+2n}{p}\right)^2}} = 0 \tag{A5.1.20}$$

or:

$$\sum_{n=1}^{n=\infty} \frac{A_n' - B_n'}{\sqrt{x^2+\left(\frac{2n}{p}\right)^2}} = \sum_{n=1}^{n=\infty} \frac{A_n'' - B_n''}{\sqrt{x^2+\left(\frac{2n}{p}\right)^2}} \tag{A5.1.21}$$

and:

$$\frac{1}{\rho_1} \sum_{n=1}^{n=\infty} \frac{\left(A_n' - B_n'\right)\frac{2n}{p}}{\sqrt[3]{x^2+\left(1+\frac{2n}{p}\right)^2}} = \frac{1}{\rho_2} \sum_{n=1}^{n=\infty} \frac{\left(A_n'' - B_n''\right)\frac{2n}{p}}{\sqrt[3]{x^2+\left(\frac{2n}{p}\right)^2}} \tag{A5.1.22}$$

From these equations, it is now possible to identify, term by term, the values of the same index n, and derive the recurrence formula by replacing n with p (cf. Eq. A5.1.19) such that:

$$B_n' = +1 \tag{A5.1.23}$$

Then by comparing the terms:

$$\frac{1+\frac{2n}{p}}{\sqrt[3]{x^2+\left(1+\frac{2n}{p}\right)^2}} \tag{A5.1.24}$$

in Eq. (A5.1.19), it follows immediately that:

$$A_n' = B_{n+p}' \tag{A5.1.25}$$

Similarly from the Eq. (A5.1.20) can be extracted:

$$A''_{n+1} = \frac{1}{2p - 1} \quad \text{(A5.1.26)}$$

and:

$$A''_{n+1} = B''_{n} \quad \text{(A5.1.27)}$$

as well as for Eqs. (A5.1.21) and (A5.1.22):

$$A'_{n} + B'_{n} = A''_{n} + B''_{n} \quad \text{(A5.1.28)}$$

and:

$$\frac{1}{\rho_1}\left(A'_{n} + B'_{n}\right) = \frac{1}{\rho_2}\left(A''_{n} + B''_{n}\right) \quad \text{(A5.1.29)}$$

From the last two relations, it expressly emerges that:

$$2A''_{n} = A'_{n}\left(1 + \frac{\rho_2}{\rho_1}\right) + B'_{n}\left(1 - \frac{\rho_2}{\rho_1}\right) \quad \text{(A5.1.30)}$$

and:

$$2B''_{n} = A'_{n}\left(1 - \frac{\rho_2}{\rho_1}\right) + B'_{n}\left(1 + \frac{\rho_2}{\rho_1}\right) \quad \text{(A5.1.31)}$$

which from Eqs. (A5.1.25) and (A5.1.27) gives:

$$2A''_{n} = A'_{n}\left(1 + \frac{\rho_2}{\rho_1}\right) + A'_{n-p}\left(1 - \frac{\rho_2}{\rho_1}\right) \quad \text{(A5.1.32)}$$

and:

$$2B''_{n} = A'_{n}\left(1 - \frac{\rho_2}{\rho_1}\right) + A'_{n-p}\left(1 + \frac{\rho_2}{\rho_1}\right) \quad \text{(A5.1.33)}$$

Yet:

$$2B''_{n-1} = A'_{n}\left(1 + \frac{\rho_2}{\rho_1}\right) + A'_{n-p}\left(1 - \frac{\rho_2}{\rho_1}\right) \quad \text{(A5.1.34)}$$

and from Eq. (A5.1.33) we have:

$$A'_n\left(1+\frac{\rho_2}{\rho_1}\right)+A'_{n-p}\left(1-\frac{\rho_2}{\rho_1}\right)=A'_{n-1}\left(1-\frac{\rho_2}{\rho_1}\right)+A'_{n-p-1}\left(1-\frac{\rho_2}{\rho_1}\right) \quad (A5.1.35)$$

Using conventional notation:

$$\frac{\rho_1-\rho_2}{\rho_1+\rho_2}=k \quad (A5.1.36)$$

thus we can write:

$$A'_n=kA'_{n-p}=-kA'_{n-1}+A'_{n-p-1} \quad (A5.1.37)$$

We then obtain the first recurrence formula:

$$A'_n=-kA'_{n-1}+kA'_{n-p}+A'_{n-p-1} \quad (A5.1.38)$$

The same identity would be found for the terms in B' by setting:

$$S'_n=A'_n+B'_n \quad (A5.1.39)$$

i.e.:

$$S'_n=-kS'_{n-1}+kS'_{n-p}+S'_{n-p-1} \quad (A5.1.40)$$

and also from Eq. (A5.1.28):

$$S''_n=-kS''_{n-1}+kS''_{n-p}+S''_{n-p-1} \quad (A5.1.41)$$

This recurrence formula can be written more generally as:

$$S_n=-kS_{n-1}+kS_{n-p}+S_{n-p-1} \quad (A5.1.42)$$

However, it does not apply to the rank lower to p. For the first terms of the series, then it is necessary to directly calculate them step by step. Thus we find:

$$\begin{cases} S_0=1+k \\ S_1=2(1+k)\,(1-k) \\ S_2=2(1+k)\,\left(1+k^2\right) \\ \text{etc.} \end{cases} \quad (A5.1.43)$$

Ultimately, if we refer to the general formula for the apparent resistivity (cf. Eq. A5.1.9) and using either the Hummel values K_n or the recurrence relations

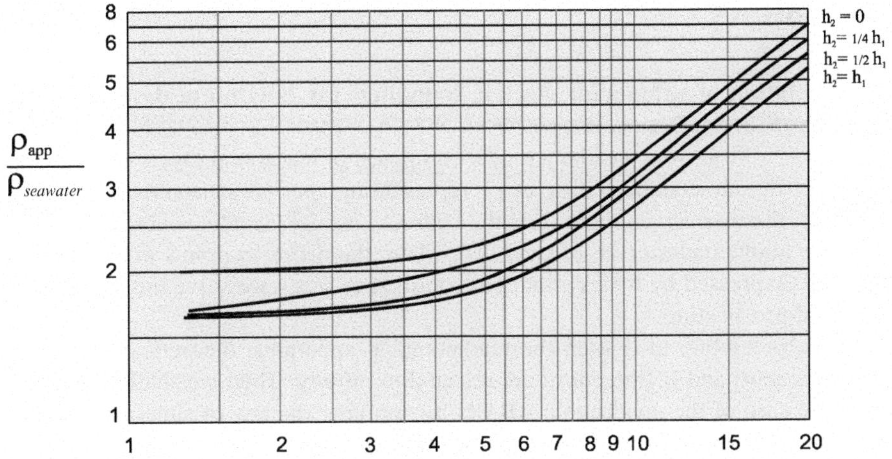

Fig. A5.19 Abacus to interpret submarine vertical electrical soundings (sediment/resistive thick layer). The use of this type of abacus has long been the cornerstone of data interpretation of electrical soundings in general. The logarithmic scale here is perfectly suited to the conductive nature of marine sediments

that allow to calculate the emissivities, then it is possible to determine the values of h_2 (marine sediments) corresponding to some given values of the ratios h_1/h_2 and ρ_1/ρ_2.

The application of the theory of electrical images is relatively well suited to unstructured acquisition streamer-type devices (and also to Wenner or Schlumberger-type arrangements, etc.) where lateral variations in resistivity can be considered negligible. This type of interpretation only allows us to approach problems in one dimension (1D) as vertical profiles.

With relatively simple programming, the technique allows with few acquisition data a rapid interpretation by abacuses (see Fig. A5.19) or any other automatic approach to successive approximations.

This technique can be effective to affirm, confirm or clarify, with relatively good accuracy, features of geological objects (lithology and sedimentary cover in particular), in a specific structural context (a tabular model on a resistive layer). However, this type of investigation is not suitable for multidimensional interpretation in two dimensions where other methods, coupled with more sophisticated technologies and acquisition systems, are then more efficient.[24]

[24]In surface prospecting, 2D and 3D models were proposed for arrangements of the dipole–dipole type (Medkour 1984).

Appendix A5.2

Interpretation of submarine electric soundings for horizontal devices by the *integrals theory* (1D modeling)

In what follows we consider a model composed of horizontal layers of thickness $h_{1,2}$ of different conductivities $\sigma_{1,2,3}$ representing, *per descensum*, the air, the seawater, the marine sediments and the bedrock or thick resistive substratum.

In the above pattern (see Fig. A5.20), where the different points in the vertical plane are expressed by their cylindrical coordinates (r, z), we solve the equation of electrical prospecting.[25]

In the preamble, to lighten the mathematical apparatus, theoretically poles N (measurement) and B (injection) are rejected to infinity. Then we shall easily pass on to the case of the quadrupole ABMN by applying the law of superimposing of the states of balance (called also superposition theorem).

If the layers 1 (seawater) and 2 (sediments) respectively fill two half-spaces ($z < 0$) and ($z > 0$), the electrode A supplying a current I thus produces at any point M on the line a primary potential of the form:

$$V_0 = \frac{I}{4\pi} \frac{2}{(\sigma_1 + \sigma_2)} \frac{1}{\sqrt{z^2 + r^2}} \tag{A5.2.1}$$

To solve the equation we can first of all:

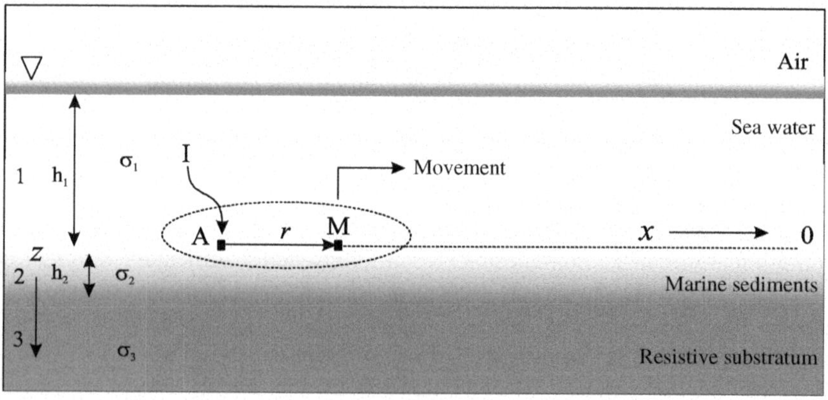

Fig. A5.20 Tabular submarine geoelectric model (three layers) corresponding to a horizontal investigation device that can be dragged at the bottom of the sea

[25]The fundamental equations of electrical prospecting were defined by the French school of geophysics (Schlumberger, Stefanescu, Kostitzine, etc.). The results of these works (prospecting equation, resolutions and solutions) which apply only to surface prospecting are summarized in Raymond Maillet's article (Maillet 1947).

Replace the term $\left(\sqrt{z^2+r^2}\right)^{-1}$ with the following Weber–Lipchitz integrals A and B (Gray and Mathew 1922):

– in seawater:

$$A = \int_0^\infty e^{\lambda z} J_0(\lambda r)\, d\lambda \tag{A5.2.2}$$

– in marine sediments:

$$B = \int_0^\infty e^{-\lambda z} J_0(\lambda r)\, d\lambda \tag{A5.2.3}$$

where J_0 is the Bessel function of the first kind and zero order and λ is an arbitrary constant

Then set in general:

$$V_i = \frac{I}{4\pi} \frac{2}{(\sigma_1+\sigma_2)} v_i \tag{A5.2.4}$$

Finally, knowing that the thicknesses of the water layer and marine sediments are not unlimited, an additional potential, disturbing, is then added to the previous potential for each of these horizons. It is also known as secondary potential v_i which is written using the functions $A_{1,2,3}$ and $B_{1,2}$:

– In seawater (1):

$$v_1' = \int_0^\infty \left[A_1(\lambda)\, e^{-\lambda z} + B_1(\lambda)\, e^{\lambda z}\right] J_0(\lambda r)\, d\lambda \tag{A5.2.5}$$

– In marine sediments (2):

$$v_2' = \int_0^\infty \left[A_2(\lambda)\, e^{-\lambda z} + B_2(\lambda)\, e^{\lambda z}\right] J_0(\lambda r)\, d\lambda \tag{A5.2.6}$$

Then the potential in the substratum (3) is written as follows:

$$v_3 = \int_0^\infty A_3(\lambda)\, e^{-\lambda z} J_0(\lambda r)\, d\lambda \tag{A5.2.7}$$

as the latter has to nullify when z increases indefinitely.

If considering firstly the boundary and frontier conditions and borders (×5) at the interfaces of the different media, such as:

– On the water surface ($z = -h_1$), we have:

$$(1)\quad \left.\frac{\partial V_1}{\partial z}\right|_{z=-h_1} = 0 \tag{A5.2.8}$$

– On the surface of separation between the seawater and sediment ($z = 0$), we have:

$$\begin{cases} (2) & V_1 = V_2 \\ (3) & \sigma_1 \dfrac{\partial V_1}{\partial z} = \sigma_2 \dfrac{\partial V_2}{\partial z} \end{cases} \tag{A5.2.9}$$

– On the surface of separation between marine sediments and substratum ($z = h_2$), we have:

$$\begin{cases} (4) & V_2 = V_3 \\ (5) & \sigma_2 \dfrac{\partial V_2}{\partial z} = \sigma_3 \dfrac{\partial V_3}{\partial z} \end{cases} \tag{A5.2.10}$$

and secondly the equality (A5.2.4), which can replace V_i by v_i, the functions $v_1 = v_0$ and $v_2 = v_0$ now satisfy conditions (2) and (3) because:

$$\begin{cases} v_0 = \dfrac{1}{\sqrt{z^2 + r^2}} & \text{(A5.2.11)} \\ \dfrac{\partial v_0}{\partial z} = -\dfrac{z}{\sqrt[3]{z^2 + r^2}} & \text{(A5.2.12)} \end{cases}$$

hence, for $z = 0$:

$$v_0 = \frac{1}{r} \quad \text{therefore} \quad v_1 = v_2 \tag{A5.2.13}$$

with among others:

$$\frac{\partial V_0}{\partial z} = 0 \quad \text{therefore} \quad \sigma_1 \frac{\partial V_1}{\partial z} = \sigma_2 \frac{\partial V_2}{\partial z} \tag{A5.2.14}$$

Then we just submit to conditions (2) and (3) the secondary potentials v_1' and v_2'.

In the other equations v_0 is replaced by one of integrals A and B. To find the unknown functions A_1, B_1, A_2, B_2 and A_3, we can form the following system:

$$\begin{cases} e^{-\lambda h_1} - A_1 e^{\lambda h_1} + B_1 e^{-\lambda h_1} = 0 \\ A_1 + B_1 = A_2 + B_2 \\ \sigma_1(-A_1 + B_1) = \sigma_2(-A_2 + B_2) \\ e^{-\lambda h_2} + A_2 e^{-\lambda h_2} + B_2 e^{-\lambda h_2} = A_3 e^{-\lambda h_2} \\ \sigma_2\left(e^{-\lambda h_2} + A_2 e^{-\lambda h_2} - B_2 e^{-\lambda h_2}\right) = \sigma_3 A_3 e^{-\lambda h_2} \end{cases} \tag{A5.2.15}$$

Then by setting:

$$e^{-2\lambda h_1} = p \quad \text{and} \quad \frac{\sigma_1 - \sigma_2}{\sigma_1 + \sigma_2} = k_1 \tag{A5.2.16}$$

as well as:

$$e^{-2\lambda h_2} = q \quad \text{and} \quad \frac{\sigma_2 - \sigma_3}{\sigma_2 + \sigma_3} = k_2 \tag{A5.2.17}$$

the previous system (A5.2.15) becomes:

$$\begin{cases} p\,(1 + B_1) - A_1 = 0 \\ A_1 + B_1 = A_2 + B_2 \\ (1 + k_1)\,(B_1 - A_1) = (1 - k_1)\,(B_2 - A_2) \\ q\,(1 + A_2) + B_2 = qA_3 \\ (1 + k_2)\,[q(1 + A_2) - B_2] = (1 - k_2)\,qA_3 \end{cases} \tag{A5.2.18}$$

The resolution of this system with five equations thus allows us to obtain the potential values at any point in the bottom of the water, located at a distance r from the current injection electrode of intensity I such that:

$$V = \frac{I}{4\pi}\frac{2}{(\sigma_1 + \sigma_2)}\left\{\frac{1}{r} + \int_0^{\infty} [A_1(\lambda) + B_1(\lambda)]\, J_0(\lambda r)\, d\lambda\right\} \tag{A5.2.19}$$

where:

$$A_1 + B_1 = A_2 + B_2 = \frac{(1 + k_1)\,p + k_2(1 - k_1)\,q + 2k_2pq}{1 - k_1p + k_1k_2q - k_2pq}.... \qquad \text{(A5.2.20)}$$

In these circumstances, the apparent resistivity measured by a dipole or a Schlumberger-type quadrupole, for example, will have finally for value:

$$\rho_a = \frac{4\pi r^2}{I}\frac{\partial V}{\partial r} \qquad \text{(A5.2.21)}$$

Appendix A5.3

Interpretation of submarine electric soundings for vertical devices by the *integrals theory* (1D modeling)

We still consider here a model composed of horizontal layers of thickness $h_{1,2}$, of different conductivities $\sigma_{1,2,3}$ presenting *per descensum*, the air, the seawater, the marine sediments and the bedrock or resistive thick layer.

In this model (see Fig. A5.21) whose injection point B is placed to infinity, the different potentials V_i in the successive layers can be respectively written with a sum of integrals comprising the functions $A_{1,2}$ and $B_{1,2}$:

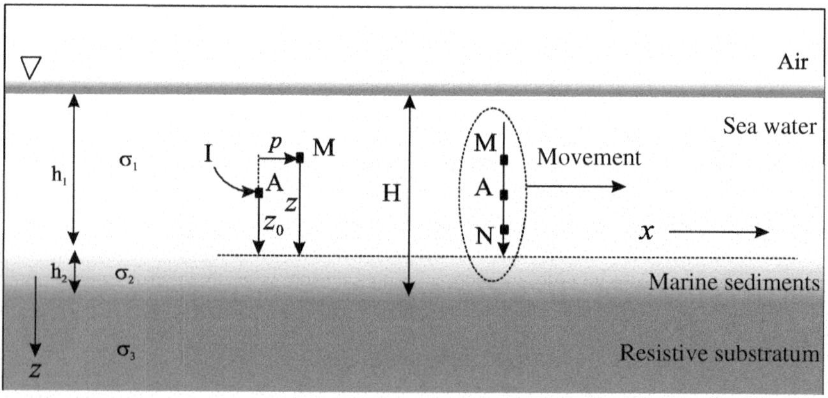

Fig. A5.21 Submarine tabular geoelectric model (three layers) corresponding to an investigation by a vertical device (measurement MN)

– In seawater (1):

$$V_1 = \frac{I}{4\pi\sigma_1}\left\{\int_0^\infty e^{-\lambda|z-z_0|} J_0(\lambda p)\, d\lambda + \int_0^\infty \left(A_1 e^{-\lambda z} + B_1 e^{+\lambda z}\right) J_0(\lambda p)\, d\lambda\right\} \quad (A5.3.1)$$

– In the marine sediments (2):

$$V_2 = \frac{I}{4\pi\sigma_1}\int_0^\infty \left(A_2 e^{+\lambda z} + B_2 e^{-\lambda z}\right) J_0(\lambda p)\, d\lambda \quad (A5.3.2)$$

where J_0 is the Bessel function of the first kind and zero order (see *Special functions* at the end of the volume) and λ is an arbitrary constant.

Considering now the following boundary conditions:

– At the surface of separation seawater/sediment ($\sigma_1 \rightarrow \sigma_2$):

$$\begin{cases} e^{-\lambda z_0} + A_1 + B_1 = A_2 + B_2 \\ \sigma_1\left(e^{-\lambda z_0} - A_1 + B_1\right) = \sigma_2(A_2 - B_2) \end{cases} \quad (A5.3.3)$$

– At the surface of separation sediments/Resistive sediment ($\sigma_2 \rightarrow \infty$):

$$A_2 e^{-\lambda h_2} + B_2 e^{\lambda h_2} = 0 \quad (A5.3.4)$$

– At the surface of separation water/air ($\sigma_1 \rightarrow \infty$):

$$-e^{-\lambda(h_1 - z_0)} - A_1 e^{-\lambda h_1} + B_1 e^{\lambda h_1} = 0 \quad (A5.3.5)$$

and setting now:

$$B_1 + e^{-\lambda z_0} = B_1' \quad (A5.3.6)$$

we have the following system:

$$\begin{cases} A_1 + B_1' - A_2 - B_2 = 0 \\ \sigma_1 A_1 - \sigma_1 B_1' + \sigma_2 A_2 - \sigma_2 B_2 = 0 \\ e^{-\lambda h_2} A_2 - e^{\lambda h_2} B_2 = 0 \\ e^{-\lambda h_1} A_1 - e^{\lambda h_1} B_1' = -e^{-\lambda(h_1 - z_0)} - e^{\lambda(h_1 - z_0)} = -2\cosh\lambda\,(h_1 - z_0) \end{cases} \quad (A5.3.7)$$

which is solved by achieving the ratio $\frac{\sigma_1 - \sigma_2}{\sigma_1 + \sigma_2} = k$:

- Such that we have for the numerator of A_1:

$$^{N}A_1 = \left[e^{-\lambda(h_1 - z_0)} + e^{\lambda(h_1 - z_0)}\right]\left[ke^{-\lambda h_2} + e^{\lambda h_2}\right] \quad (A5.3.8)$$

- Such that we have for the numerator of B_1':

$$^{N}B_1' = \left[e^{-\lambda(h_1 - z_0)} + e^{\lambda(h_1 - z_0)}\right]\left[ke^{\lambda h_2} + e^{-\lambda h_2}\right] \quad (A5.3.9)$$

- Such that for their common denominator we have:

$$D_{com} = e^{\lambda h_1}\left(ke^{\lambda h_2} + e^{-\lambda h_2}\right) - e^{-\lambda h_1}\left(ke^{-\lambda h_2} + e^{\lambda h_2}\right) \quad (A5.3.10)$$

In practice, potential difference measurements or optionally field difference measurements are carried out between two points M and N situated on the vertical on both sides of the injection point A.

By introducing as a variable $Z = z - z_0$, equal in absolute value to the length AM, we can calculate the difference of potential such that:

$$\Delta V_{MN} = V_1(Z) - V_1(-Z) \quad \text{for } \rho = 0 \quad (A5.3.11)$$

Setting $\frac{I}{4\pi\sigma_1} = K$ then we find:

$$\frac{\Delta V_{MN}}{K} = \int_0^{\infty}\left(B_1 e^{\lambda z_0} - A_1 e^{-\lambda z_0}\right)\left(e^{\lambda z} - e^{-\lambda z}\right) d\lambda \quad (A5.3.12)$$

As we have the equality $B_1 e^{\lambda z_0} = B_1' e^{\lambda z_0} - 1$ and according to the previous notations, the first bracket can be written:

$$\frac{^{N}B_1' e^{\lambda z_0} - {}^{N}A_1 e^{-\lambda z_0} - D_{com}}{D_{com}} \quad (A5.3.13)$$

which gives when developing:

$$-\frac{\left(e^{\lambda(h_1+h_2-2z_0)} - e^{-\lambda(h_1+h_2-2z_0)}\right) + k\left(e^{\lambda(h_1-h_2-2z_0)} - e^{-\lambda(h_1-h_2-2z_0)}\right)}{k\left(e^{\lambda(h_1+h_2)} - e^{-\lambda(h_1-h_2)}\right) + \left(e^{\lambda(h_1-h_2)} - e^{-\lambda(h_1-h_2)}\right)} \quad (A5.3.14)$$

Now, to simplify the writing, we can set:

$$z_1 = h_1 + h_2 - 2z_0 \quad \text{and} \quad z_2 = -(h_1 - h_2 - 2z_0) \quad (A5.3.15)$$

as well as:

$$H = h_1 + h_2 \quad \text{and} \quad h = h_1 - h_2 \quad (A5.3.16)$$

we finally obtain:

$$\frac{\Delta V_{MN}}{2K} = \int_0^\infty \frac{k \sinh\lambda\, z_2 - \sinh\lambda\, z_1}{k \sinh\lambda\, H - \sinh\lambda\, h} \sinh\lambda\, Z\, d\lambda \tag{A5.3.17}$$

a formula that is suitable for numerical calculation by approached integration, the integral rapidly converging, since H is always greater than $z_1 + Z$ and than $z_2 + Z$.

Assuming now that $z_1 > z_2$ and reintroducing the exponential functions, we can literally write:

$$\frac{\Delta V_{MN}}{K} = \frac{1}{k}\int_0^\infty e^{-\lambda(H - z_1 - Z)} \times \frac{1 - e^{-2\lambda Z} - k\left[e^{-\lambda(z_1 - z_2)} - e^{-\lambda(z_1 - z_2 + 2Z)} - e^{-\lambda(z_1 + z_2)} - e^{-\lambda(z_1 + z_2 + 2Z)}\right] - e^{-2\lambda z_1} + e^{-2\lambda(z_1 + Z)}}{1 + k^{-1}e^{-\lambda(H-h)} - e^{-2\lambda H} - e^{-2\lambda(H-h)}} d\lambda \tag{A5.3.18}$$

For more convenience, we can choose the parameters a_n so that the powers of exponential function e admit a common divisor α as large as possible such that we have $e^{-\alpha\lambda}$. We obtain then:

$$\frac{1}{k}\int_0^\infty e^{-\lambda(H - z_1 - Z)} \sum_{n=0}^{\infty} a_n e^{-n\lambda\alpha} d\lambda \tag{A5.3.19}$$

whose value is:

$$\frac{1}{k}\frac{a_n}{(H - z_0 - Z) + n\alpha} \tag{A5.3.20}$$

We just have then to multiply a_n by an inverse series to find the researched values (Fig. A5.22).

Appendix A5.4

Interpretation of isometric anomalies (sphere type) for submarine vector electrical devices (transverse fields) by solving the Laplace equation (3D modeling)

This type of model in low frequency approximation can only describe the galvanic effects of a sphere on the currents (distribution of potentials and electric fields around it). However, it is possible to get in some way a phase term comparable to a periodic investigation (≈) in opposition phase for injection, by choosing an arrangement such that the points of the entrance and exit of current are then

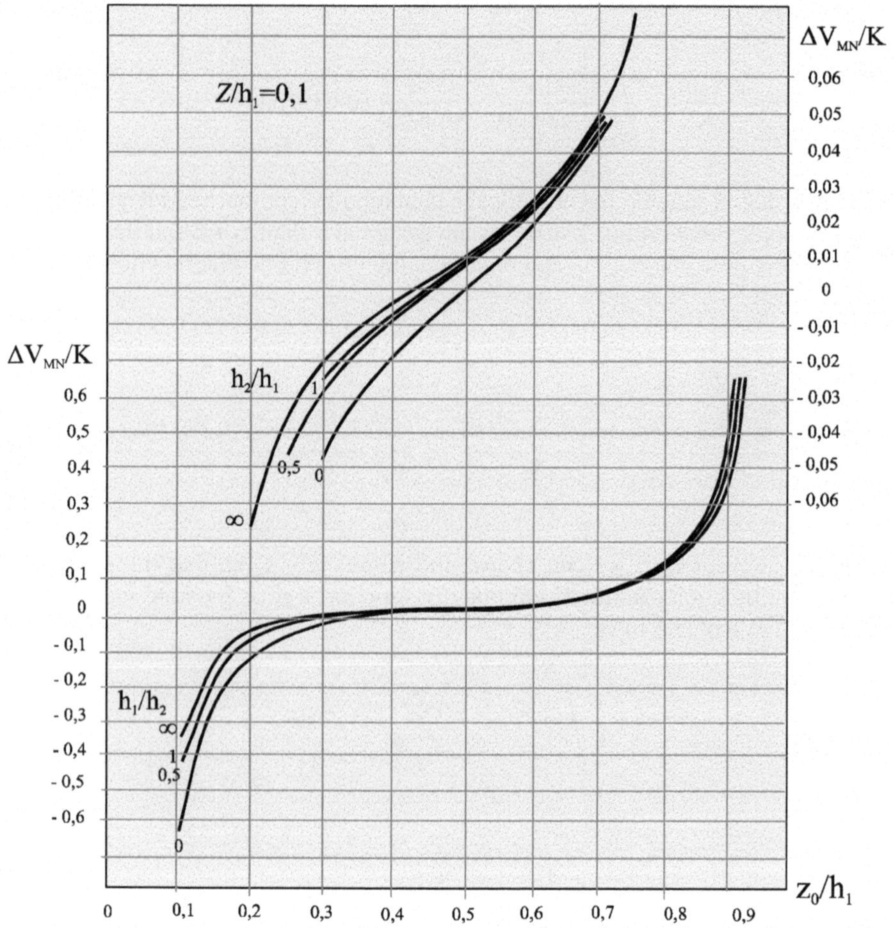

Fig. A5.22 Abacus for the interpretation of submarine soundings (marine sediment/resistive thick layer) for a vertical acquisition device

alternately at opposite potentials (+/−). This modeling can have theoretical interest for the study of the mechanisms of electromagnetic detection (respective contributions of galvanic phenomena and vortex), and applications in:

- Interpretation of data acquisition
- Development of in situ correction devices
- Calibration of acquisition systems
- Development of field sensors
- Confrontation with measures made by means of analog models (rheostatic tanks for example), etc (Fig. A5.23).

The analytical expression of anomalous fields caused by the presence of a body with sufficient resistivity contrast with the surrounding grounds is given by solving the wave equation. For its resolution, this expression can be put in the form of a partial differential equation that takes into account, for its simplification, the

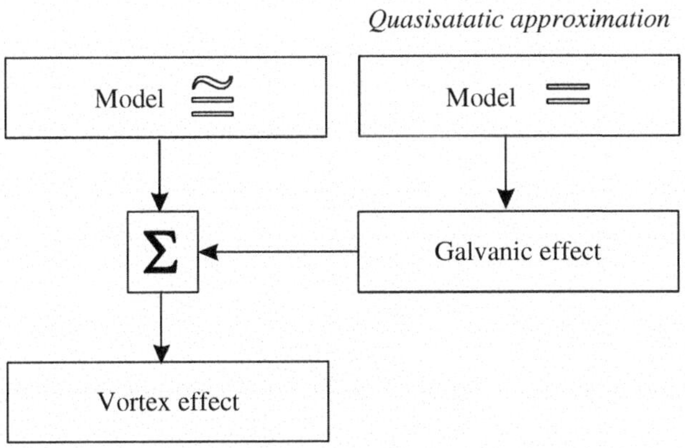

Fig. A5.23 Interest of stationary models for differentiation of vortex and galvanic effects by extraction by calculating the galvanic effect (quasistatic approximation)

geometry of the problem.[26] For example, for a spherical anomaly, the potential V at a distance r from the center of the sphere is obtained by solving the equation in spherical coordinates as described in Fig. A5.24:

$$\left[\frac{1}{r^2}\frac{\partial}{\partial r}\left(r^2\frac{\partial}{\partial r}\right)+\frac{1}{r^2\sin\theta}\frac{\partial}{\partial\theta}\left(\sin\theta\frac{\partial}{\partial\theta}\right)+\frac{1}{r^2\sin\theta}\frac{\partial^2}{\partial\varphi^2}+k^2r^2\right]V(r,\theta,\varphi,\omega)=0 \tag{A5.4.1}$$

A4.1. Laplace Equation

Using low frequencies also authorizes the use of the Laplace equation ($\nabla^2 V = 0$) so that the frequency ω ($k^2r^2 = 0$) is not required. Then we resolve, as for a direct current (quasistatic approximation), the equation in three dimensions:

[26]The theory of electrical images can also be used (Grant and West, 1965). The problem of the influence of a sphere on the potential distribution by this theory was discussed for the first time (sphere in a uniform current field) by Hummel (Hummel 1928). The use of spherical functions for solving the Laplace equation was proposed a few years later (Boursian 1933). The distribution of potential caused by a punctual current injection was then calculated by Zaborovsky (Zaborovsky, 1936). Finally, the anomalous field, on its transverse components caused by a symmetrical dipolar injection, was determined in the 1980s (Sainson, 1984). These theoretical investigations had mining goals at that time (detection and location of massive sulphides around a drill hole).

The study of the potential distribution around a sphere, for example, reduced to that of a curvilinear coordinate system (spherical coordinates) returns to characterize this system by differential invariants of the functions, i.e., by calculating the Laplacian (denoted Δ or ∇^2). It then remains to find the separated variable solutions of equations associated with ∇^2 on the surfaces of coordinates and finally to solve the problem by series of such solutions.

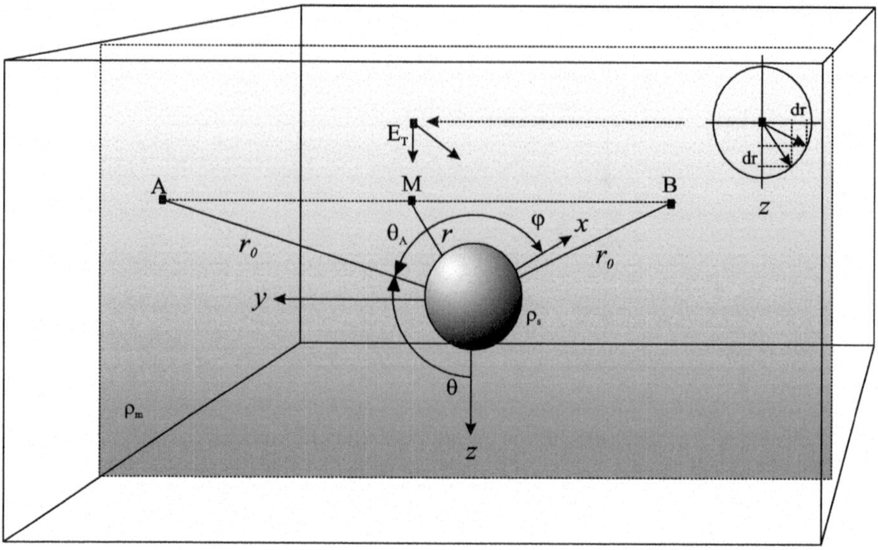

Fig. A5.24 Geoelectric model of the conducting sphere. For small angles, small variations dr are more important on the transverse component of the field along z. Points AB (injection Tx), M (measurement Rz) and the center of the sphere are in the same plane

$$\nabla^2 V = \frac{1}{r^2}\frac{\partial}{\partial r}\left(r^2\frac{\partial V}{\partial r}\right) + \frac{1}{r^2 \sin\theta}\frac{\partial}{\partial \theta}\left(\sin\theta\frac{\partial V}{\partial \theta}\right) + \frac{1}{r^2 \sin\theta}\frac{\partial^2 V}{\partial \varphi^2} = 0 \quad \text{(A5.4.2)}$$

which allows us to obtain the potential distribution and then the variations of the electric field around a sphere contained in a medium of different resistivity.

If we now consider that the potential created by the injection of a current out of the sphere only depends on two coordinates, once one of the axes is passed through the source point, we obtain by symmetry around the axis Oz the simplified expression which no longer depends on the angle φ:

$$\nabla^2 V = \frac{\partial}{\partial r}\left(r^2\frac{\partial V}{\partial r}\right) + \frac{1}{r^2 \sin\theta}\frac{\partial}{\partial \theta}\left(\sin\theta\frac{\partial V}{\partial \theta}\right) = 0 \quad \text{(A5.4.3)}$$

The determination of solutions to this equation (cf. Eq. A5.4.3) is firstly accompanied by the decomposition of the potential function V(r,θ) in a product of two functions R and H with a single variable and respectively dependent on the distance r and the angle θ such that:

$$V(r, \theta) = R(r)\, H(\theta) \quad \text{(A5.4.4)}$$

Replacing the function V by the functions R and H in (A5.4.3), then it becomes:

$$\frac{\partial}{\partial r}\left(r^2\frac{\partial \mathrm{R}(r)\,\mathrm{H}(\theta)}{\partial r}\right)+\frac{1}{\sin\theta}\frac{\partial}{\partial\theta}\left(\sin\theta\frac{\partial \mathrm{R}(r)\,\mathrm{H}(\theta)}{\partial\theta}\right)=0 \qquad \text{(A5.4.5)}$$

Since R and H do not depend respectively either of θ and of r, we have:

$$\begin{cases}\dfrac{\partial \mathrm{R}(r)\,\mathrm{H}(\theta)}{\partial r}=\mathrm{H}(\theta)\dfrac{\mathrm{dR}(r)}{\mathrm{d}r}\\[2ex] \dfrac{\partial \mathrm{R}(r)\,\mathrm{H}(\theta)}{\partial\theta}=\mathrm{R}(r)\dfrac{\mathrm{dH}(\theta)}{\mathrm{d}\theta}\end{cases} \qquad \text{(A5.4.6 and A5.4.7)}$$

Dividing the Eq. (A5.4.5) by R and H thus we arrive at:

$$\frac{1}{\mathrm{H}(\theta)\sin\theta}\frac{\mathrm{d}}{\mathrm{d}\theta}\left(\sin\theta\frac{\mathrm{dH}(\theta)}{\mathrm{d}\theta}\right)+\frac{1}{\mathrm{R}(r)}\frac{\mathrm{d}}{\mathrm{d}r}\left(r^2\frac{\mathrm{dR}(r)}{\mathrm{d}r}\right)=0 \qquad \text{(A5.4.8)}$$

We then find a solution if each member of the equation is constant (λ), i.e., if we have:

$$\begin{cases}\dfrac{1}{\sin\theta}\dfrac{\mathrm{d}}{\mathrm{d}\theta}\left(\sin\theta\dfrac{\mathrm{dH}(\theta)}{\mathrm{d}\theta}\right)=-\lambda\mathrm{H}(\theta) & \text{(A5.4.9)}\\[3ex] \dfrac{\mathrm{d}}{\mathrm{d}r}\left(r^2\dfrac{\mathrm{dR}(r)}{\mathrm{d}r}\right)=\lambda\mathrm{R}(r) & \text{(A5.4.10)}\end{cases}$$

By setting now, $x=\cos\theta$, we get:

$$\frac{\mathrm{d}}{\mathrm{d}\theta}=\frac{\mathrm{d}x}{\mathrm{d}\theta}\frac{\mathrm{d}}{\mathrm{d}x}=-\sin\theta\frac{\mathrm{d}}{\mathrm{d}x} \qquad \text{(A5.4.11)}$$

Under these conditions, the Eq. (A5.4.9) becomes:

$$-\frac{\mathrm{d}}{\mathrm{d}x}\left(-\sin^2\theta\frac{\mathrm{dH}}{\mathrm{d}x}\right)=-\lambda\mathrm{H} \qquad \text{(A5.4.12)}$$

or:

$$-\frac{\mathrm{d}}{\mathrm{d}x}\left[\left(1-x^2\right)\frac{\mathrm{dH}}{\mathrm{d}x}\right]=-\lambda\mathrm{H} \qquad \text{(A5.4.13)}$$

which ultimately leads to the expression:

$$\left(1 - x^2\right)\frac{\mathrm{d}^2\mathrm{H}}{\mathrm{d}x^2} - 2x\frac{\mathrm{dH}}{\mathrm{d}x} - \lambda\mathrm{H} = 0 \tag{A5.4.14}$$

Furthermore, by setting $\lambda = \mathrm{n}(\mathrm{n}+1)$, we find:

$$\left(1 - x^2\right)\frac{\mathrm{d}^2\mathrm{H}}{\mathrm{d}x^2} + 2x\frac{\mathrm{dH}}{\mathrm{d}x} - \mathrm{n}(\mathrm{n}+1)\,\mathrm{H} = 0 \tag{A5.4.15}$$

a differential equation whose solutions can be expressed using Legendre polynomials[27] P_n such that:

$$\mathrm{H}(\theta) = \mathrm{P}_\mathrm{n}(\cos\theta) \tag{A5.4.16}$$

The Eq. (A5.4.10) becomes now:

$$2r\frac{\mathrm{dR}(r)}{\mathrm{d}r} + r^2\frac{\mathrm{d}^2\mathrm{R}(r)}{\mathrm{d}r^2} = \lambda\mathrm{R}(r) \tag{A5.4.17}$$

Setting $r\,\mathrm{R}(r) = \mathrm{U}(r)$, we obtain:

$$2r\frac{\mathrm{d}\left[\frac{\mathrm{U}(r)}{r}\right]}{\mathrm{d}r} + r^2\frac{\mathrm{d}^2\left[\frac{\mathrm{U}(r)}{r}\right]}{\mathrm{d}r^2} = \lambda\frac{\mathrm{U}(r)}{r} \tag{A5.4.18}$$

which gives, neglecting the terms in $1/r^2$ and $1/r^3$:

$$r\frac{\mathrm{d}^2\mathrm{U}(r)}{\mathrm{d}r} = \lambda\frac{\mathrm{U}(r)}{r} \tag{A5.4.19}$$

or alternatively:

$$\frac{\mathrm{d}^2\mathrm{U}(r)}{\mathrm{d}r^2} - \frac{\lambda}{r^2}\mathrm{U}(r) = 0 \tag{A5.4.20}$$

Considering always $\lambda = \mathrm{n}(\mathrm{n}+1)$, the general solution of (A5.4.20) is of the form:

$$\mathrm{U}(r) = Ar^{-\mathrm{n}} + Br^{\mathrm{n}+1} \tag{A5.4.21}$$

which also gives:

$$\mathrm{R}(r) = Ar^{-\mathrm{n}-1} + Br^{\mathrm{n}} \tag{A5.4.22}$$

The general solution of the Laplace equation with two separate variables r and θ is therefore a linear combination of the solutions corresponding to the different values of n such that:

---

[27] See the definition at the end of the Appendix (formulas and curves).

$$V(r,\theta) = \sum_{n=0}^{\infty} \left(A_n r^{-n-1} + B_n r^n\right) P_n(\cos\theta) \quad (A5.4.23)$$

The constants A, B, C and D can be determined by the boundary conditions imposed by the model, i.e.:

- In the sphere:

$$V_S = \sum_{n=0}^{\infty} \left(A_n r^{-n-1} + B_n r^n\right) P_n(\cos\theta) \quad (A5.4.24)$$

- Out of the sphere:

$$V_M = \sum_{n=0}^{\infty} \left(C_n r^{-n-1} + D_n r^n\right) P_n(\cos\theta) \quad (A5.4.25)$$

As the potential V_M must be zero to infinity, this means that:

$$D_n = 0 \quad (A5.4.26)$$

Similarly V_S cannot tend to infinity if we make r tend to 0. We have then:

$$A_n = 0 \quad (A5.4.27)$$

Consequently, this leads to:

- In the sphere:

$$V_S = \sum_{n=0}^{\infty} B_n r^n P_n(\cos\theta) \quad (A5.4.28)$$

- Out of the sphere:

$$V_M = \sum_{n=0}^{\infty} C_n r^{-n-1} P_n(\cos\theta) \quad (A5.4.29)$$

The potential near the source A must be in $1/r$, that is, more exactly in $1/|\vec{r}_0 - \vec{r}|$ which, considering the resistivity ρ_m of the medium, leads to (Fig. A5.4.25):

$$V_M = \frac{\rho_m I}{4\pi} \frac{1}{|\vec{r}_0 - \vec{r}|} + \sum_{n=0}^{\infty} C_n r^{-n-1} P_n(\cos\theta) \quad (A5.4.30)$$

This equation is coherent if $1/|\vec{r}_0 - \vec{r}|$ is turned into Legendre polynomials.

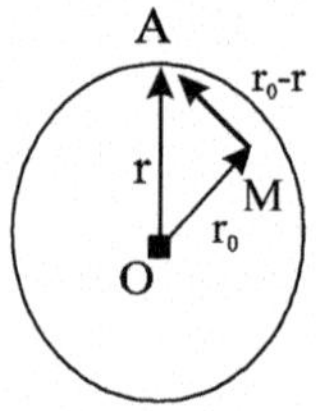

Fig. A5.25 Descriptive diagram positioning OAM, r and r_o

Then by setting the norm:

$$|\vec{r}-\vec{r}_0| = \sqrt{r_0^2 - 2rr_0\cos\theta + r^2} \tag{A5.4.31}$$

or considering its inverse:

$$\frac{1}{r_0} = \frac{1}{\sqrt{h^2 - 2h\cos\theta + 1}} \tag{A5.4.32}$$

where h is the ratio $r/r_0 < 1$, then we obtain:

$$\frac{1}{|\vec{r}_0 - \vec{r}|} = \frac{1}{r_0}\sum_{n=0}^{\infty} h^n P_n(\cos\theta) \tag{A5.4.33}$$

Accordingly, the expression of the potential is:

$$V_M = \frac{\rho_m I}{4\pi}\frac{1}{r_0} + \sum_{n=0}^{\infty}\left(h^n + \frac{4\pi r_0}{\rho_m I}\right) C_n r^{-n-1} P_n(\cos\theta) \tag{A5.4.34}$$

or even:

$$V_M = \frac{\rho_m I}{4\pi}\frac{1}{r_0} + \sum_{n=0}^{\infty}\left[\left(\frac{r}{r_0}\right)^n + \frac{4\pi r_0}{\rho_m I} C_n r^{-n-1}\right] P_n(\cos\theta) \tag{A5.4.35}$$

A4.2. Boundary Conditions

First condition: potentials continuity

The condition of continuity of the potential on the surface of the sphere symbolized by the relation:

$$V_m(P) = V_s(P) \tag{A5.4.36}$$

allows us to write:

$$\frac{\rho_m I}{4\pi}\frac{1}{r_0}+\sum_{n=0}^{\infty}\left[\left(\frac{r}{r_0}\right)^n+\frac{4\pi r_0}{\rho_m I}C_n r^{-n-1}\right]P_n(\cos\theta)=\sum_{n=0}^{\infty}B_n r^n P_n(\cos\theta) \tag{A5.4.37}$$

with the radius of the sphere being R ($r = R$), this means:

$$\frac{\rho_m I}{4\pi}\frac{1}{r_0}\left(\frac{R^n}{r_0^n}+\frac{4\pi r_0}{\rho_m I}C_n R^{-n-1}\right)=B_n R^n \tag{A5.4.38}$$

and thus after calculations:

$$B_n=\frac{\rho_m I}{4\pi r_0^{n+1}}+C_n R^{-2n-1} \tag{A5.4.39}$$

Second condition: continuity of the normal component of the current density

Now, taking into account the resistivity ρ_s of the sphere, the condition of continuity of the normal component of the current density vector required at the interface by the relation[28]:

$$\frac{1}{\rho_m}\frac{\partial V_M}{\partial r}=\frac{1}{\rho_s}\frac{\partial V_S}{\partial r} \tag{A5.4.40}$$

leads to:

$$\frac{1}{\rho_m}\frac{\rho_m I}{4\pi r_0}\frac{n r^{n-1}}{r_0^n}+\frac{1}{\rho_m}(-n-1)\,C_n r^{-n-2}=\frac{1}{\rho_s}nB_n r^{n-1} \tag{A5.4.41}$$

and to:

$$B_n=\frac{\rho_s}{\rho_m}\left[\frac{\rho_m I}{4\pi r_0^{n+1}}+\frac{(-n-1)}{n}C_n r^{-2n-1}\right] \tag{A5.4.42}$$

and, according to the relations (A5.4.39) and (A5.4.40), to:

$$C_n=\frac{\rho_m I}{4\pi r_0^{n+1}}\frac{n(\rho_s-\rho_m)\,r^{2n+1}}{[n(\rho_m+\rho_s)+\rho_s]} \tag{A5.4.43}$$

[28]The potential V varies continuously when the current lines cross the parting surface of the two media. Due to the current conservation law it is the same for $1/\rho\ \partial V/\partial r$ when r denotes the normal to the separation surface. These conditions govern the refraction of the current lines and equipotential surfaces (thus the fields) at the crossing of resistivity discontinuity surfaces.

The continuity conditions used to determine C_n being valid on the surface of the sphere, with $r = R$, there is finally:

$$C_n = \frac{\rho_m I}{4\pi r_0^{n+1}} \frac{n(\rho_s - \rho_m) R^{2n+1}}{[n(\rho_m + \rho_s) + \rho_s]} \tag{A5.4.44}$$

The potential anomaly created by the sphere outside the latter is equal to the potential difference with (V_s) and without the sphere (V_{wos}) such that:

$$V_a = V_s - V_{wos} \tag{A5.4.45}$$

suggesting an anomalous potential equal to:

$$V_a = \frac{\rho_m I}{4\pi r_o} \sum_{n=1}^{\infty} \frac{n(\rho_s - \rho_m) P_n(\cos\theta) R^{2n+1}}{r_o^n [n(\rho_m + \rho_s) + \rho_s] r^{n+1}} \tag{A5.4.46}$$

The electric field deriving from a scalar potential such as:

$$\vec{E} = -\vec{\nabla} V \tag{A5.4.47}$$

can also be expressed in Cartesian coordinates in a coordinate system $\left(O, \vec{i}, \vec{j}, \vec{k}\right)$ in the form:

$$\vec{E} = -\left(\frac{\partial V}{\partial x}\vec{i} + \frac{\partial V}{\partial y}\vec{j} + \frac{\partial V}{\partial z}\vec{k}\right) \tag{A5.4.48}$$

A4.3. Choice of the Transverse Component

For field variations following small angles, the expression of the transverse component of the electric field (along z) in the direction of the anomaly is by far the most significant for the detection (see Fig. A5.4.24). We note as well:

$$E_{T(z)} = -\frac{\partial V}{\partial z} \tag{A5.4.49}$$

A4.4. Injection by Electric Dipole

Referring now to two relatively close current injection points A and B (along the antenna), such that we have $[A,B]_{(y)} \perp E_{T(z)}$ (in-line configuration, for example), the transverse component of the anomalous field is then equal to the sum of the transverse components of the fields induced by each point A and B, which is the equivalent of writing:

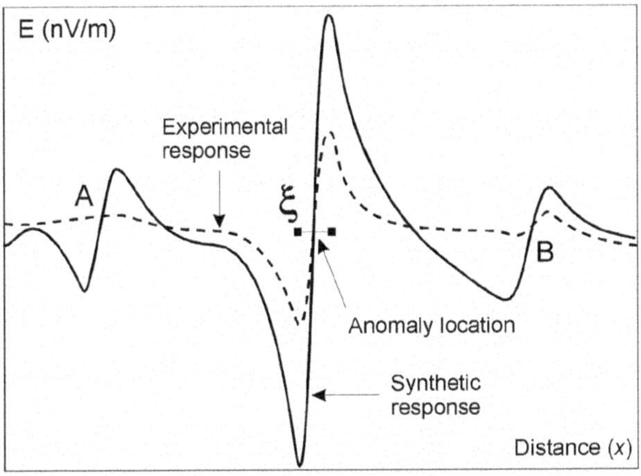

Fig. A5.26 Synthetic responses (mathematical model) for a conducting sphere, confronted with experimental responses in a rheostatic tank (see Chap. 5, Sect. 6.2) showing dispersive curves. For a resistive sphere we have absorption curves (not represented). The current injection electrodes A and B (at low frequency) placed on either side of the measuring device (in line) allow us to measure with the phase change the variations of the transverse electric field component. The electrokinetic equivalence is obtained by an LF current injection symmetrical system, whose injection points provide an antiphase signal, similar to a current dipole +/− (Sainson, 1984)

$$E_{T(z)} = E^{A}_{T(z)} + E^{B}_{T(z)} \tag{A5.4.50}$$

or literally:

$$E_{T(z)} = \overbrace{\frac{-\rho_m I}{4\pi r_{0A}} \sum_{n=1}^{\infty} \frac{n(\rho_s - \rho_m)\, R^{2n+1}}{r^{n}_{0A}[n(\rho_m - \rho_s) + \rho_s]} \left[\frac{\partial P_n(\cos\theta)_A}{\partial(\cos\theta)_A} \frac{\partial(\cos\theta)_A}{\partial z} \frac{1}{r^{n+1}} + P_n(\cos\theta)_A \frac{\partial}{\partial z} \frac{1}{r^{n+1}} \right]}^{E^{A}_{T(z)}} + E^{B}_{T(z)} \tag{A5.4.51}$$

This expression gives, after calculations of the derivatives (∂), the values of variations of the anomalous transverse field ($-\partial V/\partial z$), directly usable for the comparative interpretation of field or experimental data available from the electrometers after correction and calibration.

When there is no anomaly (angle ≈ 0), the transverse component is zero.

$$E_{T(z)} = 0 \tag{A5.4.52}$$

However, these models have limitations. If we consider, for example, a sphere of infinite radius, i.e., a flat surface seen from the top, it could then be detected at an infinite depth, which of course is not true. At this level then we join the problems relative to horizontal strata mentioned in the previous appendices (tabular model).

This type of modeling is all the more effective when the resistivity contrast is important. In absolute terms, the resolution is at maximum when the contrast is considered infinite, i.e., more exactly when the resistivity of the anomaly is zero or infinite (a perfect insulator or conductor). With an equivalent resistivity contrast, in a conductive medium, the conductive sphere has a greater anomaly than the resistant sphere. For the latter, the detectability threshold lies approximately at a depth two times lower than that of the conductive sphere (Fig. A5.4.26).

A4.5. Legendre Polynomials

Legendre polynomials, introduced in the past to study the Newtonian potential,[29] are defined by a series:

$$P_n(z) = \frac{1.3.5\ldots(2n-1)}{n!}\left\{z^n - \frac{n(n-1)}{2(2n-1)}z^{n-2} + \frac{n(n-1)(n-2)(n-3)}{2.4.(2n-1)(2n-3)}z^{n-4} - \ldots\right\}$$

where in most cases the variable z, which is the colatitude of a point in spherical coordinates, is equal to cosθ.

This gives for the values of n = 0, 1, 2, 3...

$$P_0(z) = 1$$
$$P_1(z) = z$$
$$P_2(z) = \frac{1}{2}(3z^2 - 1)$$
$$P_3(z) = \frac{1}{2}(5z^3 - 3z)$$
$$P_4(z) = \frac{1}{8}(35z^4 - 30z^2 + 3)$$
$$P_5(z) = \frac{1}{8}(63z^5 - 70z^3 + 15z)$$

....

$$P_0(\cos\theta) = 1$$
$$P_1(\cos\theta) = \cos\theta$$
$$P_2(\cos\theta) = \frac{1}{4}(3\cos 2\theta + 1)$$
$$P_3(\cos\theta) = \frac{1}{8}(5\cos 3\theta - 3\cos\theta)$$
$$P_4(\cos\theta) = \frac{1}{64}(35\cos 4\theta + 20\cos 2\theta + 9)$$
$$P_5(\cos\theta) = \frac{1}{128}(63\cos 5\theta + 35\cos 3\theta + 30\cos\theta)$$

[29]Legendre A M (1785). Researches on the attraction of homogeneous spheroids. Mem. Math. Phys., presented to the Academy of Sciences. pp. 411–434. Legendre A. M. (1787). Researches on the figure of planets. Mem. Math. Phys., presented to the Academy of Sciences. pp. 370–389.

For the calculations and more specifically for programming (see subroutine XLEG of program P5.1 titled ETRAN®), a recurrence relation (Brillouin 1933) is preferably used such that for three consecutive polynomials we have:

$$(n+1)\,P_{n+1}(z) - (2n+1)\,z\,P_n(z) + nP_{n-1}(z) = 0$$

Appendix A5.5

Interpretation of magnetotelluric submarine soundings by solving Maxwell's equations (1D modeling)

In Chap. 2 on the physical principles, the apparent resistivity was calculated generally for a homogeneous medium. In prospecting, the subsoil is of course more complex. Analytically only simple structures can be modeled. The following demonstration is a n layers tabular model topped with a layer of seawater (see Fig. A5.27).

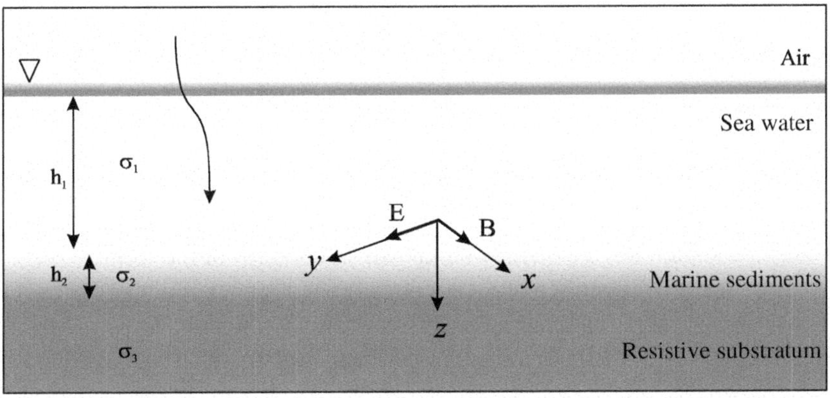

Fig. A5.27 Geoelectric submarine tabular model (three layers) corresponding to a device comprising an electrometer measuring E (along O*y*) and a magnetometer measuring B (along O*x*) placed at the bottom of the sea (an MT underwater survey, commonly called mMT). The apparent resistivity is given along the direction O*z*

From Maxwell's equations and considering that the displacement currents are negligible, considering the electric $\vec{E}$ and magnetic $\vec{B}$ fields, we form:

$$-\frac{\partial \vec{E}_y}{\partial z} = -\frac{\partial \vec{B}_x}{\partial t} \tag{A5.5.1}$$

and:

$$\vec{\nabla} \wedge \vec{B} = \mu_0 \vec{J} = \mu_0 \sigma \vec{E} \tag{A5.5.2}$$

Now we get:

$$\frac{\partial \vec{B}_x}{\partial z} = \mu_0 \sigma \vec{E}_y \tag{A5.5.3}$$

If we now consider that B_x is of the form (frequency domain: wave of pulsation ω) $B_x = B_{xo}\, e^{i\omega t}$, from (A5.5.1) we have:

$$-\frac{\partial \vec{E}_y}{\partial z} = -i\omega t B_{x_o} e^{i\omega t} = -i\omega B_x \tag{A5.5.4}$$

and from (A5.5.2):

$$-i\omega B_x = -\frac{1}{\mu_0 \sigma} \frac{\partial^2 B_x}{\partial z^2} \tag{A5.5.5}$$

where finally:

$$\frac{\partial^2 B_x}{\partial z^2} = i\omega \mu_0 \sigma B_x \tag{A5.5.6}$$

The above equation has a general well known solution (β) of the form:

$$B_x = A e^{i\beta z} + B e^{-i\beta z} \tag{A5.5.7}$$

that is to say that:

$$\frac{\partial^2 B_x}{\partial z^2} = -\beta^2 e^{i\beta z} + \beta^2 e^{-i\beta z} = -\beta^2 B_x \tag{A5.5.8}$$

In essence, it follows that:

$$-\beta^2 B_x = i\omega \mu_0 \sigma B_x \tag{A5.5.9}$$

or that:

$$\beta = i\sqrt{i\omega \mu_0 \sigma} \tag{A5.5.10}$$

By replacing β by its value, the expression (A5.5.7) is written:

$$B_x = Ae^{-\sqrt{i\omega\mu_o\sigma z}} + Be^{\sqrt{i\omega\mu_o\sigma z}} \tag{A5.5.11}$$

By setting $k = \sqrt{i\omega\mu_0\sigma}$, it immediately becomes at layer i:

$$B_x = A_i e^{-kiz} + B_i e^{kiz} \tag{A5.5.12}$$

and from (A5.5.3):

$$\begin{aligned} E_y &= \frac{1}{\mu_0\sigma}\frac{\partial \vec{B}_x}{\partial z} \\ &= \frac{1}{\mu_0\sigma}\left(-k_i A_i e^{-kiz} + k_i B_i e^{kiz}\right) \\ &= \frac{k_i}{\mu_0\sigma}\left(-A_i e^{-kiz} + B_i e^{kiz}\right) \end{aligned} \tag{A5.5.13}$$

Now considering the boundary conditions:

- When z tends to infinity: $E_y = 0$ and $B_y = 0$, it imposes for layer n, $B_n = 0$
- at the interface of layers (n−1) and (n−1)−n, we obtain the equations for the horizontal fields E_y and B_x at:
- the layer (n−1) at the depth z_{n-1}:

$$\begin{cases} B_x = (A_{n-1}e^{-k_{n-1}z_{n-1}} + B_{n-1}e^{k_{n-1}z_{n-1}}) & \text{(A5.5.14)} \\ E_y = \dfrac{k_{n-1}}{\mu_0\sigma_{n-1}}(-A_{n-1}e^{-k_{n-1}z_{n-1}} + B_{n-1}e^{k_{n-1}z_{n-1}}) & \text{(A5.5.15)} \end{cases}$$

- the layer (n−1)-n:

$$\begin{cases} B_x = A_n e^{-k_n z_{n-1}} & \text{(A5.5.16)} \\ E_y = -\dfrac{k_n}{\mu_0\sigma_n} A_n e^{-k_n z_{n-1}} & \text{(A5.5.17)} \end{cases}$$

Setting $A_n = 1$ for the layer n, thus we normalize the ratio E_y/B_x. The expressions (A5.5.14) and (A5.5.16) as well as (A5.5.15) and (A5.5.17) then form a system of equations with two unknown variables which are the coefficients A_n and B_n. Solving this system for n layers is done going back to the upper layers, each time setting the Eqs. (A5.5.14, A5.5.15, A5.5.16 and A5.5.17), and can be calculated numerically (see program P5.2) (Fig. A5.28).

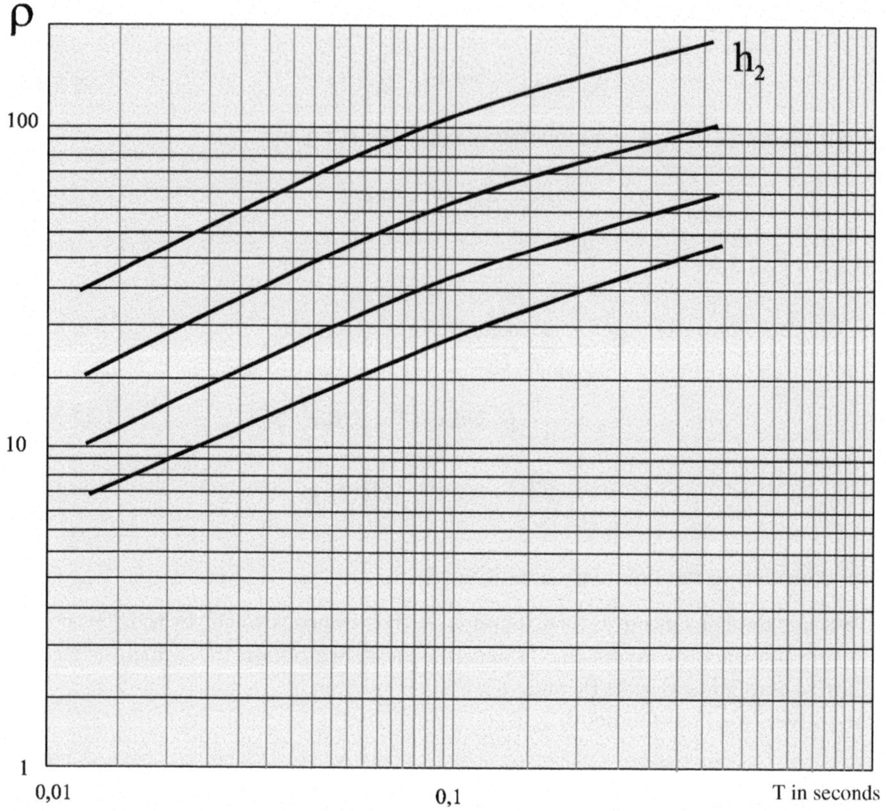

Fig. A5.28 Abacus for interpreting mMT underwater soundings to obtain the characteristics of the layer thickness h_2 corresponding to marine sediments (1–10 m) resting on a resistive thick layer (see Fig. A5.27)

Appendix A5.6

Interpretation of isometric anomalies (cylinder) for vector magnetotelluric devices (transverse fields) by the method of the coefficients of reflection (2D modeling).

The distribution of an anomalous field caused by the presence of an isometric anomaly can be calculated analytically (cf. Appendix A5.4). For more complex structures with multiple axes of radial symmetry such as disks or cylinders, special techniques are proposed. This is the case, for example, of the following one (a cylinder) which takes into account the reflections of waves on the different interfaces (Jegen and Edwards 2000). We shall also find in the scientific literature the technique for example concerning the distribution of fields around an insulating disk (Bailey 2008).

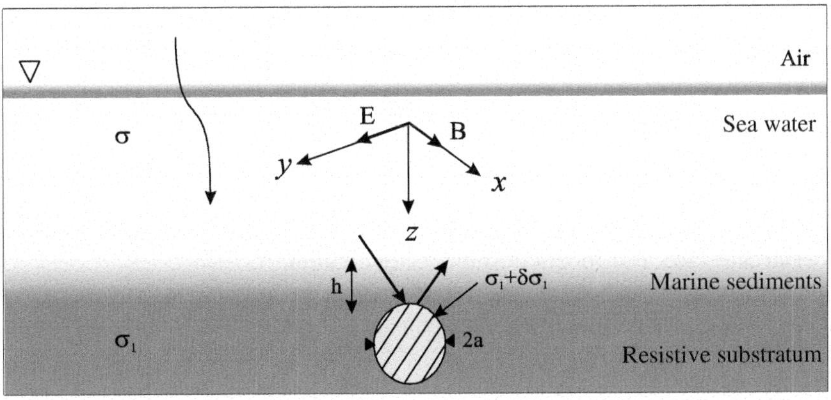

Fig. A5.29 Geoelectric model mMT used in this section (sectional cylinder)

The transverse component, along y, of the electric field (TE mode), at the bottom of the sea, directly above an anomaly of conductivity $\sigma_1+\delta\sigma_1$, buried at a depth h in a medium of conductivity σ_1 (see Fig. A5.29), as a function along y of the initial field (sea surface), is equal to:

$$E_y^1(x,h) = E_y^1(x,0)\, e^{-k_1 h} \qquad \text{(A5.6.1)}$$

where the wave number k_1 is equal to $\sqrt{i\omega\mu_0\sigma_1}$ and ω the wave pulsation.

In this case, the current in the anomaly of radius a can be put in the form:

$$I_y^a = \delta\sigma_1 E_y^1(x,h)\, \pi a^2 \qquad \text{(A5.6.2)}$$

The electric and magnetic field intensities are respectively equal to:

$$E_y^a(x,0) = \frac{I_y^a \mu_0}{2} \int_{-\infty}^{\infty} -\frac{i\omega e^{-\theta_1 h}}{\theta_1}(1 - R_{TE})\, e^{-ipx} dp \qquad \text{(A5.6.3)}$$

$$B_y^a(x,0) = \frac{I_y^a \mu_0}{2} \int_{-\infty}^{\infty} e^{-\theta_1 h}(1 + R_{TE}) e^{-ipx}\, dp \qquad \text{(A5.6.4)}$$

where the incident angle θ_1 is defined as $\theta_1^2 = i\omega\mu_0\sigma_1 + p^2\theta \approx i\omega\mu_0\sigma_1 + p^2$ with p the wave number of the anomalous field in the x direction.

The anomalous signal comes from the superposition of the upgoing field (from the anomaly) and the downgoing field, from the reflections with the water/air interface. The reflection coefficient (TE mode), depending on the different incidence wave angles θ, θ_1 and θ_{air}, is then equal to:

$$R_{TE} = \frac{\frac{\theta}{\theta_1} - R'_{TE}}{\frac{\theta}{\theta_1} + R'_{TE}} \quad \text{with} \quad R'_{TE} = \frac{e^{2\theta d} - \left(\frac{\theta_{air}}{\theta} - 1\right)/\left(\frac{\theta_{air}}{\theta} + 1\right)}{e^{2\theta d} + \left(\frac{\theta_{air}}{\theta} - 1\right)/\left(\frac{\theta_{air}}{\theta} + 1\right)} \qquad \text{(A5.6.5)}$$

where $\theta^2_{air} = \varepsilon\mu_0\omega^2 + p^2$ and where R'_{TE}, that is the reflection coefficient of the air/ocean interface, is a function of the water column.

Now, forming the ratio of the disturbed fields to the initial fields measured on the seabed directly above the anomaly along the x, y directions, thus we form the sensitivity functions:

– For the electric fields (y):

$$\frac{E^a_y(x,0)}{E^1_y(x,0)} = \frac{\mu_0}{2}\delta\sigma_1\pi a^2 e^{-k_1 h}\int_{-\infty}^{\infty}\left[-\frac{i\omega}{\theta_1}e^{-\theta_1 h}\left(1 - R_{TE}\right)e^{-ipx}\right]dp \qquad \text{(A5.6.6)}$$

– For the magnetic fields (x):

$$\frac{B^a_x(x,0)}{B^1_x(x,0)} = \frac{\mu_0}{2}\delta\sigma_1\pi a^2 e^{-k_1 h}\int_{-\infty}^{\infty}\left[-\frac{i\omega}{\theta_1}e^{-\theta_1 h}\left(1 + R_{TE}\right)e^{-ipx}\right]dp \qquad \text{(A5.6.7)}$$

representative of the sensitivity to resistivity variations between the anomaly and its surroundings depending on its burial depth.

Appendix A5.7

Interpretation of anomalies of any shape by the numerical method of integral equations (3D modeling)

The theoretical calculations that follow arise from parts of the works led by the Department of Geology and Geophysics at the University of Utah (Holmann 1989).

Numerical models do not escape the simplicity of the geometric configuration of the geological features they are supposed to represent. The traditional model in use is generally that of the heterogeneous conductivity of any shape placed in a horizontal layered medium. The latter can also be modeled by any other techniques including those using analytical methods (Fig. A5.30).

Modeling an anomalous field caused by 3D heterogeneity found in a laminate ground corresponds to the superposition of the *primary field* (answer of the laminate ground) and the *secondary field* (answer of the body in the stratified ground).

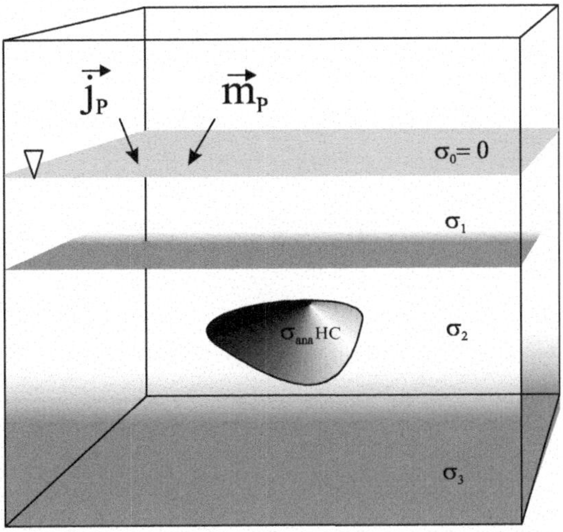

Fig. A5.30 Generally used Geoelectric model: any form (reservoir) placed for example in a layered medium (horizontal geological layers of variable conductivity σ function of the depth)

$$\vec{E} = \vec{E}_P + \vec{E}_S \tag{A5.7.1}$$

Primary and secondary fields are calculated separately. The real and imaginary parts (corresponding to amplitudes and phases) of the different components of the field are expressed in Cartesian coordinates.

A7.1. Modeling of the Primary Field

If the displacement currents are neglected, and setting $\mu = \mu_0$ (geological materials), the electric $\vec{e}$ and magnetic $\vec{h}$ fields are described in the time domain (according to the space $\vec{r}$ and time t) by Maxwell's equations such that we have:

$$\vec{\nabla} \wedge \vec{e}\,(\vec{r}, t) = -\mu_0 \frac{\partial \vec{h}(\vec{r}, t)}{\partial t} - \mu_0 \frac{\partial \vec{m}_P}{\partial t}(\vec{r}, t) \tag{A5.7.2}$$

and:

$$\vec{\nabla} \wedge \vec{h}(\vec{r}, t) = \sigma \vec{e}(\vec{r}, t) + \vec{j}_p(\vec{r}, t) \tag{A5.7.3}$$

where m_p and j_p are respectively the electric and magnetic primary currents and σ the electrical conductivity in the medium.

Taking the rotational of the Eq. (A5.7.2) and replacing it in the Eq. (A5.7.3) we form the equality for the field $\vec{e}$:

$$\vec{\nabla}\wedge\vec{\nabla}\wedge\vec{e}+\mu_0\sigma\frac{\partial\vec{e}}{\partial t}=-\mu_0\frac{\partial\vec{j}_p}{\partial t}-\mu_0\vec{\nabla}\wedge\frac{\partial\vec{m}_p}{\partial t}\qquad\text{(A5.7.4)}$$

Also using the rotational of the Eq. (A5.7.3) and replacing it in the Eq. (A5.7.2) we get the diffusion equation for the field $\vec{h}$:

$$\vec{\nabla}\wedge\left(\frac{\vec{\nabla}\wedge\vec{h}}{\sigma}\right)+\mu_0\frac{\partial\vec{h}}{\partial t}=\vec{\nabla}\wedge\left(\frac{\vec{j}_p}{\sigma}\right)-\mu_0\frac{\partial\vec{m}_p}{\partial t}\qquad\text{(A5.7.5)}$$

Using the vector identity $\vec{\nabla}\wedge\vec{\nabla}\wedge\vec{A}=-\nabla^2\vec{A}+\vec{\nabla}(\vec{\nabla}.\vec{A})$ the Eq. (A5.7.4) becomes:

$$-\nabla^2\vec{e}+\vec{\nabla}\left(\vec{\nabla}\cdot\vec{e}\right)+\mu_0\sigma\frac{\partial\vec{e}}{\partial t}=-\mu_0\frac{\partial\vec{j}_p}{\partial t}-\mu_0\vec{\nabla}\wedge\frac{\partial\vec{m}_p}{\partial t}\qquad\text{(A5.7.6)}$$

Expressly using the divergence of the Eq. (A5.7.3), $\vec{\nabla}.(\sigma\vec{e})=\sigma\vec{\nabla}.\vec{e}+\vec{\nabla}\sigma.\vec{e}=-\vec{\nabla}.\vec{j}_p$, and substituting $\vec{\nabla}.\vec{e}$ in the Eq. (A5.7.6), we arrive at:

$$\nabla^2\vec{e}+\vec{\nabla}\left(\vec{e}\cdot\frac{\vec{\nabla}\sigma}{\sigma}\right)-\mu_0\sigma\frac{\partial\vec{e}}{\partial t}$$
$$=\mu_0\frac{\partial\vec{j}_p}{\partial t}-\frac{1}{\sigma}\vec{\nabla}\left(\vec{\nabla}\cdot\vec{j}_p\right)+\mu_0\vec{\nabla}\wedge\frac{\partial\vec{m}_p}{\partial t}\qquad\text{(A5.7.7)}$$

considering that the source is in a medium with a homogeneous conductivity.

Now assigning the following identity, $\vec{\nabla}\wedge\varphi\vec{A}=\varphi\vec{\nabla}\wedge\vec{A}-\vec{A}\wedge\vec{\nabla}\varphi$, the Eq. (A5.7.5) can be written now as:

$$\frac{1}{\sigma}\vec{\nabla}\wedge\vec{\nabla}\wedge\vec{h}-\left(\vec{\nabla}\wedge\vec{h}\right)\wedge\vec{\nabla}\left(\frac{1}{\sigma}\right)+\mu_0\frac{\partial\vec{h}}{\partial t}=\frac{1}{\sigma}\vec{\nabla}\wedge\vec{j}_p-\mu_0\frac{\partial\vec{m}_p}{\partial t}\qquad\text{(A5.7.8)}$$

or:

$$-\nabla^2\vec{h}+\vec{\nabla}\left(\vec{\nabla}\cdot\vec{h}\right)-\sigma\left(\vec{\nabla}\wedge\vec{h}\right)\wedge\vec{\nabla}\left(\frac{1}{\sigma}\right)+\mu_0\sigma\frac{\partial\vec{h}}{\partial t}$$
$$=\vec{\nabla}\wedge\vec{j}_p-\mu_0\sigma\frac{\partial\vec{m}_p}{\partial t}\qquad\text{(A5.7.9)}$$

Knowing besides that the divergence of the magnetic field is nonzero at the source, the divergence of the expression (A5.7.2) shows that $\vec{\nabla}.\vec{h}=-\vec{\nabla}.\vec{m}_p$ and that:

$$\nabla^2\vec{h} + \sigma\left(\vec{\nabla}\wedge\vec{h}\right)\wedge\vec{\nabla}\left(\frac{1}{\sigma}\right) - \mu_0\sigma\frac{\partial\vec{h}}{\partial t}$$

$$= \mu_0\sigma\frac{\partial\vec{m}_p}{\partial t} - \vec{\nabla}\left(\vec{\nabla}\cdot\vec{m}_p\right) - \vec{\nabla}\wedge\vec{j}_p \tag{A5.7.10}$$

The above Eqs. (A5.7.7) and (A5.7.10) are the general equations of the total electric and magnetic fields at any point in the propagation medium. The primary electric $\vec{e}_P$ and magnetic $\vec{h}_p$ fields that correspond to the fields in a stratified medium without anomaly can be calculated from the equations:

$$\vec{\nabla}\wedge\vec{e}_P = -\mu_0\frac{\partial\vec{h}_p}{\partial t} - \mu_0\frac{\partial\vec{m}_p}{\partial t} \tag{A5.7.11}$$

and:

$$\vec{\nabla}\wedge\vec{h}_p = \sigma_{wob}\vec{e}_p + \vec{j}_p \tag{A5.7.12}$$

where σ_{wob} is the electrical conductivity in the absence of any foreign body (or without body).

These equations, in integral form, can then be calculated numerically. By application, for example, for each function of time $e^{i\omega t}$, the Fourier transforms (Papoulis 1962; Bracewell 1986; Wijewardena 2007) such that:

$$F(\vec{r},\omega) = \int_{-\infty}^{\infty} f(\vec{r},t)\, e^{-i\omega t}dt$$

and

$$f(\vec{r},t) = \frac{1}{2\pi}\int_{-\infty}^{\infty} F(\vec{r},\omega)\, e^{i\omega t}d\omega \tag{A5.7.13}$$

whose representations are equivalent, $f(\vec{r},t) \Leftrightarrow F(\vec{r},\omega)$, we get the Eqs. (A5.7.7) and (A5.7.10) in the *frequency domain* $\left(\vec{e},\ \vec{h}\rightarrow\vec{E},\ \vec{H}\right)$:

$$\nabla^2\vec{E} + \vec{\nabla}\left(\vec{E}\cdot\frac{\vec{\nabla}\sigma}{\sigma}\right) + k^2\vec{E}$$

$$= i\omega\mu_0\vec{j}_p - \frac{1}{\sigma}\vec{\nabla}(\vec{\nabla}\cdot\vec{J}_p) + i\omega\mu_0\vec{\nabla}\wedge\vec{M}_p \tag{A5.7.14}$$

and:

$$\nabla^2\vec{H} + \sigma(\vec{\nabla}\wedge\vec{H})\wedge\vec{\nabla}\left(\frac{1}{\sigma}\right) + k^2\vec{H}$$

$$= i\omega\mu_0\vec{M}_p - \overleftarrow{\nabla}(\vec{\nabla}\cdot\vec{M}_p) - \vec{\nabla}\wedge\vec{J}_p \tag{A5.7.15}$$

with $k^2 = -i\omega\mu_0\sigma$

A7.2. Modeling of the Secondary Field

In the *time domain*, by subtracting the Eq. (A5.7.9) from the Eq. (A5.7.2) and the Eq. (A5.7.10) from the Eq. (A5.7.3), we get the secondary fields such that $\left(\vec{e}_S, \vec{h}_S\right)$:

$$\vec{\nabla} \wedge \vec{e}_S = -\mu_0 \frac{\partial \vec{h}_S}{\partial t} \tag{A5.7.16}$$

and:

$$\vec{\nabla} \wedge \vec{h}_S = \sigma \vec{e}_S + \sigma_{an} \vec{e}_P \tag{A5.7.17}$$

or:

$$\vec{\nabla} \wedge \vec{h}_S = \sigma_{wob} \vec{e}_S + \vec{j}_S \tag{A5.7.18}$$

where $\vec{j}_S = \sigma \vec{e}$ and $\sigma_{an} = \sigma - \sigma_{wob}$

The above equation is similar to Eq. (A5.7.7) without the magnetic source term and where $\vec{j}_S$ is substituted by $\sigma_{an} \vec{e}_P$ which finally gives:

$$\nabla^2 \vec{e}_S + \vec{\nabla}\left(\vec{e} \cdot \frac{\vec{\nabla}\sigma}{\sigma}\right) - \mu_0 \sigma \frac{\partial \vec{e}_S}{\partial t} = \mu_0 \sigma_{an} \frac{\partial \vec{e}_P}{\partial t} - \vec{\nabla}\left(\vec{e}_P \cdot \frac{\vec{\nabla}\sigma_{an}}{\sigma}\right) \tag{A5.7.19}$$

Similarly, the secondary magnetic field is obtained in the same manner such as (cf. Eq. A5.7.10):

$$\begin{aligned} \nabla^2 \vec{h}_S + \vec{\nabla} \wedge \vec{h}_S \wedge \vec{\nabla}\left(\frac{1}{\sigma}\right) - \mu_0 \sigma \frac{\partial \vec{h}_S}{\partial t} \\ = \mu_0 \sigma_{an} \frac{\partial \vec{h}_P}{\partial t} - \sigma \vec{\nabla}\left(\frac{\vec{\nabla}\sigma_{an}}{\sigma}\right) \wedge \vec{e}_P \end{aligned} \tag{A5.7.20}$$

Solving the Eqs. (A5.7.19) and (A5.7.20), which correspond to each secondary field, is preferable to solving the equations for the total field, which would numerically require a larger complete discretization of the model (anomaly + grounds) and therefore would lead to longer calculation times.

In the *frequency domain*, the above Eqs. (A5.7.19) and (A5.7.20) are respectively obtained through the Fourier integrals $f(\vec{r}, t) \Leftrightarrow F(\vec{r}, \omega)$ where $\vec{e}_S, \vec{h}_S \xrightarrow{FT} \vec{E}_S, \vec{H}_S$ such that:

$$\nabla^2 \vec{E}_S + \vec{\nabla}\left(\vec{E}_S \cdot \frac{\vec{\nabla}\sigma}{\sigma}\right) + k^2 \vec{E}_S = -k_{an}^2 \vec{E}_P - \vec{\nabla}\left(\vec{E}_P \cdot \frac{\vec{\nabla}\sigma_{an}}{\sigma}\right) \tag{A5.7.21}$$

and:

$$\nabla^2\vec{H}_S + \sigma\left(\vec{\nabla} \wedge \vec{H}_S\right) \wedge \vec{\nabla}\left(\frac{1}{\sigma}\right) + k^2\vec{H}_S = -k_{an}^2\vec{H}_P - \sigma\vec{\nabla}\left(\frac{\vec{\nabla}\sigma_{an}}{\sigma}\right) \quad (A5.7.22)$$

with $k^2 = -i\omega\mu_0\sigma$ and $k_{an}^2 = -i\omega\mu_0\sigma_{an}$

Solutions for secondary fields are finally calculated from Eqs. (A5.7.16) and (A5.7.18) in the frequency domain, such that:

$$\vec{\nabla} \wedge \vec{E}_S = -i\omega\mu_0\vec{H}_S \quad (A5.7.23)$$

and:

$$\vec{\nabla} \wedge \vec{H}_S = \sigma_{wob}\vec{E}_S + \vec{J}_S \quad (A5.7.24)$$

To formulate then the integral equation, we consider $\vec{J}_S$ as a current source. In this empty space, the secondary electric field is then given by the expression:

$$\vec{E}_S = -i\omega\mu_0\vec{A}_S - \vec{\nabla}V_S \quad (A5.7.25)$$

where $\vec{A}_S$ and V_S are respectively the vector and the scalar potential for the Lorentz gauge.

The integral equations on the volume v for $\vec{A}_S$ and V_S as a function of $\vec{r}$ are then formulated by the following integrals:

$$\vec{A}_S(\vec{r}) = \int_v \vec{J}_S(\vec{r}')\, G(\vec{r}, \vec{r}')\, dv' \quad (A5.7.26)$$

and:

$$V_S(\vec{r}) = -\frac{1}{\sigma_{wob}}\int_v \nabla \cdot \vec{J}_S(\vec{r}')\, G(\vec{r}, \vec{r}')\, dv' \quad (A5.7.27)$$

where G is the scalar Green's function such that:

$$G(\vec{r}, \vec{r}') = \frac{e^{-ik_{wob}|r-r'|}}{4\pi\left|\vec{r} - \vec{r}'\right|} \quad (A5.7.28)$$

with $k_{wob} = \sqrt{-i\omega\mu_0\sigma_{wob}}$

For a conductivity anomaly present in a homogeneous (Holmann 1975) or stratified half-space (Vannamaker et al. 1984), an additional term is added to Eq. (A5.7.25).

A7.3. The Total Field

Finally, by adding the primary and secondary fields (cf. Eq. A5.7.1), we obtain the integral equation of the total field so that:

$$\vec{E}(\vec{r}) = \vec{E}_P(\vec{r}) + \int_v \underline{G}(\vec{r}, \vec{r}') \cdot \sigma_{an}(\vec{r}')\, \vec{E}(\vec{r}')\, dv' \qquad (A5.7.29)$$

where $\underline{G}$ is the tensor of Green's function.

This integral equation is limited to a 3D anomaly in a 1D environment, stratified or not. This corresponds to the majority of cases encountered in petroleum geology where the sedimentary layers remain relatively monotonous laterally.

Today, one of the main purpose of 3D modeling is also to include the broadside EM field data taking into account TE and TM modes, the amplitude and the phase measurements.

Program P5.1

ETRAN Fortran program for the evaluation of TE transverse field. (Model of the sphere/Laplace, cf. Appendix A5.4).

Preamble

The f2c software available for free at AT&T Bell Labs is used to transcribe the FORTRAN 77 programs in C. The following program was written in FORTRAN 4, whose syntax is almost identical to that of Fortran 77 or 90.

```
C*********************** PROGRAM ETRAN ***************************
C          THIS PROGRAMM CALCULATES THE TRANSVERSE ELECTRIC FIELD
MODIFIED BY AN
C         IMMERGED SPHERE PLACED AT A DISTANCE ALONG THE MEASUREMENT
DETECTION
C          DEVICE COMPOSED BY A DIPOLE PLACED IN THE CENTER OF AN
TRANSMITTING DIPOLE
C           LINE WHERE THE TWO INJECTION ELECTRODES ARE IN PHASE
OPPOSITION (≈ +/-).
C***************************************************************
0001 FTN4,L
0002        PROGRAM ETRAN. AUTHOR: STEPHANE SAINSON
0003        DOUB1E PRECISION XLEG, ZA, ZB, EA, EB, ET, XPA, XPAP,
XPB, XPBP
0004        DIMEMSION IPAR(5)
0005        CA11 RNPAR(IPAR)
```

```
      IN=IPAR(1)
      NOUT=IPAR(2)
      IF(IH.EQ.O) IN=1
      IF(NOUT.EQ.O) HOUT=1
      WRITE (NOUT.100)
100    FORMAT("ENTER THE RESISTIVITY OF THE GROUND, OF THE SPHERE")
       READ(IN,*)ROM.ROS
       WRITE(NOUT 200)
200    FORMAT("ENTER THE RADIUS OF THE SPHERE, THE LENGTH OF THE DEVICE")
       READ(IN,*) RS.ZL
       WRITE(NOUT 300)
300    FORMAT("ENTER THE CURRENT INTENSITY,/
       2 "THE HORIZONTAL DISTANCE FROM THE SPHERE TO THE DEVICE")
       READ(IN,*) XI.YDS
       WRITE(NOUT,400)
400    FORMAT("ENTER THE RANGE N")
       READ(IN,*) NW
       WRITE(NOUT,900)
900    FORMAT("ENTER THE VERTICAL DISTANCES MINI AND MAXI OF THE
1     SPHERE TO THE DEVICE",/," AND THE STEP DZ")
       READ(IN,*) ZMS1, ZMS2 ZMS0
       INBS=IFIX((S2-ZKS1)/ZKSO)
       DO 20 IMZS =1,INBMS+1
       IZKSP=IZMS-1
       ZMS=ZMS1+IZMSP*ZMS0
       ZAS=ZMS-ZL
       ZBS=ZMS+ZL
       DMS=SQRT(YDS**2+ZMS**2)
       DAS=SQRT(YDS**2+ZAS**2)
       DBS=SQRT(YDS**2+ZBS**2)
       ZA= (YDS**2+ZAS*ZMS)/(SQRT(YDS**2+ZAS**2)*SQRT(YD5**2+ZMS**2)
       ZB=(YDS**2+ZBS*ZMS)/(SQRT(YDS**2+ZBS**2)*SQRT(YDS**2+ZMS**2)
       WK=-ROM*XI*(ROS-ROM).YDS/(4.3.14159)
       ET=O
       XPAP=XLEG(0,ZA,IN,NOUT)
       XPBP=XLEG(0,ZB,IN,NOUT)
       DO 10 N=1, NN
       XPA=XLEG(N,ZA,IN,NOUT)
       XPB=XLEG(N,ZB,IN,NOUT)
C      WRITE(NOUT,800) ZA,ZB,XPA,XPB,XPAP,XPBP
800    FORMAT("COS TETA FOR A , B "F5,2,3X,F5,2,/,
```

```
      1, "P LEG, FOR A, B (RG N)",E8.3,3XE8.3)
      1, "P LEG, FOR A, B (RG N-1)",E8.3,3X,E8.3)
      WKN=(N*RS**(2*n+1))/(N*(ROM-ROS)+ROS)
      EA=(N*ZA*XPA+XPAP)/(DMS**(N+1)*(-(1-ZA**2)))
      EB=(N*ZB*XPB+XPBP)/(DMS**(N+1*(-(1-ZB**2)))
C     WRITE(NOUT,700) EA,ED
700   FORMAT("EA,EB INTERMEDIAIRE",E8.3,3X,E8.3)
      EA=EA*(2-(YDS**2+ZAS*ZMS)*(1/DAS**2-1/DMS**))/(DAS*DMS)
      EB=EB*(2-(YDS**2+ZBS*ZMS)*(1/DBS**2-1/DHS**2))/(DBS*DMS)
C     WRITE(NOUT,700) EA,EB
      EA=EA+XPA*(-(N+1)*DMS**(-N-3)
      EB=EB+XPB*(-(N+1)*DMS**(-N-3)
C     WRITE(NOUT,700) EA,EB
      EA=EA/DAS**(N+1)
      EB=EB/DBS**(N+1)
C     WRITE(NOUT,700) EA,EB
      EABN=(EA-EB)*WKN
      ET=ET+EABN
C     WRITE(NOUT,600) N,ET
600   FORMAT("AT RANGE N=",I3, "E TRAN= ",E8.3)
      XPAP=XPA
      XPBP=XPB
10    CONTINUE
      ET=ET*WK
      WRITE(NOUT,500) ZMS,ET
500   FORMAT("FIELD IN DZ=",E8.3,"-----E=")
20    CONTINUE
      END
      END$
      FTN4,L
C   ***** X1EG (LEGENDRE POLYNOMIALS CALCULATION)***
      FUNCTION XLEG(N,Z,IN,NOUT)
      DOUB1E PRECISION X,Y,Z,XLEG
C     WRITE (NOUT 333) Z
333   FORMAT("COS TETA AT THE ENTRANCE OF XLEG ", D20.15)
      X=0
      XN=H/2
      IFIN=IFIX(XN)
C     WRITE(MOUT,800) IFIN,XN
800   FORMAT("E(N/2)=", "13", XN=",F5.2)
      DO4K=0,IFIN,1
C     WRITE(MOUT,500) K
500   FORMAT("K=", 13)
1     Y=(-1)**I
      DO3L=1,N
```

```
C     WRITE(NOUT,600) L
600   FORMAT("L="',I3)
2     J1=N -(2*K)+L
      IF(J1,GT,(H-K)*2)GOTO3
C     WRITE(NOUT, 1000) J1
1000  FORMAT ("J1="',I3)
      Y=Y*J1
      L1=L-1
      IF(L1,GT,(H-K-1)GOTO3
      J2=N-K-L1
C     WRITE(NOUT, 300) J2
300   FORMAT ("J2="',E8.3)
      Y=Y/J2
      IF(L1,GE,K)GOTO3
      J3=K-L1
C     WRITE(NOUT,400) J3
400   FORMAT ("J3="',E8.3)
      Y=Y/J3
3     CONTINUE
      KN=N-2*K
      WRITE(NOUT,900) KN,Y
900   FORMAT ("COEF OF RANG N-2K=",13----",15.3)
      Y=Y*(Z**(N-2*K)
      X=X+Y
4     CONTINUE
C     WRITE(NOUT,1100) X
1100  FORMAT("X=" ,20.3)
      X=X /(2,**N)
      XLEG=X
C     WRITE(NOUT,200) N,Z,X
200   FORMAT("P (n="13, z=",D20.15,"),=, D20.15)
      RETURN
      END
      END$
```

Program P5.2

mMT program in HP Basic for the calculation of apparent resistivity for marine magnetotelluric method. (Tabular model, cf. Appendix A5.5).

```
! ------------------- PROGRAM mMT ----------------------------
! Programm using marine Magneto-Telluric method for the
! calculation of apparent resistivity and thicness of different
! layers for a tabular model
! -----------------------------------------------------------
10    ! PROGRAM mMT. AUTHOR: URBAIN RAKOTOSOA
20    ! ---- LOOP OVER PERIOD ----
30    ! CALCULATION OF APP RESISTIVITY AND
40    ! PHASE IN mMT , N STRATA
50    DIM R0(20),H0(20),T1(51)
60    DIM P1(S1),R1(51),R(2),I(2)
70    DEG
80    ! ------ INPUTS ------
90    F0=0
100   DISP "NORMALISATION?(1=YES)" @ INPUT N9
110   IF H9<>1 THEM GOTO 130
120   DISP "Rho min et Tmin" @ INP UT RS,TS
130   DISP "ABSCISSE IN 'T'(0) OR IN 'SQR(T)'" @ INPUT 01
140   DISP  "INITIAL 'T' AND FINAL 'T'"
150   INPUT T2,T3
160   DISP "NB OF PERIODS" @ INPUT P
170   ! ------ CONST INIT -------
180   M0=4*P1*.0000001
190   M=SQR(M0/2)
200   I0=1
210   IF 01=1 THEN T0=SQR(T2) @ T1=SQR(T3)
220   TO=LGT(T2) @ Tl=LGT<T3) @ T9 =(TI-T0)/P
230   DISP "NB OF STRATA" @ INPU
240   H=0
250   FOR I=1 TO N-1
260   DISP "STRATA ";I;"RHO.THICKNESS"
270   INPUT R0(I),H0(I)
280   H=H+H0(I)
290   HO(1)=H
300   NEXT I
310   DISP "STRATA ";N;"RHO" @ INPUT R0(N)
320   ! ------ CONST INIT -------
330   I0=1
340   FOR T=T0 TO Tl STEP T9
```

```
350  T7=10^T @ T1(10)=T7
360  IF 01=1 THEH T7=T7^2
370  0=2*PI/T7
380  M1=M*SQR(O)
390  A3=1 @ A4=0 @ B3=0 @ B4=6
400  ! ---- STRATA LOOP ------
410  FOR I=1 TO N-1
420  N0=N-I
430  A0=SQR(R0CN0)/R0(N0+1)
440  S=M1*(1/SQR(R0(H0))+1/SQR(R0 CN0+1)))
450  D=M1*(1/SQR(RO(NO))-1/SQR(R0(N0+1)))
460  X=D*HO(N0) @ GOSUB 1990
470  Y=1+A0
480  C1=Y*R(1)*EXP(X) @ C2=Y*I(1)*EXP(X)
490  F1=Y*R(1)*EXP(-X) @ F2=-(Y*I(1)*EXP(-X))
500  X= S*H0(N0) @ GOSUB 1990
510  Y=1-A0
520  D1=Y*R(1)*EXP(X) @ 02=Y*I(1)*EXP(X)
530  E1=Y*R(1)*EXP(-X) @ E2=-(Y*I(1)*EXP(-X))
540  A1=.5*(A3*C1-A4*C2+B3*D1-B4*D2)
550  A2=.5*(A4*C1+A3*C2+D2*B3+D1*B4)
560  B1=.5*(A3*E1-A4*E2+B3*F1-B4*F2)
570  B2=.5*(A4*E1+A3*E2+B3*F2+B4*F1)
580  A3=A1 @ A4=A2 @ B3=B1 @ B4=B
590  NEXT I
600  ! ---- RAPP AND PHI CALCULATION ----
610  R(2)=A3 œ I(2)=A4@ R(1)=B3 @ 1(1)=B4
620  GOSUB 1670
630  E1=X @ E2=V
640  GOSUB 1710
650  R(1)=X @ I(1)=Y
660  H1=X @ H2=Y
670  R(1)=E1 @ I(1)=E2 @ R(2)=H1 @ I(2)=H2
680  GOSUB 1850
690  Z1=X @ Z2=Y
700  R1(IO)=(Z1^2+Z2^2)*R0(1)
710  R(1)=H1 @ I(1)=H2
720  R(2)=1 @ 1(2)=-1
730  GOSUB 1750
740  H1=X @ H2=Y
750  R(1)=E1 @ I(1)=E2 @ R(2)=H1 @ I(2)=H2
760  GOSUB 1850
770  Z1=X @ Z2=Y
780  ! DISP "Re(E)=H";E1 @ DISP "I m(E)=";E2
790  ! DISP "Re(H)=";H1 @ DISP "I m(H)=";H2
```

```
800  P1(I0)=ATN(Z2/Z1)
810  ! DISP "Re(E/H)=";Z1
820  ! DISP "Im(E/H)=";Z2
830  I0= I0+1
840  IF 01#1 THEH D1SP "T=";T1(IO-1) @ GOTO 860
850  DISP "rac(T)=";T1(I0-1)
860  DISP "RAPP= ";R1(I0-1)
870  DISP "PHI=";P1(IO-1)
880  DISP
890  NEXT T
900  ! ---- TRACES ------
910  IF N9=1 THEN GOTO 990
920  T0=100000000000
930  R0=100000000000
940  FOR 1=1 TO P
950  IF T1(I)<=T0 THEN T0=T1(1)
960  IF R1(I)<=R0 THEN R0=R1(I)
970  NEXT 1
980  GO1O 1010
990  T0=T5
1000 R9=R5
1010 T0=10^INT(LGT(T0))
1020 R0=10^INT(LGT(R0))
1030 ! ---- AXES ------
1040 PLOTTER 1S 705
1050 CSIZE 4
1060 MSCALE 220,20
1070 IF F0<>0 THEM GOTO 1560
1080 XAXIS 0,0,-195,0
1090 YAXIS O,0,0,180
1100 PLOT -210,-10,-2
1110 LDIR 90 @ LABEL "Ra( .m)"
1120 FOR E=0 TO 1
1130 K0=10^E
1140 FOR K=K0 TO 10*K0 STEP K0
1150 PLOT 0,LGT(K)*75,-2
1160 LORG 5 @ LDIR 0
1170 LABEL "-"
1180 IF K#K0 THEN 1210
1190 PLOT 10,LGT(K)*75.-2
1200 LDIR 180 @ LABEL K0*T0
1210 HEXT K
1212 HEXT E
1214 FOR J=0 TO 2
1216 K0=10^J
```

```
1220 FOR K=K0 TO 10*K0 STEP K0
1230 PLOT -(LGT(K)*65),0,-2
1240 LORG 5 @ LOIR 90
1250 LABEL "-"
1260 IF K#K0 THEM 1290
1270 PLOT -(LGT(K)*65),-10,-2
1280 LABEL K0*R0
1290 HEXT K
1300 HEXT J
1310 PLOT 10,135,-2 @ LOIR 90
1320 IF 01=1 THEN LABEL "rac(T)" ELSE LABEL "T(Sec.)"
1560 MSCALE 220,20 @ P0=-2
1570 P0=-2
1580 FOR 1=1 TO P
1590 PLOT -(65*LGT(Rl(I)/R0)),75*LGT(Tl(I)/T0),P0 @ P0=-1
1600 NEXT 1
1610 PENUP
1620 OISP "OTHER CURVE?(1=yes)" @ INPUT F0
1630 IF F0=0 THEN GOTO 2110
1640 GOTO 220
1650 ! SUBROUTINES COMPLEXES NB
1660 ! ------------------------
1670 ! ADDITION
1680 X=R(1)+R(2)
1690 Y=I(1)+I(2)
1700 RETURN
1710 ! SOUSTRACTION
1720 X=R(1)-R(2)
1730 Y=I(1)-I(2)
1740 RETURN
1750 ! MULTIPLICATION
1760 X=R(1)*R(2)-I(1)*I(2)
1770 Y=R(1)*I(2)+R(2)*I(1)
1780 RETURN
1790 ! CONJUGATE COMPLEXE (1)
1800 X=R(1) @ Y=-1(1)
1810 RETURN
1820 ! CONJUGATE COMPLEXE (2)
1830 X=R(2) @ Y=-1(2)
1840 RETURN
1850 ! DIVISION
1860 R1=R(1) @ I1=I(1)
1870 X2=R(2) @ Y2=I(2)
1880 GOSUB 1820
1890 R(2)=X @ 1(2)=Y
```

```
1900 GOSUB 1750
1910 X1=X @ Y1=Y
1920 R(1)=X2 @ I(1)=Y2
1930 GOSUB 1750
1940 D=X
1950 X=X1/D @ Y=Y1/D
1960 R(1)=R1 @ I(1)=I1
1970 R(2)=X2 @ I(2)=Y2
1980 RETURN
1990 ! NOTATION PASSAGE EXP
2000 ! A NOT "POLAIRE"
2010 ! FIRST VAR
2020 RAD
2030 R(1)=COS(X) @ I(1)=SIN(X)
2040 DEG
2050 RETURN
2060 ! SECONDE VAR
2070 RAD
2080 R(2)=COS(X) @ I(2)=SIN(X)
2090 DEG
2100 RETURN
2110 END
```

Appendix A6.1

Simplified geological time scale in thousands of years (ky), millions of years (My) and billions of years (Gy)

Cenozoic	Quaternary	Neogene	Holocene	10 ky
			Pleistocene	1.8 My
			Pliocene	5.3 My
	Tertiary		Miocene	23.8 My
		Paleogene	Oligocene	33.7 My
			Eocene	54.8 My
			Paleocene	65 My
Mesozoic	Cretaceous			142 My
	Jurassic			206 My
	Triassic			248 My
Paleozoic	Permian			292 My
	Carboniferous			354 My
	Devonian			417 My

(continued)

	Silurian	443 My
	Ordovician	495 My
	Cambrian	545 My
Precambrian	Proteozoic	2.5 Gy
	Archean	4.5 Gy

Glossary

Basic Notations and Use of Units

The meaning of symbols and units has not been systematically recalled in the text. The tables below are here as a reminder regarding terms. Certain symbols may occasionally have more than one name. These are usually specified in the text.

Symbols	Parameters	Units
I, i	Electrical current	A
V, *V*, u, U	Electrical potential, voltage, scalar	V
U_{AB}	Voltage difference $U_A - U_B$	V
U_{eff}	Efficient electromotive force	V
u	Potential	V
ζ	Chemical potential	V
R, R_c, R_r	Electrical resistance	Ω
η	Noise at 1 Hz	$V/A.m^2$
L	Inductance	H
C	Capacitance	F
C	Conductance	S
X, X_0	Reactance	Ω
σ	Electrical conductivity	S/m
ρ, ρ_a	Electrical resistivity, apparent resistivity	Ω.m
ε, ε_0, ε_r	Dielectric permittivity, free space, relative	F/m
μ, μ_0, μ_r	Magnetic permeability, free space, relative	H/m
χ	Magnetic susceptibility	–
s	Salinity	mg/l
$\boldsymbol{\sigma}$	Conductivity matrix	–
$\vec{E}$, $\vec{e}$	Electrical field	V/m ($/Hz^{1/2}$)
$\vec{H}$, $\vec{h}$	Magnetic field	A/m
$\vec{B}$, $\vec{b}$	Magnetic induction	T, $V.s/m^2$

(continued)

S. Sainson, *Electromagnetic Seabed Logging*, DOI 10.1007/978-3-319-45355-2

Symbols	Parameters	Units
$\vec{D}, \vec{d}$	Electric displacement	$A.s/m^2$
$\vec{F}$	Varying magnetic field force	A/m
$\vec{J}, \vec{J}^S$	Current density	A/m^2
$\vec{n}$	Normal vector, or *Poynting* vector	$V.A/m^2$
$\vec{r}$	Directional vector	–
$\vec{A}$	Potential vector	V.s/m
TE_x, TE_y	Transverse component	V/m, A/m
q	Charge or charge density	C or C/m^3
Φ	Magnetic flux	Wb
Z	Impedance	Ω
$\tilde{\zeta}$	Complex impedance	Ω
$\tilde{\varepsilon}$	Complex permittivity	F/m
$\tilde{c}$	Complex speed	m/s
f, Δf	Frequency and frequency bandwidth	Hz
T	Period	s
ω	Angular frequency or pulsation	rad/s
k, k_z,	Wave number, in z direction	–
p, p_{min}, p_{max}	Sample, extreme axes	–
$\vec{k}$	Wave vector	–
K	Geometric factor	–
k,ε,	Factor, scale factor	–
k_B	*Boltzmann*'s constant ($=1.38 \cdot 10^{-23}$)	Joule/°K
h_P	*Planck*'s constant ($=663 \cdot 10^{-34}$)	J.s
π	Number Pi (≈3.14159)	–
δ	Skin depth	m
λ	Wave length	m
R_{TE}	Reflection coefficient	–
R_L^{TE} R_L^{TM}	Interactions coefficient	–
p	Pressure, hydrostatic pressure	Pa
T	Temperature	°C or °K
l, L	Length, offset	m or km
α, β, γ	Attenuation	dB
n, m	Index, mode, number or parameter	–
α, β, θ, ϕ, φ	Angles	° or rad
F, Π	Plan	–
ϕ_0, ϕ_a, $\Delta\phi$	Phase and phase shift	° or rad
ϕ, dϕ	Flux, flux variation	–
(x, y, z)	Cartesian or rectangular coordinates	–
(r, θ)	Spherical or polar coordinates	–
(r, θ, z)	Cylindrical coordinates	–
t, τ, Δt, $\Delta\tau$	Times, time shift and delay time	s
a, R, r, d	Radius or distance, diameter	m or km
h, h, p_R, t_h	Water, depth and layer thickness	m

(continued)

Symbols	Parameters	Units
S, s, a, V, v, β_v	Surface, volume	m^2, m^3
h_a	Altitude	m or km
g	Gravity acceleration (earth: 9.81)	m/s^2
A, B, M, N, O, P	Points (measurement, observation, etc.)	–
(C)	Curve	–
A, S_n, D	Area, plan, surface, domain or travel	–
v, v, c	Velocity, celerity, speed	m/s
P	Dipole moment	A.m
W	Power	W
P_D	Spectral density power	–
ϕ^m	Porosity	%
R^N	Digital resolution	%
D	Dynamic range	dB

Principal Unit Denominations

Unit abbreviation	Unit name
A	Ampere
V	Volt
Ω	Ohm
W	Watt
dB	Decibel
Hz	Hertz
S	Siemens
T, γ, G	Tesla, gamma, Gauss
F	Farad
H	Henry
C	Coulomb
Wb	Weber
Pa	Pascal
J	Joule
°C, °K	°Celsius, °Kelvin
°, rd	Degree, radian
m	Meter
s	Second
l	Liter
g	Gram
%	Percent

Prefixes in the International System of Units and Corresponding Values

Prefix (SI)		Value
Tera	T	10^{12}
Giga	G	10^{9}
Mega	M	10^{6}
Kilo	K	10^{3}
Milli	m	10^{-3}
Micro	µ	10^{-6}
Nano	n	10^{-9}
Pico	p	10^{-12}

Mathematical Symbolism

Notations and mathematical operators	
$\tilde{c}$, c	Complex
$\underline{a}$, $\underline{G}$	Tensor, Green tensor
m, M, $\boldsymbol{\chi}$	Matrix
$\mathbf{C}_X$	Covariance matrix
$\mathbf{J}^{T}$	*Jacobian* matrix
I	*Identity* matrix
D, **W**	*Regularization* matrix
C	*Covariance* matrix
ϑ, $G^{\mathrm{E,H},D,S}$	Operator, integration operator
s, n, n	Scalar, number
λ, A_{n}, B_{n}, C_{n}	Constant, coefficient
D, S	Domain, surface
$\vec{a}$, $\vec{A}$, 0	Vectors (temporal and frequency domain)
Re, Re{ }	*Real* part
i, Im	*Imaginary* part
$\hat{z}$	*Director* vector
$\vec{\Pi}$	*Poynting* vector
$+, -, \pm$	Addition, subtraction, more or less
$\times, \div, /$	Multiplication, division
$=, \approx, \equiv$	Equal, approximately equal to, equivalent
$< >$	Average or distribution

(continued)

Notations and mathematical operators	
$\perp$, //	Perpendicular to, parallel to
$\Leftrightarrow$, $\rightarrow$	Equivalent to, tends to
Δ	Difference
d, ∂	Differential, partial derivative
$\int$, $\iint$, $\iiint$	Simple, double and triple integral
Σ, $\bar{a}$	Sum, average
$\sqrt{}$	Square root
!	Factorial
$\in$	Is an element of
$\propto$	Infinity
(x, y, z)	Deterministic variables
a, b, Γ	Independent random variables
F(), f(),	Functions
$\psi\ (x, y, z, t)$	*Potential* function
$F(x, y, z).e^{j\omega t}$	*Harmonic* function
$P^{\alpha}(m)$	*Parametric* function
$S(m)$	*Stabilization* function
$\phi(m)$	*Predicted* function
δ, $\delta(t)$	*Dirac* function, *Shah* function
G	Scalar *Green* function
J_m	*Bessel* function of m-th order
N_m	*Neumann* function of m-th order
H_m	*Hankel* function of m-th order
P_n	*Legendre* polynomial of order n
$\lim\limits_{\rightarrow}$	Limit
$\wedge$	*Vectorial* product
.	*Scalar* product
$\vec{\nabla}\wedge$	Rotational
$\vec{\nabla}.$	Divergence
$\vec{\nabla}$	Gradient of
∇^2	*Laplacian*
cos, cosh	Cosine, hyperbolic cosine
sin, sinh	Sine, hyperbolic sine
tan, cotg	Tangent, cotangent
log	Decimal logarithm
ln	*Neperien* logarithm
e, exp	Exponential
e	Base of natural logarithms
| |	Absolute value
|| ||	Norm
P()	Probability distribution
d, g, I	Measured and calculated data, information

(continued)

Notations and mathematical operators	
e, $\vec{e}$	Error, parameter, error vector
$P\{\}$	Probability
$f(m\|I)$, $P(X)$	Probability distribution
*	Convolution product

Definition of the Vector Operators (Gradient, Divergence, Rotational, Laplacian)

- The *gradient* operator transforms a scalar function into a vector function. The gradient vector $\vec{\nabla}\psi$ of a scalar field ψ in a given direction corresponds to the partial derivative of ψ in this direction.

In a system of rectangular coordinates (x, y, z), axes of unit vectors $\left(\vec{\mathrm{i}}, \vec{\mathrm{j}}, \vec{\mathrm{k}}\right)$, it is defined by its projections such that:

$$\vec{\nabla}\psi = \frac{\partial\psi}{\partial x}\vec{\mathrm{i}} + \frac{\partial\psi}{\partial y}\vec{\mathrm{j}} + \frac{\partial\psi}{\partial z}\vec{\mathrm{k}}$$

This vector is then the variations of a scalar quantity in a given direction.

- The *divergence* operator $\vec{\nabla}.\vec{\mathrm{a}}$ of a vector $\vec{\mathrm{a}}$ at any point in the space corresponds to the flux variation dϕ of this vector relative to an elementary volume dv containing the point. This flux passes through the closed surface which delimits this volume such that:

$$\vec{\nabla}.\vec{\mathrm{a}} = \lim_{\mathrm{dv}\to 0} \frac{\mathrm{d}\phi}{\mathrm{dv}}$$

In a system with rectangular coordinates (x, y, z), its expression as a function of the vector projections $\vec{\mathrm{a}}$ becomes:

$$\vec{\nabla}.\vec{\mathrm{a}} = \frac{\partial\ \vec{\mathrm{a}}_x}{\partial x} + \frac{\partial\ \vec{\mathrm{a}}_y}{\partial y} + \frac{\partial\ \vec{\mathrm{a}}_z}{\partial z}$$

- The *rotational* operator $\vec{\nabla} \wedge \vec{a}$ of a vector $\vec{a}$ in a given direction $d\vec{l}$ corresponds to the limit of the movement $\vec{a}$ (c) around the surface $d\vec{s}$ normal at $d\vec{l}$ when $d\vec{s}$ approaches zero, such that:

$$\vec{\nabla} \wedge \vec{a} = \lim_{ds \to 0} \oint_{(c)} \vec{a}.d\vec{l}$$

In a system with rectangular coordinates (x, y, z), axes of unit vectors $\left(\vec{i}, \vec{j}, \vec{k}\right)$, its expression is equivalent to:

$$\vec{\nabla} \wedge \vec{a} = \left(\frac{\partial\,\vec{a}_z}{\partial y} - \frac{\partial\,\vec{a}_y}{\partial z}\right)\vec{i} + \left(\frac{\partial\,\vec{a}_x}{\partial z} - \frac{\partial\,\vec{a}_z}{\partial x}\right)\vec{j} + \left(\frac{\partial\,\vec{a}_y}{\partial x} - \frac{\partial\,\vec{a}_x}{\partial y}\right)\vec{k}$$

- The *Laplacian* operator ∇^2 is applied to scalar and vectorial fields and is equal to $\nabla.\nabla$. It is used to resolve the Laplace equation ($\nabla^2 = 0$), the Poisson equation or the wave equation. It is given in Cartesian or rectangular coordinates (x, y, z) by the expression:

$$\nabla^2 = \frac{\partial^2}{\partial x^2} + \frac{\partial^2}{\partial y^2} + \frac{\partial^2}{\partial z^2}$$

Depending on the case to solve, it can be also expressed in polar, cylindrical or spherical coordinates.

For vectorial analysis, see, for example, the work of Professor H. Skilling (1942): Fundamentals of Electric Waves, Ed Wiley (pp. 10–36) or that of Professor E. Durand (1964): Electrostatique. Volume 1. Distributions. Masson Ed. Chap. 2, or that of G. Goudet (1956): Electricity. Masson Ed.

Special Functions

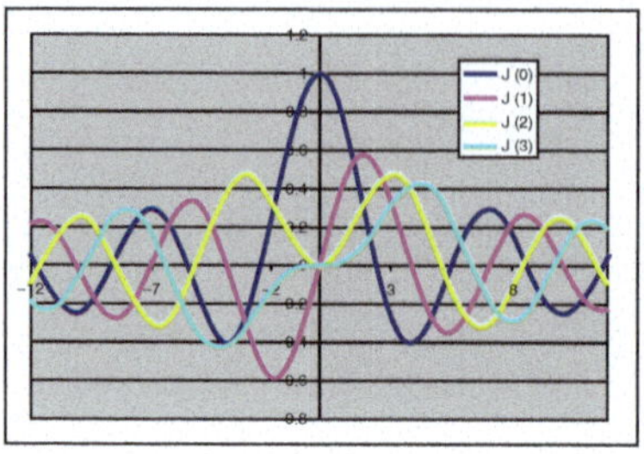

Bessel Function of first kind (J_m or $J_{(m)}$)

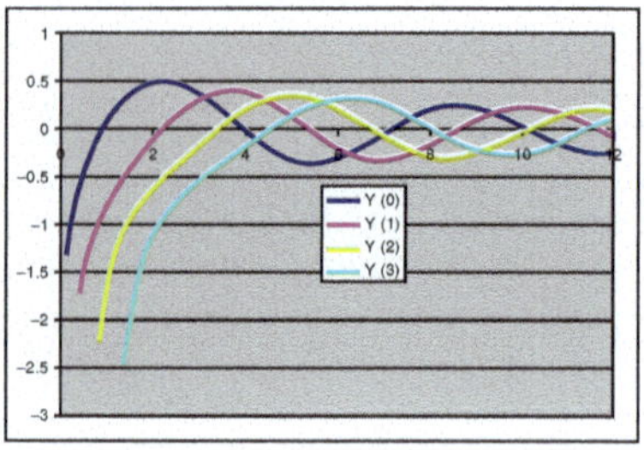

Bessel Function of second kind (Y_m) or *Neumann* function (N_m)

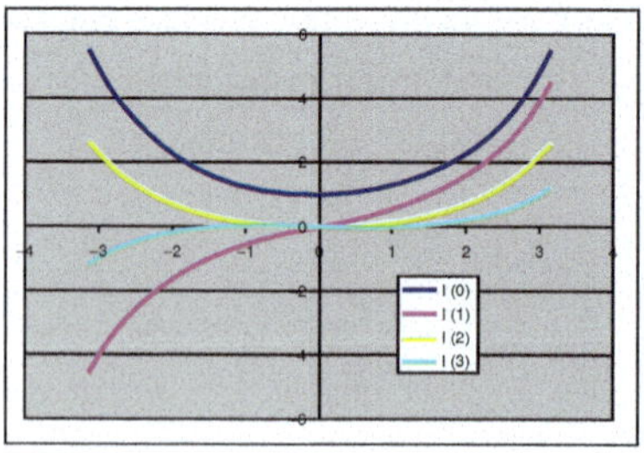

Modified Bessel Function of first kind (I_m)

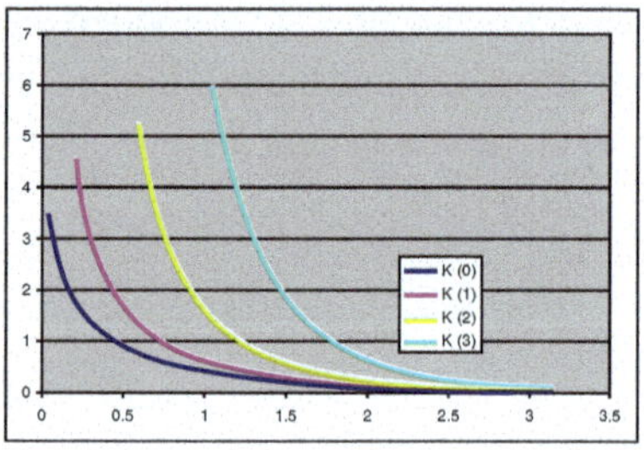

Modified Bessel Function of second kind (K_m)

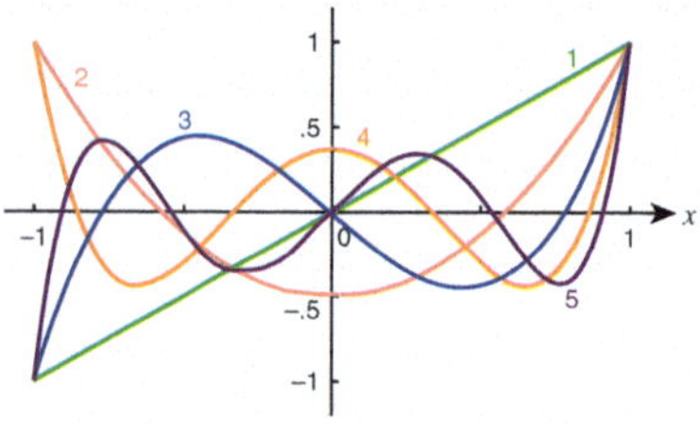

Legendre polynomials (P_n)

Special functions and particularly cylindrical *Bessel* functions said of different orders (m = 1, 2 and 3), used in solving wave equations a priori as part of the design of sensors and a posteriori in the interpretation of field data. The functions of the third kind, commonly called *Hankel* functions (H_m), are linear combinations of *Bessel* functions of the first kind (J_m) and second kind (Y_m). The orthogonal *Legendre* polynomials (P_n) are used in solving the *Laplace* equation in spherical symmetry.

Index

S. Sainson, *Electromagnetic Seabed Logging*, DOI 10.1007/978-3-319-45355-2

H

I

J

K

L

T

The manufacturer's authorised representative in the EU is Springer Nature Customer Service Centre GmbH, Europaplatz 3, 69115 Heidelberg, Germany. If you have any concerns regarding our products, please contact ProductSafety@springernature.com

Printed and bound by CPI Group (UK) Ltd, Croydon, CR0 4YY
05/10/2026
02228133-0011